The Biology of Scorpions

Contributors

JOHN L. CLOUDSLEY-THOMPSON

NEIL F. HADLEY

JOHN T. HJELLE

SHARON J. MCCORMICK

GARY A. POLIS

THOMAS M. ROOT

J. MARC SIMARD

W. DAVID SISSOM

MICHAEL R. WARBURG

DEAN D. WATT

The BIOLOGY of *Scorpions*

EDITED BY

Gary A. Polis

STANFORD UNIVERSITY PRESS
STANFORD, CALIFORNIA

Stanford University Press
Stanford, California

Printed in the United States of America

Published with the assistance of the
Vanderbilt University Committee for
Monograph Subvention

CIP data appear at the end of the book

Preface

This book was conceived out of need. In the middle 1970's, I started to conduct research on scorpions and wanted to relate my findings to what was already known about these animals. I had visions of going to a few sources to find this information, but I soon discovered that this was impossible. I encountered a myriad of reports scattered in all types of sources, including several quite obscure journals. There were many facts but little cohesion, few common themes, and seldom any synthesis. In some cases, authors were apparently unaware of similar research. Colleagues informed me that this was the situation in many areas of scorpion biology. It became obvious that we needed a single source for all our information on scorpions.

It is the purpose of this book to present both what we know and what we don't know about scorpions. There is plenty of each. In this sense, the book's scope is broad, and the chapters attempt to be comprehensive. Various information was collected and collated, and, we hope, common themes were extracted and presented systematically. Some specific themes may change or mature as more research is conducted. This is undoubtedly true for our knowledge of scorpion systematics, ecology, and toxicology: especially in these areas much work has to be performed. We hope that this book will aid future researchers with this task. The bibliography contains almost all pertinent literature on scorpion biology and should be particularly useful to researchers and arachnologists. I anticipate that the book will be a standard supplement to courses in arachnology or in the biology of arthropods.

However, the book itself is intended for scientists and interested lay persons alike. I believe that it bridges whatever gap may exist between these two groups. The reason for this is found in scorpions themselves. They are both an intrinsically interesting subject and one that we humans view with a combination of curiosity, fascination, and fear. As the book communicates just how amazing scorpions are, it becomes not just an edifying scholarly work but one that also fascinates the reader. At least this is the hope of each author.

This book is composed of twelve chapters written by ten different authors. Because of the nature and pace of current scorpion research, it was inconceivable that any one specialist could have done justice to the subject working alone. Thus the usual problems of inconsistency, repetition, and so forth, of a multiauthor work, carefully attended to but doubtless still in evidence, are more than offset by the volume's depth and authority. Under different conditions or in different times, an additional chapter on paleontology would be appropriate. However, with the recent deaths of Erik Kjellesvig-Waering and Leif Størmer, there are no living experts on fossil scorpions. A limited treatment of paleontology (see Chapters 1 and 3) is included in the present book.

I would like to dedicate my work in this book to my mom and dad, Marie and Sam Polis, and to Märie Turner, Ken Sculteure, Larry Pomeroy, and Sharon Polis. These people have been the major influences in my growth as a person and a biologist. I salute each of them, and give them my heartfelt love and gratitude. I would also like to acknowledge with gratitude the friends and co-workers who have helped me during the past ten years of field work during the many long, often lonely and scary, but usually enjoyable evenings: Lars Carpelin, Denise Due, Roger Farley, Rick Fleck, Sharon McCormick, Chris Myers, Dan Polis, Sharon Lee Polis, Steve Polis, Mike Quinlan, Dave Sissom, Joe Vehige, and Tsunemi Yamashita. Stanley Williams, Oscar Francke, Wilson Lourenço, William Shear, Richard Bradley, Denise Due, and David Lightfoot allowed me to use unpublished information. Robert Mitchell, Edward Ross, Philip Brownell, and Dave Sissom provided photographs. Dave Sissom, Sharon McCormick, Wilson Lourenço, Herbert Levi, and William Carver were particularly helpful with their suggestions on how to improve both my writings and the entire book.

Dave Sissom performed the yeoman's chore of updating synonymies and nomenclature. My continuing thanks to Sherrie Hughes for the fine job of typing the never-ending manuscripts, and to Lou Gorman for organizing my files and the bibliography of this book. All these people generously, cheerfully, and (usually) uncomplainingly have given me their time and energy. I sincerely thank every one of these *amigos* for their help, ideas, moral support, camaraderie, and companionship. My research has been sponsored by funds from the Vanderbilt University Research Council, Vanderbilt Natural Science Committee, Sigma Xi, and the National Science Foundation.

G. A. P.

Contents

Tables and Figures

Tables

Figures

Contributors

JOHN L. CLOUDSLEY-THOMPSON is Emeritus Professor of Zoology in the University of London. After war service in North Africa and Normandy he returned to Cambridge University with the honorary rank of Captain, receiving the degrees of M.A. and Ph.D. From 1950 to 1960 he was Lecturer in Zoology at King's College, London, and obtained the degree of D.Sc. (London). He became Professor of Zoology, University of Khartoum, and Keeper of the Sudan Natural History Museum in 1960, and was Professor of Zoology, Birkbeck College, London, from 1972 to 1986. He was awarded an honorary D.Sc. (Khartoum) and Gold Medal in 1981. He is now working at University College, London, where he holds a Leverhulme Emeritus Fellowship.

NEIL F. HADLEY received his Ph.D. in biology from the University of Colorado and is currently Professor of Zoology at Arizona State University. His research interests include the thermal, water, and metabolic relations of arthropods, with special emphasis on the structure and function of arthropod cuticle. He recently authored *The Adaptive Role of Lipids in Biological Systems* (New York: Wiley, 1985).

JOHN T. HJELLE received his Ph.D. in entomology from the University of California, Berkeley. His primary research interests are the systematics, ecology, and morphology of scorpions. He is currently Research Associate in the Department of Entomology at the California Academy of Sciences.

SHARON J. MC CORMICK received her bachelor's degree from University of California, Riverside, and her master's degree from California State University, San Diego. She has held research positions at Vanderbilt University and (currently) at Valdosta State University. She is now Sharon J. McCormick-Carter. Her latest research is centered on parental effects of family raising.

GARY A. POLIS is Associate Professor of Biology at Vanderbilt University. He received his bachelor's degree in biology and philosophy from Loyola University, Los Angeles, and his doctorate in biology from the University of California, Riverside. His academic interests include evolution, behavioral and community ecology, desert biology, and natural history. He is interested in inducing general evolutionary and ecological principles from empirically based research on model organisms such as scorpions and spiders.

THOMAS M. ROOT received his Ph.D. from the University of Wyoming and is currently Associate Professor of Biology at Middlebury College. His primary research interest is the neural control of locomotion, burrowing, and stinging behavior of scorpions. His other research interests include studies of arachnid neurobiology and behavior, and comparative studies of arthropod sensory and motor systems.

J. MARC SIMARD is Assistant Professor of Surgery and of Physiology and Biophysics in the Department of Surgery, Division of Neurosurgery, at the University of Texas Medical Branch, Galveston. His main research activity is in the physiology and biophysics of ion channels, and has led him to study the effects of scorpion toxins.

W. DAVID SISSOM received his Ph.D. in general biology from Vanderbilt University and is now Assistant Professor of Biology at Elon College. His primary interests are systematics and phylogeny of all scorpion families. He is currently engaged in revisionary studies of the species groups of the genus *Vaejovis* in mainland Mexico and is evaluating phylogenetic relationships among the various vaejovid taxa. He is also interested in scorpion natural history and ecology.

MICHAEL R. WARBURG received his Ph.D. from Yale University and is now Professor of Zoology at the Israel Institute of Technology, Haifa. His main research activities are in physiological ecology of xeric-inhabiting animals and problems of adaptation to terrestrial life.

DEAN D. WATT is Professor of Biochemistry at Creighton University School of Medicine. His principal research activity has been the isolation and chemical characterization of the toxins from the venom of the scorpion *Centruroides exilicauda*.

The Biology of Scorpions

1

Introduction

GARY A. POLIS

Scorpions are fascinating animals. Unfortunately, the interest shown by most people stems from the scorpion's reputation as a deadly scourge, a killer of man and his animals. Indeed, there are about 25 species whose venom is capable of causing human death, in some cases within seven hours. However, almost 1,500 other species are no more than efficient predators of insects and other small animals—the sting of these scorpions is usually far less painful than that of a honeybee.

Scientists interested in scorpions are captivated by their great antiquity and the amazing suite of biochemical, physiological, behavioral, and ecological adaptations that have combined to ensure their continued success over the past 450 million years. Though these "living but sophisticated fossils" have changed little in their morphology since their first appearance, their adaptations nevertheless range from superbly efficient behavioral repertoires to the maternal care of offspring, ultrasensitive tactile and visual fields, and complex venoms that are a precise mixture of different toxins, each with its own action. Scorpions thus represent a model system for research ranging from biochemistry to evolutionary ecology.

The evolutionary history of scorpions dates back to their appearance in the middle Silurian (about 425–450 million years ago). They almost certainly evolved from the Eurypterida, or "water scorpions" (Wills 1966; Kjellesvig-Waering 1966a, 1986; Manton 1977; Savory 1977; Rolfe and Beckett 1984; Rolfe 1985). Paleozoic scorpions and eurypterids share several common features, including external book gills, flaplike abdominal appendages, large, multifaceted compound eyes, and similar chewing structures on the coxae of the first pair of appendages (Kjellesvig-Waering 1966a, 1986; Størmer 1970; Manton 1977; Shear 1982, pers. comm., 1983). Kjellesvig-Waering's (1986) monograph indicates that scorpions and eurypterids are sister groups, and confirms that the traditional division of chelicerates into the mainly aquatic Merostomata and the mainly terrestrial Arachnida is an artificial one, based largely on environment.

It is generally accepted that early scorpions were aquatic, marine or amphibious, like many of the living crabs (this point is controversial: see Chapter 3). The earliest scorpions apparently not only possessed gills but also had legs adapted to a benthic existence. The fact that many of the early scorpions were relatively large (see below) also strongly suggests that these species would need water to support their bodies. Finally, the earliest fossils are often associated with the remains of marine organisms (Kjellesvig-Waering 1966a, 1986).

Marine and amphibious scorpions probably persisted well into the Carboniferous (about 250–300 million years ago; see Chapter 3). The first decidedly terrestrial (air-breathing) scorpions probably appeared on land during the late Devonian or early Carboniferous (325–350 million years ago). The earliest unequivocally terrestrial scorpion is *Palaeopisthacanthus*, from the Upper Carboniferous, in which stagmata are preserved (Rolfe 1980). The evolution of enclosed book lungs in place of external book gills was the major change associated with the transition from water to land. Early scorpions radiated into several now-extinct families and superfamilies (Kjellesvig-Waering 1986). They ranged in size from a few centimeters (e.g., *Palaeophonus nuncius*) to an estimated length of up to a meter (e.g., *Gigantoscorpio willsi, Brontoscorpio anglicus, Praearcturus gigas*). Although *Brontoscorpio* and *Praearcturus* occur in terrestrial sediments, they are too large to have molted on land, and must have been amphibious, if not aquatic.

Evidently several other groups of terrestrial arthropods were already established on land before scorpions first appeared (Manton 1977; Savory 1977; Rolfe 1980, 1985; Shear 1982, pers. comm., 1983). Fossils of various arachnids (Acari, Amblypygi, Trigonotarbi), myriapods (Chilopoda, Diplopoda), and possibly even insects occur in the late Silurian–early Devonian, approximately 380 million years ago. These fossils indicate that land was colonized repeatedly by arthropods and that scorpions were probably not the first arthropods to establish themselves successfully on land.

The occurrence of these fossils in combination with different interpretations of embryological and morphological data has produced a controversy over the origin of the arachnids and the relationship of scorpions to the arachnids (Chapters 3 and 4). One view considers that scorpions are a group in the Arachnida and may even be the ancestor of the other arachnids. The alternative view contends that scorpions are not arachnids at all, but modern, terrestrial merostomes: by this reckoning, the horseshoe crab *Limulus* is the closest living relative, and Merostomata (including scorpions) is a group distinct from the arachnids (Bergström 1979). The evidence for these competing hypotheses, pro and con, is presented by Sissom (Chapter 3). Whatever their exact taxonomic relationship, it is clear that scorpions form a distinct group that has been consistently separated by taxonomists from other arachnids.

Modern scorpions are generally similar in appearance to Paleozoic forms. Except for the changes in locomotion and respiration necessitated by the migration to land, the basic body plan is externally similar to that of scorpions that lived 425 million years ago (see Chapters 2 and 3). The earliest scorpions possessed a segmented opisthosoma with the mesosoma and metasoma clearly differentiated, well-formed chelate pedipalps and chelicerae, eight walking legs, pectines (unique to scorpions), and a terminal telson. Evidently this early body plan was a particularly successful one, sufficiently general and well adapted that major changes did not subsequently occur. Thus, no great architectural revolution in external morphology accompanied the taxonomic diversification of scorpions into the various extinct and extant families. Neither has there been extensive modification during radiation into different habitats: tropical species are similar to desert species; intertidal scorpions resemble those from high altitudes. Although some species show ecomorphological adaptation to different soils and microhabitats (see Chapter 6), similarities in appearance among scorpions far overshadow morphological differences. All scorpions look generally alike.

Today, scorpions have a wide geographical distribution and live on all major land masses except Antarctica (they were accidentally introduced into New Zealand and England by man in recent times). They range from Canada and central Europe to the tips of South America and Africa. This tropical-to-temperate distribution is generally similar to that of amblypygids, uropygids, and theraphosid spiders. Of the arachnids, only spiders, mites, pseudoscorpions, and Opiliones have a wider distribution, extending into subpolar areas (Savory 1977).

Scorpions have radiated into all nonboreal habitats, including desert, savanna, grasslands, temperate forests, tropical forests, rain forests, the intertidal zone, and snow-covered mountains over 5,500 m in altitude. Several species live in caves, and one species, *Alacran tartarus,* is found at depths of more than 800 m (Francke 1982a).

Although some species are quite specific in (micro)habitat requirements, many exhibit a high degree of plasticity in habitat use. For example, *Vaejovis janssi* is the only scorpion found on Socorro Island (southwest of the tip of Baja California Sur); it is found there in jungle, heavy brush, rocky terrain, and sand, on the ground, in vegetation, and near the surf (Williams 1980). *Euscorpius carpathicus* lives in caves, above the ground, and in the intertidal zone. *Scorpio maurus* occurs below sea level in Israel and above 3,000 m in the Atlas Mountains, several thousand kilometers to the west.

In some habitats scorpions appear to be one of the most successful and important predators in terms of density, diversity, standing biomass, and role in community energetics and structure (see Chapters 6 and 7). As many as 13 species occur sympatrically, and in most locales 3 to 6 species can be found. Several species occur at densities exceeding $0.5/m^2$. *Vaejovis littoralis,* an intertidal scorpion of Baja California, exhibits the highest density of any scorpion, $2.0–12.0+/m^2$ along the high-tide mark. Since adults of various species commonly weigh 0.5–10.0 g, the standing biomass of some populations of the more dense species is remarkably high.

A number of reasons account for scorpions' success. Although they are morphologically conservative, scorpions are quite adaptable and plastic in their ecology, behavior, physiology, and life history (see Chapters 4, 5, 6, and 8). They are well adapted to extreme physical conditions: some species can be supercooled below the freezing point for several weeks and yet return to normal levels of activity within a few hours. Other species can survive total immersion under water for as long as one to two days. Desert species can withstand temperatures several degrees higher than most other desert arthropods are able to tolerate. Thermoregulation by scorpions is primarily behavioral—nocturnal activity, retreat to burrows during periods of stressful temperatures, stilting—rather than strictly physiological via evapotranspiration of water from the body surface.

Desert scorpions conserve water more efficiently than any other arthropod (see Chapter 8). They excrete nearly insoluble nitrogenous waste products (xanthine, guanine, uric acid), and their feces are likewise extremely dry. There is minimal water loss through the spiracles of the book lungs or through the integument. The exoskeleton is coated with a thin layer of lipids that is extremely impermeable to water. Some scorpions can live indefinitely without drinking water; sufficient water is normally contained in their food.

Scorpions share the record with many spiders for the lowest arthropod metabolic rates ever recorded (see Chapter 7). And because scorpions normally exhibit low levels of activity and movement, they have been known to survive without food for over a year. Many species spend 92 to 97 percent of their entire existence relatively inactive in their burrows. Most species forage by waiting motionless until the prey happens upon them. These opportunistic predators are capable of gaining one-third or more of their body weight from one meal, and little is wasted. They exhibit a remarkably high conversion rate of prey biomass into scorpion biomass; external digestion is a major factor in such efficient assimilation.

They are adept predators with an extremely varied diet (see Chapter 7). Insects and various arachnids (especially spiders, solifuges, and other scorpions) are the most frequent prey. However, scorpions also eat isopods, gastropods, and vertebrates (snakes, lizards, rodents).

The neural and behavioral mechanisms used to locate prey are exact and sophisticated (see Chapters 7 and 9). By accurately monitoring the wave characteristics of substrate vibrations, some species of scorpion are able to locate and capture prey that are on or below the surface (Brownell 1977a). This method of prey location is basically the same as that seismologists use to detect the location and magnitude of earthquakes. Some species actually catch prey flying in the air; they probably detect such prey with pedipalpal trichobothria.

The scorpion nervous system, although morphologically primitive, may exhibit other relatively sophisticated sensory capabilities (see Chapters 5 and 9). For example, according to Fleissner (1977c) the scorpion eye is among the most sensitive of the arthropod eyes and may even allow scorpions to orient by starlight.

Many species of scorpion exhibit a number of life-history traits that are quite different from those of most other terrestrial arthropods (Polis and Farley 1980; Chapters 4 and 6). In fact, these traits are more characteristic of long-lived vertebrates. Although some species are small (the smallest scorpion, *Typhlochactas mitchelli* is 8.5 to 9 mm in total length), most species are relatively large. Among terrestrial arthropods, only myriapods, some mygalomorph spiders, and a few insects rival the size of larger scorpions. The larger species of scorpion are among the biggest of all terrestrial invertebrates and are bigger than one-third to one-half of all terrestrial vertebrate species (McCormick and Polis 1982). For example, nongravid females of the African *Hadogenes troglodytes* weigh 32 g and may reach 21 cm in length (Newlands 1972a). By comparison, hummingbirds vary in length from 8 to 25 cm and in weight from 2 to 20 (average = 5) g. Developmental times are relatively long. Gestation periods range from 3 to 18 months; time to maturity varies from 6 to 96 months. Most scorpions live 2 to 10 years, and at least some species live 25 years or longer. Although some other arachnids show equivalent developmental times, the vast majority complete their life cycles far more rapidly than scorpions.

Scorpion reproduction is also uncharacteristic of terrestrial arthropods (see Chapter 4) in that the mother invests a great amount of time and energy in her offspring. All scorpions are unlike most other terrestrial arthropods in being viviparous, and gestation is long. Embryos are nourished *in utero* or via a "placental" connection with the mother. The young are born large; for example, the second instar of *Paruroctonus mesaensis* (a medium to large species) is greater in size (0.03 g) than the adults of a large proportion of North American insects (Polis and Farley 1980). After birth, maternal care lasts from two days to several months. Mother–offspring relationships may even involve such intermediately subsocial behaviors as cooperative feeding and burrow construction (see Chapter 6). Overall, these life-history characteristics combine to produce long generation times and very low values for other reproductive parameters. The maximum rate of population increase (r_{max}) for several species of scorpion is the lowest reported for any animal, including long-lived vertebrates (see Chapter 6).

Overall then, scorpions are biologically interesting. They exhibit a curious but obviously successful mix of primitive and advanced characteristics. They are a very old group, one that has survived over 400 million years of geological and environmental change. Great plasticity in physiology, behavior, life history, and response to environmental factors—rather than plasticity in gross morphology—is the major adaptive feature that best explains their continued success.

Apart from their interesting biology, it is also clear that scorpions themselves have long fascinated human beings (see Chapter 12). For most people, they are an object of fear and a subject of folklore. Myths about scorpions were created by cultures as different as the ancient Greeks and Egyptians and the modern cowboys and gauchos of the New World (Baerg 1961, Cloudsley-Thompson 1965, Parrish 1966). Occasionally scorpions are associated with good. The gods placed Scorpio in the zodiac because a scorpion killed Orion the Hunter after Orion bragged that he would kill all the animals on earth. More commonly, however, scorpions are associated with death and evil. The scorpion has appeared repeatedly in religious cults of ancient and modern history as an agent of the night, the devil, or the gods of the underworld.

Such morbid interest in scorpions is probably generated by the fact that a few species possess potent toxins capable of killing human beings (see Chapter 10). With the exception of snakes and bees, scorpions are responsible for more human deaths every year than any other nonparasitic group of animals. Although such statistics are largely educated guesses, Bücherl (1971) estimates that over 5,000 people die each year by scorpion envenomation. In Mexico alone, scorpions of the genus *Centruroides* caused an average of 1,696 deaths per year for the 12 years reported by Mazzotti and Bravo-Becherelle (1963). In the city of Durango, Mexico (pop. 40,000), there was an average of 45 deaths per year between 1890 and 1926 (Baerg 1961); yet this astonishing rate ranked Durango only fourteenth among all the states in the Mexican Republic (Mazzotti and Bravo-Becherelle 1963). Scorpions are also a major health hazard in India, North Africa, and parts of South America.

Potentially lethal species produce a complex neurotoxin that causes both local and systemic effects (see Chapter 10). Paralysis, severe convulsions, and cardiac irregularities precede death. In some cases, death occurs within a few hours after envenomation. Fortunately, death can be avoided if antivenins are administered.

But most scorpions are not deadly. Although there are over 1,500 described species in more than 110 genera (see Chapter 3 and Francke 1985), fewer than 25 (in eight genera) are lethal. All the deadly species belong to the family Buthidae. Most scorpions produce toxins that cause mild to strong local effects, including edema, discoloration, and pain; victims recover fully in a matter of minutes or days.

Considering the medical and biological importance of scorpions, it is unfortunate that we are still relatively ignorant about them. Obviously, more research needs to be conducted on almost all aspects of scorpion biology. For example, we know practically nothing about the natural history or field behavior of any of the deadly species.

This lack of information is paradoxical, because scorpions exhibit many characteristics that make them well suited as research organisms (see Chapter 11). They occur in high densities in many habitats and survive well in captivity. Their cuticle fluoresces under ultraviolet light (Lawrence 1954, Pavan and Vachon 1954), and they are thus readily detected in the field with portable lights (Stahnke 1972a). This feature allows the rapid collection of animals for laboratory research and facilitates field studies in behavior or ecology. Scorpions live for many years and are large enough for easy study. Because their anatomy is fairly simple and relatively well known, they are ideal for various physiological and neurobehavioral studies. Finally, comparative research is easily conducted because scorpions are often locally diverse and there is considerable interspecific variation in behavior and physiology.

Scorpions thus present a model research vehicle for addressing many basic questions, and they are of course extraordinarily interesting in their own right.

2

Anatomy and Morphology

JOHN T. HJELLE

Anatomical and morphological investigations of any group of organisms commonly constitute a large proportion of our knowledge about the group. The reasons are numerous, but basically one must know something of these topics before embarking on studies of such other areas as systematics, physiology, and behavior. The literature on scorpion anatomy and morphology began in earnest in the early 1800's. Beginning with general description and gross dissection, investigations have become more and more closely focused with the advent of new equipment and techniques. With the development of the scanning electron microscope (SEM), microanatomical investigations have become more common in the scorpion literature.

From the many individual studies undertaken over the last hundred years or more, several works have attempted to bring at least some of the widely scattered literature together. Six relatively recent publications serve such a purpose: Millot and Vachon 1949, Snodgrass 1952, Vachon 1952, Savory 1977, Kästner 1968, and Legendre 1968. Two more recent publications devote sections to this topic and include some excellent illustrations: Keegan 1980 and Levy and Amitai 1980.

The six volumes provide a basic knowledge of scorpion morphology and anatomy, but none is comprehensive, and they suffer from the same problem that plagues the many other publications on specific morphological topics—the lack of a standard nomenclature. In addition, a review of the literature reveals that there are significant gaps in our knowledge in several areas, not to mention differences in interpretation. As a result, my primary purpose is to provide a more comprehensive review of the current state of knowledge of scorpion morphology and anatomy. Preferred and alternative terminologies are given, and the preferred terms are italicized on first use. In some instances, the same structure bears two different names that are so widely used and accepted that no particular preference is indicated.

Gross Morphology and Segmentation

The scorpion body is divided into two tagmata, the cephalothorax and the abdomen (Figs. 2.1 and 2.2). The cephalothorax, also known as

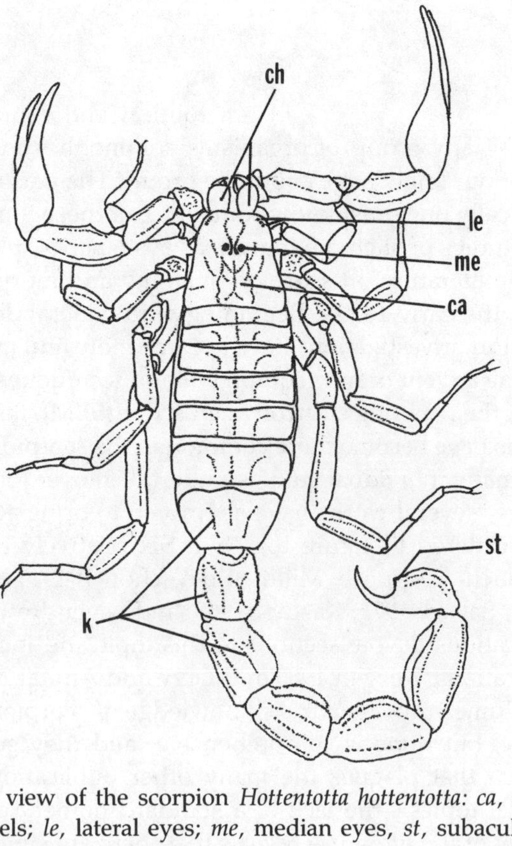

Fig. 2.1. Dorsal view of the scorpion *Hottentotta hottentotta: ca,* carapace; *ch,* chelicerae; *k,* keels; *le,* lateral eyes; *me,* median eyes, *st,* subaculear tubercle. (After Vachon & Stockmann 1968)

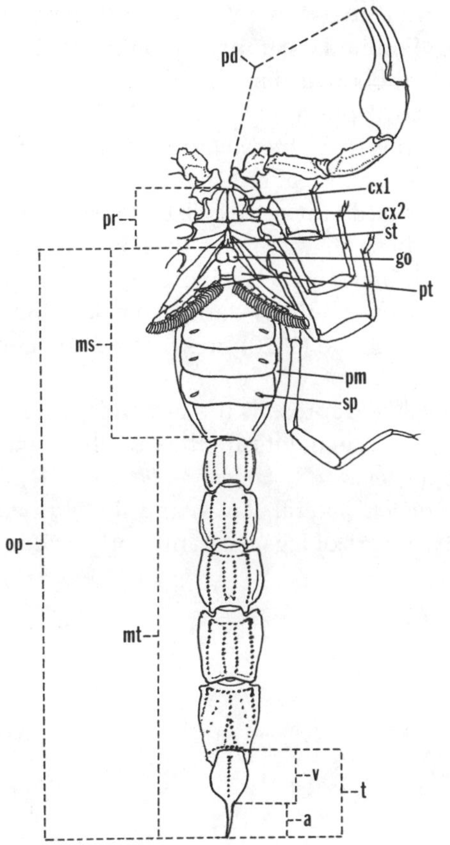

Fig. 2.2. Ventral view of the scorpion *Androctonus australis: a*, aculeus; *cx1*, coxapophysis I; *cx2*, coxapophysis II; *go*, genital operculum; *ms*, mesosoma; *mt*, metasoma; *op*, opisthosoma; *pd*, pedipalp; *pm*, pleural membrane; *pr*, prosoma; *pt*, pectines; *sp*, spiracle; *st*, sternum; *t*, telson; *v*, vesicle. (After Lankester 1885a)

the prosoma, is covered dorsally by the carapace, which bears the median and lateral eyes. The abdomen, or opisthosoma, is subdivided into a broad anterior mesosoma, or preabdomen, and a narrow taillike metasoma, or postabdomen. The cephalothorax consists of seven visible postoral somites (III–IX). The preoral clypeolabral and antennal somites I and II are absent, as in other arachnids, but their neuromeres remain as part of the cephalothoracic ganglia (Henry 1949, Anderson 1973). Thus, the first visible somite, somite III, is represented by the chelicerae, which are three-segmented appendages modified for feeding and grooming. Somite IV comprises the pedipalps, which are six-segmented, chelate appendages modified for prey immobilization, defense, and sensory perception (Figs. 2.1–2.3). The segments, from the most proximal, are the *coxa, trochanter, femur* (humerus), *patella* (brachium), *tibia* (*manus,* or hand, and its fixed finger), and the *tarsus* (movable finger). The tibia and tarsus are collectively referred to as the *chela*. On the surfaces of these segments, as well as on many other areas of the body, are raised, linear structures called *keels* or carinae (Fig. 2.1). These may be represented only by darkened coloration, or they may be heavily granulated. Stahnke (1970) presented a nomenclature for scorpion keels.

The four pairs of legs represent the appendages of somites V–VIII (Figs. 2.4 and 2.15). The segments are the *coxa, trochanter, femur, patella, tibia, basitarsus* (= *tarsomere I*), *tarsus* (= *tarsomere II*), and the *apotele* (see below) with *ungues* (lateral claws) and *dactyl* (median claw). The coxae of the first two pairs of legs bear anteriorly extended *coxapophyses*

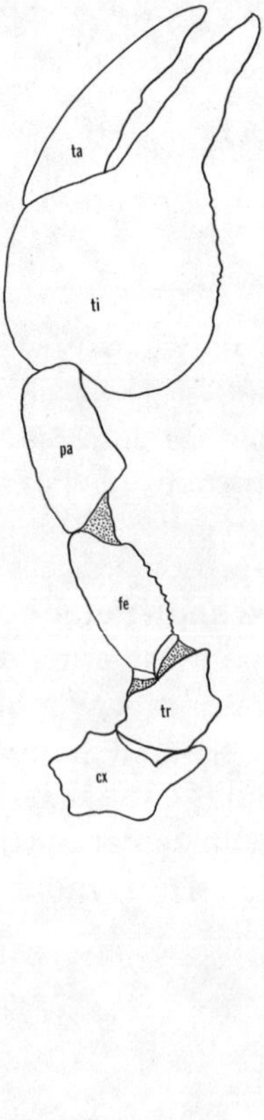

Fig. 2.3. Representative scorpion pedipalp: *cx,* coxa; *fe,* femur; *pa,* patella; *ta,* tarsus; *ti,* tibia; *tr,* trochanter.

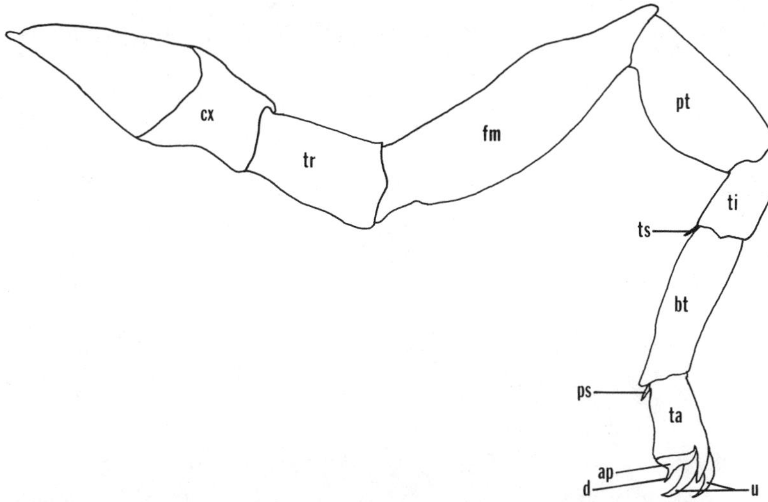

Fig. 2.4. Representative scorpion leg: *ap*, apotele; *bt*, basitarsus; *cx*, coxa; *d*, dactyl; *fm*, femur; *ps*, pedal spur; *pt*, patella; *ta*, tarsus; *ti*, tibia; *tr*, trochanter; *ts*, tibial spur; *u*, ungues.

(mesal lobes, endites) (Fig. 2.2). These substitute for the sternum and serve to close the preoral cavity ventrally.

The sternum (the fused sterna of somites VII and VIII) lies between the coxae of legs III and IV (Figs. 2.2 and 2.5). Because of its varied shape (pentagonal, triangular, or transverse), the sternum has been considered an important taxonomic character at the family level of classification. According to Pocock (1894) and Petrunkevitch (1916), the pentagonal shape is primitive, since the late embryo and first instar of all scorpions exhibit a pentagonal sternum.

Somite IX bears embryonic appendages but otherwise disappears in later stages. This ephemeral segment, first discovered by Brauer (1895) in *Euscorpius carpathicus*, was confirmed more recently by Abd-el-Wahab (1952) in *Leiurus quinquestriatus*.

The mesosoma consists of seven segments (somites X–XVI), the seventh narrowing posteriorly to its junction with the metasoma (Figs. 2.2 and 2.5). Dorsally, each segment is covered by a sclerotized plate, or tergum. The ventral sclerotized plates, or sterna, are found only on mesosomal segments 3 to 7 (somites XII–XVI). Ventrally, mesosomal segment 1 bears paired genital opercula, which cover the gonopore.

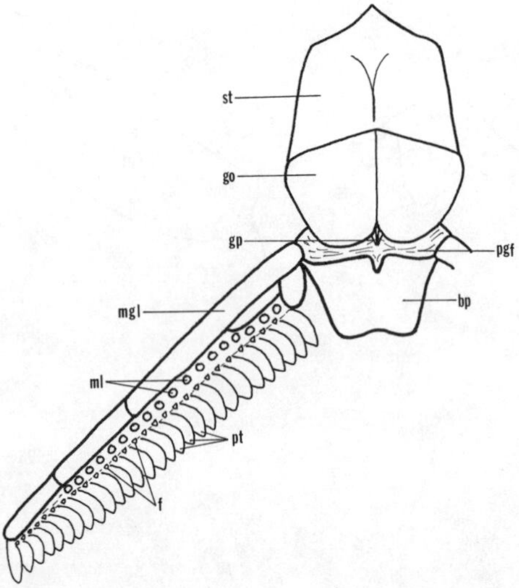

Fig. 2.5. Representative sternum, genital operculum, and pecten (body somites VI–XI): *bp*, basal piece; *f*, fulcra; *go*, genital operculum; *gp*, genital papillae; *mgl*, marginal lamella; *ml*, median lamellae; *pgf*, postgenital fold; *pt*, pectinal teeth; *st*, sternum.

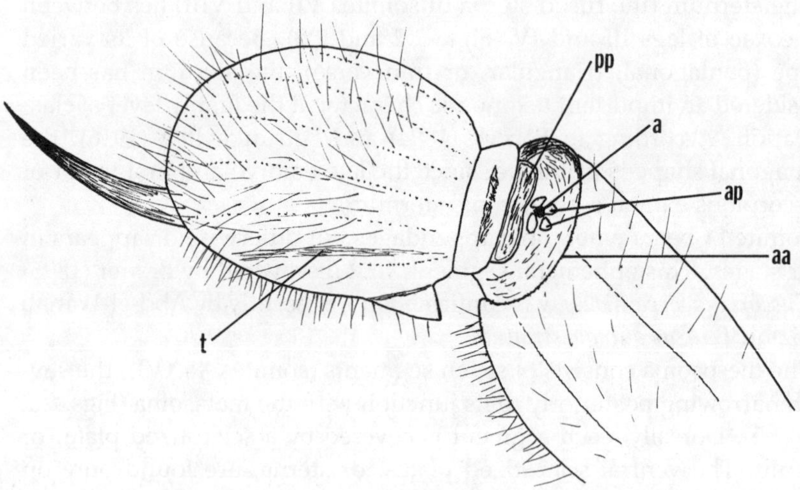

Fig. 2.6. Representative scorpion anal area, ventrolateral view: *a*, anus; *aa*, anal arch; *ap*, anal papillae; *pp*, pedicular plate; *t*, telson.

These are usually fused medially in the female and are partially or completely separate in the male. The males of some taxa also bear two genital papillae, which may be partially or completely covered by the genital opercula. Somite XI (mesosomal segment 2) comprises the basal piece and the pectines. Somites XII–XV (mesosomal segments 3–6) each bear a pair of openings called *spiracles,* or stigmata (Figs. 2.2, 2.20, and 2.24), which are the external apertures of the respiratory organs called *book lungs* (lung books). The apertures may be slitlike, elliptical, oval, or circular. Mesosomal segment 7 (somite XVI) bears no appendages or other taxonomically significant external structures.

All the connective membranes between segments are called intersegmental membranes. Those membranes connecting the sterna and terga laterally are the *pleural membranes* (Fig. 2.2). The membrane between the genital operculum and the basal piece is called the *postgenital fold* (Fig. 2.5).

The metasoma consists of five segments (somites XVII–XXI) plus the telson, or "sting," which is not considered a true segment. Metasomal segments I–V are simple body rings without evident sterna or terga. The segments become progressively longer distally, with segment V always being the longest, and they bear various keels, setae, and bristles, which are of taxonomic importance (see Stahnke 1970). The anus opens at the ventrodistal end of segment V and is bordered by the *anal arch* and surrounded by four *anal papillae* (Fig. 2.6). The telson is subdivided into the bulbous *vesicle* and the needlelike *aculeus.* For purposes of measurement, the *subaculear tubercle* (Fig. 2.1) is considered to be the boundary between these two areas, but many species have the subaculear tubercle much reduced or lacking, which makes a precise distinction less certain.

Muscular System

At least 150 pairs of muscles have been mentioned or described by various authors (e.g., Ch'eng-Pin 1940, Millot and Vachon 1949). The most obvious body muscles are the dorsoventral muscles. These attach to a tergum dorsally and to a sternum ventrally within the same segment. The mesosomal dorsoventral muscles function as compressors and serve a vascular function in regulating abdominal blood pressures (Firstman 1973; see also "Respiratory System," below). Specific muscle nomenclature is beyond the scope of this chapter, although the general musculature of various structures is covered in the appropriate sections.

Appendages

The appendages of the scorpion, as of other arthropods, are extensions of the body proper that serve various functions necessary to the survival of the organism. As will be discussed below, however, scorpions have evolved appendages unique to the order, such as the pectines, and the pedipalps have been modified to grasp and crush prey. All the appendages are liberally endowed with various sensory receptors, some of which have yet to be fully studied and described.

Chelicerae

The chelicerae (Figs 2.1 and 2.7) are three-segmented chelate appendages that may be extended anteriorly from underneath the anterior margin of the carapace. There is no apparent way of identifying the segments (Snodgrass 1952), but Vyas (1971) dubbed them the coxal, the deutomerite, and the terminal sclerites. It would be preferable to retain some degree of consistency in nomenclature, and the following terminology is recommended to approximate more closely that applied to

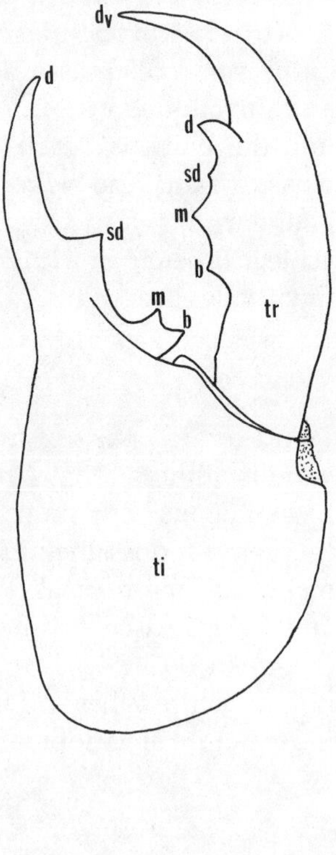

Fig. 2.7. Representative scorpion chelicera, dorsal view (coxa and coxal endosclerite omitted): *b*, basal tooth; *d*, distal tooth; *m*, median tooth; *sd*, subdistal tooth; *ti*, tibia; *tr*, tarsus. The teeth of the movable digit are further delineated as dorsal and ventral (or external and internal, respectively). By convention, the ventral tooth is indicated with the subscript "*v*."

the pedipalps: that the basal segment be called the coxa, and that the second and third segments be called the tibia (with its fixed finger) and the tarsus (movable finger), respectively. The coxa is ringlike, with its proximal margin being drawn into a process called the coxal endosclerite, to which many of the cheliceral muscles attach. The tibia is a barrel-shaped structure that is produced distally into the fixed finger. The inner ventral edge of the tibia is set with a brush of long, stout hairs, and has three functions: straining food before ingestion, grooming the pedipalps and legs, and contact chemoreception (Abushama 1964).

The fixed and movable fingers are provided with various teeth, denticles, and granules that are used to break down the harder tissues of prey items. A definite nomenclature has been set up for these structures (Vachon 1956, 1963), and their presence or absence, position, and shape have proved to be of considerable taxonomic importance (see Fig. 3.1; Vachon 1963, San Martín 1972, San Martín and Cekalovic 1972).

Vyas (1974a) determined that the movements of a scorpion chelicera are brought about by 18 muscles. Using *Heterometrus fulvipes,* he found and named 5 intrinsic muscles (originating and inserting within the appendage) and 13 extrinsic muscles (originating from outside the appendage). The movable finger has a powerful adductor (closing) muscle and a relatively weak abductor (opening) muscle. This is in contrast to the pedipalps, where abductor muscles are lacking.

Pedipalps

These second prosomal appendages consist of the coxa, trochanter, femur, patella, tibia, and tarsus (Figs. 2.1–2.3). The shapes and textures of the segments vary from one species to another (Stahnke 1970, Vyas 1971), and these appendages and their attendant structures are taxonomically very important. The sensory setae called trichobothria, discussed below, are present only on the pedipalps.

Barrows (1925), noting the lack of an abductor muscle to the movable pedipalp finger, hypothesized that the chela is opened by hydraulic pressure, that is, by an increase in blood pressure. Mathew (1965, 1966) removed portions of the cuticle from the pedipalp of *Heterometrus scaber,* thereby creating a major reduction in blood pressure. These scorpions were still able to open their chelae, and Mathew, as well as Dubale and Vyas (1968), accordingly proposed that the adductor muscles apply tension on the corium (the hinge area at the base of the movable finger), causing it to infold upon itself, and that when the adductor

TABLE 2.1

Nomenclature applied to scorpion leg segments by various authors

<table>
<tr><td>Börner 1903</td><td>coxa</td><td>trochanter</td><td colspan="2">femur</td><td colspan="2">tibia</td><td colspan="2">basitarsus</td><td colspan="3">tarsus 1</td><td>tarsus 2</td><td>pretarsus</td></tr>
<tr><td>Hansen 1930</td><td>precoxa</td><td>transcoxa</td><td>prefemur</td><td>femur</td><td>patella</td><td>tibia</td><td colspan="6">tarsus</td><td>transtarsus</td></tr>
<tr><td>Millot & Vachon 1949</td><td>coxa</td><td>trochanter</td><td>prefemur</td><td>femur</td><td colspan="2">tibia</td><td colspan="3">basitarsus</td><td colspan="3">tarsus</td><td>posttarsus</td></tr>
<tr><td>Vachon 1957</td><td>coxa</td><td>trochanter</td><td>prefemur</td><td>femur</td><td colspan="2">tibia</td><td colspan="3">basitarsus</td><td colspan="3">tarsus</td><td>apotele</td></tr>
<tr><td>Snodgrass 1952</td><td>coxa</td><td>trochanter</td><td colspan="2">femur</td><td colspan="2">patella</td><td>tibia</td><td colspan="3">tarsomere I</td><td colspan="2">tarsomere II</td><td>pretarsus</td></tr>
<tr><td>Stahnke 1970, Vyas 1971</td><td>coxa</td><td>trochanter</td><td colspan="2">femur</td><td colspan="2">patella</td><td>tibia</td><td colspan="3">tarsomere I</td><td colspan="2">tarsomere II</td><td>pretarsus</td></tr>
<tr><td>Couzijn 1976</td><td>coxa</td><td>trochanter</td><td colspan="2">femur</td><td colspan="2">patella</td><td>tibia</td><td colspan="3">basitarsus</td><td colspan="2">telotarsus</td><td>apotele</td></tr>
</table>

muscles relax, the tension on the corium is released and the chela opens. Further, Govindarajan and Rajulu (1974) showed that the corium contains resilin or a protein resembling it. Since a property of resilin is that it regains its original shape or condition after having been compressed or stretched, this would account for the capability of extending the movable finger without benefit of abductor muscles. But hydraulic pressure may yet be found to play some role in this mechanism.

The fixed and movable fingers of the pedipalp chelae are set along their cutting edge with granules that are found in various patterns. Specific nomenclature was given by Stahnke (1970). The chelae also occasionally exhibit sexual dimorphism, for example, heavy crenulations of the cutting edges of the fingers of the male that are lacking in the female. The male may also exhibit processes or depressions on the manus that aid in holding the pedipalp fingers of the female during courtship (Fig. 4.20; Maury 1975a).

Legs

The nomenclature applied to the segments of the legs has varied considerably (Fig. 2.4). All authors appear to agree that there are eight segments, but they have applied different names to them or have subdivided segments differently. Table 2.1 compares the nomenclature used by several authors, and that presented in this chapter is a blend of this information.

The most recent discussion of scorpion leg terminology and homology is included in a study of the legs' functional anatomy (Couzijn 1976). In addition to providing a glossary and list of synonyms, Couzijn introduced or revived several terms in an attempt to provide "a homogeneous, . . . precisely defined, terminology." One of these terms is *apotele,* which refers to the pretarsus, posttarsus or transtarsus of other authors. This term was introduced by Grandjean (1952) for mites and was first used for scorpions by Vachon (1957). Couzijn continued the usage because the segment possesses its own elevator and depressor tendons, a condition he describes as eudesmatic. Other new or reactivated terms in Couzijn's paper include *coxapophyses* (coxal endites of legs I and II); calcar (*pedal spur,* between the basitarsus and the tarsus, originally used in Birula 1905; see Figs. 2.4 and 2.15); pseudonychium (*dactyl,* or median claw of the apotele); and ambulacrum (a functional term referring to the terminal part of the leg that is in regular contact with the substrate). It remains to be seen whether or not these terms, as well as other functional terms not mentioned here, become accepted and commonly used.

Functionally, the legs are ambulatory, none of them being modified for any other purpose as are those of some other arachnids. However, they are occasionally used for non-ambulatory purposes, such as digging, or by the female for catching the emergent young at birth. The legs are liberally endowed with proprioceptors, bristles, and sensory setae, which are described in the next section. A tibial spur is found at the tibio-basitarsal joint of some Buthidae. Only three species of scorpions are known to lack pedal spurs (Stahnke 1970, Mitchell and Peck 1977).

Snodgrass (1952), Manton (1958), Vyas (1971), Couzijn (1976), and Bowerman and Root (1978) provided extensive analyses of the functional morphology and musculature of the legs, which will not be repeated here. Suffice it to say that the relative positions of the leg segments, the directions of movement, and the articulation joints are complexly interrelated as an adaptation to life in burrows and under surface objects. Snodgrass found it "remarkable that 26 muscles should be required to operate a single leg of the scorpion, in addition to those inserted in the coxa." Root (1985), however, indicated that each leg was controlled by 15 muscles.

Pectines

The pectines (sing., pecten) are comblike sensory structures unique to scorpions (Figs. 2.2, 2.5, 2.8, and 2.9), which some authors (e.g., Savory 1977) regard as having been derived from the gill lamellae of the Xiphosurida. They consist of three marginal lamellae and a variable number of median lamellae, fulcra, and pectinal teeth. Median lamellae or fulcra are absent in some taxa, and the number of pectinal teeth varies widely between taxa and to some extent between the sexes of the same species, with males nearly always having a greater number. W. Lourenço (pers. comm. 1982) has found that the number of pectinal teeth is statistically the same for both sexes in some species.

Various functions have been attributed to the pectines, including those as external respiratory organs, external genitalia, and tactile organs. After discussing the various interpretations of function, Cloudsley-Thompson (1955) concluded that their main function is the perception of ground vibrations. Carthy (1966, 1968), using his determination of the fine structure of the pectines and the behavioral evi-

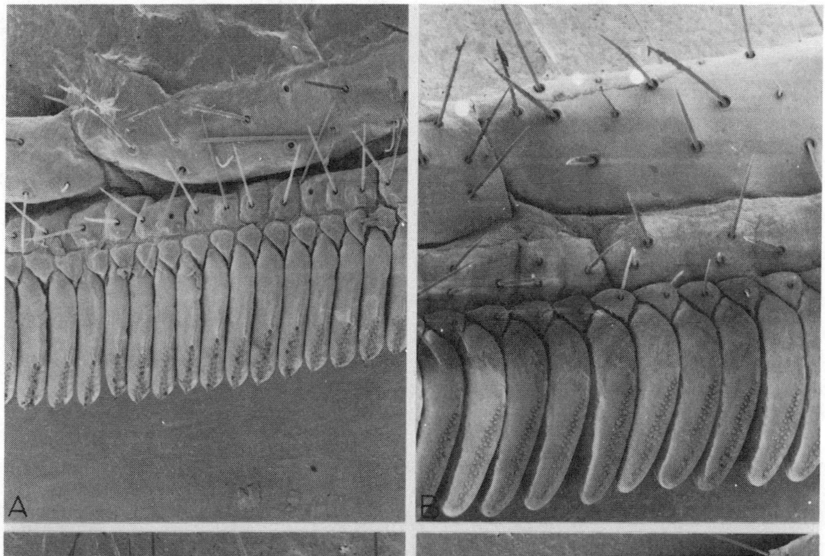

Fig. 2.8. Series of views of the pectines of juvenile to adult male *Paruroctonus mesaensis*. Note the increase in the number of setae and pectinal sensilla with age. *A*, Carapace length 1.99 mm, ×89; *B*, carapace length 2.98 mm, ×77; *C*, carapace length 3.83 mm, ×62; *D*, adult male, carapace length 8.24 mm, ×36 (Photos courtesy of Maury Swoveland, M.A. thesis, San Francisco State University)

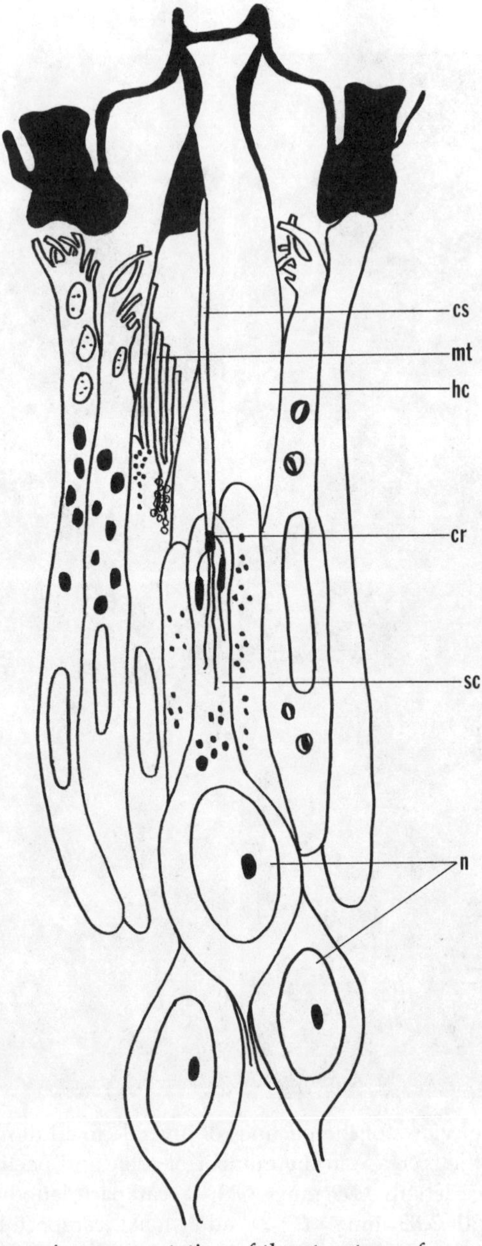

Fig. 2.9. Diagrammatic representation of the structure of a pectinal sensillum: *cr*, ciliary root; *cs*, ciliary strand; *hc*, hypodermal cell; *mt*, microtubules; *n*, nuclei of sensory cells; *sc*, sensory cell. (After Carthy 1968)

dence of Abushama (1964), concluded that the primary function of the pectines is to detect the nature of the substrate in order to select preferential areas for spermatophore deposition. Swoveland (1978) suggested three functions performed by different sensilla: habitat selection, proprioception, and chemoreception. Finally, the results of studies by Ivanov and Balashov (1979) and by Foelix and Müller-Vorholt (1983) strengthened the conclusion that the pectines function primarily as mechanoreceptors and contact chemoreceptors.

The pectines are generally set with various macrosetae and microsetae, but the most distinctive sensory structures are those found on the anteroventral margin of each pectinal tooth, called *peg sensilla* (sensilla basiconia of Sreenivasa Reddy 1959; sensory pegs of Carthy 1966, 1968). Carthy (1966, 1968), using the electron microscope on the pectines of *Leiurus quinquestriatus,* was the first to describe these sensilla in detail. He found that each pectinal tooth had about 400 sensilla, each sensillum being "blunt-ended, about 2 μm in diameter and 2 μm long, mounted on a cylindrical base 5 μm in diameter inserted into a pit in the cuticle." Swoveland (1978) compared the fine structure of the pectines of nine species of New World scorpions. He found a small sclerite, which he called the *basidens,* at the base of each pectinal tooth, on the dorsal surface of the pectines; this sclerite was found in all species examined. He also found two types of circular sensilla on two species. The first, which he called Type One circular sensilla, were found only on *Anuroctonus phaiodactylus* (an obligate burrower) and appeared to be typical stretch receptors. These were found on both sexes over all parts of the pecten. Type Two circular sensilla were found on both sexes of *Didymocentrus comondae* but were restricted to the pectinal teeth, and a chemoreceptive function was suggested. Swoveland also noted that the number of peg sensilla increases as the scorpion increases in size (see Fig. 2.8), and that, in the chactoid species studied, males have a greater number of peg sensilla than females, while the peg sensilla of females are longer than those of the male. The sexual dimorphism was less apparent in the single buthoid studied.

The most recent study of the fine structure of the peg sensilla is that by Foelix and Müller-Vorholt (1983). Significantly, they found that the tip of each peg sensillum is perforated by a slit that leads to a fluid-filled chamber separated from the lower part of the peg shaft by a cuticular platelet. This platelet has an open connection to the lumen of the shaft below, allowing the dendrites in the lumen to communicate with the outside. They also found that each peg sensillum is dually innervated, providing the dual function of mechanoreception and contact chemoreception. Finally, they determined that the axons of the peg sensilla form synaptic connections with each other, giving rise to a ganglionic structure located within each pecten.

Integument

The integument of scorpions, as in other terrestrial arthropods, serves several purposes: it provides a protective covering for the internal organs and tissues; it is resistant to water loss, particularly important for those species inhabiting xeric areas; it provides points of attachment for muscles, making locomotion and other movement possible; and, through its respiratory openings and sensory structures, it is the organism's primary connection to its environment. The basic structure of the scorpion integument is essentially the same as that detailed for other arthropods by Wigglesworth (1947), Richards (1951), and Chapman (1971), although terminology differs and there are some unique features.

A diagrammatic representation of the integument of *Euscorpius italicus* was presented by Pavan (1958; see Fig. 2.10). He indicated that the integument consists of five layers: epicuticle (4 μm), exocuticle (5 μm), a layer intermediate between exocuticle and endocuticle (1 μm), various layers of endocuticle separated by membranes (35 μm), and the epidermis (15 μm). Pavan's unnamed layer between exocuticle and endocuticle has since been referred to as the colorless exocuticle (Kennaugh 1959) and the outer endocuticle (Malek 1964). The preferred name, however, seems to be *mesocuticle* (Dennell 1975, following the nomenclature of Schatz 1952 and Lower 1956).

Dennell (1975) examined the mesocuticle of the pedipalps, mesosomal terga and sterna, and legs of *Pandinus imperator.* He found that the ducts of integumental glands were abundant in some areas but absent in others, and that clearly defined laminae were closely spaced in the regions devoid of ducts and more widely spaced where ducts were present. Although the mesocuticle of the pedipalps and mesosomal sclerites were similar, he found striking differences in the legs, in that the fibers were aligned in parallel with the long axis of the podomere. He suggested that this axial arrangement of fibers is related to the incidence of mechanical stress and that it makes the cylindrical podomeres more resistant to bending. The more random distribution of fibers in

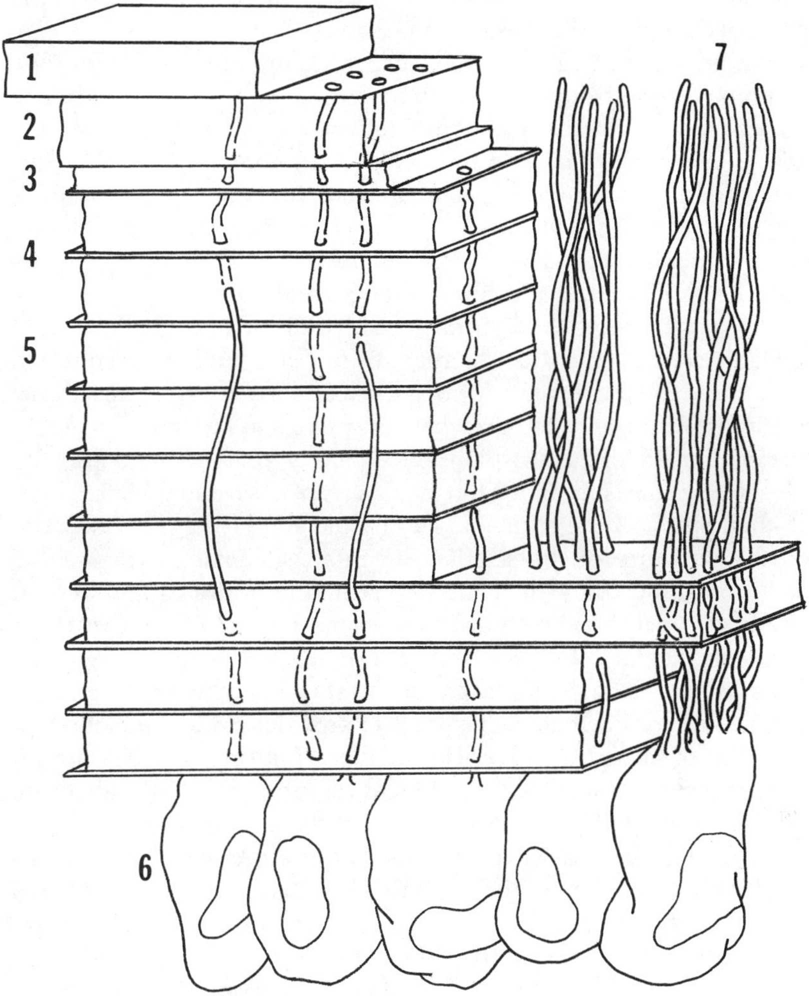

Fig. 2.10. Diagrammatic representation of the cuticle of *Euscorpius italicus: 1*, epicuticle; *2*, esocuticle (hyaline exocuticle and mesocuticle); *3*, intermediate layer (inner exocuticle); *4*, lamellae of endocuticle; *5*, interlamellar endocuticular membranes; *6*, epidermis; *7*, pore canals. (After Pavan 1958)

other parts of the body, he indicated, "may be more appropriate to the requirements imposed by less localized stress."

In the pedipalp of *Heterometrus fulvipes,* Dalingwater (1980) found a nonlaminate zone below the fourth or fifth lamina of the mesocuticle that begins near the bases of both fixed and movable fingers and extends to their tips. Similar to Dennell's (1975) findings about fiber arrangement in the legs, the nonlaminate zone contained sheets of fibers in parallel alignment with the long axis of the fingers. He concluded that these nonlaminate zones of mesocuticle appear in those regions where the cuticle needs additional strength and rigidity.

Using transmission electron microscopy, Filshie and Hadley (1979) and Hadley and Filshie (1979) studied the fine structure of the cuticle in general, and of the epicuticle in particular, of *Hadrurus arizonensis.* Their interest concentrated on what contributes to the impermeability of the integument of this scorpion, since it has one of the lowest cuticular permeabilities measured for a terrestrial organism (Hadley 1970a).

In their study of the entire cuticle, Filshie and Hadley (1979) examined the structure of the sclerite and intersegmental cuticles of the opisthosoma and found that the fine structure of dorsal and ventral sclerotized cuticle was essentially the same, but the fine structure of the intersegmental cuticle differed in several respects from that of the sclerite cuticle. Since their study is the most recent on the subject, a résumé of their results follows; Figure 2.11 provides a diagrammatic representation of the cuticle. (See also Chapter 7 for further discussion.) If the figures for the thickness of the cuticle are compared with those given by Pavan (1958), it will be seen that the cuticle of *Hadrurus arizonensis* is more than twice as thick as that of *Euscorpius italicus* (omitting epidermis), whereas the epicuticle of *Euscorpius* is many times thicker than that of *Hadrurus.* Although the possibility of mismeasurement exists, it is logical that the cuticle of a desert-dwelling arthropod like *Hadrurus* would be thicker than that of *Euscorpius,* which lives in more mesic areas.

The overall thickness of the sclerite cuticle of *Hadrurus arizonensis* is about 100 μm (compared to about 45 μm for *Euscorpius italicus*):

1. The epicuticle is about 0.3 μm thick and consists of four layers: the outer membrane (cement layer of Wigglesworth 1947; tectocuticle of Richards 1951), about 3 mμ, irregular in thickness; the outer epicuticle, about 3 mμ; the cuticulin layer, about 5 mμ; and the dense homogeneous layer making up the bulk of the thickness.

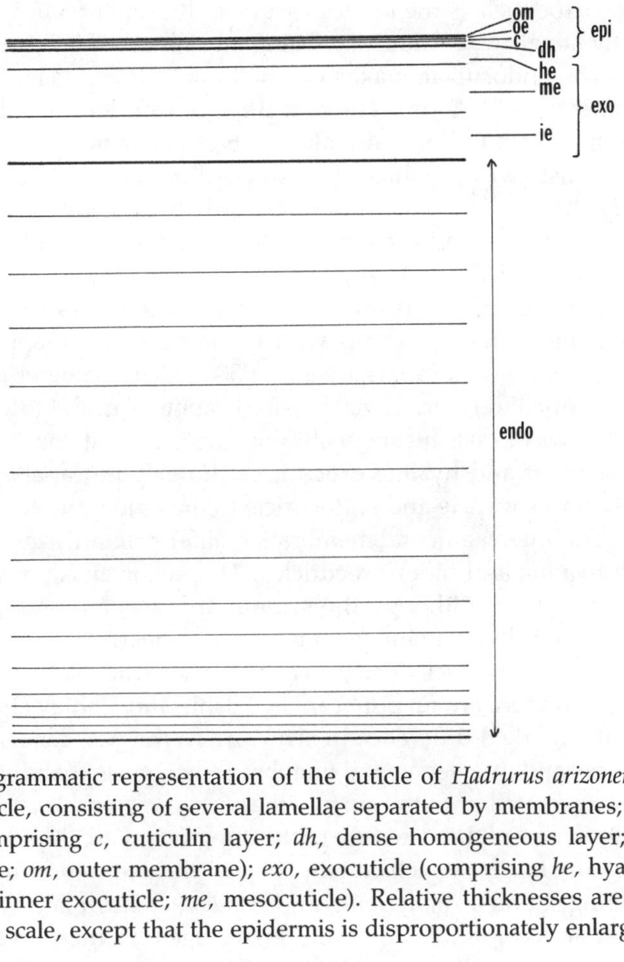

Fig. 2.11. Diagrammatic representation of the cuticle of *Hadrurus arizonensis: endo,* endocuticle, consisting of several lamellae separated by membranes; *epi,* epicuticle (comprising *c,* cuticulin layer; *dh,* dense homogeneous layer; *oe,* outer epicuticle; *om,* outer membrane); *exo,* exocuticle (comprising *he,* hyaline exocuticle; *ie,* inner exocuticle; *me,* mesocuticle). Relative thicknesses are approximately to scale, except that the epidermis is disproportionately enlarged.

2. The exocuticle consists of three sublayers. The first, the hyaline exocuticle (Kennaugh 1959), is the outermost fibrous component of the sclerotized cuticle and apparently does not have a counterpart in other arthropod groups. It is about 7.5 μm thick and consists of two sublayers: the outer layer, about 2 μm thick, is lamellate, with about 20 lamellae of differing thicknesses; the inner layer, about 5.5 μm and nonlamellate, corresponds to the mesocuticle of Dennell (1975) and Dalingwater (1980). I am inclined to retain the term "mesocuticle" to describe this layer, owing to its descriptive nature. The innermost layer

of the exocuticle is the inner exocuticle. It is about 5 μm thick and distinctly lamellate, the lamellae being an average of 0.2 μm.

3. The endocuticle makes up the bulk of the cuticle, having a total thickness of 65–85 μm. The first three or four lamellae deposited (the outermost) are uniform and about 5.5 μm. The next couple of lamellae are almost twice as thick, but succeeding lamellae become progressively thinner, until the innermost are about 1 μm. The lamellae are clearly defined, and a membrane separates the inner exocuticle from the endocuticle.

4. A series of interconnecting channels traverses the entire cuticle and connects the epidermis with the surface of the epicuticle. In the epicuticle, the wax canals are about 500 mμ in diameter and penetrate to the cuticulin layer. Their density is about $1/\mu m^2$. Outer pore canals join the wax canals just beneath the epicuticle and pass to the junction of the inner and hyaline exocuticles. Inner pore canals pass through the inner exocuticle and endocuticle to the epidermis.

5. The intersegmental membranes differ primarily in the absence of both hyaline and inner exocuticles. The major divisions are epicuticle and procuticle. Although the same four layers are present in the epicuticle, the dense homogeneous layer is nearly seven times thicker, averaging 2 μm thick. The procuticle varies from 30 to 40 μm in thickness, and there are no pore canals. Filshie and Hadley (1979) indicated that the marked difference in structure of the wax canals between the sclerites and intersegmental membranes suggests that different materials are secreted onto the surface in the two regions.

Also a part of the structure of the integument are the so-called stridulatory organs, by which certain species are able to produce buzzing or scraping sounds. *Androctonus* and *Parabuthus* stridulate by scraping the telson across granules or transverse ridges on the dorsal surfaces of metasoma I and II (Pocock 1902b, 1904; Millot and Vachon 1949; Kästner 1968; Newlands 1974a). According to Pocock (1896, 1904), Millot and Vachon (1949), and Kästner (1968), *Opistophthalmus* creates a sound by extending and retracting the chelicerae, causing dorsal bristles or chitinous platelets to rub against the anterior edge of the carapace. Alexander and Ewer (1957), however, ascribed a sensory rather than a stridulatory function to these structures. Other stridulatory granules or bristles, or both, have been described from the basal segments of the pedipalps and of the first pair of legs of *Heterometrus* and *Pandinus* (Pocock 1896, 1904; Millot and Vachon 1949) and from the pectines of *Rhopalurus* (Millot and Vachon 1949). Constantinou and Cloudsley-

Thompson (1984) reviewed the stridulatory structures found in the families Scorpionidae and Diplocentridae.

Two topics relate to a discussion of the integument of scorpions: the recently recognized extraocular light sensitivity of the metasoma, and the well-known phenomenon of fluorescence of the sclerotized portions of the cuticle. Zwicky (1968, 1970a) first noted the existence of extraocular photoreceptors in the metasoma of *Urodacus*. Geetha Bali and Pampathi Rao (1973) later found similar photoreceptors in two species of *Heterometrus*. Their findings differed from Zwicky's in two important respects: first, only a single peak, at 480 nm, was found in *Urodacus*, whereas a large peak at 568 nm and a smaller peak at 440 nm were found in *Heterometrus;* and second, cutting the telsonic nerves did not completely abolish the light sensitivity of the metasoma, unlike the situation in *Urodacus*.

Pavan (1956) demonstrated that the fluorescent material is concentrated in the surface epicuticle. Stahnke (1972a) mentioned that during ecdysis the exuviae fluoresce brightly but the emergent animal does not, with full fluorescence being attained about 48 hours after emergence. He also found that if the emergent scorpion is at once frozen and lyophilized, the fluorescent quality never develops. Thus, the fluorescent material is probably part of the complex mixture of mucopolysaccharides and lipids produced by dermal and epidermal glands that are brought to the epicuticle via pore and wax canals to provide the impermeable quality of the integument. Filshie and Hadley's (1979) suggestion that different materials are secreted onto the surface of the intersegmental membranes than onto the sclerites may indicate why the intersegmental membranes do not fluoresce.

Koehler (1979) studied the physical properties of scorpion fluorescence. He found that the general form of the emission spectrum was that of a skewed single-peak Gaussian curve with the peak at approximately 472 nm. All specimens tested, whether living or dead, had the same peak position, regardless of species, locality, coloration, or method of preservation. Although the peak position remained invariant, the bandwidth varied from 70 to 78 nm owing to coloration. The bandwidth varied appreciably between light and dark specimens, with lighter scorpions emitting over a broader spectrum. The shapes and positions of the emission and excitation spectra indicated mirror fluorescence, in which emission and excitation curves form approximate mirror images of one another. At the excitation maximum, the fluorescence-emission intensity was less than 2 percent of the intensity of the exciting radiation. Using Zwicky's (1968, 1970a) data on extraocular light sensitivity, Koehler determined that it is possible for a scorpion to detect very low intensities of light at certain wavelengths. The large neural response detected at 480 nm (by Zwicky) was possibly produced by incident light of an intensity not greater than 40 $\mu W/cm^2$.

To determine the intensities of fluorescence a scorpion could produce under natural ultraviolet excitation, Koehler used the solar energy distribution, with the sun at its zenith, at Tucson, Arizona, for *Centruroides exilicauda*. He found that, taking into account only the exciting radiation in the ultraviolet region, scorpions could conceivably radiate energy in the visible range, with emission intensities on the order of tens of microwatts per square centimeter of epicuticle; consideration of the exciting wavelengths in the visible region would increase the emission-intensity figures. He concluded that it is possible for fluorescence to serve as an ultraviolet-sensitivity mechanism in scorpions, although more study is needed.

Incidentally, other arthropods are known to fluoresce, including fluorescent sowbugs in the Mojave Desert of California and moth pupae with fluorescent cocoons in northern California. There have also been other unpublished reports of fluorescence of the intersegmental membrane areas of some species of solpugids (rather than the sclerotized cuticle as in scorpions).

Sense Organs

Eyes

The single pair of median eyes (Fig. 2.1) is situated on either side of the median plane of the carapace, usually just anterior to the midline but in some species considerably anterior or posterior to the midline. Along each anterolateral margin of the carapace are from zero to five lateral eyes, which vary in number among species. These are usually paired, that is, with the same number of eyes on each side, although specimens are occasionally found with, for example, three lateral eyes on one side and two or four on the other. In troglobitic species the eyes are much reduced or absent. The median eyes are sometimes referred to as diplostichous because they are derived from two layers of hypodermis, whereas the lateral eyes may be termed monostichous (Savory 1977).

Bedini (1967) determined that the dioptric apparatus of the median

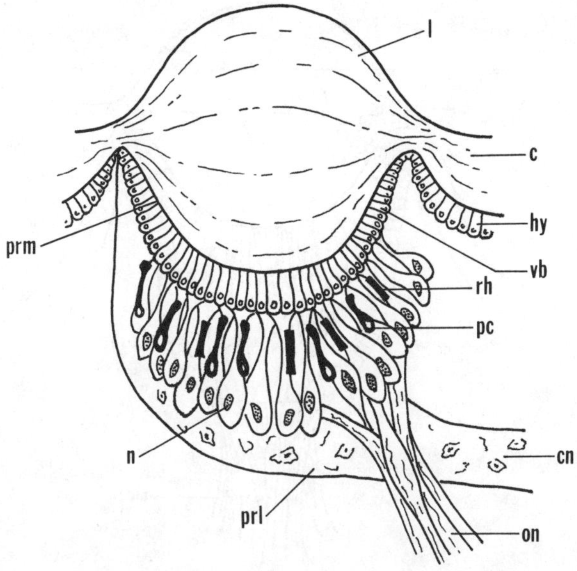

Fig. 2.12. Diagrammatic representation of a median eye, with only a few pigment cells included: *c*, cuticle; *cn*, connecting nerve to other eye; *hy*, hypodermis; *l*, lens; *n*, nucleus of retinula cell; *on*, optic nerve; *pc*, pigment cell; *prl*, postretinal layer; *prm*, preretinal membrane; *rh*, rhabdomere; *vb*, vitreous body. (After Machan 1967)

eyes of *Euscorpius carpathicus* is composed of a lens and a vitreous body, which consists of transformed epidermal cells. The complete eye (Fig. 2.12) is made up of a lens, a vitreous body, a preretinal membrane, a retina, and a postretinal membrane, which continues on to form the optic nerve sheath. A single layer of pigment cells surrounds the whole retina. The lateral eyes (Fig. 2.13) consist of a lens, a preretinal membrane, a retina, a layer of pigment cells, and a postretinal membrane, without a vitreous body.

Schliwa and Fleissner (1980) undertook a more detailed examination of the lateral eyes of *Androctonus australis*. Their work basically agreed with that of Bedini (1967), but they provided further information. In the lateral eyes, the retinula cells—the main cellular constituent of the retina—bear a distal receptive segment characterized by rhabdomeral microvilli that interdigitate with microvilli of neighboring cells to form a rhabdom. The rhabdomeres of one cell are contiguous with those of neighboring cells, so that the entire retina forms a contiguous rhabdom

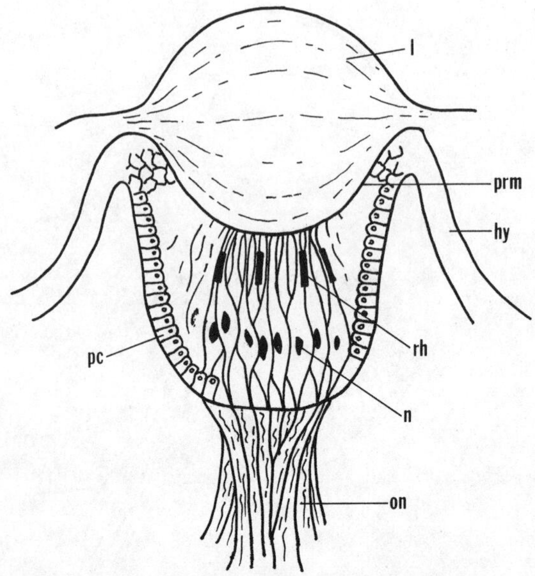

Fig. 2.13. Diagrammatic representation of a lateral eye, with a portion of the hypodermal cells omitted to show the proximal ends of retinula cells: *hy*, hypodermis; *l*, lens; *n*, nucleus of retinula cell; *on*, optic nerve; *pc*, pigment cell; *prm*, preretinal membrane; *rh*, rhabdomere. (After Machan 1967)

network, unlike the discrete rhabdom arrangement in the median eyes. This lack of insulation between rhabdomeres suggests the existence of an optical and electrical coupling between neighboring retinula cells resulting in electrical cross talk. Arhabdomeric cells are probably an important part of this system. These cells, present in both median and lateral eyes, do not contribute to the structure of the rhabdom, but they are in intimate contact with the retinula cells via extensions of the dendrite. It appears that the arhabdomeric cells serve as secondary neurons involved in the processing of visual information. The ratio of retinula cells to arhabdomeric cells is approximately 2:1 in the larger lateral eyes, 5:4 in the smaller lateral eyes, and 5:1 in the median eyes.

In a study of albino individuals of *Urodacus yaschenkoi*, Locket (1986) found that the rhabdoms were abnormal, consisting of reduced and disorganized microvilli, although all components of the eye were present.

Retinula cells of both median and lateral eyes display characteristic movements of pigment granules within their cell bodies. In the light-adapted state, pigment granules accumulate in the distal portion of the cell, close to the preretinal membrane. In the dark-adapted state, pigment granules gather near the nucleus. When scorpions are maintained in constant darkness, retinula cells of the lateral eye show only weak circadian pigment movements, in contrast to median eyes, where displacements of the screening pigment are dramatic under even these conditions.

Schliwa and Fleissner (1980) concluded that the importance and biological role of the lateral eyes differ from those of the median eyes. They state that the lateral eyes are highly sensitive and can detect differences in brightness even at very low light intensities, and that they assume a role as light detectors of *Zeitgeber* stimuli for the synchronization of the circadian clock (see also Fleissner 1975). Their description of the scorpion visual system is that of two distinct types of eyes: one (the median eyes) provided with relatively high acuity and good spatial discrimination, and the other (the lateral eyes) with high absolute sensitivity. (See also Chapter 9 for a more complete discussion.)

Proprioceptors and Sensory Projections

Various kinds of proprioceptors in arthropods serve to measure the onset, direction, and velocity of movement and determine the position of one segment or subsegment in relation to another. In scorpions, the slit sensilla (Fig. 2.14), first described by Bertkau (1878), are proprio-

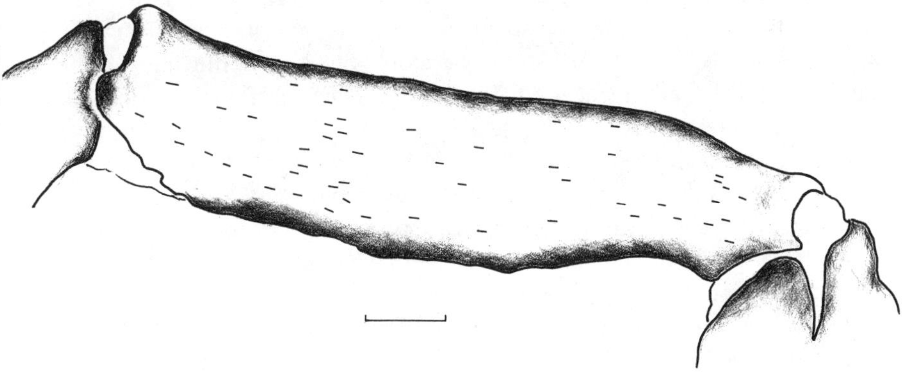

Fig. 2.14. Anterior view of the patella of the right fourth leg of a scorpion showing the many isolated slits (–) shorter than 30 μm. Bar = 1 mm. (After Barth & Wadepuhl 1975)

ceptors. In spiders, slit sensilla occur in characteristic groups called lyriform organs, but these are lacking in scorpions, in which only isolated and grouped slits are found (Barth and Wadepuhl 1975). Pringle (1961) discussed the homologies of different arthropod proprioceptors and how a knowledge of their structure and physiology can contribute to a better understanding of the patterns of arthropod movement. The slit sensilla detect deformations in the exoskeleton resulting from loads to which the cuticle is exposed (Barth 1978). They form elongate holes in the cuticle, which are covered by a membrane to which the dendritic process of a sensory cell attaches. An axon from the sensory cell unites with small nerves located just below the cuticle.

The slit sensilla of the legs have been most closely examined, but Pampathi Rao and Murthy (1966) described two pairs of proprioceptors on each metasomal segment. Labeled Organ I (anterior) and Organ II (posterior), these receptors monitor the bending movements of the metasoma; the organs of metasoma V also monitor the movements of the telson.

Barth and Wadepuhl (1975) and Barth (1978, 1985), studying the slit sensilla of the legs, found that slit sensilla occur on all segments of the leg, including the coxapophyses of legs I and II and the apotele. They defined two types of sensilla groupings: an isolated organ, which measures 30 μm or more in length and lies at least 100 μm from its closest neighbor; and a group of single slits, which comprises at least one slit measuring 30 μm or more plus a minimum of one additional slit less than 100 μm away. Most of the grouped slits as well as the large isolated slits occur on the lateral surfaces of the leg and are concentrated near the joints (Fig. 2.15). Small isolated slits and some large slits can be found on all surfaces of the leg. The slits are usually oriented roughly parallel to the long leg axis (less than 45° angle of deviation), one exception being a group on the anterior surface of the basitarsus. A slit sensillum is deformed most easily by loads perpendicular to its long axis, and since slit length varies from about 5 μm to more than 200 μm, a broad spectrum of sensitivities is probable (Barth 1978).

Brownell and Farley (1979a) discovered another function for certain of the slit sensilla on the basitarsus of *Paruroctonus mesaensis*, that of prey detection. This psammophilic species locates its prey by orienting to substrate vibrations produced by movements of the prey in sand. The sense organ, which they called the basitarsal compound slit sensillum (BCSS), consists of eight slits on the thick cuticle of the basitarsal

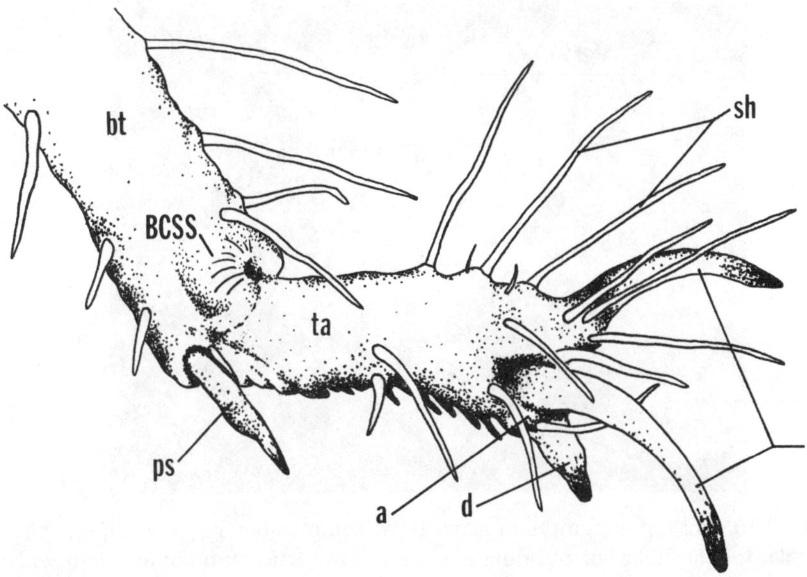

Fig. 2.15. Anterodorsal view of the right fourth leg of *Paruroctonus mesaensis: a*, apotele; *BCSS*, basitarsal compound sense sensilla; *bt*, basitarsus; *d*, dactyl; *ps*, pedal spur; *sh*, sensory hairs; *ta*, tarsus; *u*, ungues. (After Brownell & Farley 1979a)

condyle, just above the tarsal hinge. Two anterior long slits are oriented perpendicular to the long axis of the leg, and the shorter, posterior ones are oriented parallel to the leg axis. Interposed between these are slits of intermediate length and orientation, the entire group forming a fan-shaped array about 200 μm wide and 300 μm long (Fig. 2.15). The BCSS is sensitive to surface vibration waves and can respond to insects moving more than 15 cm away. Although the BCSS is necessary for determining the direction of the vibration source, it appears that the integration of input from receptors on two or more legs is necessary for target localization.

In general, one of the least-studied areas of scorpion morphology and behavior has been that of the sensory projections. This has resulted in a somewhat spotty knowledge of the various structures involved and a lack of standard nomenclature. Although some of the structures discussed below are considered nonsensory by some, they are included because the case is not certainly proved. These projections

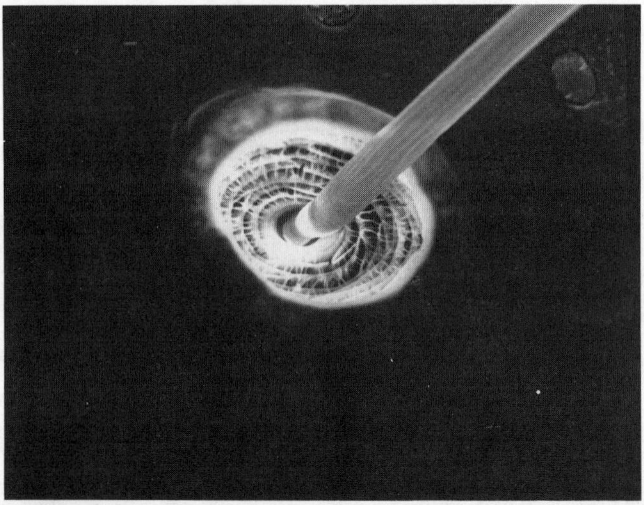

Fig. 2.16. SEM photograph of a trichobothrium, showing the cup-shaped areola, the meshlike surrounding cristae, and the articulation of the shaft with the base. Patella of *Anuroctonus phaiodactylus*, ×720.

may be divided into two groups: immovable projections, or spines, which are found at various locations on the body, particularly the legs; and movable projections.

Stahnke (1970) considered these movable projections to be setae and defined three types: first, microchaetes—small, whitish, fine setae with a base attached to a poorly developed areolar cup, found on the fulcra of the pectines; second, macrochaetes—large, colored setae, each arising from a somewhat cup-shaped areola that is completely filled by the base of the seta, found on various areas of the body, but particularly on the metasoma, on sternum 7 of the mesosoma, on the intercrestal area of the anal arch, and on the pectines (he considered these setae nonsensory); and third, trichobothria—long, thin setae that react to the slightest movement of air, each arising from a cup-shaped areola that is not completely filled by the base of the seta, found only on the femur, patella, and tibia of the pedipalps (Fig. 2.16; see Reissland and Görner 1985).

The trichobothria are known to respond to the movements and vibrations of air, and their location on the pedipalps results in efficient detection and location of other scorpions and of prey items as well as recognition of a threatening situation. The number and placement of trichobothria varies between taxa, and Vachon (1973) instituted a detailed nomenclature in an attempt to standardize terminology at the order level. The trichobothrial patterns have proved to be important characters and have been used in many systematic studies (e.g., Gertsch and Soleglad 1972, Maury 1973a, Soleglad 1975).

Abushama (1964) divided sensory "hairs" into four types (excluding trichobothria, which she called "thin hairs"):

TYPE 1: thin, about 0.2 mm long, not visible to the naked eye; found widely over the body, covering the vesicle of the telson extensively, the pedipalps, and, to a lesser extent, the legs and intersegmental regions of the metasoma.

TYPE 2: thin, about 1 mm long, just visible to the naked eye; found in all hirsute parts, but mainly covering the pectines, intersegmental membranes of the metasoma, pedipalps, telson, and legs.

TYPE 3: about 5 or 6 in number, 0.5–1 mm long, not visible to the naked eye; found only on the dorsal side of the distal tarsi of the legs.

TYPE 4: strong, curved, and spiny, 1–1.5 mm long, covering the tarsi of the legs.

Abushama, by observing scorpion behavior after painting or amputating various body parts, considered the Type 1 hairs to be thermal receptors and the Type 3 hairs to be humidity and tactile receptors. No sensory function was indicated for hairs of Types 2 or 4, although the latter are probably immovable spines.

Brownell and Farley (1979a) found that tarsal sensory hairs are excited by substrate vibrations along with the slit sensilla (Fig. 2.15). These setae respond best to compressional waves, and the receptors consist of a single bundle of cell bodies below each hair socket, which give rise to dendritic processes terminating on a cuticular fold at the base of each shaft.

Foelix and Schabronath (1983) and Foelix (1985) investigated the fine structure of mechanoreceptor hairs, slit sensilla, chemoreceptor hairs, and tarsal organs of the tarsus and basitarsus. They found that the structure of the various receptors closely resembles that of spider sensilla and, unlike insect sensory organs, always has multi-innervated mechanoreceptor hairs with a higher number of sensory cells per sensillum. Their conclusions concerning mechanoreceptor hairs and slit sensilla generally follow the results of the other workers cited above. They describe the chemoreceptor sensilla as being 100 μm long and 4

μm in diameter (compared to 200–500 μm long and 15–25 μm in diameter for mechanoreceptor hair sensilla). The tip is blunt, opens to the outside, and is innervated by 22 or 23 bipolar neurons. The authors also found tarsal organs on all legs, situated dorsally behind the claws, for which a hygroreceptive function was suggested.

Occasionally, collections of bristles or setae have specialized functions. Examples of these would be the bristlecombs on the distal leg segments of psammophilous species, which aid in walking on sand, the hairbrush of the chelicerae, and the maxillary brush of coxapophysis I (see "Digestive System," below).

Nervous System

In the numerous publications on various aspects of the scorpion nervous system, the nomenclature applied to the structures involved has varied considerably. What appears to be the most widely followed terminology is used in the following description of the nervous system and draws most heavily on the following publications: Ch'eng-Pin 1940, Tembe and Awati 1944, Henry 1949, Millot and Vachon 1949, Babu 1965, Bullock and Horridge 1965, Babu 1985, and Legendre 1985. For a histological discussion, see Babu 1965.

The central nervous system of the scorpion (Figs. 2.17 and 2.18) consists of the *brain* (supraesophageal ganglionic mass, dorsal cerebral ganglion, cerebral nerve mass, protocerebrum and tritocerebrum); the *subesophageal ganglion* (thoracic nerve mass), which is connected to the brain by a pair of thick, short *circumesophageal commissures*; three free *mesosomal ganglia*; four free *metasomal ganglia*; and associated nerves.

The brain is situated above the esophagus in the anterior part of the cephalothorax and gives rise to the following nerves: (1) a pair of *median ocellar nerves* innervating the median eyes, along with an associated pair of *tegumentary nerves*; (2) a pair of *lateral ocellar nerves*, which branch to innervate the lateral eyes; (3) three pairs of *cheliceral nerves*, which innervate the associated muscles and glands of the chelicerae; (4) a pair of small nerves that unite into a single median nerve, which extends into the rostrum and is called the *rostral nerve*; (5) a pair of tegumentary nerves arising from the posterolateral angles of the brain, which extend posteriorly, innervating the integument along the lateral and posterior borders of the carapace; and (6) two other nerves arising close to the base of the rostral nerve, which unite to form the *frontal ganglion*. From the frontal ganglion, a single nerve extends posteriorly along the dorsal

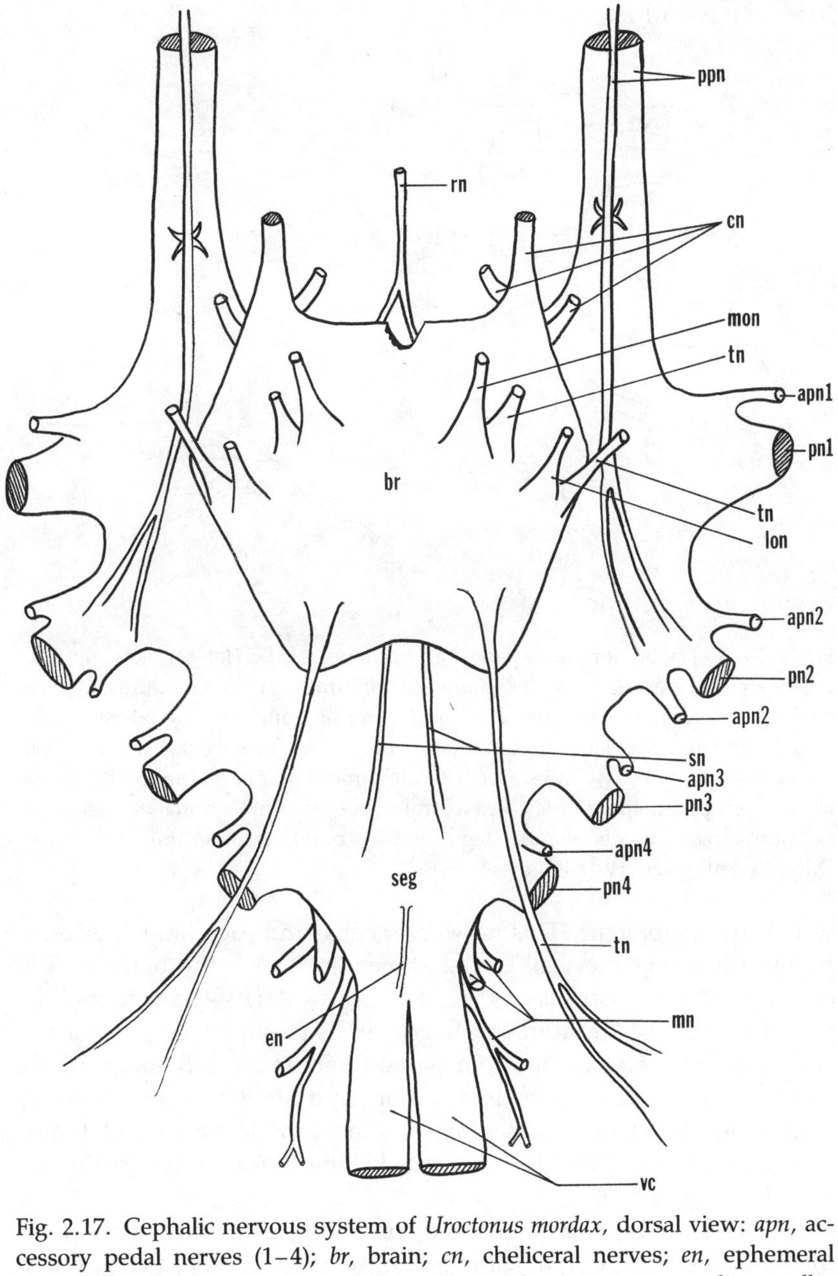

Fig. 2.17. Cephalic nervous system of *Uroctonus mordax*, dorsal view: *apn*, accessory pedal nerves (1–4); *br*, brain; *cn*, cheliceral nerves; *en*, ephemeral nerve; *lon*, lateral ocellar nerve; *mn*, mesosomal nerves; *mon*, median ocellar nerve; *pn*, pedal nerves (1–4); *ppn*, pedipalpal nerves; *rn*, rostral nerve; *seg*, subesophageal ganglion; *sn*, stomatogastric nerves; *tn*, tegumentary nerves; *vc*, ventral connectives. (Modified after Henry 1949)

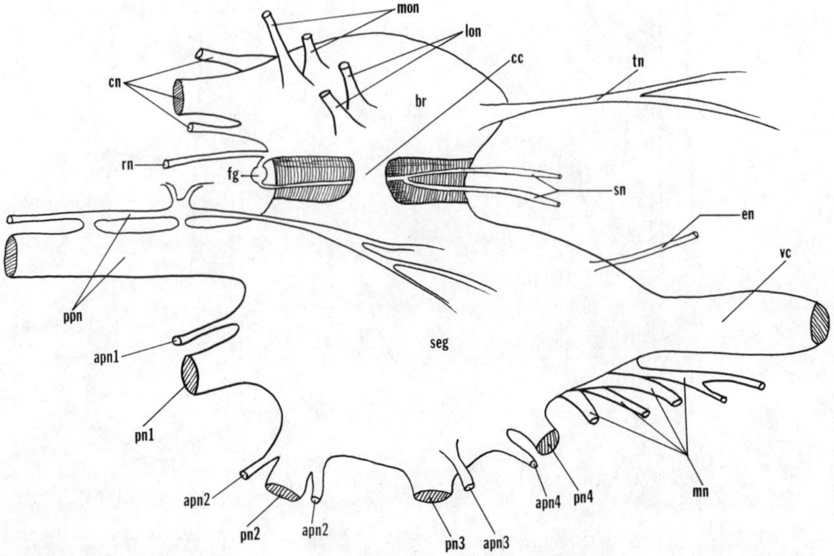

Fig. 2.18. Cephalic nervous system of *Uroctonus mordax*, lateral view: *apn*, accessory pedal nerves (1–4); *br*, brain; *cc*, circumesophageal commissure; *cn*, cheliceral nerves; *en*, ephemeral nerve; *fg*, frontal ganglion; *lon*, lateral ocellar nerve (with its associated tegumentary nerve); *mn*, mesosomal nerves; *mon*, median ocellar nerve (with its associated tegumentary nerve); *pn*, pedal nerves (1–4); *ppn*, pedipalpal nerves; *rn*, rostral nerve; *seg*, subesophageal ganglion; *sn*, stomatogastric nerves; *tn*, tegumentary nerve; *vc*, ventral connective. (Modified after Henry 1949)

wall of the esophagus. This nerve branches, and each branch extends posteriorly along the wall of the alimentary canal; the branches are called the *stomatogastric nerves*. Tembe and Awati (1944) considered the brain to consist of the supraesophageal and subesophageal ganglia and subdivided these areas into four parts: forebrain and midbrain (these two considered the brain by this author), and hindbrain and accessory brain (considered the subesophageal ganglion). The area that Tembe and Awati termed the midbrain was called the tritocerebrum by Bullock and Horridge (1965).

The subesophageal ganglion is a compound ganglion that supplies the following nerves: (1) a pair of *pedipalpal nerves*; (2) four pairs of *pedal nerves* (crural nerves; Henry 1949 also mentioned one small *accessory pedal nerve* to legs I, III, and IV, and two accessory pedal nerves to leg II);

and (3) four pairs of *mesosomal nerves* (vagus nerves), the destinations of which are uncertain. Ch'eng-Pin (1940) indicated that the first pair goes to the area around the pectines, the second pair to the genital opercula, and the third and fourth pairs to the first two pairs of book lungs. Henry (1949) said that the first pair goes to the opercula, the second are tegumentary nerves to the area around the pectines, the third innervates the pectines, and the fourth branches again, sending one branch to the area around the gonopore and a longer branch into the gonads. Babu (1965) described an unpaired, dorsal *ephemeral nerve*, which innervates the endosternite and surrounding muscles; anterior and posterior genital nerves; large pectinal nerves; and third and fourth mesosomatic segmental nerves, which comprise dorsal branches to the dorsum of the respective segment and ventral branches to the book lungs and ventral muscles.

A pair of *ventral connectives* extends posteriorly from the subesophageal ganglion and enlarges into seven free ganglia, three in the mesosoma and four in the metasoma. The three free ganglia located in the mesosoma are actually the fifth, sixth, and seventh mesosomal ganglia, and, according to Millot and Vachon (1949), their exact position varies between families. The fifth ganglion gives off a pair of nerves to the third book lung along with an unpaired visceral nerve. The sixth mesosomal ganglion innervates the fourth pair of book lungs and also extends an unpaired visceral nerve. The seventh mesosomal ganglion gives off a pair of nerves to its own segment, along with posterior branches. Yellamma et al. (1982) found that this last mesosomal ganglion consists of a larger number of cells compared to other ganglia and hypothesized a role in the stinging reflex.

The free metasomal ganglia are situated in the anterior portions of the first four segments. Each gives off a pair of branched nerves to innervate the respective segment. The fourth ganglion is a fusion of the fourth and fifth metasomal ganglia; from its posterior border, a pair of nerves extends posteriorly, each dividing into two branches, one innervating the fifth metasomal segment, and the other, consisting of giant fibers, innervating the telson (*telsonic nerves*). According to Babu (1965), giant fibers control the movements of the telson via the terminal ganglion and seem to connect with other syncytial giant fibers to the pedipalps and legs, coordinating attack movements.

Finally, two singular reports merit further study. In *Leiurus quinquestriatus*, Abd-el-Wahab (1957) described a well-developed ganglionic mass situated at the end of the lumen of the testicular cylindrical glands, from which lateral and median nerves extended to the wall of the gland and to the paraxial organ. In *Centruroides* species, Keegan and Lockwood (1971) noticed that the dorsal gland, an unpaired structure dorsal to the muscles surrounding the venom glands of the telson, grossly resembled a ganglion of the nervous system. These observations have been neither corroborated nor disputed by other researchers.

Respiratory System

The respiratory system consists of four pairs of respiratory organs, called *book lungs*, that are located in the mesosoma. They open to the surface through apertures called *spiracles* (stigmata), which are located on the sterna of mesosomal segments 3 through 6 (Figs. 2.2 and 2.23) and may be slitlike, oval, elliptical, or circular in shape. The description of the book lungs in Snodgrass (1952) is still basically accurate, with a few additional terms and structures introduced by Mill (1972), Vyas and Laliwala (1972a), and Vyas (1974b), from whom this discussion is taken.

Each book lung is enclosed in a pulmonary sinus and is covered by a sheath of connective tissue. The book lung contains two chambers: a dorsal *pulmonary chamber* (covered dorsally by the connective tissue sheath) and a ventral *atrial chamber*, which tapers posteroventrally into the spiracle (Figs. 2.19 and 2.20). The pulmonary chamber contains a variable number of thin, hollow, leaflike structures called *lamellae*. The scorpions examined thus far appear to have about 140 to 150 lamellae in each book lung. The lamellae are attached to the posterior wall of the pulmonary chamber, and each is a tubular extension of the epithelium of the posterior wall. The distal free ends are covered by a thin layer of cuticle that, along with the cuticular layer of the atrial chamber, is cast off and renewed with each molt. The space between two adjacent lamellae is called the *interlamellar space* (lumina of Mill 1972), and these spaces communicate with the atrial chamber via slitlike openings in the roof of the atrial chamber. Adjacent lamellae are held apart by bristles (cuticular bars of Mill 1972), although Pavlovsky (1926) mentioned that in *Chaerilus*, "the free margin of the laminae [lamellae] is devoid of spinelets." The hollow interior of each lamella, called the *lumen*, is partitioned toward its atrial end, forming a small space called the *epithelial sinus* (Fig. 2.21; Vyas 1974b).

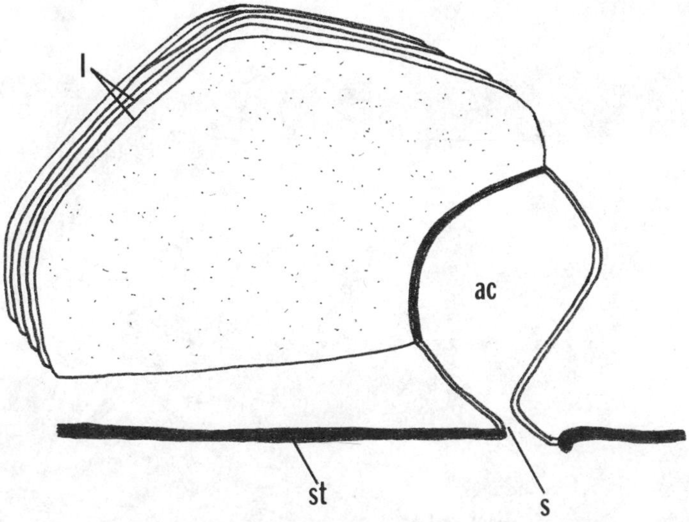

Fig. 2.19. Diagrammatic representation of the scorpion book lung, vertical section: *ac*, atrial chamber; *l*, lamellae within the pulmonary chamber; *s*, spiracular opening; *st*, sternite. (After Snodgrass 1952)

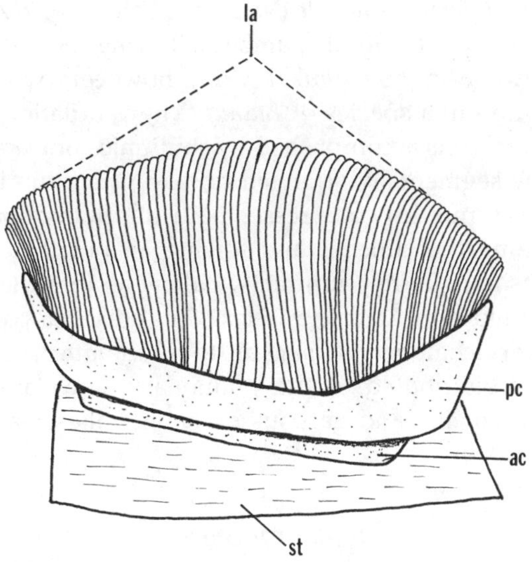

Fig. 2.20. Diagrammatic representation of the scorpion book lung, dorso-posterior view: *ac*, atrial chamber; *la*, lamellae; *pc*, pulmonary chamber; *st*, sternite. (After Snodgrass 1952)

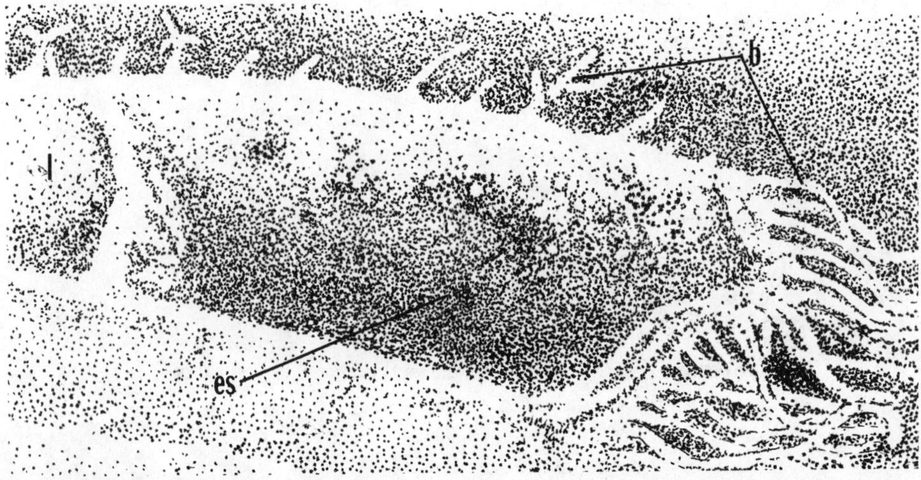

Fig. 2.21. Atrial end of book-lung lamella: *b*, bristles; *es*, epithelial sinus; *l*, lumen. (After Vyas 1974b)

Respiration is effected by two groups of muscles. The contraction of the *ventralis poststigmaticus muscle* (Vyas and Laliwala 1972a) opens the spiracle and expands the atrial chamber, allowing the entry of air into the chamber. According to Fraenkel (1929), however, two muscles perform this function in a species of *Buthus*. The ventilation of air in the pulmonary chamber is accomplished by rhythmic contractions and relaxations of the segmentally arranged dorsoventral muscles. The contraction of these muscles compresses both the pulmonary and atrial chambers, folding the lamellae and forcing air out the spiracle. The poststigmaticus muscle then contracts, pulling air into the atrial chamber, and the dorsoventral muscles relax. This creates a partial vacuum in the pulmonary chamber, causing air to fill the interlamellar spaces. Hemolymph is distributed to the pulmonary sinus and enters the lumen of each lamella, and gaseous exchange takes place by simple diffusion.

Circulatory System

The endosternite (Fig. 2.22) is an internal skeleton that occurs in all chelicerates except solpugids and some groups of mites (Firstman 1973). Although serially metamerized, it is centralized in the cephalothorax and serves as an attachment site for many skeletal muscles. In scor-

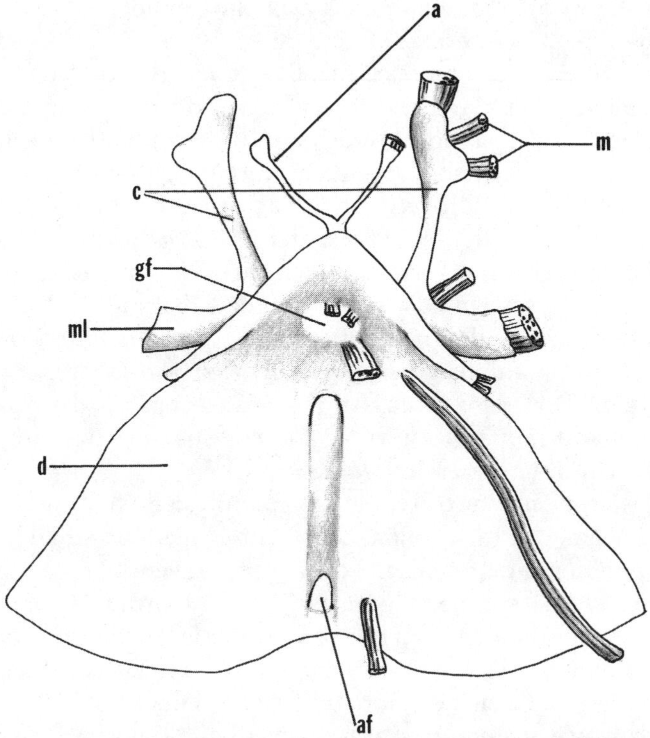

Fig. 2.22. Representative scorpion prosomal endosternite, ventral view, showing various muscle attachments: *a*, anterior process; *af*, arterial foramen; *c*, cornua; *d*, diaphragm; *gf*, gastric foramen; *m*, muscles; *ml*, mediolateral process. (After Lankester 1885a)

pions, the endosternite lies horizontally above the subesophageal ganglion and consists of a pair of longitudinal rods that join each other posteriorly. At this point they join the diaphragm, which separates the cephalothoracic and abdominal cavities. At the diaphragm, the endosternite is circumneural, forming a complete transverse ring around the posterior end of the subesophageal ganglion. It is also perforated to allow passage of the alimentary canal, the dorsal blood vessel, and various muscles.

The endosternite has been described and discussed by numerous authors, most notably Schimkewitsch (1893, 1894) and Pocock (1902a). It has generally been regarded as ectodermal in origin and primarily skeletal in function, and its morphological significance led to controversies in the recognition of *Limulus* as a chelicerate. The reason for mentioning this structure in a discussion of the circulatory system is that Firstman (1973) demonstrated that the endosternite is "continuous with a perineural vascular membrane which encloses a periganglionic arterial sinus" in the apulmonate arachnids (those without book lungs). In the pulmonate arachnids, including scorpions, there is no longer any anatomical connection between the adult arterial system and the endosternite except in the caponiid spider *Orthonops gertschi*. Firstman described the endosternites of all the arachnid orders, as well as those of xiphosurids and pycnogonids, and concluded that although the primary function is skeletal in modern groups, the original function may have been vascular, and that the embryonic origin is mesodermal rather than ectodermal. He also discussed the related topic of the origin of dorsoventral and transverse muscles and concluded with a proposed classification of the chelicerates.

An endosternite may also occur independently in each of one or more abdominal segments. In scorpions, an abdominal endosternite is present at the anterior end of the mesosoma, which is ventral to the ventral connectives of the subesophageal ganglion. According to Firstman (1973), this is morphologically a transverse muscle because it is contractile at either end, it originates on the body wall, and it is fused with the connective tissue of a single pair of dorsoventral muscles.

The circulatory system of scorpions has been described by many authors, including Newport (1843), Schneider (1892), Petrunkevitch (1922), Ch'eng-Pin (1940), Tembe and Awati (1942), Millot and Vachon (1949), Awati and Tembe (1956), Randall (1966), Vyas and Laliwala (1972b), Firstman (1973), and Tjonneland et al. (1985). The following description of the circulatory system, including terminology, is taken from these sources, and specific citations will not be given other than to indicate significant differences from the general pattern.

The *heart* is situated in the mesosoma and extends from the diaphragm to the seventh mesosomal segment (Fig. 2.23). It is limited anteriorly by a flaplike valve located slightly in front of the diaphragm. It is suspended in a *pericardial sinus* (pericardium, dorsal sinus) defined by a continuous pericardial membrane, which merges anteriorly with the diaphragm. Most authors indicate that the heart is divided into seven chambers by small transverse constrictions, but Randall (1966) found no interostial partitions in either *Uroctonus mordax* or *Centruroides exilicauda* that could divide the heart lumen into chambers. There

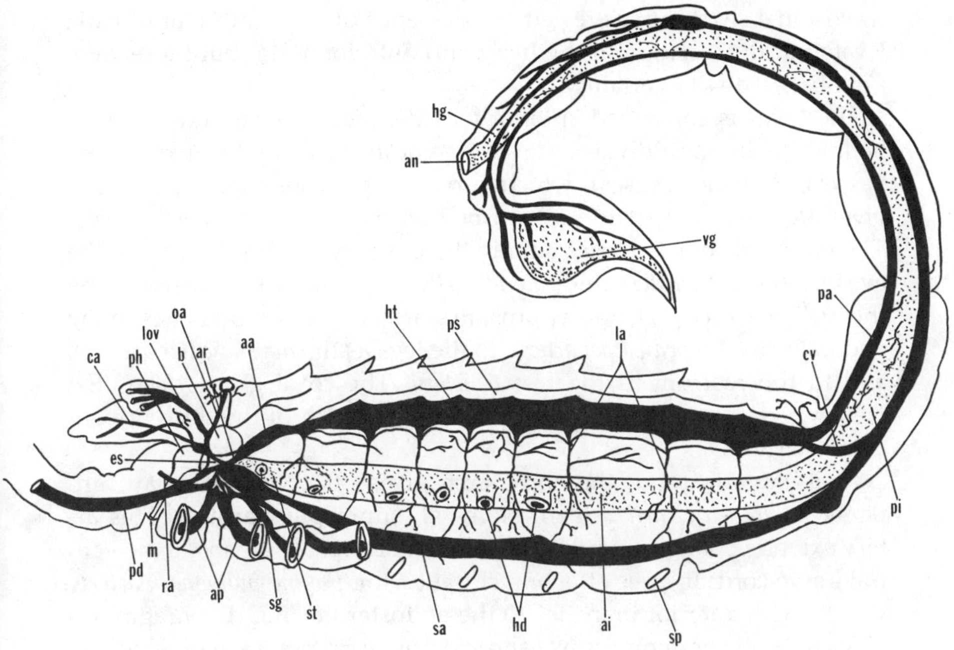

Fig. 2.23. Diagrammatic representation of scorpion, lateral view, showing the circulatory and digestive systems and the venom gland: *aa*, anterior aorta; *ai*, anterior intestine; *an*, anus; *ap*, appendicular arteries; *ar*, aortic arch; *b*, brain; *ca*, cheliceral artery; *cv*, commissural vessel; *es*, esophagus; *hd*, hepatopancreatic gland ducts; *hg*, hindgut; *ht*, heart; *la*, lateral arteries; *lov*, lateral ocellar vessel; *m*, mouth; *oa*, optic artery; *pa*, posterior aorta; *pd*, pedipalpal artery; *ph*, pharynx; *pi*, posterior intestine; *ps*, pericardial sinus; *ra*, rostral artery; *sa*, supraneural artery; *sg*, stomach gland duct; *sp*, spiracle; *st*, stomach; *vg*, venom gland. (After Vachon 1952)

are seven distinct pairs of *ostia* provided with internal valves situated at the anterior end of each chamber, each ostium opening to a *lateral artery* (systemic artery), which supplies the viscera. The heart is attached to the body wall by seven pairs of *alary muscles* (alae cordis).

Regarding the ostia, Randall (1966) found distinct differences between *Uroctonus mordax* and *Centruroides exilicauda*. In *Uroctonus* the ostia of each pair lie directly opposite each other and are perpendicular to the long axis of the heart. In *Centruroides* they are diagonally oriented and slightly offset from one another. She also noted that the occurrence of a pair of lateral arteries and hypocardial muscles in the first mesosomal segment suggested the presence of an eighth pair of ostia located at the junction of the heart and anterior aorta, but these were not observed with certainty.

The heart is continued anteriorly as the *anterior aorta*. Immediately behind the brain, it divides into two *aortic arches* (lateral trunks, lateral vessels, thoracic sinuses), which surround the esophagus. Each arch gives off two principal arteries, termed *inner* and *outer prosomal arteries*. The inner prosomal arteries supply the brain by cerebral branches, the eyes by the *optic artery* and *lateral ocellar vessel*, and the chelicerae by the *cheliceral artery*. The outer prosomal artery gives off branches to the pedipalps as the *pedipalpal artery*, to the legs as the *appendicular arteries*, and to the rostrum as the *rostral artery*. The coxal glands (see "Excretory System," below) are supplied by a branch of the third appendicular artery.

At its anterior end, the aortic arch gives off a ventromedian, unpaired *supraneural artery* (ventral vessel, supraspinal artery). This artery extends posteriorly below the alimentary canal and above the ventral nerve cord. It gives off a branch called the *pectino-pulmonary artery*, which, with anterior branches to the endosternite and diaphragm, divides to form common *pectinal* and *pulmonary arteries*, each of which bifurcates and supplies hemolymph to the pectines and the first pair of book lungs (Vyas and Laliwala 1942b, *Heterometrus fulvipes*). The second through fourth pairs of pulmonary arteries are given off separately from the supraneural artery. Ch'eng-Pin (1940, *Mesobuthus martensi*) described a situation in which, in mesosoma 2–7 and metasoma I–III, the supraneural artery gives rise to a single, median subneural vessel anterior to each free abdominal ganglion; each subneural vessel then divides into two lateral subbranches, the commissural vessels, which extend outward and upward to join the heart or posterior aorta, or both. Tembe and Awati (1942) indicated that the supraneural (supraspinal) artery of *M. tamulus* gives off single short branches at both the anterior and posterior ends of each free ganglion, with an additional pair of median lateral arteries arising from the posterior aorta of metasoma V. Either additional dissections are indicated or different species within the same genus have morphologically disparate supraneural arterial branches.

According to Ch'eng-Pin (1940), the heart gives off eight pairs of lateral arteries (hepatic vessels), one pair in each of the first six segments and two pairs in the last segment. Tembe and Awati (1942) indicated that a single lateral artery (hepatic vessel) arises from the ventral surface of the heart in mesosoma 7, which divides into three branches, two of them serving the muscles of metasoma I, while the third joins the supraneural artery. Randall (1966) noted that no lateral arteries were associated with the sixth and seventh pairs of ostia in *Uroctonus mordax;* in *Centruroides exilicauda,* however, a pair of lateral arteries were found at the sixth pair of ostia, and a pair of arteries left the ventral surface of the heart in the terminal segment. There appears to be agreement that at least a single commissural vessel from the heart in mesosoma 7 extends ventrally to join the supraneural vessel in metasoma I.

In dissection and dye-infusion studies of *Paruroctonus mesaensis,* Farley (1987) described blood-flow patterns and pacemaker activity, and he added to our knowledge of the morphology of the circulatory system. Of particular interest was his finding of valves at the origins of the posterior and communicating arteries.

There are also some discrepancies in descriptions of the metasomal blood vessels. Depending on the source, there may be one, two, or four pairs of lateral arteries arising from the *posterior aorta* (caudal artery, posterior artery), which serve the metasomal muscles and alimentary canal. There is agreement that the posterior aorta terminates in the telson to serve the poison glands and muscles of the aculeus.

The circulatory system of scorpions is an open system: after passing through the heart and arterial system, the blood collects in a *ventral sinus,* which encloses the supraneural vessel and its branches, the ventral nerve cord, and the book lungs. From there, it passes through the book lungs and is brought to the pericardial sinus via *pulmonary veins.* According to Vyas and Laliwala (1972b), there are four pairs of pulmonary veins in *Heterometrus fulvipes,* whereas Tembe and Awati (1942) recorded seven pairs of pulmonary veins entering the pericardial sinus in *M. tamulus.* The blood then enters the heart to be recirculated. Vyas and Laliwala also mentioned dorsal (other than pericardial) and paired lateral sinuses without further discussion or description.

Ravindranath (1974) described the hemocytes of *Heterometrus swammerdami* and divided them into six classes using the nomenclature of Jones (1962). These classes are: prohemocytes (with four subclasses), plasmatocytes, granular hemocytes, cystocytes, spherule cells, and adipohemocytes. Scorpions do not have a discrete fat body, and the presence of adipohemocytes, which are similar to the fat body cells of insects, is of particular interest. The hematopoietic tissues of scorpions consist of a series of cellular masses attached to the nerve cord in the region of the cephalothorax and mesosoma (lymphatic glands of Millot and Vachon 1949; see also "Excretory System," below).

Digestive System

The alimentary canal is a simple, slightly differentiated tube lying ventral to the heart and aorta and dorsal to the ventral nerve cord (Fig. 2.23). It consists of three main divisions: the *stomodeum* (foregut; anterior intestine of Millot and Vachon 1949), *mesenteron* (midgut; middle intestine of Millot and Vachon 1949, including the thoracenteron, chylenteron, and intestinum), and *proctodeum* (hindgut, posterior intestine). The following description is taken from Bardi and George (1943), Ch'eng-Pin (1940), Millot and Vachon (1949), Snodgrass (1952), Srivastava (1955), and Tembe and Awati (1942), and specific citations will not be given except in cases of discrepancy or unique description.

The stomodeum consists of the *mouth, pharynx,* and *esophagus.* The mouth is hidden in a large preoral cavity formed by the pedipalp coxae laterally, the chelicerae dorsally, and the coxapophyses of the first two pairs of legs ventrally. The mouth opens at the base of the *rostrum* (labrum), a large, laterally compressed structure that extends forward from between the bases of the pedipalp coxae. The opposing median edges of coxapophyses I leave a groove that leads directly to the mouth. The ventral surface of each coxapophysis I is closely set with short bristles forming an oval *maxillary brush* (Fig. 2.24; Srivastava 1955). These oppose the dorsal surfaces of coxapophyses II, each of which has a median channel leading to the mouth. On either side of the median channel of each coxapophysis II lies an irregular network of lateral channels that is absent in younger instars and develops gradually through successive molts. Both coxapophyses I and II contain numerous *maxillary glands*—simple, alveolar structures formed as invaginations of the hypodermis. During feeding, rhythmic movements of coxapophyses I allow the juices of the prey to pass between the opposing surfaces of both pairs of coxapophyses and fill the lateral and median channels of coxapophyses II. The maxillary brush probably serves two purposes: to filter prey juices and to clean the opposing surface of coxapophysis II to prevent the maxillary channels from becoming clogged. The secretions of the maxillary glands begin the digestive process.

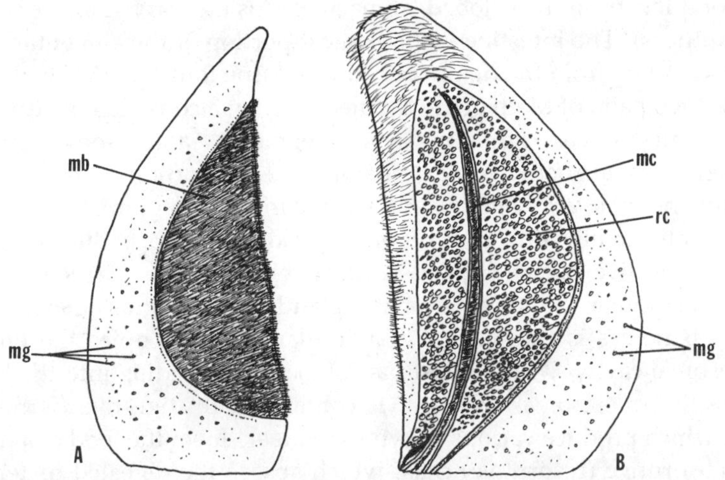

Fig. 2.24. *Heterometrus bengalensis: A*, ventromedial surface of coxapophysis I; *B*, dorsolateral surface of coxapophysis II. *mb*, maxillary brush; *mc*, median maxillary channel; *mg*, maxillary gland openings; *rc*, raised cuticular surface. (After Srivastava 1955)

The mouth leads into the pharynx, a small pear-shaped chamber that has been modified into a sucking organ. The broad, dorsal end lies between the two anterior processes of the endosternite. It is provided with numerous muscles, which are divided into two groups: dilators and compressors (sphincters). Five groups of dilator muscles (Tembe and Awati 1942) attach to various surfaces within the anterior cephalothorax and rostrum; the compressors consist of muscle bundles surrounding the pharynx.

The esophagus, which extends from the lower portion of the pharynx and proceeds posteriorly to near the base of the fourth coxae, is a thin tube slightly dilated posteriorly. According to Tembe and Awati (1942), this dilated area forms a valvular structure, which may prevent the return flow of the stomach contents. These authors also mentioned a postcerebral sucking apparatus in the esophagus of *Mesobuthus tamulus* consisting of two groups of muscles.

The mesenteron consists of the *stomach* and *stomach glands*, the *intestine*, and *the hepatopancreas*. The stomach (crop of Ch'eng-Pin 1940) is a dilated tube in the cephalothorax extending from the esophagus to the diaphragm. It lacks the circular lining of the stomodeum and receives a pair of ducts from the trilobed stomach glands (salivary glands of various authors). The intestine is the longest portion of the alimentary canal, extending from the diaphragm through the fourth metasomal segment. Two pairs of Malpighian tubules insert at a constriction either in the last mesosomal or first metasomal segment, leading some authors to refer to an *anterior intestine* (hepatopancreatic portion, pars-tecta intestini) and a *posterior intestine* (pars-nuda intestini, ileum).

The anterior intestine gives off on each side five pairs of ducts, which lead to the five pairs of glands collectively called the *hepatopancreas* (liver, glandular intestine, stomach glands of Snodgrass 1952). The hepatopancreas is the main digestive gland and occupies the entire mesosomal cavity, with the last pair of glands extending into the base of the first metasomal segment. It contains both digestive (ferment) cells, which produce various enzymes to break down the food material, and absorptive (resorptive) cells, which absorb the digested material. When the absorptive cells become filled with excretory granules, the granules are discharged into the intestine. The hepatopancreas thus acts in part as a storage organ and allows scorpions to fast for long periods of time. The posterior intestine is a thin-walled tube with a uniformly cylindrical epithelium surrounded by muscle pads.

Having found that the stomach glands and hepatopancreas are histologically identical, Bardi and George (1943) regarded any major distinction between them to be without foundation. Their interpretation is that the stomach glands are the anteriormost pair of hepatopancreatic glands even though they are separated from the rest of the hepatopancreas by the diaphragm.

The proctodeum consists of the *hindgut* and the *anus*. Their cuticular lining is continuous with the intersegmental membrane of metasoma V and the telson. The anal opening is surrounded by four chitinous anal papillae (Fig. 2.6).

Excretory System

In addition to the excretory function performed by the absorptive cells of the hepatopancreas, as mentioned above, the excretory system consists of the *Malpighian tubules*, the *coxal glands*, *nephrocytes*, and the *lymphatic glands*. Two pairs of Malpighian tubules enter the intestine in the area of the last mesosomal segment and are free and closed at their branched distal ends. A longer pair extends to the anterior end of the stomach in the cephalothorax, and a much shorter pair extends into the lobes of the hepatopancreas (Tembe and Awati 1944). Their epithelial cells absorb waste matter from the hemolymph and excrete it in the form of guanine.

The coxal glands are a pair of white, subtriangular glands situated posterolaterally in the prosoma (Fig. 2.25). Expanding on the original view of them as digestive organs, Lankester (1884) first considered them to be excretory in function. Each gland consists of a saccule, through which extend numerous ramifications of a branch of the third appendicular artery, and a tortuously coiled labyrinth, which terminates in a short exit duct. Various authors (e.g., Laurie 1890, Bernard 1893, Ch'eng-Pin 1940, Millot and Vachon 1949) have determined that the ducts open to the exterior through minute apertures on the posterior margins of the third and fourth coxae of the legs. These glands remove wastes from the hemolymph, but it is not known whether they are selective in removing particular substances.

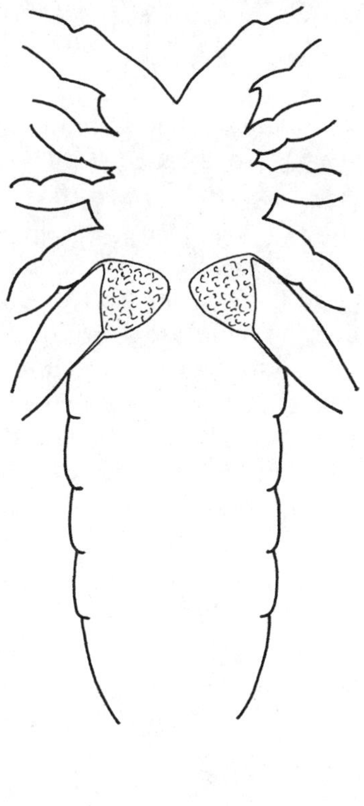

Fig. 2.25. Diagrammatic representation of scorpion, dorsal view, showing the position of the coxal glands. (After Ch'eng-Pin 1940)

Nephrocytes are large excretory cells found in the ventral prosoma and mesosoma, particularly around the nervous system and book lungs, at the base of the pectines, and into the coxae of the legs. They often possess two nuclei and contain waste products in the form of either small yellowish granules or round, oily vacuoles.

The lymphatic glands consist of oddly shaped cellular strands found along the dorsal surface of the ventral nerve cord in the mesosoma. These glands were mentioned above in reference to their hematopoietic function. In addition, Millot and Vachon (1949) mentioned that they contain many phagocytic cells, which perform a major role in the elimination of foreign substances. Further, the same authors ascribed the same functions to a pair of "lateral lymphoid glands" found in the anterior mesosoma directly behind the diaphragm in scorpions other than the Buthidae. Nayar (1966) described the latter glands from *Heterometrus scaber*. He suggested that they are endocrine in nature, because an extract produced a pronounced hyperglycemic effect when administered to another scorpion, raising the blood-sugar level by nearly 500 percent.

Venom Glands

A pair of venom glands, one on each side of the midline, is found in the vesicle of the telson (Fig. 2.26). Each gland is invested mesally and dorsally with compressor muscles that press the gland against the cuticle along its exterior lateral and ventral surfaces. The glands are separated by a median vertical muscular septum. Each gland is provided with an exit duct, and these communicate to the exterior via two minute, closely apposed apertures just before the tip of the aculeus.

Each gland is enclosed in a basement membrane, within which is a layer of connective tissue; between the connective tissue and lumen is a single layer of glandular secretory epithelium (Halse et al. 1980). In most scorpions, the secretory epithelium is thrown into more or less extensive foldings that limit the size of the lumen, and variation in the complexity of these foldings is of some significance at both the generic and family levels. In the Chactidae, the glands are relatively simple structures with little or no epithelial folding; the Buthidae show extensive folding, and even lateral ramifications, of the epithelium; and the remaining families exhibit variable intermediate levels of complexity

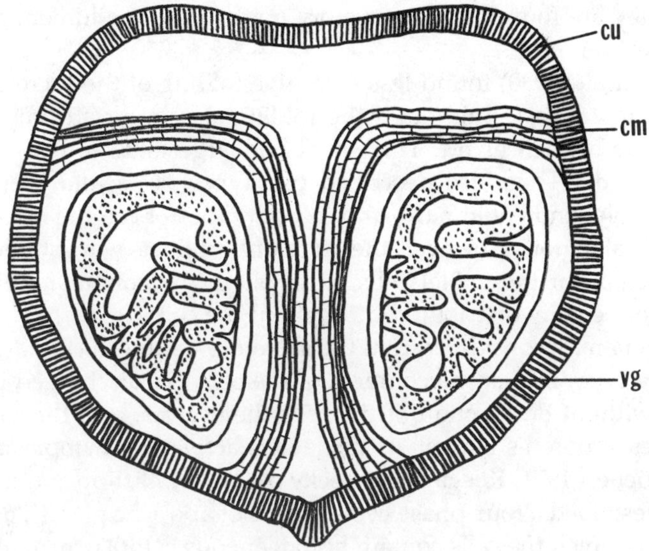

Fig. 2.26. Diagrammatic representation of scorpion venom glands: *cm,* compressor muscles; *cu,* cuticle; *vg,* venom gland, showing secretory epithelial folding typical of the Buthidae. (After Snodgrass 1952)

(Millot and Vachon 1949, Bishop and de Ferriz 1964, Keegan and Lockwood 1971). Rosin (1965) described a hitherto unique condition in *Nebo hierichonticus* (Diplocentridae), in which each gland is divided into two lobes by a longitudinal septum in addition to the secretory epithelium being thrown into folds.

Ultrastructurally, the venom glands of *Centruroides exilicauda* and *Urodacus novaehollandiae* have been examined (Mazurkiewicz and Bertke 1972, Halse et al. 1980). Mazurkiewicz and Bertke found that the secretory epithelium of *Centruroides* is extensively folded on the mesal side (i.e., the side away from the cuticle), and that the surface of the epithelial cells is invested with microvilli (also noted in Keegan and Lockwood 1971 in other *Centruroides* species), which may serve to increase the secretory surface. They also noted many membrane-bound vesicles within the lumen; these vesicles segregate morphologically different secretion products, which probably mix and rupture upon injection. Mazurkiewicz and Bertke interpreted these different morphologies of secretory products to be discrete entities, not stages of synthesis, because they are found in the secretory epithelium, the lumen, and the expressed venom.

Halse et al. (1980) found less extensive folding of the secretory epithelium in *Urodacus*, although the folding also occurred only on the mesal side of each gland. There was one large fold, and two to four smaller folds. They also described two types of secretory epithelial cells—goblet and columnar—and in both types of cell the nucleus lay in the basal region. However, the secretory products were stored in the apical region of the goblet cells, whereas in the columnar cells they were more evenly distributed throughout the cell.

It is generally considered that the secretory epithelial cells are of the apocrine type, in that the contents of the cells are discharged into the lumen without destruction of the cells themselves, and the same cell produces venom as long as its nucleus is active and cytoplasm exists (e.g., Bücherl 1971, Keegan and Lockwood 1971). Bishop and de Ferriz (1964) described a four-phase cycle typical of apocrine glands: a resting phase, in which the cells contain homogeneous cytoplasm; an elaboration phase, in which they contain fine uniform granules; an accumulation phase, in which larger granules are seen; and an expulsion phase, in which the venom granules and a portion of the cytoplasm of the cell are extruded into the lumen of the gland.

On a related though not strictly morphological topic, Newlands (1974a) described the venom-squirting ability of seven species of *Parabuthus*. All these species have telsons with large vesicles, and the venom emerges as a fine jet or spray that may reach distances of up to a meter when squirted at a low trajectory and at least 50 cm when squirted vertically. The spray does not appear to be aimed in any definite direction, but its direction is generally toward the front of the scorpion.

Reproductive System

Female

The *ovariuterus* (ovaries) consists of a reticular network of longitudinal and transverse ovarian tubes, which lie parallel and ventral to the alimentary canal and extend from the third to the sixth or seventh mesosomal segments (body somites X through XIII or XIV; Fig. 2.27). The *longitudinal ovarian tubes* are either three or four in number; if three,

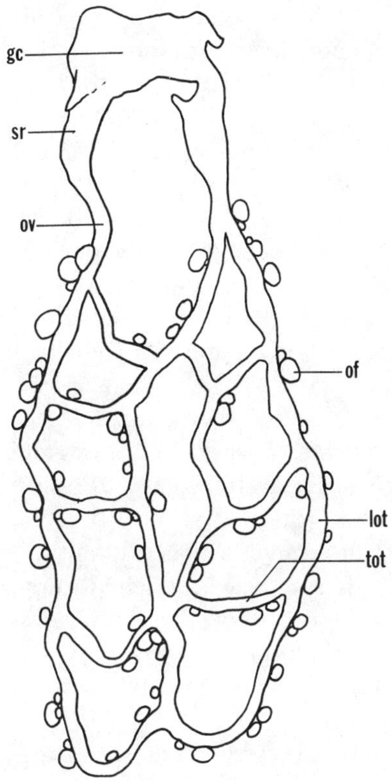

Fig. 2.27. Ovariuterus of *Rhopalurus rochae*, dorsal view: *gc*, genital chamber; *lot*, longitudinal ovarian tube; *of*, ovarian follicle; *ov*, oviduct; *sr*, seminal receptacle; *tot*, transverse ovarian tube. (After Matthiesen 1970)

they result from a fusion of the median tubes (O. Francke, pers. comm. 1982). The longitudinal ovarian tubes are connected by either four or five *transverse ovarian tubes*. According to most authors, female Buthidae have an eight-celled reticular ovariuterus, that is, five transverse ovarian tubes (or anastomoses, referring to the connections between longitudinal and transverse tubes), whereas all other families have a six-celled ovariuterus with four transverse ovarian tubes (Birula 1917a, Millot and Vachon 1949, Mathew 1956, Francke 1979a). Piza (1939a, b), however, showed that *Tityus bahiensis* and *T. serrulatus* (Buthidae) do not have five transverse tubes; only the anterior and posterior transverse ovarian tubes are present, forming a two-celled ovariuterus. Matthiesen (1970) confirmed Piza's observations and included *Tityus cambridgei* and *T. stigmurus* as buthids with a two-celled ovariuterus. More recently, O. Francke (pers. comm. 1982) has indicated that he considers the transverse ovarian tubes to be lacking in these species, the cells being formed by the longitudinal tubes alone. The transverse ovarian tubes are normally found near the middle of each of the third through sixth or seventh mesosomal segments.

From the anterior angle of each lateral longitudinal ovarian tube, the *oviducts* proceed anteriorly, forming the dilated *seminal receptacles* (spermathecae). The seminal receptacles open into an expanded *genital chamber* (vaginal part), which in some species may be produced posteriorly to form what Millot and Vachon (1949) call a spermatophore pouch. The genital chamber opens to the exterior through the *gonopore* (genital aperture) on the first mesosomal segment (body somite X), which is covered externally by the *genital opercula*.

After scorpions mate, the spermatozoa, leaving the seminal receptacles apparently at the volition of the female (Bücherl 1971), migrate along the oviducts and fertilize the oocytes. The fertilized eggs develop in place and are fixed inside *ovarian follicles* (diverticula, oval bodies), which are normally found only on one side of any particular longitudinal or transverse ovarian tube segment. Each follicle is attached to the tube by a single *pedicel* (stalk, appendix), and, upon dissection, embryos may be found in various stages of development. Ultimately, the embryo becomes detached from the pedicel, moves into the ovarian tube, and migrates to the genital operculum.

As Mathew (1962) has described, the reticular ovarian tubes are functionally both ovaries and oviducts. In studying *Heterometrus scaber*, he found that as the ovarian tubes undergo changes to serve as the oviducts for mature embryos, only small amounts of germinal epithelium are retained for the next brood. Physiologically, then, the ovarian tubes serve as ovaries, giving rise to ova in the early stages; they then change into oviducts to serve as the passageway for the embryos. After the extrusion of the embryos, the ovarian tubes again regain their germinal epithelium to serve as ovaries. Mathew equated this changing role of the ovarian tubes to the reconstitution of the uterine mucosa of the mammalian uterus after every birth.

Laurie (1896b) distinguished three types of ova and embryonic development:

1. Apoikogenic. Large eggs with much yolk; further embryonic development takes place in the oviduct after leaving the follicle; no blind outgrowths from the oviducts; the embryo is surrounded by embryonic

membranes. Found in all buthids; *Broteochactas, Euscorpius, Brotheas* (Chactidae); *Chaerilus* (Chaerilidae); and *Bothriurus* and *Cercophonius* (Bothriuridae).

2. Katoikogenic. Very small eggs with no yolk; development in blind outgrowths of the oviducts; no embryonic membranes. Found in the Scorpionidae and Diplocentridae.

3. An unnamed intermediate condition with eggs comparatively poor in yolk. Found in *Iurus, Caraboctonus, Vaejovis, Scorpiops, Anuroctonus,* and *Hadrurus* (Vaejovidae).

Nearly a hundred years later, Francke (1982d), in a review of the literature on scorpion parturition, determined that scorpions may be separated into two groups, apoikogenic and katoikogenic, with Laurie's (1896a) unnamed condition considered apoikogenic. Further, he discussed the question of oviparity, concluding that all Recent scorpions should be considered viviparous, because all embryos derive some nutrients directly from the mother during development. Makioka and Koike (1984, 1985) found natural populations of *Liocheles australasiae* in which no males were found and assumed that the females reproduced by thelytokous parthenogenesis.

Male

The following description of the male reproductive system, excluding the spermatophore, is taken from the following authors: Tembe and Awati (1944, *Mesobuthus*), Millot and Vachon (1949, *Buthus* and *Heterometrus*), and Abd-el-Wahab (1957, *Leiurus*). The male organs consist of paired *testes*, each of which is formed by four *longitudinal tubules* (trunks) united by four *transverse tubules* (trabeculae, anastomoses), thus forming a six-celled reticular network (Fig. 2.28). Previous authors have described the system as having two longitudinal tubules but have always shown four tubules in their illustrations (e.g., Millot and Vachon 1949, p. 417). Occasionally, deviations from this basic pattern have been found, for example, missing tubule segments, blind outgrowths, or additional transverse branches. In *Scorpio maurus* (Pavlovsky 1921), each testis is made up of only a single longitudinal tubule provided with four blind diverticula projecting medially, although the first one or two diverticula may partially unite with one another.

The anterolateral loop of each testis gives rise to a *vas deferens*, into which the *seminal vesicles* and two pairs of *accessory glands* empty. There

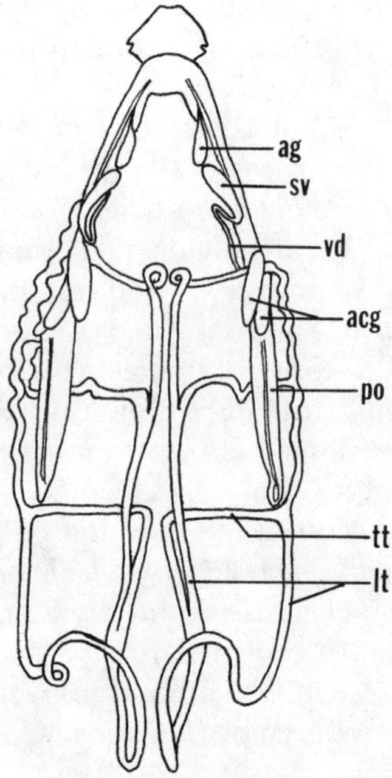

Fig. 2.28. Paired testes of *Hottentotta hottentotta: acg,* accessory glands; *ag,* anterior gland; *lt,* longitudinal tubules; *po,* paraxial organ; *sv,* seminal vesicle; *tt,* transverse tubule; *vd,* vas deferens. (After Vachon 1952)

is considerable confusion in the literature about the terminology to be applied to the accessory glands, and there is no definite knowledge about their functions, although inferences have been made. Two pairs of accessory glands have been described by several authors: the *cylindrical glands* and the *oval glands* (described as seminal receptacles in Millot and Vachon 1949). Birula (1917a) said that the oval gland (ovate vesicle) is posterior to the cylindrical gland, whereas Millot and Vachon (1949) and Abd-el-Wahab (1957) showed the reverse situation in their illustrations. Tembe and Awati (1944) mentioned that the cylindrical glands are dorsal to the oval glands, but they did not indicate longitudinal placement. The hypothesized function of the accessory glands is the secretion of the various components of the hemispermatophore, but this, as well as accurate comparative descriptions of the morphology of the male reproductive system, is an area that needs considerable study. In addition, Abd-el-Wahab (1957) noted a pair of minute, many-branched glands of unknown function that emerge from the junction of the vasa deferentia.

For many years, it was not realized that male scorpions produce a spermatophore. Because of this, various authors referred to paired ejaculatory sacs, ejaculatory organs, or a penis, and inferred that these structures formed tubular organs for transmitting sperm. It is now considered that the *ejaculatory sacs* (elongate, hollow tubules behind the seminal vesicles), the seminal vesicles, and the accessory glands constitute the paired *paraxial organs* (terminology after Pavlovsky 1917). Each paraxial organ produces half the spermatophore, called the hemispermatophore. The sperm is introduced into the *capsule* of the hemispermatophore by the seminal vesicle; the accessory glands produce components of the hemispermatophore or the material to glue the two halves together, or both; and the ejaculatory sac serves to push the two halves out of the gonopore, forming the spermatophore. Spermatophores are ejected one at a time (Francke 1979a), and males may produce several spermatophores during a mating cycle.

Basically, a spermatophore (Fig. 2.29) consists of the *pedicel* (base, foot), *trunk* (basal portion, stem), *capsule*, and *lamina* (distal lamina). There is usually a *pedal flexure* between the pedicel and trunk and a *truncal flexure* (basal flexure) between the trunk and lamina. All these structures may exhibit varying degrees of development in different taxa. The capsule contains the sperm and is quite variable in shape, structure, and ornamentation. Alberti (1983) reviewed the current knowledge of the structure of scorpion spermatozoa.

Two basic types of spermatophore have been described: *flagelliform*, found only in the Buthidae and so called because of the elongate, elastic *flagellum* at the distal end of the lamina; and *lamelliform*, found in all other families. Francke (1979a) suggested that the lamelliform spermatophore is homologous to the spermatophores of atemnid pseudoscorpions.

It must be realized that the spermatophores of only about 30 species in 20 genera have been described (see Francke 1979a), out of approximately 1,400 described species of scorpions. To complicate matters, preinsemination and postinsemination spermatophores differ in appearance, and descriptions of both conditions have been presented in the literature. In addition, the hemispermatophores of over 60 species of bothriurids and buthids have been examined (Koch 1977, Maury 1980), many of which have been described. However, there is no published evidence that the morphology of the internal hemispermatophore and the external pre- and postinsemination spermatophore is the same, and comparisons among the three forms may not be valid.

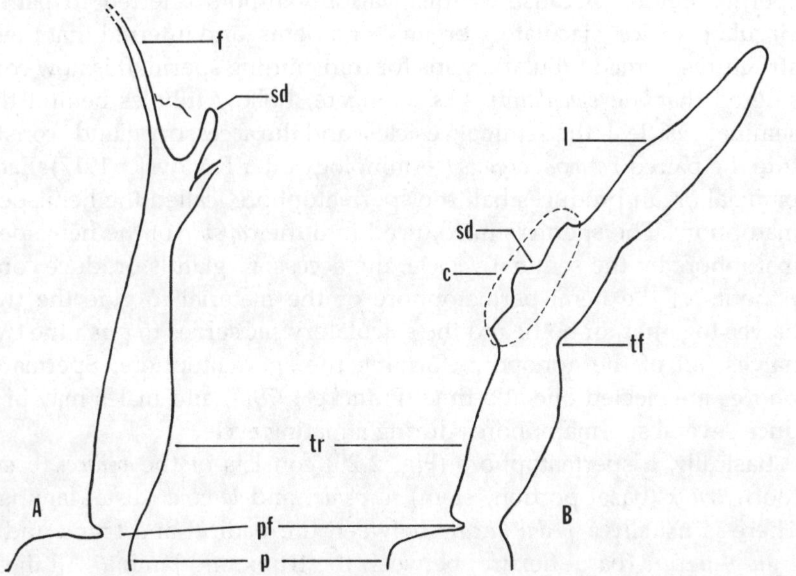

Fig. 2.29. Diagrammatic representation of the spermatophore: *A*, flagelliform; *B*, lamelliform. *c*, capsule; *f*, flagellum; *l*, lamina; *p*, pedicel; *pf*, pedal flexure; *sd*, sperm duct; *tf*, truncal flexure; *tr*, trunk. (After Francke 1979a)

As Maury (1980) indicated, it is not always a simple matter to obtain preinsemination spermatophores for study, but it is probable that these will have the greatest significance for taxonomic purposes, not only because of their consistent rigidity, but because it can be confidently assumed that all pertinent structures are present and unaltered.

Because the spermatophores of so few scorpions have been described, and because there has been such a profusion of terms used to describe various parts of the capsule of the spermatophore, I have elected to provide only a basic nomenclature in Figure 2.29. For further discussion regarding detailed capsular structure and terminology, and the general taxonomic importance of the spermatophore, the following publications are recommended: San Martín and Gambardella 1967, Maury 1975c, Koch 1977, Francke 1979a, and Maury 1980. Spermatophore morphology is one area that requires much more comparative study and the application of a uniform nomenclature before meaningful conclusions may be drawn.

Teratology

Developmental anomalies are well known in scorpions, the most common being the duplication of various posterior body segments. For example, specimens have been found with two telsons (Williams 1971a); duplication of the metasoma from segment I (Vachon 1952) and segment II (Berland 1913); and complete distal duplication from mesosoma 3 (Armas 1977) and mesosoma 4 (Balboa 1937, Vachon 1952). Duplication of the anterior part of the body is less common, having been reported in *Euscorpius* by Brauer (1917) and in *Tityus* by Matthiesen (1980).

Other anomalies of buthids reported by Armas (1977) include deformation of the patella of the pedipalp, with keels and trichobothria lacking; various walking legs with deformed, fused, shortened, or missing tarsomeres; and a malformed telson. These particular abnormalities appear only on one side of any particular specimen, although they may occur on either side of the body.

A singular sexual anomaly was described by Maury (1983), in which a hermaphroditic *Brachistosternus pentheri* was found with both embryos and hemispermatophores.

With the exception of reports by Pavesi (1881) and Brauer (1917) concerning *Euscorpius* (Chactidae), and by Maury (1983) concerning a bothriurid, all developmental anomalies have been reported from the family Buthidae. Matthiesen (1980) provided a more extensive listing of reported anomalies.

ACKNOWLEDGMENTS

My sincerest thanks and gratitude to several people who contributed their time and energy to this chapter. Frank Koehler, Jr., and Dennis Koehler provided several of the illustrations; Stanley C. Williams and Maury Swoveland supplied the SEM photographs; Edward L. Smith contributed to interpretation of the segmentation of various parts of the body; and Oscar Francke, Herbert Levi, and W. David Sissom provided comments and suggestions on the manuscript.

3

Systematics, Biogeography, and Paleontology

W. DAVID SISSOM

The order Scorpiones, with nine living families and about 1,400 described species and subspecies, is a rather small group of chelicerate arthropods. Scorpions have nevertheless received considerable taxonomic attention through the years, probably as a result of three factors: their medical importance, their antiquity and obvious importance to the analysis of chelicerate phylogeny, and their local abundance and wide geographical distribution. Despite this attention, many taxa (even higher categories) are poorly understood, and the relationship of scorpions to other chelicerates remains a topic of controversy.

This chapter is divided into two main sections. The first section deals exclusively with the characterization, identification, and biogeography of Recent taxa. The second section deals with the origin and evolution of the order and includes a discussion of the fossil taxa and their relationships to Recent forms. At the end of the chapter is a brief discussion of the role of scorpions in the evolution of the subphylum Chelicerata.

This chapter reflects our current understanding of scorpion systematics, and is not intended as a systematic revision or a phylogenetic reappraisal of the order Scorpiones. Undoubtedly, many taxonomic changes will be made as intergeneric relationships become clarified.

Morphological Characters of Recent Scorpions

The classification of living scorpions is based almost entirely on internal and external morphology, although some researchers are now exploring the value of hemolymph electrophoresis and karyology (e.g., Goyffon et al. 1973a, b; Koch 1977; Vachon and Goyffon 1978; Lamoral 1979; Newlands and Martindale 1979). These techniques have been applied in only a few taxa, but the results appear promising. The following discussion reviews the morphological characters that have traditionally been used to distinguish the scorpion families and subfamilies.

Chelicerae

Vachon (1956, 1963) recognized that within families and genera the cheliceral dentition is rather consistent, and devised a nomenclatural system for the teeth. The fixed finger (with only two exceptions) has four teeth: distal, subdistal, medial, and basal. The medial and basal teeth are almost always fused to form a bicusp. The movable finger demonstrates considerable variability in morphology between families (Fig. 3.1). There are four or five teeth on the external (or dorsal) margin, occupying distal, subdistal, medial, and basal positions. Of these, the distal and medial teeth are the largest in the row, and this facilitates identification of the teeth, particularly when there are two subdistals or two basals. The distal external tooth usually apposes a large distal internal tooth (i.e., the two form a fork). In addition to the distal tooth, the internal margin may possess one or two medial teeth and/or a row of serrations, or it may be completely smooth.

The consistent and distinct morphology of cheliceral dentition in scorpion families is arranged according to four basic types. The Buthidae, Chaerilidae, and Iuridae each have unique dentition, and the remaining families have the fourth basic type. Dentition of each family is discussed below.

Trichobothria

Trichobothrial numbers and patterns are among the most useful external characters currently used in scorpion systematics, primarily as a result of the work of Dr. Max Vachon. Vachon (1973) culminated years of research with a definitive study of 97 genera, in which a nomenclatural system for the various trichobothria was completed, and this

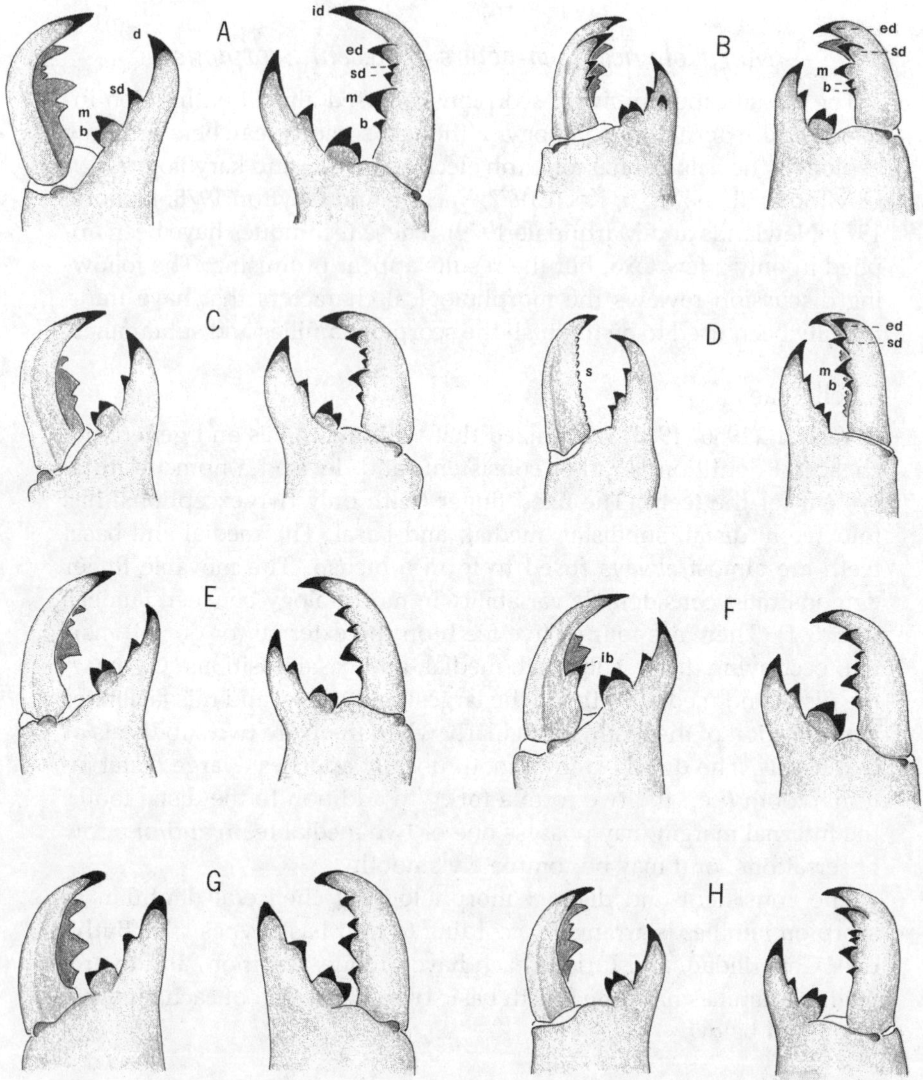

Fig. 3.1. Cheliceral dentition of scorpions by family, showing internal (*left*) and external (*right*) views of right chelicera: *A*, Bothriuridae (*Brachistosternus*); *B*, Buthidae (*Centruroides*); *C*, Chactidae (*Superstitionia*); *D*, Chaerilidae (*Chaerilus*); *E*, Diplocentridae (*Diplocentrus*); *F*, Iuridae (*Hadruroides*); *G*, Scorpionidae (*Opistophthalmus*); *H*, Vaejovidae (*Vaejovis*). *b*, basal; *d*, distal; *ed*, external distal; *ib*, internal basal; *id*, internal distal; *m*, medial; *s*, serrated internal margin of movable finger; *sd*, subdistal.

system has been adopted by almost all scorpion systematists. Stahnke (1974) also devised a nomenclatural system for trichobothria, but his system is not widely accepted.

Vachon's (1973) nomenclatural system is actually quite simple. Each trichobothrium on the pedipalpal femur, patella, and tibia (manus and fixed finger) is assigned a code, consisting of letters and, if necessary, a number, according to its position. The first letter of the code refers to the surface on which a trichobothrium is located: *d*, dorsal; *e*, external; *i*, internal; and *v*, ventral. The next letter (or letters) refers to the specific area on the surface in which a trichobothrium is situated: *b*, basal; *sb*, subbasal; *m*, medial; *st*, subterminal; and *t*, terminal. When more than one trichobothrium is found in a given area, each is assigned a number. For example, in *Diplocentrus whitei*, five trichobothria lie on the external basal surface of the patella (Fig. 3.5C); they are named patellar trichobothria eb_1, eb_2, eb_3, eb_4, and eb_5. To prevent confusion, capital letters are assigned to the surface designations of the chela manus (*D*, *E*, *I*, and *V*) and lowercase letters to those of the fixed finger.

In general, three basic trichobothrial patterns occur among living scorpions (Table 3.1). Each pattern possesses a fundamental number of trichobothria, but deviations from this number are common. The Type A pattern (Fig. 3.2), found only in the Buthidae, is characterized by the absence of trichobothria on the ventral surface of the patella and by the presence of four or five trichobothria on the dorsal surface of the femur. Two different configurations, α and β, exist for the femur in the Type A pattern (Vachon 1975; Fig. 3.3). The Type B pattern is found only in the Chaerilidae (Fig. 3.4), and the Type C pattern occurs in the remaining six families (Fig. 3.5).

Considerable variation exists within these basic patterns, particularly in Type C. First, the number of trichobothria deviates in many genera from the fundamental number (Vachon 1973). In cases where the total number is higher, the extra trichobothria are called accessory trichobothria (Fig. 3.6). Sometimes there are so many accessory trichobothria that the basic Type C pattern is obscured. Losses of trichobothria are less common, and usually involve only one or a few trichobothria. Second, the positions of homologous trichobothria are not fixed, although they do occur within generally predictable limits called territories (Vachon 1973). In some cases, however, the variability in position among presumably homologous trichobothria is so drastic that Vachon suggested it is the result of trichobothrial "emigration." Stahnke (1974)

TABLE 3.1

Numbers, patterns, and nomenclature of scorpion trichobothria

Segment	Surface: Dorsal	Ventral	Internal	External	Number	
TYPE A BASIC TRICHOBOTHRIAL PATTERN						
Femur	d_1–d_5		i_1–i_4	e_1, e_2	11	39
Patella	d_1–d_5		i	eb_1, eb_2, esb_1, esb_2, em, est, et	13	
Chela manus		V_1, V_2		Eb_1–Eb_3, Esb, Est, Et	8	
Fixed finger	db, dt		it	eb, esb, est, et	7	
TYPE B BASIC TRICHOBOTHRIAL PATTERN						
Femur	d_1–d_4		i	e_1–e_4	9	37
Patella	d_1, d_2	v_1–v_3	id, iv	eb_1, eb_2, esb, em, est_1, est_2, et	14	
Chela manus		V		Eb_1–Eb_3, Est, Et	6	
Fixed finger	db, dt		ib, it	eb, esb, est, et	8	
TYPE C BASIC TRICHOBOTHRIAL PATTERN						
Femur	d		i	e	3	48
Patella	d_1, d_2	v_1–v_3	i	eb_1–eb_5, esb_1, esb_2, em_1, em_2, est, et_1–et_3	19	
Chela manus	Db, Dt	V_1–V_4		Eb_1–Eb_3, Esb, Est, Et_1–Et_5	16	
Fixed finger	db, dsb, dst, dt		ib, it	eb, esb, est, et	10	

SOURCE: Vachon 1973.

likewise postulated some form of migration to account for such cases. Francke and Soleglad (1981), however, proposed that the drastic differences in position result from trichobothrial loss or gain, and that the trichobothria in question are not actually homologues.

Lamoral (1979) and Francke and Soleglad (1981) agreed that caution should be used in evaluating the phylogeny of scorpions and trichobothrial data. The variability in number and pattern is great, both intraspecifically and interspecifically, and this variability is not well understood for many taxa.

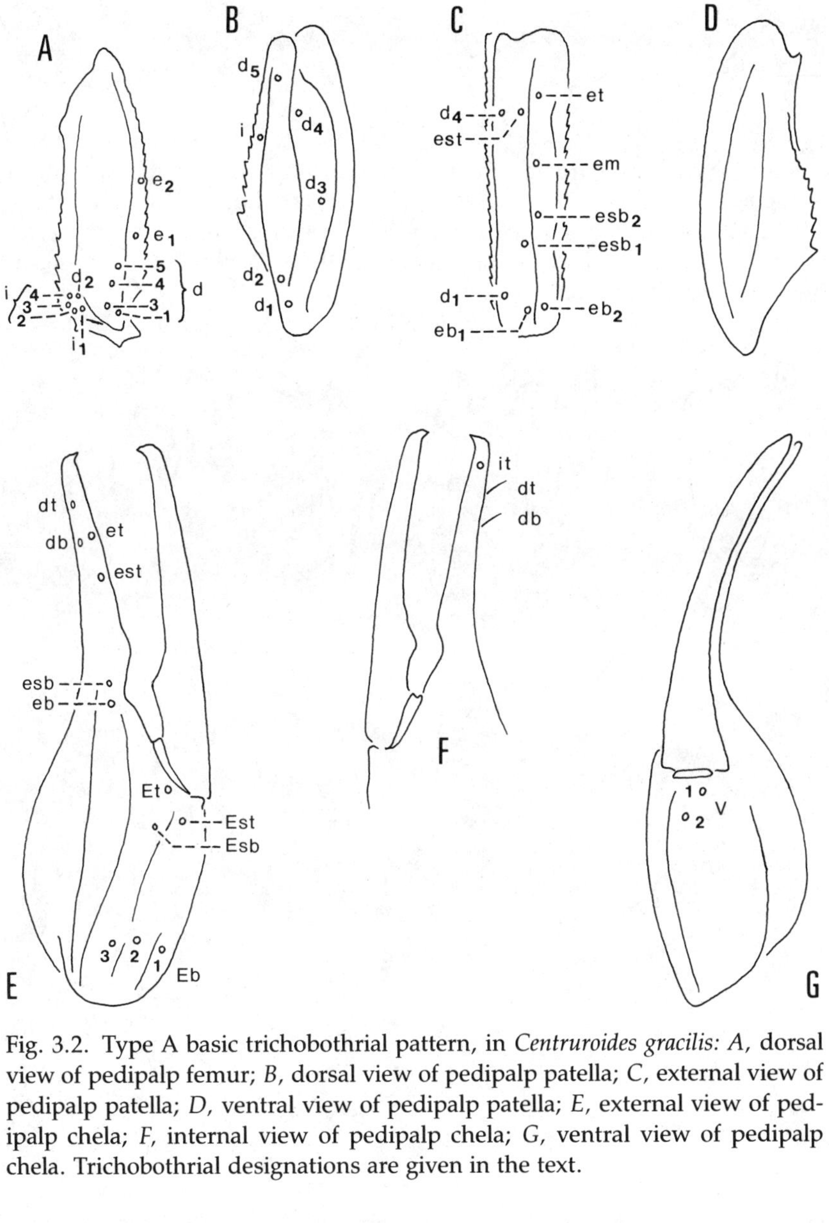

Fig. 3.2. Type A basic trichobothrial pattern, in *Centruroides gracilis: A,* dorsal view of pedipalp femur; *B,* dorsal view of pedipalp patella; *C,* external view of pedipalp patella; *D,* ventral view of pedipalp patella; *E,* external view of pedipalp chela; *F,* internal view of pedipalp chela; *G,* ventral view of pedipalp chela. Trichobothrial designations are given in the text.

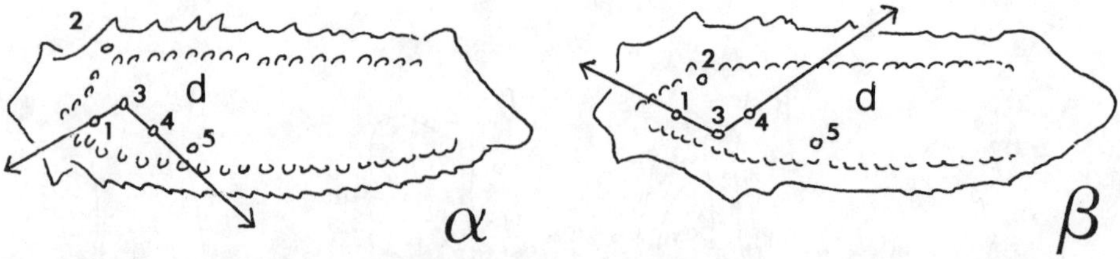

Fig. 3.3. Type A basic trichobothrial pattern, α and β configurations. In the α configuration (shown in *Rhopalurus*), the angle formed by dorsal trichobothria d_1, d_3, and d_4 opens toward the external face; in the β configuration (shown in *Ananteris*), the angle opens toward the internal face. Trichobothrial designations are given in the text.

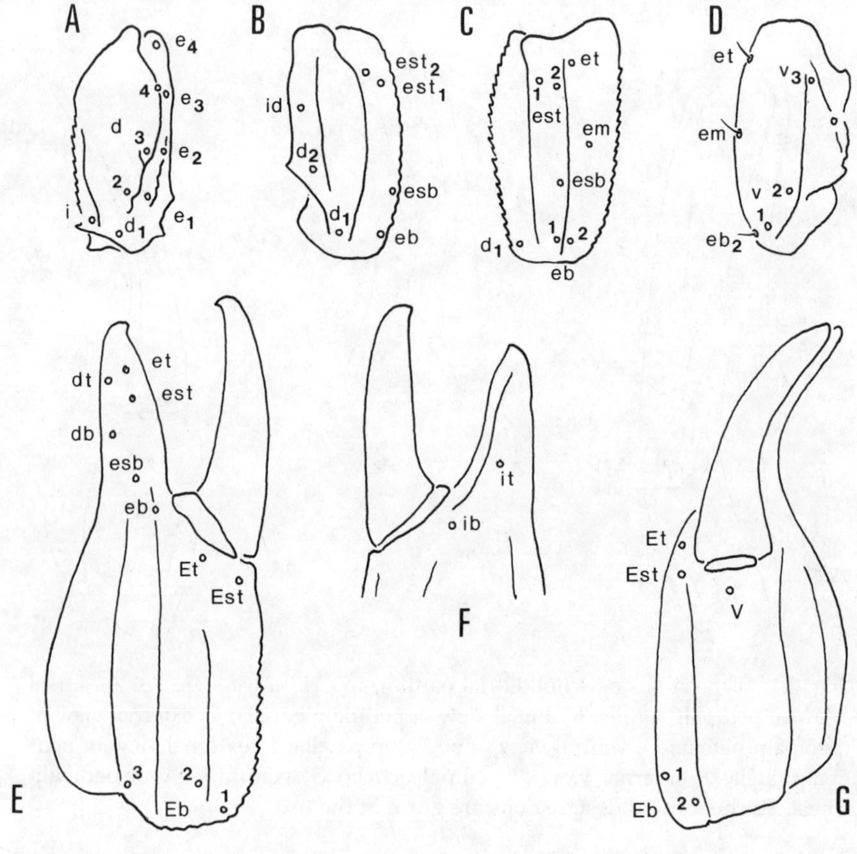

Fig. 3.4. Type B basic trichobothrial pattern, in *Chaerilus: A*, dorsal view of pedipalp femur; *B*, dorsal view of pedipalp patella; *C*, external view of pedipalp patella; *D*, ventral view of pedipalp patella; *E*, external view of pedipalp chela; *F*, internal view of pedipalp chela; *G*, ventral view of pedipalp chela. Trichobothrial designations are given in the text.

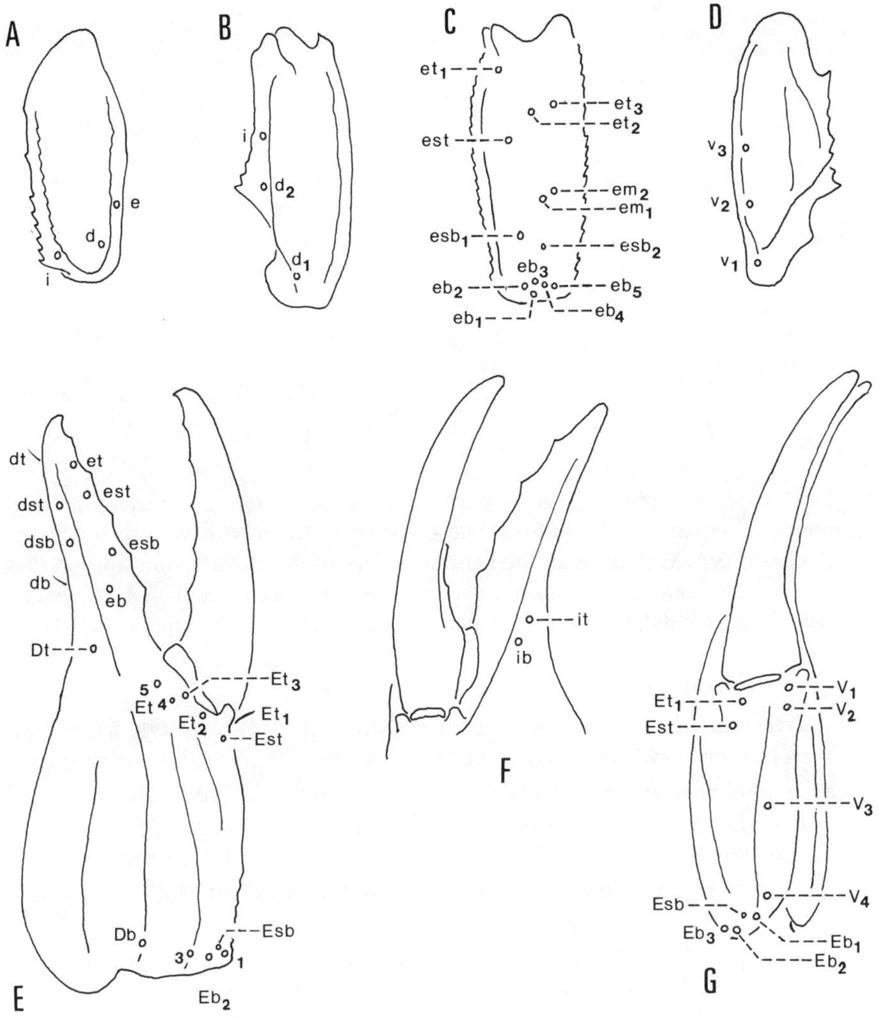

Fig. 3.5. Type C basic trichobothrial pattern, in *Diplocentrus whitei: A,* dorsal view of pedipalp femur; *B,* dorsal view of pedipalp patella; *C,* external view of pedipalp patella; *D,* ventral view of pedipalp patella; *E,* external view of pedipalp chela; *F,* internal view of pedipalp chela; *G,* ventral view of pedipalp chela. Trichobothrial designations are given in the text.

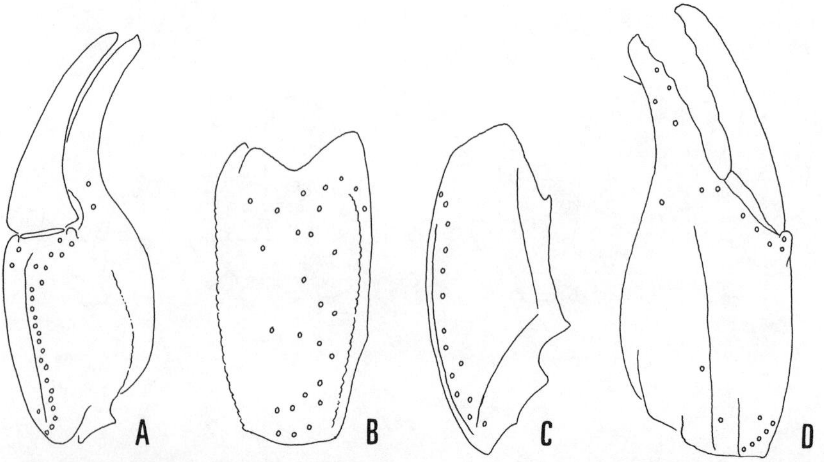

Fig. 3.6. Trichobothrial pattern of *Anuroctonus phaiodactylus*, a species with numerous accessory trichobothria. These accessory trichobothria have obscured the basic Type C pattern on the ventral surface of the chela manus and on the external and ventral surfaces of the patella. *A*, ventral aspect of chela; *B*, external aspect of patella; *C*, ventral aspect of patella; *D*, external aspect of chela.

Coxosternal Region

Two features of the coxosternal region (Fig. 3.7) have been used in the characterization of scorpion families: the shape of the coxapophyses (coxal endites or gnathobases) and the shape of the sternum.

The shape of the coxapophyses of leg I serves to separate the Chaerilidae from the remaining families. In the Chaerilidae, the anterior margin of the endite is very broad, producing anterior "lobes" (Pocock 1900). In the other families, the margin is rounded.

Among Recent scorpions the sternum is either subpentagonal, subtriangular, or broadly slitlike. The subpentagonal sternum is found in most scorpions, and in at least one genus in each family (but only rarely in the Bothriuridae and Buthidae). The subtriangular sternum, characterized by having the lateral edges strongly convergent anteriorly, is known only in the Buthidae, and the slitlike sternum only in the Bothriuridae. The observation that all first-instar buthids possess a subpentagonal sternum has led to the hypothesis that this shape represents the primitive condition. It also appears that the sternum of bothriurids (which sometimes appears to be composed of two small and

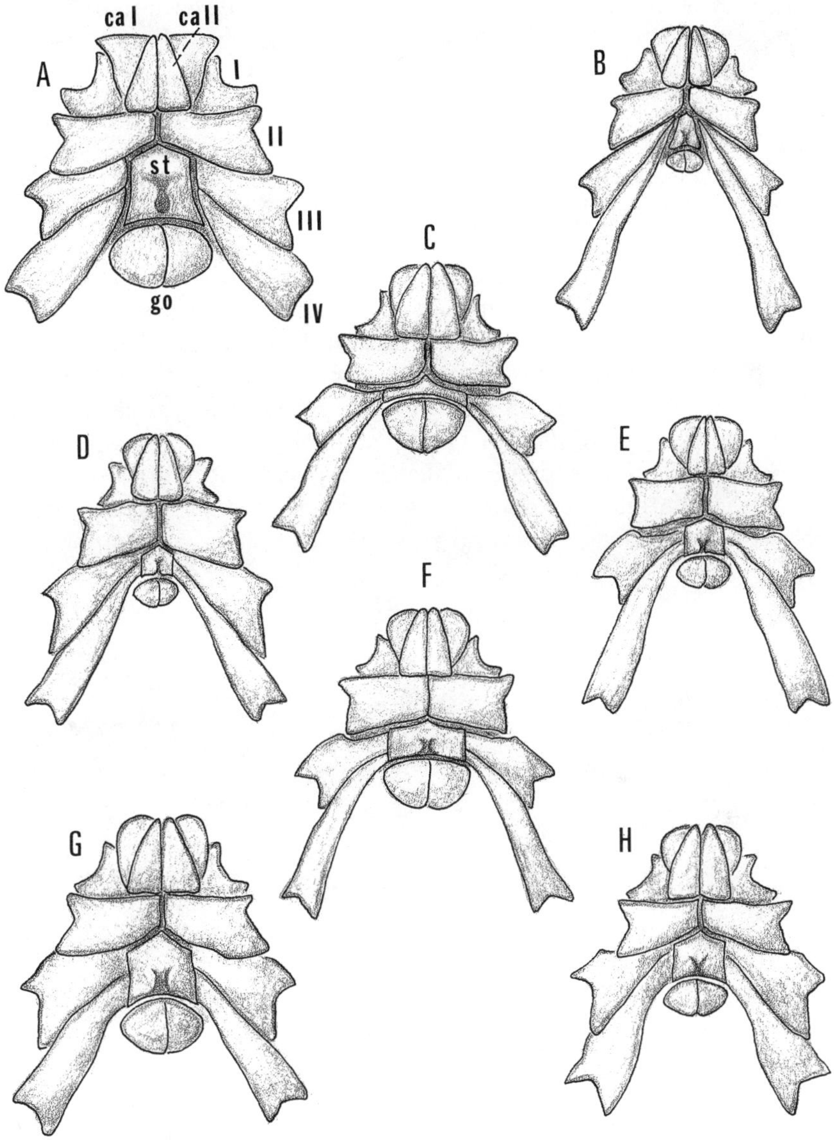

Fig. 3.7. Coxosternal regions of scorpions: *A*, Chaerilidae (*Chaerilus*); *B*, Buthidae (*Centruroides*); *C*, Bothriuridae (*Bothriurus*); *D*, Vaejovidae (*Vaejovis*); *E*, Iuridae (*Hadruroides*); *F*, Chactidae (*Superstitionia*); *G*, Scorpionidae (*Opistophthalmus*); *H*, Diplocentridae (*Diplocentrus*). *caI*, *caII*, coxapophyses of legs I and II; *I–IV*, coxae of legs I–IV; *go*, genital operculum; *st*, sternum.

separate sclerites) is actually a folded subpentagonal sternum. If boiled in potassium hydroxide, the sternum flattens and assumes a somewhat subpentagonal shape (Petrunkevitch 1953).

Leg Spination

The presence or absence of tibial and pedal spurs on the legs of scorpions (Fig. 3.8) has been important in classifying the higher categories (Pocock 1893a, Birula 1917a, Vachon 1952, Mitchell 1968). Tibial spurs, if present, occur singly on the leg (always on legs III and/or IV) in the arthrodial membrane between the tibia and tarsomere I. Pedal spurs, which are found between tarsomeres I and II, occur either singly or in pairs on each leg. If paired, there is a prolateral (interior) and a retrolateral (exterior) pedal spur; if single, the spur is always prolateral.

The morphology of the tibial and pedal spurs is often of taxonomic value at the generic and specific levels, especially in bothriurids and buthids. In buthids, the spurs are sometimes bifurcate and bear variable numbers of setae. The relative sizes of the spurs is also important.

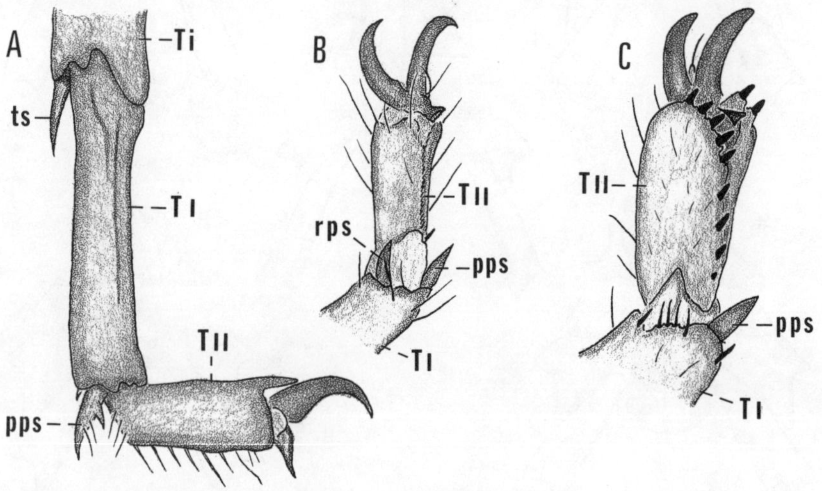

Fig. 3.8. Tibial and pedal spurs in scorpions: *A*, leg of *Buthus* (Buthidae), showing tibial spur (*ts*) and prolateral pedal spur (*pps*); *B*, tarsus of *Vaejovis* (Vaejovidae), showing both prolateral (*pps*) and retrolateral (*rps*) pedal spurs; *C*, tarsus of *Opistophthalmus* (Scorpionidae), showing the characteristic single prolateral pedal spur (*pps*). (View *A* after Vachon 1952)

For example, *Vachonia* is distinguished from other bothriurids by having unequal pedal spurs. In some buthids (e.g., *Anomalobuthus* and *Apistobuthus*), the tibial spurs are often vestigial if present.

Venom Glands and Telson

The venom glands of scorpions were extensively studied by Pavlovsky (1913, 1924b) and Lourenço (1985). The glands are paired oval sacs that lie inside the telson, and each has a secretory duct that empties into the aculeus. In order to observe the venom gland, a transverse section can be made through the middle of the telson vesicle. The glands can be examined without further preparation. The epithelium of the venom gland may be simple or folded (Fig. 3.9). In bothriurids and some ischnurids, folding is limited to the medial side of the venom gland. In other taxa, folding is more complex and occurs over the entire surface of the epithelium. The types of venom glands associated with each family are discussed further below.

The telson itself often bears a raised protuberance underneath the curvature of the aculeus. This protuberance may be either a rounded tubercle, a sharp "tooth," or a more complex structure (Fig. 3.10). Buthids display considerable variability in the size and shape of the tubercle (Fig. 3.10*D–F*); W. D. Sissom, pers. obs.; W. Lourenço, pers. comm. 1983), but the other families show a more consistent structure. The possession of a subaculear tubercle is one of the characteristics

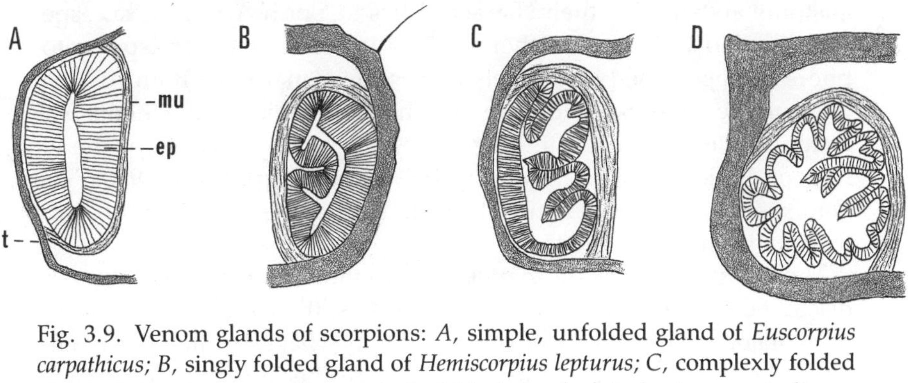

Fig. 3.9. Venom glands of scorpions: *A*, simple, unfolded gland of *Euscorpius carpathicus*; *B*, singly folded gland of *Hemiscorpius lepturus*; *C*, complexly folded gland of *Scorpio maurus*; *D*, complexly folded gland of *Androctonus australis*. *ep*, epithelium of venom gland; *mu*, muscle; *t*, wall of telson. (After Werner 1934 from Pavlovsky)

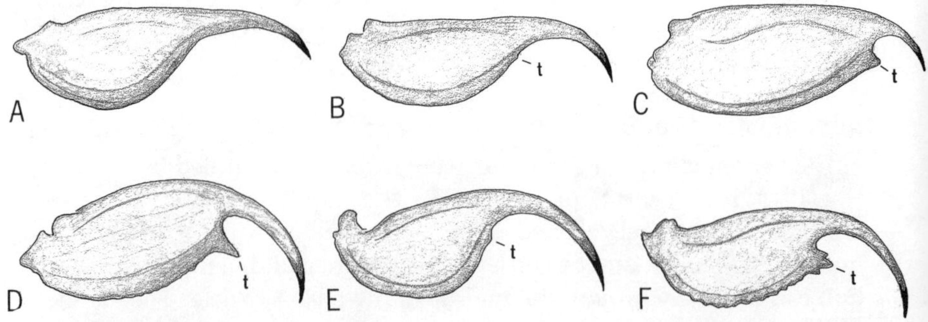

Fig. 3.10. Lateral views of telsons of scorpions, showing development of subaculear protuberances: *A*, *Opistophthalmus* (no subaculear tubercle); *B*, *Vaejovis intermedius* (very subtle subaculear tubercle); *C*, *Diplocentrus whitei* (subaculear tubercle); *D*, *Centruroides gracilis* (subaculear tooth); *E*, *C. exilicauda* (very subtle subaculear tubercle); *F*, *Tityus* sp. (complex tubercle, with accessory subdistal tubercles). *t*, tooth, tubercle.

commonly used in keys to separate the Diplocentridae from the closely related Scorpionidae.

Male Reproductive System

Pavlovsky (1924c) should be credited with recognizing the importance of male genitalia in scorpion taxonomy. Unfortunately, the structures were largely ignored until Vachon (1940, 1952) reviewed their anatomy and applied their characteristics to North African taxa, especially buthids. The only other families in which the hemispermatophore has been used extensively in classification are the Bothriuridae, Ischnuridae, and Scorpionidae. In the Bothriuridae, the hemispermatophore is now a useful diagnostic tool, largely because of the efforts of Abalos (1955), San Martín (1969), and Maury (1971, 1980). The hemispermatophores of almost three-fourths of the species in this family have been examined.

The morphology of the hemispermatophore and paraxial organ is discussed by Hjelle (Chapter 2). As he states, there are two basic types of hemispermatophore (and spermatophore): flagelliform and lamelliform. In a flagelliform hemispermatophore the important taxonomic characteristics are the relative size and shape of the trunk, lobes, and flagellum (Vachon 1952, Lamoral 1979, Levy and Amitai 1980). The paraxial organ associated with the flagelliform type of hemispermatophore possesses a number of accessory glands. The lamelliform type often has a heavily sclerotized capsule, which provides numerous useful characters for differentiating taxa. In this type, too, the relative size and shape of the lobes, trunk, and lamina are important. The paraxial organ associated with the lamelliform hemispermatophore usually does not bear accessory glands. For details of hemispermatophore and paraxial organ morphology specific to familial characterization, see Figures 3.11 and 3.12, and below.

Female Reproductive System and Embryonic Development

The female reproductive system in scorpions consists of the ovariuterus and associated structures. Variation in the morphology of the ovariuterus is strongly linked to the mode of embryonic development, and the following discussion includes both subjects. Female reproductive morphology, embryology, and parturition were recently reviewed by Francke (1982d), and I have drawn heavily on his paper and references.

Basically, the ovariuterus is constructed of a reticulate system of ovarian tubes arising from the genital atrium. Longitudinal ovarian tubes are connected by transverse ovarian tubes (anastomoses), giving the reticulate appearance. The medial longitudinal tubes may be partially or completely fused. Of taxonomic importance are the number of transverse tubes present (reflected by the number of cells making up the ovariuterus), the position of the follicles in relation to the ovariuterus, and the presence or absence of diverticula.

In the family Buthidae, five pairs of transverse ovarian tubes (Fig. 3.13*A*) generally divide the ovariuterus into eight cells. The ovariuterus of *Tityus* (at least in the species studied) has only two cells (Fig. 3.13*B*), because all except the distal anastomoses are absent (Matthiesen 1970). In all other families, the ovariuterus possesses four pairs of transverse tubes, forming six cells.

The position of the follicles in relation to the ovariuterus provides further distinctions. In the Bothriuridae, Buthidae, Chactidae, Chaerilidae, Iuridae, and Vaejovidae the oocytes are contained in follicles that are in direct contact with the ovariuterus (as in Fig. 3.13*C–E*). The iurids and vaejovids differ from this trend somewhat by having the follicles connected to the ovariuterus by a short pedicel. Nevertheless, development of the embryos still occurs in the lumen of the ovariuterus, just as it does in the other families. The developmental type to which these six families belong is referred to as apoikogenic (Laurie 1896a, b).

The Diplocentridae, Ischnuridae, and Scorpionidae differ consider-

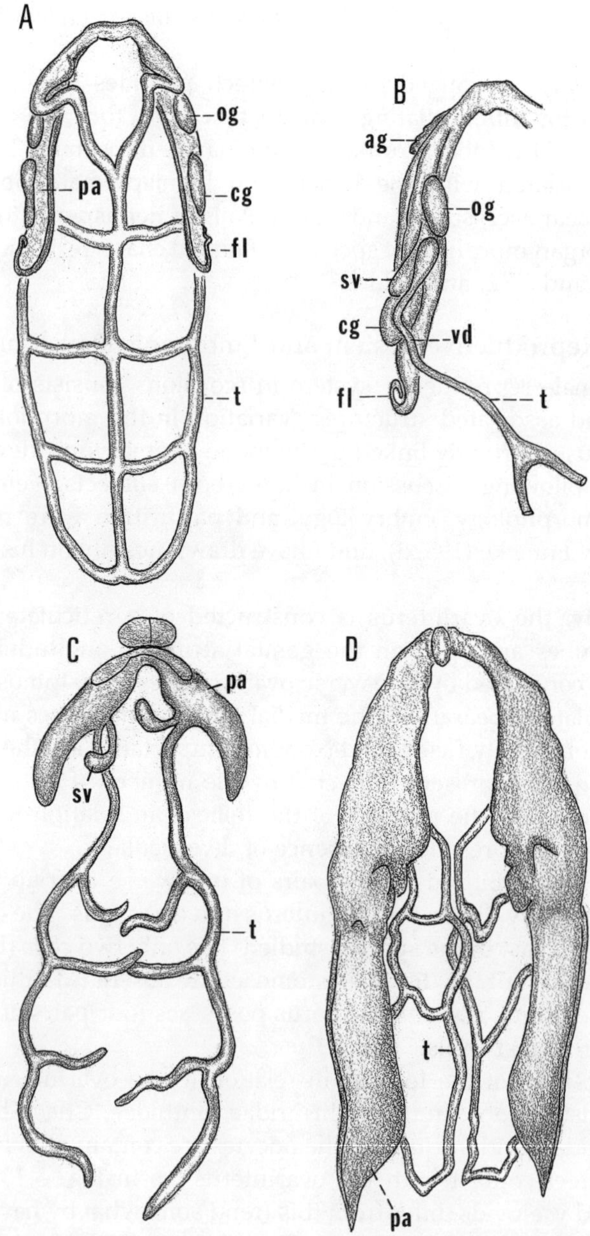

Fig. 3.11. Paraxial organs of scorpions: *A*, paraxial organ and testes of *Centruroides vittatus* (Buthidae); *B*, paraxial organ of *Isometrus maculatus* (Buthidae); C, paraxial organ and testes of *Scorpio maurus* (Scorpionidae); *D*, paraxial organ and testes of *Urodacus manicatus* (Scorpionidae). *ag*, anterior gland; *cg*, cylindrical gland; *fl*, flagellum; *og*, oval gland; *pa*, paraxial organ; *sv*, seminal vesicle; *t*, testes; *vd*, vas deferens. (Views *B–D* after Werner 1934 from Pavlovsky)

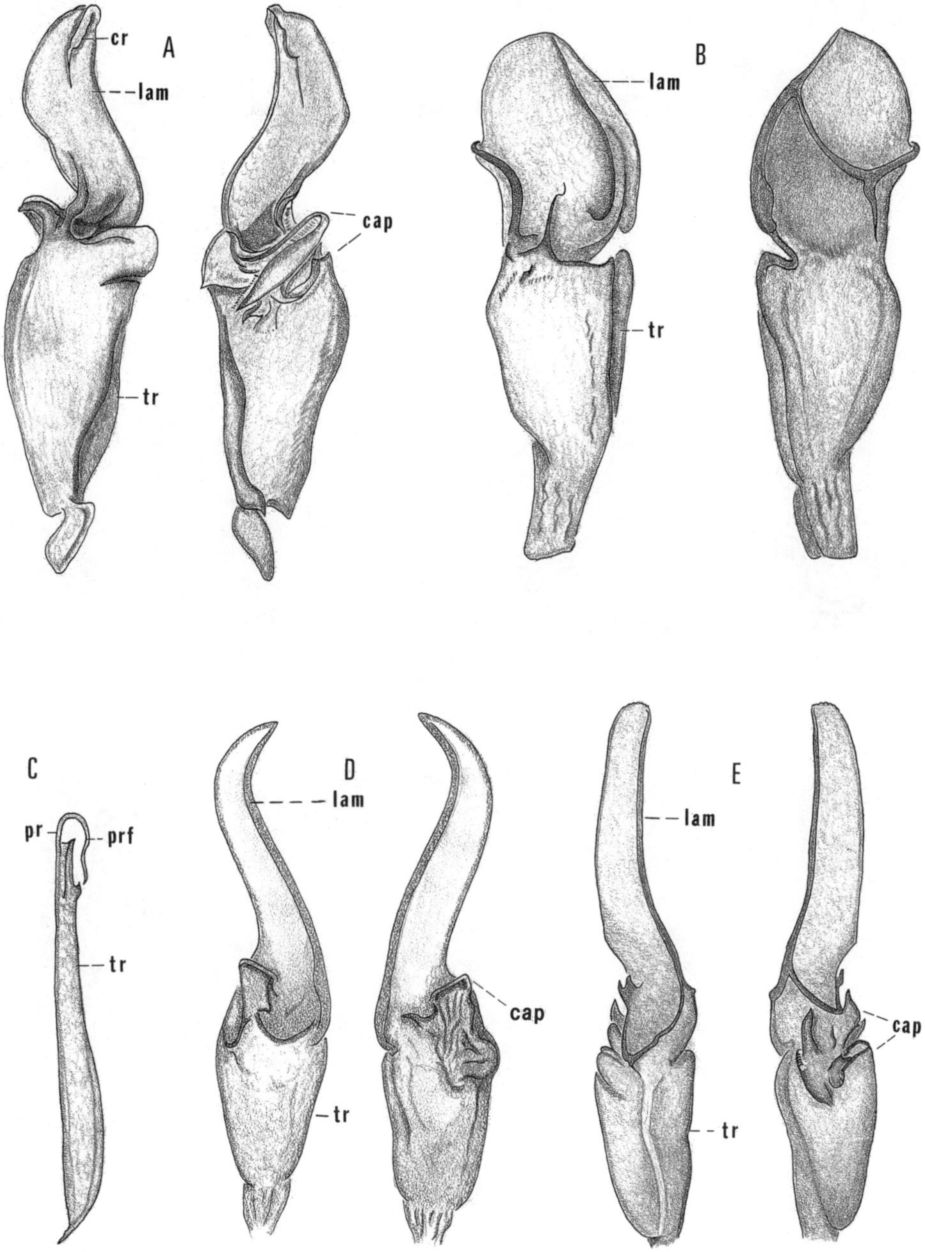

Fig. 3.12. Left hemispermatophores of scorpions, lateral (*left*) and medial (*right*) views: *A*, *Bothriurus vachoni* (Bothriuridae); *B*, *Hadrurus arizonensis* (Iuridae); *C*, *Centruroides vittatus* (Buthidae), lateral view; *D*, *Diplocentrus whitei* (Diplocentridae); *E*, *Vaejovis intermedius* (Vaejovidae). *cap*, capsule; *cr*, crest; *lam*, distal lamina; *pr*, pars recta; *prf*, pars reflecta; *tr*, trunk. (View *A* after San Martín 1968)

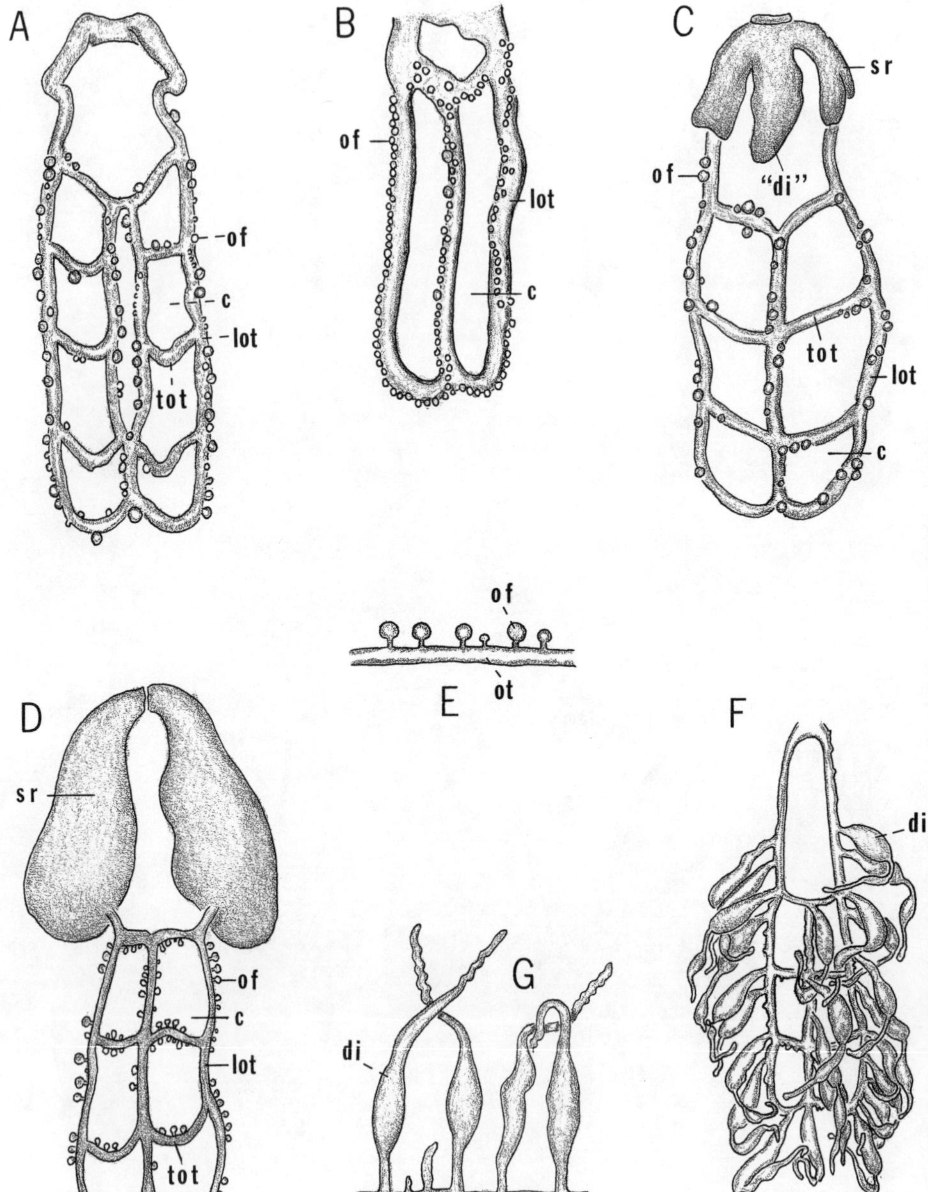
A
of
c
lot
tot
B
of
lot
c
C
sr
of
"di"
tot
lot
c
of
E
ot
D
sr
of
c
lot
tot
G
di
ot
F
di

ably from the other families. The oocytes are located within numerous lateral diverticula arising from the branches of the ovariuterus (Fig. 3.13*F*, *G*), with one embryo developing in each diverticulum. This type of development is referred to as katoikogenic (Laurie 1896a, b) and is a synapomorphic feature for these taxa. Although some bothriurids also possess a "diverticulum" (Fig. 3.13C), it is connected to the genital atrium and does not resemble the diverticula of katoikogenic scorpions; its function is not clear.

Classification of Recent Scorpions

The classification scheme resulting from the morphological characters discussed in the preceding section is given below. Currently, nine families are recognized in the Scorpiones: the Bothriuridae, Buthidae, Chactidae, Chaerilidae, Diplocentridae, Ischnuridae, Iuridae, Scorpionidae, and Vaejovidae. A key to these scorpion families is provided, and then each family is characterized in detail. Keys to subfamilies and genera are given. Following each familial characterization, the distributions of all genera are given, with estimates of the numbers of species.

The information in this section is derived from a large number of sources, including regional studies, generic revisions, and species descriptions, most of which are listed in the bibliography. Doubtful taxa and synonyms are excluded; for a complete list of generic synonymies through 1982, consult Francke's (1985) *Conspectus Genericus Scorpionum*. Several changes made in the *Conspectus* should be mentioned here, because they reflect changes in well-established names. Two names were changed because they were discovered to be homonyms;

Fig. 3.13. Female reproductive systems of scorpions: *A*, eight-celled ovariuterus of *Parabuthus planicauda* (Buthidae); *B*, two-celled ovariuterus of *Tityus bahiensis* (Buthidae); *C*, six-celled ovariuterus of *Brachistosternus intermedius* (Bothriuridae); *D*, ovariuterus of *Hadrurus arizonensis* (Iuridae); *E*, section of ovariuterus of *Paruroctonus mesaensis* (Vaejovidae), showing follicles attached to *ot* by short pedicels; *F*, ovariuterus of *Scorpio maurus* (Scorpionidae), showing numerous lateral diverticula; *G*, section of ovariuterus of *Heterometrus fulvipes* (Scorpionidae), showing diverticula. *c*, cell of ovariuterus; *di*, diverticulum; *"di,"* saclike "diverticulum" extending from genital atrium; *lot*, longitudinal ovarian tubule; *of*, ovarian follicle; *ot*, ovarian tubule; *sr*, seminal receptacle; *tot*, transverse ovarian tubule. (Views *A*, *C*, *G* after Werner 1934 from Pavlovsky; view *B* after Matthiesen 1970; view *F* after Millot & Vachon 1949) the new names are *Pocockius* (for *Stenochirus*) and *Paraiurus* (for *Calchas*). A third generic name, *Buthotus*, was shown to be a junior synonym of *Hottentotta*.

Key to the Families of Recent Scorpiones

The following key is modified from Francke and Soleglad (1981). This key, and those that follow, will work best with later-instar scorpions or adults.

Pedipalp femur with 10 or more trichobothria, of which 4 or 5 are on the internal aspect (Figs. 3.2*A* and 3.3); pedipalp patella without ventral trichobothria (Fig. 3.2*D*) .... **Buthidae**

Pedipalp femur with 9 or fewer trichobothria, of which only 1 is on the internal aspect (Figs. 3.4*A* and 3.5*A*); pedipalp patella with 1 or more ventral trichobothria (Figs. 3.4*D* and 3.5*D*):

Coxapophyses broadly expanded anteriorly (Fig. 3.7*A*); pedipalp femur with 9 trichobothria, of which 4 are on the dorsal aspect (Fig. 3.4*A*) .... **Chaerilidae**

Coxapophyses not broadly expanded anteriorly (Fig. 3.7*C–H*); pedipalp femur with 3 or 4 trichobothria, of which only 1 is on the dorsal aspect (Fig. 3.5*A*):

Retrolateral pedal spurs absent (Fig. 3.8*C*); always 2 trichobothria (V_1, Et_1) on ventral aspect of pedipalp-chela manus adjacent to base of movable finger (as in Figs. 3.5*G*, 3.22*C*); female ovariuterus with conspicuous diverticula (Fig. 3.13*F*, *G*):

Subaculear tubercle present (Fig. 3.10*C*) .... **Diplocentridae**

Subaculear tubercle absent (Fig. 3.10*A*):

Lateroapical margin of tarsi produced into a rounded lobe (Fig. 3.28*A*); or, if margin straight (*Hemiscorpius* and *Habibiella*), then the metasoma bears a single ventromedian carina and the carapace has three pairs of lateral eyes .... **Scorpionidae**

Lateroapical margin of tarsi straight (Fig. 3.24*A*); if metasoma has single ventromedian carina (*Heteroscorpion*), then the carapace has two pairs of lateral eyes .... **Ischnuridae**

Retrolateral pedal spurs present (Fig. 3.8*B*); if retrolateral pedal spurs absent, then 3 trichobothria (V_1, Et_1, Et_2) present on ventral aspect of pedipalp-chela manus adjacent to base of movable finger (Fig. 3.15*E*, *F*); female ovariuterus without diverticula (Fig. 3.13*C*, *D*):

Sternum reduced to a transverse slitlike sclerite in adults; or, if subpentagonal, then still at least twice as wide as long . . . **Bothriuridae**

Sternum distinctly subpentagonal, as long as or longer than wide:

Cheliceral movable finger with 1 large, darkened basal tooth on the internal margin (Fig. 3.1*F*) .... **Iuridae**

Cheliceral movable finger without single large, darkened basal tooth on the internal margin (several small denticles or tubercles may be present; Fig. 3.1*C*, *H*)... **Chactidae** and **Vaejovidae**

Family Bothriuridae Simon, 1880

The Bothriuridae (Figs. 3.14, 3.15) is a small family that reaches its highest diversity in South America (10 genera, 78 species). There are also a monotypic genus, *Cercophonius*, in Australia, and another genus, *Lisposoma*, with two species in southern Africa. This distribution pattern is typical of Gondwanaland biotas.

Lisposoma was formerly considered a scorpionid, but it was known to share similarities with the bothriurids. Francke (1982b) analyzed the relationships between *Lisposoma* and the two families and found reason to transfer the genus to the Bothriuridae. *Lisposoma* has three trichobothria ventrally near the base of the movable chela finger and legs ending in truncated tarsi bearing paired ventral spines. Both characters are derived and unique to *Lisposoma* and bothriurids. Therefore, Francke (1982b) hypothesized that the characters represent synapomorphies. Two characters that were used to place *Lisposoma* in the Scorpionidae—a pentagonal sternum and the absence of a subaculear tubercle on the telson—were dismissed as plesiomorphic. A third character, the presence of only prolateral pedal spurs at the articulation between tarsomeres I and II, is derived, but it has arisen independently at least three times in the Scorpiones (including in the Bothriuridae). It is also interesting to note that females of *Lisposoma* lack the distinctive ovariuteral diverticula of the Scorpionidae. On the basis of this analysis, then, it is appropriate to regard *Lisposoma* as a bothriurid until further counterevidence accumulates.

Characterization

CHELICERAE (Fig. 3.1*A*). On the cheliceral fixed finger the medial and basal teeth are fused into a bicusp. The movable finger carries one or two subdistal teeth and a single basal tooth on the external (dorsal) margin; the distal external tooth is smaller than the distal internal tooth, and generally not apposable to it; the internal (ventral) margin is smooth.

TRICHOBOTHRIAL PATTERN. The trichobothrial pattern is Type C (as in Fig. 3.5). Only *Thestylus* and *Lisposoma* have the basic number and

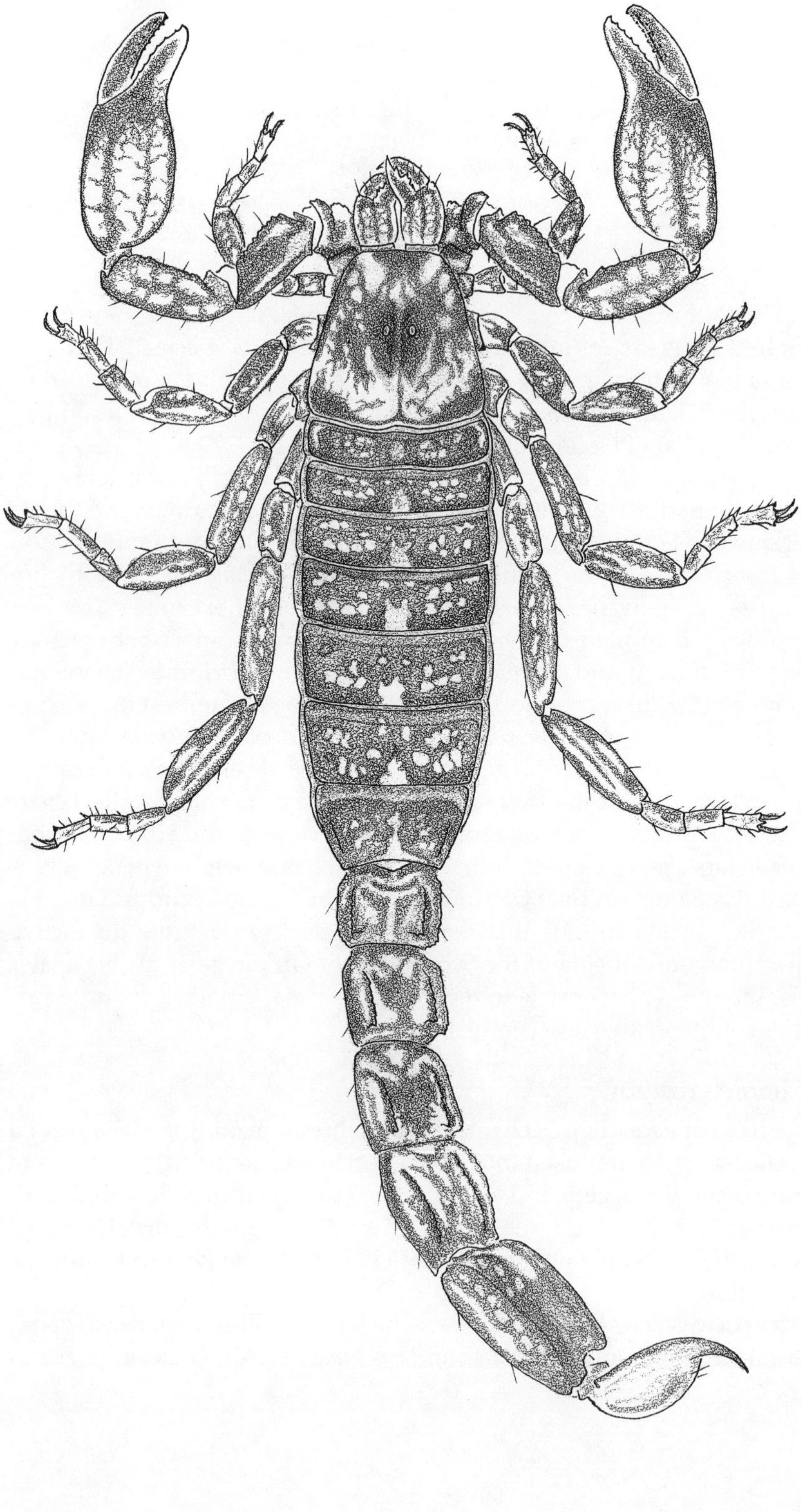

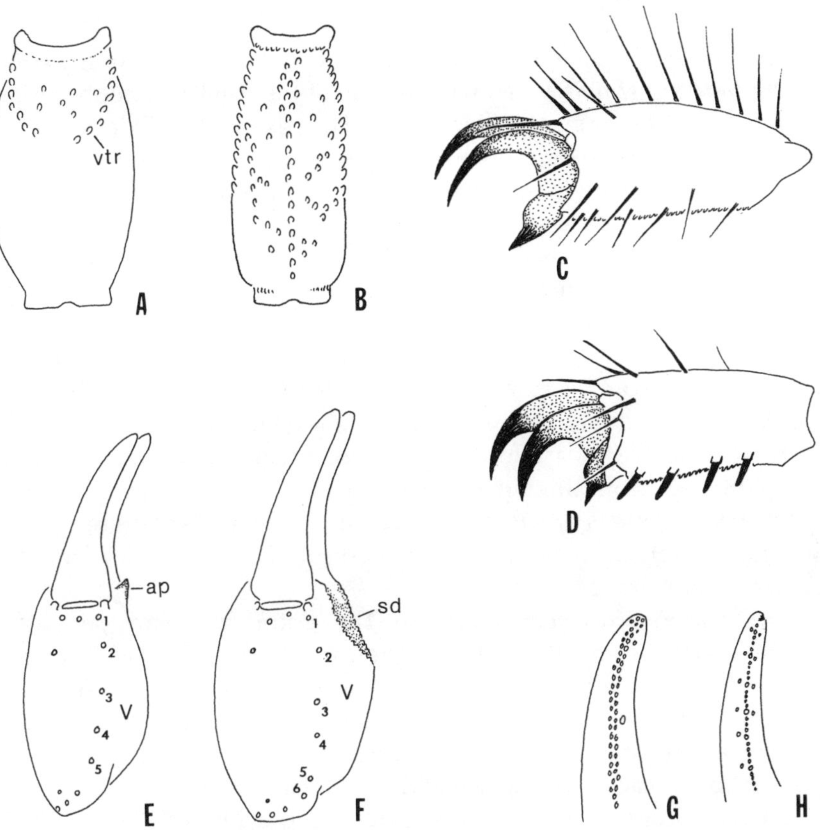

Fig. 3.15. Characteristics of bothriurid genera: *A, Bothriurus araguayae*, ventral aspect of metasomal segment V; *B, Orobothriurus alticola*, ventral aspect of metasomal segment V; *C, Brachistosternus ehrenbergi*, prolateral aspect of tarsomere II of leg III; *D, Centromachetes* sp., prolateral aspect of tarsomere II of leg III; *E, Bothriurus araguayae*, ventral aspect of male pedipalp chela, showing trichobothrial pattern and apophysis; *F, Timogenes* sp., ventral aspect of male pedipalp chela, showing trichobothrial pattern and semicircular depression; *G, Centromachetes* sp., dentition of pedipalp-chela fixed finger; *H, B. araguayae*, dentition of pedipalp-chela fixed finger. *ap*, apophysis; *vtr*, ventral transverse carina; *sd*, semicircular depression. Trichobothrial designations are as described in text (see "Trichobothria") and as shown in Figure 3.5. (View B after Maury 1975d)

Fig. 3.14 (*facing page*). *Bothriurus araguayae* from Brazil, a representative bothriurid (Bothriurinae).

pattern; all other genera have accessory trichobothria (Vachon 1973). *Brachistosternus* (*Ministernus*) has lost one of the patellar trichobothria.

COXOSTERNAL REGION (Fig. 3.7C). The coxapophyses do not possess anteriorly expanded lobes. The sternum is represented by a narrow transverse plate in all genera except *Lisposoma* and *Tehuankea* (in which it is more or less subpentagonal).

LEG SPINATION. There are no tibial spurs. Only prolateral (interior) pedal spurs are present in *Lisposoma, Phoniocercus, Thestylus,* and *Vachonia;* both prolateral and retrolateral (exterior) pedal spurs are present in the other genera. Except in *Brachistosternus* and *Vachonia,* the tarsi possess two rows of ventral spines. *Brachistosternus* and *Vachonia* inhabit sandy soils and possess bristlecombs on the tibiae and tarsi; the increased surface area afforded by these bristlecombs prevents the scorpion's legs from sinking into the sand.

VENOM GLANDS AND TELSON. The venom glands are thick-walled and folded; the venom is opalescent. The telson lacks a subaculear protuberance.

MALE REPRODUCTIVE SYSTEM. The bothriurid hemispermatophore is lamelliform (Fig. 3.12*A*). The capsular region is well developed and, in *Brachistosternus,* possesses spines (Maury 1980). The distal lamina is characterized by the presence of a strong crest on the externolateral aspect, but this is not yet confirmed in *Lisposoma.* The paraxial organ lacks accessory glands.

FEMALE REPRODUCTIVE SYSTEM. The ovariuterus (Fig. 3.13C) forms a reticulate mesh of six cells. A single saclike diverticulum is suspended from the genital atrium in some species; lateral diverticula are lacking.

DEVELOPMENT. Apoikogenic.

Key to the Subfamilies and Genera of the Bothriuridae

The following key is based primarily on those provided by Mello-Leitão (1945) and Maury (1971). In several couplets it is necessary to count the tarsal spines of the legs; these spines are usually paired, but asymmetry in number sometimes occurs (e.g., some *Cercophonius* may have two spines in the prolateral row, but one in the retrolateral row: Koch 1977).

Ventral aspects of pedipalp tibia and chela manus with numerous accessory trichobothria; total number of trichobothria exceeding 100; lateral eyes in 2 pairs .... Subfamily VACHONIANINAE ... ***Vachonia***

Ventral aspect of pedipalp patella with 6 or fewer trichobothria; ventral aspect of chela manus with 10 or fewer trichobothria; total number of trichobothria on pedipalps ranging from 48 to 56; lateral eyes in 3 pairs:

Tarsi with ventral setae, rather than spines (Fig. 3.15*C*); ventral aspect of cheliceral movable finger with 4 to 6 setae; hemispermatophore with spines in the capsular region ....Subfamily BRACHISTOSTERNINAE... ***Brachistosternus***

Tarsi armed ventrally with 2 rows of spines (Fig. 3.15*D*); ventral aspect of cheliceral movable finger with numerous setae; hemispermatophore without spines in the capsular region ... Subfamily BOTHRIURINAE:

Sternum subpentagonal or nearly so, about twice to three times as wide as long:

Tarsi of legs with prolateral pedal spurs only (Fig. 3.8*C*) .. ***Lisposoma****

Tarsi of legs with both prolateral and retrolateral pedal spurs (Fig. 3.8*B*) .... ***Tehuankea***

Sternum much wider than long, often appearing slitlike:

Dentate margin of pedipalp-chela fingers composed of a single row of granules flanked by regularly spaced accessory granules (Fig. 3.15*H*):

5 ventral trichobothria on chela manus (Fig. 3.15*E*); male chela with an apophysis on inner surface near fixed finger (Fig. 3.15*E*) or lacking specialized structures:

Legs with prolateral pedal spurs only (Fig. 3.8*C*):

Tarsomere II of leg IV with 2 to 4 pairs of spines.. ***Phoniocercus***

Tarsomere II of leg IV with 5 to 7 pairs of spines..... ***Thestylus***

Legs with both prolateral and retrolateral pedal spurs (Fig. 3.8*B*):

Ventrodistal aspect of metasomal segment V with a ventral transverse carina forming a distinct arc (Fig. 3.15*A*); longitudinal carinae may be present as well; cheliceral movable finger with 1 subdistal tooth; male telson with conspicuous dorsal gland .... ***Bothriurus***

Metasomal segment V with 3 or 5 subparallel longitudinal carinae (in one species without carinae; Fig. 3.15*B*); cheliceral movable finger with 2 subdistal teeth (Fig. 3.1*A*); male telson without conspicuous dorsal gland (only a depression instead) .... ***Orobothriurus***

6 to 10 ventral trichobothria on pedipalp-chela manus (Fig. 3.15*F*); male chela with semicircular depression on inner surface near origin of fixed finger (Fig. 3.15*F*) .... ***Timogenes***

*The genus *Lisposoma* has not yet been assigned to any subfamily. It is included here in the key because it has no known characters that distinguish it from bothriurine genera.

Dentate margin of pedipalp-chela fingers composed of numerous granules irregularly comprising 2 to 5 rows (in some cases only the basal part of the dentate margin is doubled; Fig. 3.15G) flanked by regularly spaced accessory granules:

Tarsomere II of leg IV with 2 (rarely 3) pairs of spines; Australia . . . . ***Cercophonius***

Tarsomere II of leg IV with 4 to 8 pairs of spines; South America:

Tarsomere II of legs I and II with 2 to 3 pairs and 3 to 4 pairs of spines, respectively; tarsomere II of leg IV with 4 pairs of spines . . . . ***Centromachetes***

Tarsomere II of legs I and II with 1 pair and 2 pairs of spines, respectively; tarsomere II of leg IV with 5 to 8 pairs of spines . . . . ***Urophonius***

List of Genera and Their Distributions

Subfamily Bothriurinae Simon, 1880

Bothriurus Peters, 1861 (Argentina, Bolivia, Brazil, Chile, Paraguay, Peru, Uruguay), 2 subgenera, 27 spp.

Centromachetes Lönnberg, 1897 (Chile), 3 spp.

Cercophonius Peters, 1861 (Australia, Tasmania), 1 sp.

Orobothriurus Maury, 1975 (Argentina, Chile, Peru), 8 spp.

Phoniocercus Pocock, 1893 (Chile), 2 spp.

Tehuankea Cekalovic, 1973 (Chile), 1 sp.

Thestylus Simon, 1880 (Brazil), 2 spp.

Timogenes Simon, 1880 (Argentina, Bolivia; Paraguay?), 5 spp.

Urophonius Pocock, 1893 (Argentina, Brazil, Chile, Uruguay), 13 spp.

Subfamily Brachistosterninae Maury, 1972

Brachistosternus Pocock, 1893 (Argentina, Bolivia, Brazil, Chile, Colombia, Ecuador, Paraguay, Peru), 3 subgenera, 16 spp.

Subfamily Vachonianinae Maury, 1972

Vachonia Abalos, 1954 (Argentina), 1 sp.

Genus *incertae sedis*

Lisposoma Lawrence, 1927 (South-West Africa), 2 spp.

Family Buthidae Simon, 1880

The Buthidae (Figs. 3.16, 3.17), with 48 genera and over 500 species, is the largest and most widespread of the scorpion families. Its members occupy all six faunal regions, but the greatest generic diversity occurs in the Old World, especially in the Afrotropical Region (22 genera) and in southern portions of the Palaearctic Region (23 genera).

Although there have been many attempts to recognize subfamilies in the Buthidae, none appears satisfactory (Stahnke 1972b). Not only are genera shuffled back and forth between established subfamilies, but there is little agreement even about which subfamilies should be recognized or what they should be called. For this reason, subfamilies will not be recognized here. However, the interested reader is referred to the following sources for information: Birula 1917a, Werner 1934, Mello-Leitão 1945, Probst 1972, Lamoral 1976, Koch 1977, and Levy and Amitai 1980.

Lamoral (1980) recognized the buthids as constituting one of three main evolutionary lines of scorpions, representing the plesiomorphic sister group of the other Recent scorpions. Protobuthoids were suggested to have been well established by Pangaean times, originating in Laurasia and spreading to Gondwanaland before the split of the two land masses. Subsequent breakups of Laurasia and Gondwanaland during the Mesozoic led to the current distribution of the buthids.

Characterization

CHELICERAE (Fig. 3.1*B*). The cheliceral fixed finger has the medial and basal teeth fused into a bicusp. The movable finger has a single subdistal tooth and two basal teeth (sometimes one in *Akentrobuthus*) on the external margin; the distal external tooth is larger than the distal internal tooth, and generally is apposable to it (i.e., the two teeth form a fork); there are two teeth on the internal margin.

TRICHOBOTHRIAL PATTERN. The trichobothrial pattern is Type A (Figs. 3.2 and 3.3). Most genera show the basic number and pattern; however, *Alayotityus, Karasbergia, Lissothus, Mesotityus, Microbuthus, Orthochirus, Pectinibuthus, Zabius*, and some species of *Microtityus* have lost trichobothria, whereas *Buthiscus, Liobuthus*, and *Vachoniolus* possess accessory trichobothria (Vachon 1973; W. Lourenço, pers. comm. 1983).

COXOSTERNAL REGION (Fig. 3.7*B*). The coxapophyses do not possess anteriorly expanded lobes. Although most buthids have a subtriangular sternum, all first-instar buthids possess a subpentagonal sternum, and this characteristic persists in the adult in *Akentrobuthus, Butheoloides, Charmus, Karasbergia, Microtityus*, and some *Orthochirus*.

LEG SPINATION. Tibial spurs are present on legs III and/or IV, except in *Akentrobuthus, Alayotityus, Centruroides, Isometrus, Mesotityus, Microtityus, Rhopalurus, Tityus, Vachoniolus*, and *Zabius*; in *Anomalobuthus*, the single tibial spur on leg IV is vestigial (Stahnke 1972b). In

Fig. 3.16. Representative Buthidae: *Above: top, Hottentotta conspersa,* Vila Arriaga, Angola; *bottom, Parabuthus villosus,* Gobabeb, South-West Africa (photos courtesy Edward S. Ross). *Facing page: top, Centruroides vittatus,* Comstock, Texas; *bottom, Uroplectes fischeri,* Lake Manyara, Tanzania (photo courtesy Ross).

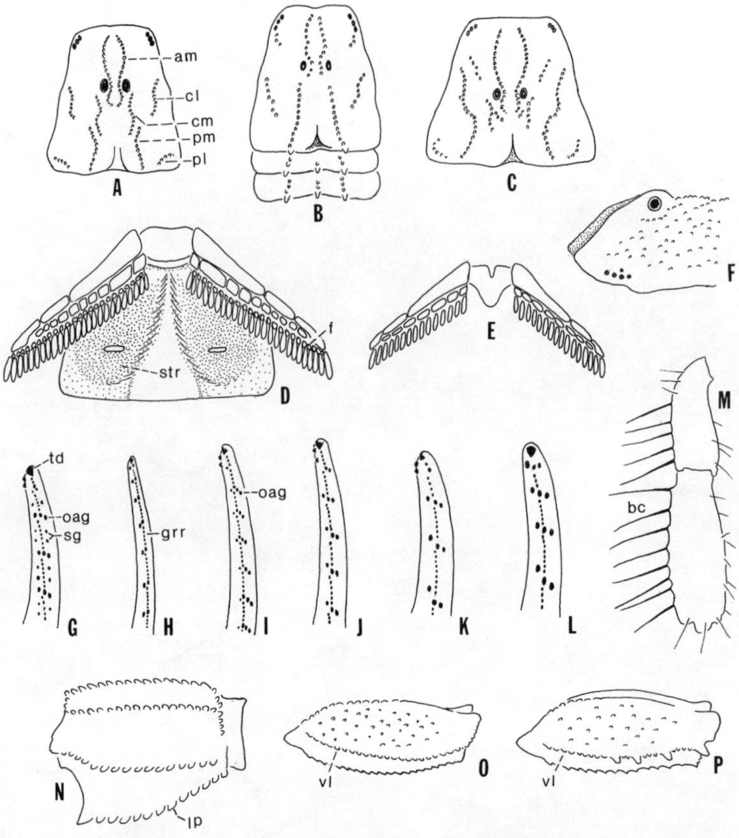

Fig. 3.17. Characteristics of buthid genera: *A*, generalized buthid carapace, showing carinal nomenclature; *B*, *Compsobuthus werneri*, carapace and anterior tergites, dorsal view; *C*, *Buthus occitanus*, carapace, dorsal aspect, showing lyre configuration of *cl* and *pm* (see below); *D*, *Rhopalurus rochae*, pectines and sternite III, ventral aspect, showing *str*; *E*, *Ananteris balzani*, pectines, ventral aspect (note fulcra absent); *F*, *Orthochirus scrobiculosus*, carapace, lateral view; *G–L*, dentition on distal part of pedipalp-chela movable finger (*G*, *Centruroides pococki* [note *sg*]; *H*, *Anomalobuthus rickmersi; I*, *Lychas variatus* [note imbricated *grr*]; *J*, *Odontobuthus doriae; K*, *Hottentotta judaica; L*, *B. occitanus); M*, *A. rickmersi*, right tibia and tarsomere I (leg III), dorsal view, showing *bc; N*, *Androctonus crassicauda*, metasomal segment IV, lateral view; *O*, *P*, metasomal segment V, lateral view (*O*, *H. judaica; P*, *Mesobuthus eupeus*). *am*, anterior median carina; *bc*, bristlecomb; *cl*, central lateral carina; *cm*, central median carina; *dl*, dorsolateral metasomal carina; *f*, fulcrum; *grr*, granular row; *oag*, outer accessory granule; *pl*, posterior lateral carina; *pm*, posterior median carina; *sg*, supernumerary granule; *str*, stridulatory area; *td*, terminal denticle; *vl*, ventrolateral metasomal carina. (View *A* after Levy & Amitai 1980)

Apistobuthus it is variable in size and sometimes even absent. Both prolateral and retrolateral pedal spurs are present. The tarsi possess ventral spines in some genera, but this condition is not common.

VENOM GLANDS AND TELSON. The venom glands are thin-walled and folded; the venom is opalescent. The telson usually bears a subaculear tubercle or tooth (Fig. 3.10*D–F*).

MALE REPRODUCTIVE SYSTEM. The hemispermatophore is flagelliform (Fig. 3.12*C*). The trunk is long and slender, and four simple lobes arise at the truncal flexure. The inner lobe is continuous with the flagellum, which is divided into the *pars recta* and *pars reflecta* (Pavlovsky 1924c; Vachon 1940, 1952; Lamoral 1979). The paraxial organ (Fig. 3.11*A*, *B*) has six glands: one cylindrical gland, one oval gland, and two pairs of anterior accessory glands.

FEMALE REPRODUCTIVE SYSTEM. The ovariuterus forms a reticulate mesh of eight cells (Fig. 3.13*A*) or two cells (in *Tityus* spp.; Fig. 3.13*B*). Diverticula are lacking.

DEVELOPMENT. Apoikogenic.

Key to the Genera of the Buthidae

The following key relies heavily on trichobothrial patterns. The trichobothria of buthids are sometimes quite small and difficult to find, but the characters that they provide are of greater quality and consistency than some characters used in the past. The reader should consult Figures 3.2 and 3.3 for trichobothrial nomenclature and approximate locations of various trichobothria.

★ Angle formed by trichobothria d_1, d_3, and d_4 opens toward external face of pedipalp femur (α configuration, Fig. 3.3):

▲ Legs III and IV lacking tibial spurs:

Pedipalp patella with only 4 dorsal trichobothria (d_2 absent):

Pedipalp femur with trichobothrium d_2 present, but displaced to internal surface; pedipalp chela lacking trichobothrium Eb_3; sternum subpentagonal, with lateral edges slightly convergent anteriorly . . . . ***Mesotityus***

Pedipalp femur with trichobothrium d_2 absent; pedipalp chela with trichobothrium Eb_3 present; sternum subtriangular, with lateral edges strongly convergent anteriorly . . . . ***Alayotityus***

Pedipalp patella with 5 dorsal trichobothria:

Tergites I–VI with 3 of 5 distinct longitudinal carinae; trichobothrium d_2 of pedipalp femur usually absent:

Small scorpions (adults less than 25 mm long); pedipalp chelae slender with elongate fingers; telson with distinct subaculear tubercle . . . . ***Microtityus***

Larger scorpions (adults to about 60 mm) with robust pedipalp chelae; telson without subaculear tubercle .... ***Zabius***

Tergites I–IV with 1 distinct median longitudinal carina (a vestigial lateral pair, if present, may be found at the posterior tergal margin); trichobothrium d_2 of pedipalp femur always present:

Dentate margins of pedipalp-chela fingers composed of 12 to 17 oblique rows of granules; adults without supernumerary granules flanking these rows:

Spiracles of book lungs small and rounded; trichobothrium d_5 of pedipalp femur situated at the level of e_1; Cuba .... ***Tityopsis***

Spiracles of book lungs large, elongate oval or slitlike; trichobothrium d_5 of pedipalp femur distinctly basal to e_1..... .... ***Tityus***

Dentate margins of pedipalp-chela fingers composed of 7 to 9 oblique rows of granules, these flanked by supernumerary granules in the adult (Fig. 3.17*G*):

Trichobothrium *db* of pedipalp-chela fixed finger usually distal to *et*; first sternite with 2 concave lateral stridulatory areas (Fig. 3.17*D*); distal segments of metasoma slightly to greatly dilated in male, but segments not longer than in female.................................. ***Rhopalurus***

Trichobothrium *db* of pedipalp-chela fixed finger basal to or at the level of *et*; first sternite lacking stridulatory areas; distal metasomal segments not dilated in male, but male segments distinctly (usually greatly) elongated .... ***Centruroides***

▲ Legs III and IV with tibial spurs (Fig. 3.8*A*):

Sternum subpentagonal:

Pedipalp chela with 2 *Eb* trichobothria on palm .... ***Karasbergia***

Pedipalp chela with 3 *Eb* trichobothria on palm:

Telson with distinct subaculear tubercle; carapace granular, but lacking distinct carinae .... ***Butheoloides***

Telson lacking subaculear tubercle; carapace possessing distinct carinae .... ***Charmus***

Sternum subtriangular:

Ventral aspect of cheliceral fixed finger smooth, lacking nodules or denticles .... ***Uroplectes***

Ventral aspect of cheliceral fixed finger with 1 or 2 denticles (Fig. 3.1*B*):

Telson with distinct subaculear tubercle:

Tergites I–VI with a single median carina; subaculear tubercle triangular, with 2 small subdistal accessory tubercles (Fig. 3.10*F*) . . . . ***Tityobuthus***

Tergites I–VI with 3 carinae (lateral pair short, restricted to posterior portion of tergite); subaculear tubercle triangular in profile, but lacking small accessory tubercles . . ***Pseudolychas***

Telson without distinct subaculear tubercle:

Granular rows of dentate margin of pedipalp-chela fingers slightly or not at all imbricated (Fig. 3.17*J*, *L*); stridulatory area located on dorsal aspect of metasomal segments I and II; pedipalp-chela movable finger with 3 accessory granules (not arranged in a row) just proximal to the terminal denticle (Fig. 3.17*L*) . . . . ***Parabuthus***

Granular rows of dentate margin of pedipalp-chela fingers distinctly imbricated (Fig. 3.17*I*); no stridulatory area on dorsal aspect of metasomal segments I and II; pedipalp-chela movable finger with 2 accessory granules along inner edge just proximal to the terminal denticle and lying adjacent to the first granular row (Fig. 3.17*J*):

Metasomal segments lacking ventrolateral and ventral submedian carinae and with others reduced; pectines with less than 18 teeth (known specimens actually with 15); females with unmodified pectinal teeth . . . . ***Pocockius***

Metasomal segments with all carinae distinct; pectines almost always with more than 18 teeth; females with proximal tooth on each pecten dilated and often elongated . . . . ***Grosphus***

★ Angle formed by trichobothria d_1, d_3, and d_4 opens toward internal face of pedipalp femur (β configuration, Fig. 3.3):

■ Legs without tibial spurs:

Sternum subpentagonal . . . . ***Akentrobuthus***

Sternum subtriangular (i.e., lateral edges strongly convergent anteriorly):

Tibia and tarsomeres of legs I–III with retrolateral row of long curved setae (bristlecombs; Fig. 3.17*M*):

Pedipalp patella with more than 7 external trichobothria; dentate margin of pedipalp-chela fingers with outer accessory denticles (Fig. 3.17*J*); pedipalp chela more robust, with movable finger less than twice as long as underhand:

Pedipalp patella with 11 external trichobothria; pedipalp femur with 6 dorsal and 4 external trichobothria . . . . ***Liobuthus***

Pedipalp patella with 8 external trichobothria; pedipalp femur with 5 dorsal and 2 external trichobothria ...... ***Vachoniolus***

Pedipalp patella with 7 external trichobothria; dentate margin of pedipalp-chela fingers lacking outer accessory denticles (Fig. 3.17*H*); pedipalp chela very slender, with long fingers (movable finger more than twice as long as underhand):

Pedipalp femur, patella, and chela with fundamental number of trichobothria; pectines with less than 35 teeth... ***Anomalobuthus***

Pedipalp femur lacking trichobothrium d_2, patella lacking d_2, and chela lacking *esb*, *Esb*, and Eb_3; pectines with more than 35 teeth .... ***Pectinibuthus***

Tibia and tarsomeres of legs I–III with setae not arranged into a bristlecomb .... ***Isometrus***

- Tibial spurs present on legs III and IV or only on leg IV (sometimes vestigial in *Anomalobuthus*):

Tibial spurs present only on leg IV:

Pedipalp femur with 3 external trichobothria .... ***Buthiscus***

Pedipalp femur with 2 external trichobothria:

Pectines without fulcra (Fig. 3.17*E*) .... ***Lychasioides***

Pectines with fulcra (Fig. 3.17*D*):

Tibiae and tarsi of legs I–III with bristlecombs; tibial spur vestigial; dentate margin of pedipalp-chela movable finger lacking outer accessory denticles (Fig. 3.17*H*) and divided into 10 slightly oblique rows (sometimes fewer if basal portion of dentate margin not distinctly divided).... ***Anomalobuthus***

Tibiae and tarsi with setae not arranged into bristlecombs; tibial spur small to well developed; dentate margin of pedipalp-chela movable finger with outer accessory denticles and divided into 6 to 9 more or less imbricated rows .. ***Babycurus***

Tibial spurs present on both legs III and IV (or, if present only on leg IV, then the second metasomal segment is much wider than the other metasomal segments):

Pectines without fulcra (Fig. 3.17*E*) .... ***Ananteris***

Pectines with fulcra (Fig. 3.17*D*):

Dentate margin of pedipalp-chela movable finger with granules indistinct, not divided into rows, and limited to distal half of finger:

Carapace heavily granulated; metasomal segment V punctate .... ***Microbuthus***

Carapace smooth; metasomal segment V not punctate.. ***Lissothus***

Dentate margin of pedipalp-chela movable finger with granules distinct, divided into rows, and occurring the length of the finger:

Carapace, in lateral view, with a distinct downward slope from median eyes to anterior margin (Fig. 3.17*F*); small scorpions (less than 30 mm long):

Carapace, tergites, sternites, and metasoma set with very dense, rounded granules producing a beaded appearance; tergites I–VI with 3 carinae, each extending posteriorly into a sharp point; metasomal segments I–IV without carinae . . . . ***Birulatus***

Carapace, tergites, sternites, and metasoma often granular, but not as above; tergites I–VI with or without carinae, but if present never extending past margin of tergite; metasomal segments I–IV carinate:

Metasomal segment V punctate; trichobothrium d_2 of pedipalp femur usually absent . . . . ***Orthochirus***

Metasomal segment V not punctate; trichobothrium d_2 of femur present . . . . ***Butheolus***

Carapace, in lateral view, with entire dorsal surface horizontal, or nearly so (possibly with slight anterior downward slope); size variable

Cheliceral fixed finger with a single ventral denticle:

Telson with distinct subaculear tubercle, ranging in size from small to very large . . . . ***Lychas***

Telson with tubercle either very subtle or absent:

Tibiae and tarsi of legs I–III with bristlecombs along retrolateral margins . . . . ***Psammobuthus***

Tibiae and tarsi of legs I–III with setae, but not arranged as above:

Tergites I–VI with 3 carinae; pedipalp-chela fingers relatively short (about as long as underhand); manus robust . . . . ***Hemibuthus***

Tergites I–VI with a single median carina; pedipalp-chela fingers elongate (more than twice as long as underhand); manus slender . . . . . ***Isometroides***

Cheliceral fixed finger with 2 ventral denticles:

Metasomal segment II widely flared, much wider than other metasomal segments . . . . ***Apistobuthus***

Metasomal segment II similar in width to other metasomal segments:

First 2 tergites with 5 carinae, the posterior ones with at least 3 . . . . ***Leiurus***

Anterior tergites without carinae, or with 1 to 3 carinae:

Carapace granular but lacking distinct carinae. . . . . ***Buthacus***

Carapace with distinct carinae:

Trichobothrium *eb* of pedipalp chela situated in the distal part of the manus rather than on the fixed finger; ventral aspect of tarsomere II with a single median row of 5 to 6 setae; pedipalp chela extremely stocky, with short fingers (movable finger about the same length as the underhand) . . . . ***Kraepelinia***

Trichobothrium *eb* always distinctly on the fixed finger; ventral aspect of tarsomere II not as above; pedipalp chela more slender, with longer fingers (movable finger longer than underhand):

Tergites I–VI with a single median carina, present at least on posterior segments; telson with denticulate subaculear tubercle . . . . ***Odonturus***

Tergites I–VI with 3 carinae (may be weak or obsolete on anterior tergites); telson with at most a subtle protuberance under aculeus, rarely with a pronounced tooth:

Pedipalp-chela movable finger with 2 distal internal granules located just proximal to terminal denticle, flanked laterally by a row of 5 to 7 smaller granules (Fig. 3.17*J*) . . . . ***Odontobuthus***

Pedipalp-chela movable finger with 3 or 4 distal granules located just proximal to terminal denticle (Fig. 3.17*K*, *L*); first row of smaller granules situated proximally to these:

- 4 granules on pedipalp-chela movable finger, just proximal to terminal denticle (Fig. 3.17*K*):

Central lateral and posterior lateral carapacial carinae joined, forming a continuous linear series of granules to posterior margin (Fig. 3.17*B*):

Trichobothrium *db* of pedipalp-chela fixed finger basal to *est*; carinae of tergites projecting beyond posterior margin as distinct spiniform processes (Fig. 3.17*B*) . . . . ***Compsobuthus***

Trichobothrium *db* distal to *est*; carinae of tergites not projecting noticeably beyond posterior margin as spiniform processes . . . . ***Darchenia***

Central lateral and posterior lateral carapacial carinae not joined as above, usually separated by a small gap, with central lateral carinae continuing distally beyond origin of posterior laterals (Fig. 3.17*A*):

Tarsomeres I and II bearing setae on ventral aspect; pedipalp chela very slender, with long, upwardly curved fingers; movable finger well over twice as long as underhand; chela trichobothria *db, dt, est,* and *et* situated distinctly in distal half of fixed finger, widely separated from *eb* and *esb* . . . . ***Cicileus***

Tarsomeres I and II with paired spines ventrally; pedipalp chela not as slender, with shorter fingers; movable finger less than twice as long as underhand (usually less than 1.5 times as long); trichobothria of chela fixed finger usually more evenly spaced along length of finger:

Ventrolateral carinae of metasomal segment V with posterior granules enlarged, often lobate (Fig. 3.17*P*); pedipalp-chela trichobothrium *db* basal to *est* on fixed finger; central, lateral, and posterior median carapacial carinae joined, forming a lyre-shaped configuration (Fig. 3.17*C*) . . . . ***Mesobuthus***

Ventrolateral carinae of metasomal segment V with all granules more or less equal in size, never lobate (Fig. 3.17*O*); pedipalp-chela trichobothrium *db* distal to *est* on fixed finger (with a few exceptions); carapacial carinae not forming a lyre-shaped configuration (Fig. 3.17*A*) . . . . ***Hottentotta***

• 3 granules on pedipalp-chela movable finger, just proximal to terminal denticle (Fig. 3.17*L*):

Scorpions of medium to small size (less than 35 mm); carapacial carinae weak; carapacial surface finely granular . . . . ***Vachonus***

Scorpions of larger size (greater than 50 mm); carapace with carinae well developed and surface coarsely granular:

Central, lateral, and posterior median carinae of carapace joined to form a lyre-shaped configuration (Fig. 3.17*C*); metasoma with all segments more or less equal in width and depth; metasomal segment IV with weakly developed dorsolateral carinae . . . . ***Buthus***

Central, lateral, and posterior median carinae of carapace not joined to form a lyre-shaped configuration (Fig. 3.17*A*); metasomal segment IV with well-developed, granulate dorsolateral carinae (Fig. 3.17*N*); metasomal segments robust, increasing in width and depth posteriorly . . . . ***Androctonus***

List of Genera and Their Distributions

Akentrobuthus Lamoral, 1976 (Zaire), 1 sp.

Alayotityus Armas, 1973 (Cuba), 4 spp.

Ananteris Thorell, 1891 (Argentina, Bissau, Bolivia, Brazil, Colombia, Costa Rica, Ecuador, French Guiana, Guinea, Guyana, Panama, Peru), 13 spp.

Androctonus Hemprich and Ehrenberg, 1829 (northern Africa; Middle East to India), 8 spp.

Anomalobuthus Kraepelin, 1900 (southwestern U.S.S.R.), 1 sp.

Apistobuthus Finnegan, 1932 (Iran, Oman Republic, Saudi Arabia), 1 sp.

Babycurus Karsch, 1886 (Africa south and east of the Sahara Desert), 16 spp.

Birulatus Vachon, 1974 (Jordan), 1 sp.

Buthacus Birula, 1908 (northern and western Africa; Middle East), 7 spp.

Butheoloides Hirst, 1925 (Ivory Coast, Mali Republic, Mauritania, Morocco, Senegal), 4 spp.

Butheolus Simon, 1883 (Oman Republic, Somalia, Sudan, Yemen), 3 subgenera, 4 spp.

Buthiscus Birula, 1905 (Algeria, Tunisia), 1 sp.

Buthus Leach, 1815 (Egypt, Ethiopia, southern France, Israel, Jordan, Libya, southern Spain), 3 spp.

Centruroides Marx, 1889 (southern United States to northern South America, including the West Indies), 40 spp.

Charmus Karsch, 1879 (India, Sri Lanka), 2 spp.

Cicileus Vachon, 1948 (Algeria), 1 sp.

Compsobuthus Vachon, 1949 (northern Africa to Middle East to India), 12 spp.

Darchenia Vachon, 1977 (Yucatán [Mexico]), 1 sp.

Grosphus Simon, 1880 (Malagasy Republic), 7 spp.

Hemibuthus Pocock, 1900 (India), 3 spp.

Hottentotta Birula, 1908 (northeastern, southern, and eastern Africa; Middle East to India), 2 subgenera, 21 spp.

Isometroides Keyserling, 1885 (Australia), 2 spp.

Isometrus Hemprich and Ehrenberg, 1829 (cosmotropical, but introduced in Neotropical and Palaearctic regions), 2 subgenera, 15 spp.

Karasbergia Hewitt, 1914 (South Africa, South-West Africa), 1 sp.

Kraepelinia Vachon, 1974 (southeastern Iran), 1 sp.

Leiurus Hemprich and Ehrenberg, 1829 (northeastern Africa through Middle East), 1 sp.

Liobuthus Birula, 1898 (Iran; Turkistan [U.S.S.R.]), 1 sp.
Lissothus Vachon, 1948 (Libya, Mauritania), 2 spp.
Lychas C. L. Koch, 1845 (Africa south of the Sahara Desert, Australia, Burma, China, India, Indonesia, Malaysia, Philippines, Thailand), 5 subgenera, about 50 spp.
Lychasioides Vachon, 1974 (Cameroon Republic), 1 sp.
Mesobuthus Vachon, 1950 (Iran and southwestern U.S.S.R. to China, India, and Mongolia), 20 spp.
Mesotityus González-Sponga, 1981 (Venezuela), 1 sp.
Microbuthus Kraepelin, 1898 (coasts of Aden, Djibouti, Ethiopia, and Mauritania), 3 spp.
Microtityus Kjellesvig-Waering, 1966 (Brazil, West Indies, Venezuela), 2 subgenera, 9 spp.
Odontobuthus Vachon, 1950 (India, Iran, Pakistan), 2 spp.
Odonturus Karsch, 1878 (Kenya, Somalia, Tanzania), 1 sp.
Orthochirus Karsch, 1891 (Algeria to Egypt; Middle East and southwestern U.S.S.R. to India; Sicily), 6 spp.
Parabuthus Pocock, 1890 (Africa south of the Sahara Desert, Red Sea coast of Arabian Peninsula), 31 spp.
Pectinibuthus Fet, 1987 (Turkistan [U.S.S.R.]), 1 sp.
Pocockius Francke, 1985 (India, Sri Lanka), 2 spp.
Psammobuthus Birula, 1911 (Turkistan [U.S.S.R.]), 2 spp.
Pseudolychas Kraepelin, 1912 (Rhodesia, South Africa), 3 spp.
Rhopalurus Thorell, 1876 (Brazil, Colombia, Cuba, Guyana, Haiti, Venezuela), 9 spp.
Tityobuthus Pocock, 1893 (Malagasy Republic), 2 spp.
Tityopsis Armas, 1974 (Cuba), 1 sp.
Tityus C. L. Koch, 1836 (Central and South America, West Indies), over 100 spp.
Uroplectes Peters, 1861 (southern and eastern Africa), 50 spp.
Vachoniolus Levy, Amitai and Shulov, 1973 (Oman Republic, Saudi Arabia), 2 spp.
Vachonus Tikader and Bastawade, 1983 (India), 2 spp.
Zabius Thorell, 1894 (Argentina; Paraguay?), 2 spp.

Families Chactidae Laurie, 1896, and Vaejovidae Thorell, 1876

As previously mentioned, the Chactidae (18 genera, about 140 species) and the Vaejovidae (12 genera, about 130 species) cannot be distinguished from one another with the current morphological evidence (Figs. 3.18, 3.19). Furthermore, the relationships between these two families and the distinct Iuridae remain obscure; however, the three families together form a monophyletic group (the "chactoids") that is the sister group of the Bothriuridae, according to Lamoral's (1980) hypothesis (see "Phylogeny of Recent Scorpion Families").

The past problems in "chactoid" systematics stem mainly from the utilization of the number of lateral eyes as a familial character. The possession of two pairs of lateral eyes was considered diagnostic for the Chactidae, whereas three pairs were diagnostic for the Vaejovidae. This character persisted in the literature despite the knowledge of chactids that do not have two pairs of lateral eyes and of vaejovids that do not have three pairs. The character is sometimes even variable at the specific level. The unreliability of the character has since been discussed at length by several authors (Mitchell 1968, Gertsch and Soleglad 1972, Soleglad 1976, González-Sponga 1977, Francke and Soleglad 1981).

Taxonomic problems are also prevalent at the subfamilial and even generic levels. The recognition of subfamilies in these two families is somewhat unsatisfactory (Soleglad 1976), although certain groups appear to be monophyletic. For example, the Megacorminae (Soleglad 1976, Francke 1981a), the Superstitioninae (Francke 1982a), and the Scorpiopsinae (Francke 1976a) each seem to consist of very closely related genera. The Chactinae also form a relatively compact group, although *Chactopsis* is problematic; Soleglad (1976) has pointed out interesting similarities between *Chactopsis* and the Megacorminae and suggested a possible relationship between them. The validity of the characters used to separate the Chactinae from the Euscorpioninae has also been questioned (Soleglad 1976).

Intergeneric relationships are very poorly understood in the Vaejovidae. The relationship of *Anuroctonus* to the other vaejovids is extremely unclear; in the past, it has been considered a member of the subfamily Vaejovinae (Kraepelin 1905), the Uroctoninae (Werner 1934), and even the Hadrurinae (Stahnke 1974; *Hadrurus* was considered a vaejovid at that time). As Francke and Soleglad (1981) implied, *Anuroctonus* is best considered a genus *incertae sedis*.

Fig. 3.18. Representative "chactoids." *Above: top, Uroctonus mordax,* Tiburon Ridge, California (photo courtesy Edward S. Ross); *center, Paruroctonus luteolus,* Palm Springs, California; *bottom, Superstitionia donensis,* Palm Canyon, California (photo courtesy Ross). *Facing page: top, Syntropis macrura,* Isla Carmen, Baja California Sur; *center, Vaejovis gravicaudus,* Puerto Escondido, Baja California Sur (photos courtesy Gary A. Polis); *bottom, Broteochactas delicatus,* El Llano, Panama (photo courtesy Ross).

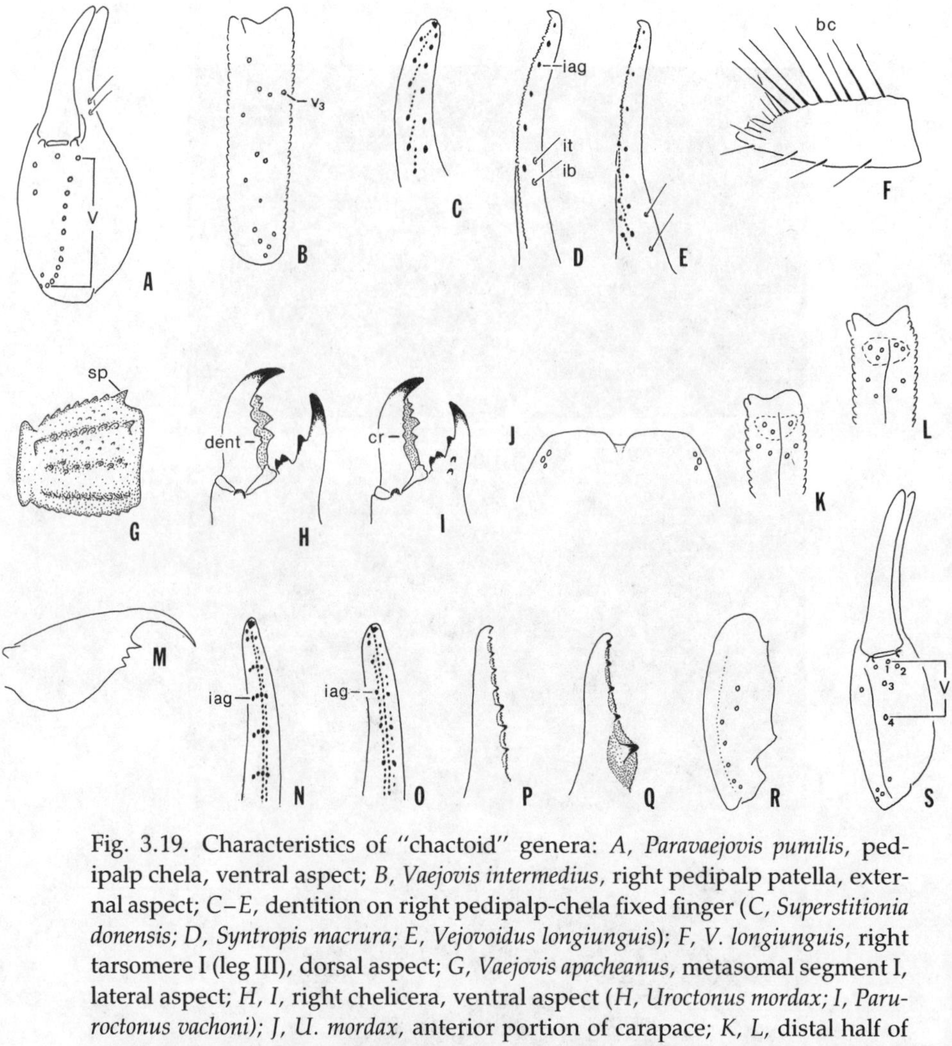

Fig. 3.19. Characteristics of "chactoid" genera: *A*, *Paravaejovis pumilis*, pedipalp chela, ventral aspect; *B*, *Vaejovis intermedius*, right pedipalp patella, external aspect; *C–E*, dentition on right pedipalp-chela fixed finger (*C*, *Superstitionia donensis; D*, *Syntropis macrura; E*, *Vejovoidus longiunguis*); *F*, *V. longiunguis*, right tarsomere I (leg III), dorsal aspect; *G*, *Vaejovis apacheanus*, metasomal segment I, lateral aspect; *H*, *I*, right chelicera, ventral aspect (*H*, *Uroctonus mordax; I*, *Paruroctonus vachoni); J*, *U. mordax*, anterior portion of carapace; *K*, *L*, distal half of patella, external aspect, showing *et* trichobothria (within dotted line: *K*, *Megacormus gertschi; L*, *Plesiochactas dilutus)*; *M*, *Broteochactas* (= *Vachoniochactas*) *lasallei*, telson, lateral aspect; *N*, *O*, dentition on distal part of pedipalp-chela movable finger (*N*, *M. gertschi; O*, *P. dilutus*); *P*, *Q*, right chela fixed finger, lateral aspect (*P*, *Chactas gestroi; Q*, *Teuthraustes rosenbergi*); *R*, *Chactopsis insignis*, right pedipalp patella, ventral aspect; *S*, *Chactas (Caribeochactas) gansi*, right pedipalp chela, ventral aspect. *bc*, bristlecomb; *cr*, crenulation; *dent*, denticle; *iag*, inner accessory granule; *sp*, spinoid denticle terminating dorsolateral carina. Trichobothrial designations are as in text (see "Trichobothria") and in Fig. 3.5. (Views *K*, *L*, *N*, *O* after Soleglad 1976; *P*, *Q* after Kraepelin 1911; *S* after González-Sponga 1978)

The inclusion of *Vejovoidus* and *Syntropis* into a separate subfamily, the Syntropinae, is another example of the taxonomic problems in the Vaejovidae. These two taxa are placed together because they share a single derived character, an unpaired ventromedian carina on metasomal segments I–IV. Several authors have stated that this character has been overemphasized at the subfamilial level (Soleglad 1976, Francke 1981a, Lourenço 1985), and it has certainly evolved independently at least four times in Recent scorpions. If this character evolved independently in *Syntropis* and *Vejovoidus*, it would be considered a convergent similarity rather than a homology. Although this is not the place to argue for the dissolution of the Syntropinae, the differences between *Syntropis* and *Vejovoidus* are outstanding and demonstrate that the two genera are not closely related (W. D. Sissom, pers. obs.).

The genus *Vaejovis* is extremely large and complex, with over 80 species and subspecies. *Vaejovis*, currently with four species groups, has been a "catchall" group for North American vaejovids. Some genera in the Vaejovidae (i.e., *Paravaejovis*, *Paruroctonus*, and *Serradigitus*) were formerly considered species groups of *Vaejovis*. In addition, some species originally described in the genus *Uroctonus* are now placed in *Vaejovis*. As a better understanding of relationships between these species groups and the other genera in the Vaejovidae is gained, it will almost certainly be possible to provide a proper characterization for the family itself (and to separate it adequately from the Chactidae).

Characterization

CHELICERAE (Fig. 3.1*C*, *H*). The number and arrangement of the teeth of the cheliceral fixed finger are variable: three teeth (in *Typhlochactas sylvestris* and *T. mitchelli*: Mitchell and Peck 1977; Sissom 1989); four teeth without a bicusp (in *T. rhodesi*, *Sotanochactas elliotti*, *Alacran tartarus*, some *Broteochactas* and *Troglotayosicus*: Mitchell 1968, 1971; González-Sponga 1978a; Lourenço 1981a; Francke 1982b); or four teeth with the medial and basal teeth fused into a bicusp (all other taxa). The movable finger has one or two subdistal teeth and a single basal tooth on the external margin; the distal external tooth is smaller than the distal internal tooth and may or may not be apposable to it; the internal margin may be smooth, or it may possess small denticles or serrations. The movable finger often has a serrula on the ventral aspect.

TRICHOBOTHRIAL PATTERN. The trichobothrial pattern of the pedipalps is Type C (as in Fig. 3.5). *Belisarius*, *Paruroctonus*, *Pseudo-*

uroctonus, Sotanochactas, Superstitionia, Syntropis, Troglotayosicus, Typhlochactas, Uroctonus, most *Vaejovis,* and *Vejovoidus* have the fundamental number and pattern. The majority of genera have accessory trichobothria.

COXOSTERNAL REGION (Fig. 3.7*D, F*). The coxapophyses are not broadly expanded anteriorly. The sternum is subpentagonal, but in some taxa (e.g., *Troglocormus* spp.) it may be quite long and narrow (Francke 1981a).

LEG SPINATION. Tibial spurs are always lacking. Pedal spurs are absent altogether in *Typhlochactas rhodesi* and *Sotanochactas elliotti*. Only prolateral pedal spurs are present in *T. reddelli, T. sylvestris, T. mitchelli,* and *Alacran tartarus*. All other genera possess both prolateral and retrolateral pedal spurs (as in Fig. 3.8*B*).

VENOM GLANDS AND TELSON. The venom glands are thin-walled and simple; the venom is opalescent. The telson, with a few exceptions, does not bear a subaculear prominence.

MALE REPRODUCTIVE SYSTEM. The hemispermatophore is lamelliform (as in Fig. 3.12*E*). The morphology of the capsular region is variable, ranging from very simple (e.g., Superstitioninae, *Anuroctonus*) to complex (most genera studied). The paraxial organ lacks accessory glands.

FEMALE REPRODUCTIVE SYSTEM. The ovariuterus forms a reticulate mesh of six cells. In *Scorpiops* and *Vaejovis* the ovarian follicles are attached to the ovariuterus by a short pedicel (as in Fig. 3.13*E*); in *Euscorpius, Brotheas,* and *Broteochactas* the ovarian follicles are in direct contact with the ovariuterus (Laurie 1896a, b). The other genera have not been studied.

DEVELOPMENT. Apoikogenic. The ova are rich in yolk in *Euscorpius, Brotheas,* and *Broteochactas;* they contain virtually no yolk in *Scorpiops* and *Vaejovis* (Laurie 1896a, b). In the latter genera, the embryo is nourished via a diffuse pseudoplacenta. The other genera have not been studied.

Key to the Genera of the Chactidae and Vaejovidae

The following key applies to all chactid and vaejovid genera currently accepted as valid without reference to familial and subfamilial assignments, some of which will undoubtedly be revised in the near future. The key is based partially on existing keys and on generic descriptions (e.g., Mello-Leitão 1945; Stahnke 1974; Williams 1974; Soleglad 1976; González-Sponga 1978a; Vachon

1980; Francke 1981, 1982, 1986). The reader should refer to Figure 3.5 for the basic nomenclature and positions of trichobothria.

The inadequate diagnoses and dubious validity of certain taxa (especially among the vaejovids) will cause some problems when using the key provided below. For example, the genus *Pseudouroctonus* (which is probably not valid) does not possess any known features that exclude it from *Vaejovis*; the characters utilized in the original description (Stahnke 1974) to separate *Pseudouroctonus* from *Vaejovis* and *Uroctonus* are not unique to the former, either alone or in combination. *Pseudouroctonus* is, therefore, not included in the key, and its single species (*P. reddelli*) will key out to *Vaejovis*. There are also a few species of *Vaejovis* in the *nitidulus* group that have the terminal spine on the dorsolateral metasomal carinae reduced, but their placement in *Vaejovis* is quite certain (Sissom and Francke 1985, Sissom 1986). I have chosen to accept the definition of *Uroctonus* proposed by Soleglad (1973), although there are others (see Stahnke 1974, Williams 1980).

★ Pedipalp patella with 3 or fewer ventral trichobothria:
  ▲ Median eyes absent; cave and forest litter inhabitants:
    No trace of lateral eyes present; tarsi of legs with prolateral pedal spurs only, or lacking pedal spurs altogether (Fig. 3.8*C*):
      Size large, with adults 60 to 70 mm in length; forms well pigmented; external face of pedipalp patella with 19 or 20 trichobothria . . . . **Alacran**
      Size small, with adults less than 20 mm in length; forms pale yellowish white to pale yellowish brown; external face of patella with 14 trichobothria:
        Pedipalp-chela fingers twice as long as palm; trichobothria *ib* and *it* of pedipalp-chela fixed finger situated distally, with *it* near midfinger; external series (*eb, esb, est, et*) of fixed finger situated in distal half of finger . . . . **Sotanochactas**
        Pedipalp-chela fingers about equal to palm length; trichobothria *ib* and *it* of pedipalp-chela fixed finger situated at the base, near its junction with the manus; trichobothrium *eb* always situated at the base of the fixed finger . . . . **Typhlochactas**
    At least vestiges of lateral eyes present; tarsi with both prolateral and retrolateral pedal spurs (Fig. 3.8*B*):
      External surface of pedipalp patella with 14 trichobothria (v_3 positioned on external surface: Fig. 3.19*B*); chela trichobothrium *Dt* displaced to fixed finger; Ecuador . . . . **Troglotayosicus**
      External surface of pedipalp patella with 13 trichobothria (v_3 in normal position, on ventral surface); chela trichobothrium *Dt* positioned basally on chela manus; Pyrenees Mountains . . . **Belisarius**

▲ Median eyes present:

Chela manus with 12 ventral trichobothria (Fig. 3.19*A*) . . . . . ***Paravaejovis***

Chela manus with 4 to 6 ventral trichobothria (Fig. 3.5*D*):

Carapace with 2 pairs of lateral eyes; dentate margin of pedipalp-chela fingers divided into 5 or 6 oblique, nonoverlapping rows of granules (Fig. 3.19*C*) . . . . ***Superstitionia***

Carapace with 3 pairs of lateral eyes; dentate margins divided into variable numbers of rows, but always arranged as a linear series, except occasionally at the base (Fig. 3.19*D*, *E*):

Metasomal segments I–IV, or only III and IV, with a single ventromedian carina:

Pedipalps and metasoma elongate, with fixed finger distinctly longer than carapace; tibiae and tarsi of legs with setae sparsely arranged; chela trichobothria *ib* and *it* situated at the level of the sixth inner accessory granule on fixed finger (Fig. 3.19*D*) . . . . ***Syntropis***

Pedipalps and metasoma more robust, with fixed finger distinctly shorter than carapace; tibiae and tarsi of legs I–III with setae arranged as bristlecombs (Fig. 3.19*F*); both *ib* and *it* situated basally to the sixth inner accessory granule (Fig. 3.19*E*) . . . . ***Vejovoidus***

Metasomal segments I–IV with paired ventral submedian carinae, or lacking carinae in that position:

Dentate margins of pedipalp-chela fingers serrated in lateral view, either undivided or divided into 2 to 4 (rarely 5) rows by slightly enlarged granules; terminal denticles of fingers enlarged, clawlike; trichobothria *ib* and *it* of pedipalp-chela fixed finger displaced to level of sixth inner accessory granule or beyond; female pectines with proximal 1 to 3 pectinal teeth enlarged . . . . ***Serradigitus***

Dentate margins usually not serrated, but if so, then the granules are divided into 5 to 7 rows; terminal denticles usually more blunt; trichobothria *ib* and *it* situated on distal portion of manus, basally on fixed finger, or at the level of the sixth inner accessory granule; female pectinal teeth never enlarged:

Cheliceral movable finger without distinct serrula on ventrodistal aspect; legs I–III with well-developed bristlecombs (Fig. 3.19*F*; except in a few species); dorsolateral carinae of metasomal segments I–IV without enlarged spinoid, terminal denticles . . . . ***Paruroctonus***

Combination of characters not seen as above: cheliceral movable finger with or without distinct serrula and legs I–III almost always without distinct bristlecombs; dorsolateral carinae almost always with enlarged terminal denticles (NOTE: if serrula absent and bristlecombs present, then dorsolateral carinae of metasomal segments I–IV terminate in enlarged spinoid denticles; Fig. 3.19*G*):

Carapace with moderate or deep anteromedian notch (Fig. 3.19*J*); ventral margin of cheliceral movable finger frequently crenulate or denticulate (Fig. 3.19*H*); pedipalp patella with 2 or 3 ventral trichobothria (Fig. 3.5*D*); ventral face of pedipalp chela distinctly flattened (owing to absence of ventromedian carina and presence of strong ventrointernal and ventroexternal carinae); serrula always present .... ***Uroctonus***

Carapace with straight margin or with weak to moderate anteromedian notch; ventral margin of cheliceral movable finger smooth; pedipalp patella with 2 ventral trichobothria; ventral face of pedipalp chela rounded or with ventrointernal carina developed; serrula absent in some species groups..................... ***Vaejovis***

★ Pedipalp patella with more than 3 ventral trichobothria:

■ Dentate margins of pedipalp-chela fingers composed of a double row, or of several indistinct rows, of denticles (Fig. 3.19*N*, *O*):

Median eyes lacking; known only from caves in Mexico ... ***Troglocormus***

Median eyes always present:

Metasomal segments I–IV with single ventromedian carina:

Only 3 to 4 trichobothria in *et* series of pedipalp patella (Fig. 3.19*K*); anterior portion of carapace distinctly tapered; inner accessory granules of pedipalp-chela fingers doubled (Fig. 3.19*N*) .... ***Megacormus***

5 trichobothria in *et* series of pedipalp patella (Fig. 3.19*L*); anterior portion of carapace not distinctly tapered; inner accessory granules single (Fig. 3.19*O*).................... ***Plesiochactas***

Metasomal segments I–IV with paired ventral submedian carinae, or lacking carinae in that position:

Patella with 7 ventral trichobothria, 5 along the ventroexternal margin and the 2 distal ones set more to the middle of the ventral face (Fig. 3.19*R*); ventral margin of cheliceral movable finger smooth; pedipalp chela lacking carinae ...... ***Chactopsis***

Patella with 6 to 23 ventral trichobothria, all positioned in a row along the ventroexternal margin (or carina); ventral margin of cheliceral movable finger with a row of serrations; pedi-

palp chela robust, strongly carinate:
2 pairs of lateral eyes .... *Parascorpiops*
3 pairs of lateral eyes:
Pedipalp patella with at most 48 trichobothria, of which 16 to 26 are on the external aspect and 6 to 19 are on the ventral aspect .... *Scorpiops*
Pedipalp patella with about 80 trichobothria, of which 59 or 60 are on the external aspect and 23 are on the ventral aspect .... *Dasyscorpiops*

- Dentate margins of pedipalp-chela fingers composed of only a single row of denticles (Fig. 3.19*D*):
Pedipalp-chela manus with over 20 ventral trichobothria (Fig. 3.6*A*); cheliceral movable finger with 2 or 3 fine basal crenulations on ventral margin (Fig. 3.19*I*) .... *Anuroctonus*
Pedipalp-chela manus with 4 to 11 ventral trichobothria; ventral margin of cheliceral movable finger smooth:
Dentate margin of pedipalp-chela fingers with paired inner accessory granules (Fig. 3.19*N*) .... *Euscorpius*
Dentate margin of pedipalp-chela fingers with unpaired inner accessory granules (Fig. 3.19*O*):
Pedipalp patella with 7 trichobothria on ventral surface:
Tarsomere II with 2 ventral submedian rows of spines; spiracles linear or oval .... *Brotheas*
Tarsomere II with irregularly placed setae on ventral surface; spiracles oval or round .... *Broteochactas*
Pedipalp patella with 5 or 6 trichobothria on ventral surface:
Pedipalp patella with 6 ventral trichobothria:
Chela fixed finger with granular rows interrupted by 7 enlarged denticles (Fig. 3.19*P*); trichobothrium V_4 situated near middle of underhand (Fig. 3.19*S*); pedipalp patella with 17 external trichobothria... ***Chactas* (*Caribeochactas*)**
Chela fixed finger with granular rows interrupted by 5 or 6 enlarged denticles; trichobothrium V_4 more basal, usually in basal third of underhand (Fig. 3.5*G*); pedipalp patella with 18 external trichobothria....... *Nullibrotheas*
Pedipalp patella with 5 ventral trichobothria:
Pedipalp-chela fixed finger with granular rows flanked laterally by 6 to 8 (usually 7) enlarged denticles, with basalmost of these not greatly enlarged (Fig. 3.19*P*):
Telson with 3 spinoid subaculear tubercles (Fig. 3.19*M*); male pedipalps similar in morphometrics to those of female ...... ***Broteochactas* ("Vachoniochactas" group)**

Telson without subaculear tubercles; male pedipalps very long and slender in comparison with those of female .... ***Chactas***

Pedipalp-chela fixed finger with granular rows flanked by 5 or 6 enlarged lateral denticles, the basalmost of which is much larger than the preceding ones (Fig. 3.19*Q*) .... ***Teuthraustes***

List of Genera and Their Distributions

The list below shows the current classification of the Chactidae and Vaejovidae, although it is generally unsatisfactory and will likely change, as explained above.

Family Chactidae Laurie, 1896

Subfamily Chactinae Laurie, 1896

Broteochactas Pocock, 1893 (Brazil, Colombia, the Guianas, Venezuela), about 30 spp.

Brotheas C. L. Koch, 1843 (Brazil, Colombia, the Guianas, Venezuela), 12 spp.

Chactas Gervais, 1844 (Colombia, Ecuador, Peru, Venezuela), 5 subgenera, 19 spp.

Chactopsis Kraepelin, 1912 (Brazil, Peru, Venezuela), 5 spp.

Teuthraustes Simon, 1878 (Ecuador, Peru, Venezuela), 15 spp.

Troglotayosicus Lourenço, 1981 (Ecuador), 1 sp.

Subfamily Euscorpioninae Laurie, 1896

Belisarius Simon, 1879 (eastern Pyrenees Mts. of France and Spain), 1 sp.

Euscorpius Thorell, 1876 (northern Africa, southern Europe; Middle East, southwestern U.S.S.R.), 3 subgenera, 5 spp. (and numerous ssp.)

Subfamily Megacorminae Kraepelin, 1899

Megacormus Karsch, 1881 (southeastern Mexico), 3 spp.

Plesiochactas Pocock, 1900 (Guatemala, southeastern Mexico), 2 spp.

Troglocormus Francke, 1981 (eastern Mexico), 2 spp.

Subfamily Superstitioninae Stahnke, 1940

Alacran Francke, 1982 (Oaxaca [Mexico]), 1 sp.

Sotanochactas Francke, 1986 (San Luis Potosí [Mexico]), 1 sp.

Superstitionia Stahnke, 1940 (Baja California [Mexico], Arizona, New Mexico, and California [United States]), 1 sp.

Typhlochactas Mitchell, 1968 (eastern and southern Mexico), 5 spp.
Family Vaejovidae Thorell, 1876
Subfamily Scorpiopsinae Kraepelin, 1905
Dasyscorpiops Vachon, 1973 (Malaya), 1 sp.
Parascorpiops Banks, 1901 (Borneo), 1 sp.
Scorpiops Peters, 1861 (Afghanistan, Southeast Asia; India, Pakistan), 4 subgenera, 17 spp.
Subfamily Syntropinae Kraepelin, 1905
Syntropis Kraepelin, 1900 (Baja California [Mexico]), 1 sp.
Vejovoidus Stahnke, 1974 (Baja California [Mexico]), 1 sp.
Subfamily Vaejovinae Thorell, 1876
Nullibrotheas Williams, 1974 (Baja California [Mexico]), 1 sp.
Paravaejovis Williams, 1980 (Baja California [Mexico]), 1 sp.
Paruroctonus Werner, 1934 (southwestern Canada, northwestern Mexico, southwestern United States), 2 subgenera, 30 spp.
Pseudouroctonus Stahnke, 1974 (Nuevo León [Mexico], Texas [United States]), 1 sp.
Serradigitus Stahnke, 1974 (northwestern Mexico, southwestern United States), 20 spp.
Uroctonus Thorell, 1876 (California [United States]), 16 spp.
Vaejovis C. L. Koch, 1836 (Mexico, United States), over 60 spp.
Genus *incertae sedis*
Anuroctonus Pocock, 1893 (Baja California [Mexico], western United States), 1 sp.

Family Chaerilidae Pocock, 1893

The Chaerilidae (Fig. 3.20) is a monotypic family consisting of the genus *Chaerilus* (15 species). The family was erected by Pocock (1900) because of the unique shape of the coxapophyses. Subsequent authors (e.g., Birula 1917a, Werner 1934, Kästner 1940, Millot and Vachon 1949) considered *Chaerilus* to be closely related to *Paraiurus* (= *Calchas*), placing the two genera in a subfamily (Chaerilinae) of the Chactidae. Vachon (1956, 1963, 1973) recognized that *Chaerilus* possesses unique cheliceral dentition and trichobothrial patterns, and the genus has since occupied its present position.

Lamoral (1980) suggested that the ancestors of the chaerilids originated in Pangaean times as an eastern Laurasian relic that moved into the Oriental Region after the Indian plate connected with Eurasia. They became isolated in the Oriental Region as the Himalayas formed.

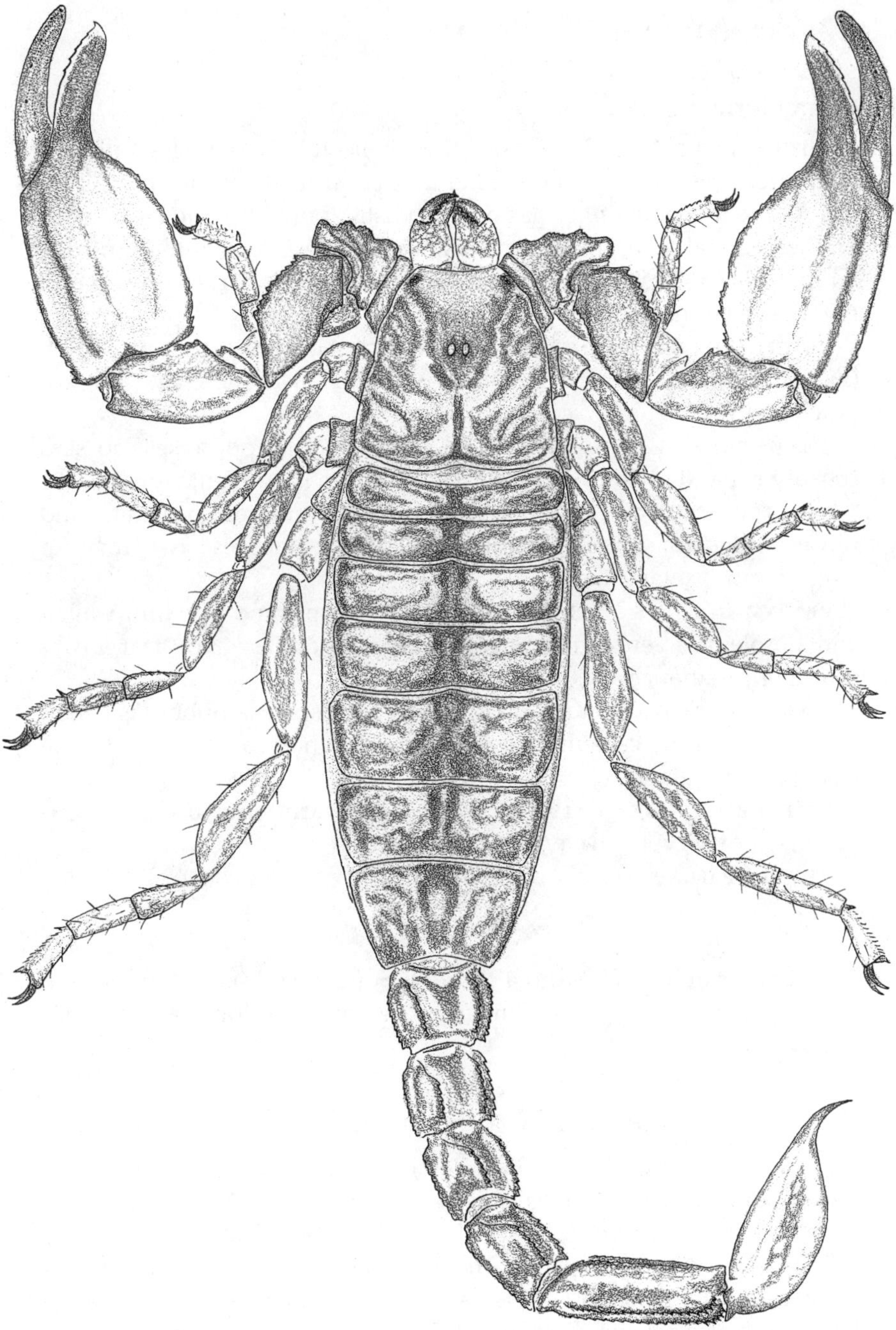

Fig. 3.20. *Chaerilus celebensis,* a representative chaerilid from the island of Luzon, Republic of the Philippines.

Characterization

CHELICERAE (Fig. 3.1*D*). On the cheliceral fixed finger, all four teeth are separate (i.e., the medial and basal teeth do not form a bicusp). The movable finger has one subdistal tooth and one basal tooth on the external margin; the distal external tooth is smaller than the distal internal tooth and apposable to it; the internal margin possesses a row of serrations or small teeth.

TRICHOBOTHRIAL PATTERN. The trichobothrial pattern is Type B (Fig. 3.4). All species possess the fundamental number and pattern (Vachon 1973).

COXOSTERNAL REGION (Fig. 3.7*A*). The coxapophyses possess broadly expanded anterior lobes. The sternum is subpentagonal.

LEG SPINATION. The legs lack tibial spurs, but both prolateral and retrolateral pedal spurs are present. The tarsi possess two rows of stout, ventral setae.

VENOM GLANDS AND TELSON. The venom glands are thin-walled and simple; the venom is opalescent. The telson does not bear a subaculear prominence.

MALE REPRODUCTIVE SYSTEM. The hemispermatophore is lamelliform. There is no capsule. The paraxial organ has a single pair of anterior accessory glands.

FEMALE REPRODUCTIVE SYSTEM. The ovariuterus forms a reticulate mesh of six cells. Diverticula are absent.

DEVELOPMENT. Apoikogenic.

Distribution

Chaerilus Simon, 1877 (India through Southeast Asia to Borneo, the Celebes, Java, Philippines, Sumatra, and other Indonesian islands), 23 spp.

Family Diplocentridae Peters, 1861

The Diplocentridae (Figs. 3.21, 3.22), a small family of seven genera and about 70 described species, is essentially a New World family, but *Nebo* and *Heteronebo* occur in the Middle East. *Heteronebo* has a peculiar distribution, with two species on the small island of Abd-el-Kuri, off the coast of the Arabian Peninsula and Somalia, and six species on islands in the Caribbean (Francke 1977a, 1978a). This distribution defies explanation.

Fig. 3.21. Representative Diplocentridae: *top, Diplocentrus anophthalmus,* a troglobite, Actún Chukum, Yucatán (photo courtesy Robert W. Mitchell); *bottom, Diplocentrus* sp., Comstock, Texas.

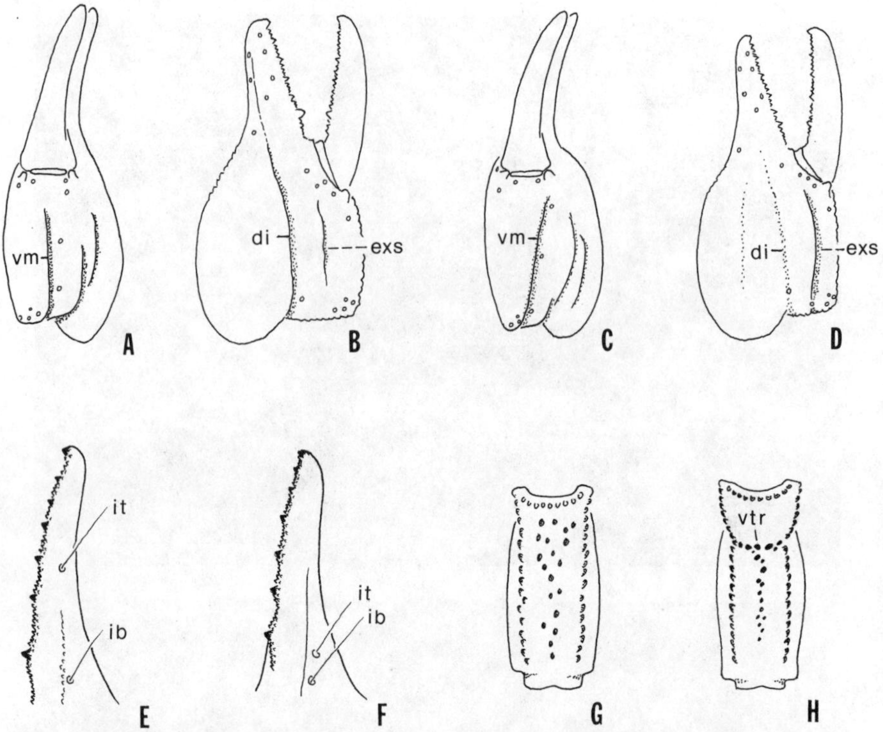

Fig. 3.22. Characteristics of diplocentrid genera: *A, Diplocentrus* sp., ventral aspect of pedipalp chela; *B, Diplocentrus* sp., external aspect of pedipalp chela; *C, Didymocentrus lesueurii,* ventral aspect of pedipalp chela; *D, D. lesueurii,* external aspect of pedipalp chela; *E, Nebo* sp., inner aspect of pedipalp-chela fixed finger, showing position of trichobothria *ib* and *it; F, Diplocentrus whitei,* inner aspect of pedipalp-chela fixed finger, showing position of trichobothria *ib* and *it; G, Heteronebo muchmorei,* ventral aspect of metasomal segment V; *H, Didymocentrus hummelincki,* ventral aspect of metasomal segment V. *di,* digital chela carina; *exs,* external secondary chela carina; *vm,* ventromedian chela carina; *vtr,* ventral transverse carina of metasomal segment V; trichobothrial designations are as given in text (see "Trichobothria"). (Views *C, D, H* after Francke 1978)

A second problem of taxonomic and zoogeographical interest arises with the four diplocentrids from Baja California Sur, Mexico. These species (and some now considered synonyms) were originally placed in two genera, *Bioculus* and *Didymocentrus,* by Stahnke (1968). Williams and Lee (1975) subsequently restudied these taxa after extensive collecting efforts and considered the diplocentrid fauna of Baja California to comprise four species, which they assigned to *Didymocentrus*. *Bioculus* was considered a junior synonym of *Didymocentrus*. Several years later, the diagnoses for the genera *Didymocentrus* and *Diplocentrus* were emended (Francke 1978), and the four diplocentrid species were considered to belong to *Diplocentrus*. This view was not addressed by Williams (1980), who retained them in *Didymocentrus*, but the case certainly warrants further study.

Characterization

CHELICERAE (Fig. 3.1*E*). The cheliceral fixed finger has the medial and basal teeth fused into a bicusp. The movable finger has a single subdistal and a single basal tooth on the external margin; the distal external tooth is smaller than the distal internal tooth and more or less apposable to it; the internal margin is smooth.

TRICHOBOTHRIAL PATTERN. The trichobothrial pattern is Type C (Fig. 3.5). All genera have the basic number and pattern (Vachon 1973).

COXOSTERNAL REGION. The coxapophyses do not possess anteriorly expanded lobes (Fig. 3.7*H*). The sternum is subpentagonal.

LEG SPINATION. Tibial spurs are lacking. Only prolateral pedal spurs are present (except in *Oiclus*, in which there are no spurs). The tarsi possess two rows of ventral spines.

VENOM GLANDS AND TELSON. The venom glands are thick-walled and folded; the venom is tinged with reddish pigments. The telson bears a subaculear tubercle (Fig. 3.10*C*).

MALE REPRODUCTIVE SYSTEM. The hemispermatophore is lamelliform (Fig. 3.12*D*). The capsular region is well developed, and the paraxial organ lacks accessory glands.

FEMALE REPRODUCTIVE SYSTEM. The ovariuterus forms a reticulate mesh of six cells. Numerous diverticula arise from the lateral and mesial branches of the ovariuterus.

DEVELOPMENT. Katoikogenic.

Key to Subfamilies and Genera of the Diplocentridae

Previous authors (Kraepelin 1905, Birula 1917a, Kästner 1940) considered the Diplocentridae to comprise two subfamilies: the Nebinae and the Diplocentrinae. *Heteronebo* and *Nebo* were placed in the Nebinae primarily on zoogeographical grounds; however, the New World species of *Heteronebo* were unknown at that time. Francke (1977a) demonstrated that *Heteronebo* has closer morphological affinities to the diplocentrine genera than to *Nebo* and reassigned the former to the Diplocentrinae. The following key is derived from the studies of Francke (1977a, 1978).

Pedipalp-chela trichobothrium *it* situated in distal half of fixed finger (Fig. 3.22*E*); subaculear tubercle often fingerlike and narrow at base .... Subfamily NEBINAE .... ***Nebo***

Pedipalp-chela trichobothrium *it* situated basally on fixed finger (Fig. 3.22*F*); subaculear tubercle rounded and broad at the base .... Subfamily DIPLOCENTRINAE:

- Metasomal segment V without ventral transverse carina (Fig. 3.22*G*).... .... ***Heteronebo***
- Metasomal segment V with distinct ventral transverse carina (Fig. 3.22*H*):
  - Pedal spurs absent from all legs .... ***Oiclus***
  - Prolateral pedal spurs present on all legs:
    - Metasomal segments dorsoventrally compressed, at least 1.5 times wider than deep; tarsomere I on all legs bearing numerous conspicuous pores ventrally and prolaterally; Venezuela .... ***Tarsoporosus***
    - Metasomal segments not dorsolaterally compressed, but more or less subcylindrical; tarsomere I on all legs lacking pores:
      - Pedipalp chela with ventromedian carina oblique, distinctly directed toward internal condyle of movable finger articulation (Fig. 3.22*C*); external secondary carina of outer chela palm most highly developed (Fig. 3.22*D*) .... ***Didymocentrus***
      - Pedipalp chela with ventromedian carina parallel to longitudinal axis of palm, directed toward external condyle (Fig. 3.22*A*); if carinae of outer chela palm developed, then digital carina most highly developed (Fig. 3.22*B*):
        - Pedipalp femur always deeper than wide; cheliceral fixed finger shorter than cheliceral manus width; cheliceral movable finger shorter than cheliceral chela (manus only) length; West Indies .... ***Cazierius***
        - Pedipalp femur wider than deep, or, if deeper than wide, then cheliceral proportions not as above; continental North America .... ***Diplocentrus***

List of Genera and Their Distributions

Subfamily Diplocentrinae Kraepelin, 1905

Cazierius Francke, 1978 (Barbados, Cuba), 3 spp.

Didymocentrus Kraepelin, 1905 (Greater and Lesser Antilles, eastern Central America; Baja California [Mexico]),* 13 spp.
Diplocentrus Peters, 1861 (northern Central America; Mexico, southwestern United States), 26 spp.
Heteronebo Pocock, 1899 (Greater and Lesser Antilles, Abd-el-Kuri I. [Yemen, P.D.R.]), 11 spp.
Oiclus Simon, 1880 (Lesser Antilles), 1 sp.
Tarsoporosus Francke, 1978 (Colombia, Venezuela), 3 spp.

Subfamily Nebinae Kraepelin, 1905
Nebo Simon, 1878 (Middle East), 10 spp.

Family Ischnuridae Pocock, 1893

The family Ischnuridae (Figs. 3.23, 3.24) includes eight genera that were formerly assigned to the subfamilies Ischnurinae and Heteroscorpioninae, in the Scorpionidae. The familial name Ischnuridae was actually used by several authors in the past (e.g., Pocock 1900, Tikader and Bastawade 1983), but it was in a recent cladistic analysis that Lourenço (1985) demonstrated why these genera should be separated from the Scorpionidae and placed in their own family. The members of the Ischnuridae differ from the Scorpionidae in the possession of smooth to moderately folded venom glands and tarsi without rounded lateroapical lobes. These characters are hypothesized to be primitive, so in fact there are as yet no synapomorphies to define the Ischnuridae (i.e., the group is paraphyletic, not monophyletic). Although Lourenço did not recognize subfamilies within the Ischnuridae, its genera can be separated into two groups based on the number of trichobothria on the pedipalps (see below).

The Ischnuridae contains some of the world's largest living scorpions. *Hadogenes troglodytes*, for example, is known to reach lengths of 21 cm and weights up to 32 g when not gravid (Newlands 1972b).

There is some question whether or not the genus *Chiromachetes* should be regarded as valid. This is a monotypic genus for the species *C. fergusoni*; it is known only from the female holotype collected at the Mala-

*Williams and Lee (1975) considered the four species of diplocentrids from Baja California to belong to *Didymocentrus*. Francke (1978a) suggested that they be placed in *Diplocentrus*, but Williams (1980), in his monograph on the scorpions of Baja California, continued to recognize them as *Didymocentrus*. Pending further investigation, I have followed the most recent treatment of these species.

Fig. 3.23. Representative Ischnuridae: *top, Hadogenes* sp., near Warmbad, South Africa; *bottom, Liocheles waigiensis*, southwest of Sarina, Australia. (Photos courtesy Edward S. Ross)

bar Coast, in southern India. This genus was recognized by Tikader and Bastawade (1983), but Lourenço (1985) suggests that the characters used to separate it from *Iomachus* are probably not legitimate at the generic level. I have chosen to accept Lourenço's view and will not consider this genus in the discussion below.

The Ischnuridae represents a typical Gondwanaland faunal element, with its members inhabiting Africa, South America, and India. The genus *Liocheles*, which hypothetically arose in India, secondarily invaded southeastern Asia and Indonesia, and eventually Australia (Koch 1977, Lourenço 1985). Although Australia was a portion of Gondwanaland, it has been suggested that *Liocheles* dispersed there via land connections between southeastern Asia and New Guinea and between New Guinea and Australia at various times during the Cenozoic (Koch 1977). This is supported by the fact that the three species of *Liocheles* in Australia are not endemic and are restricted to the northeastern portion of that continent (Koch 1977).

Another genus, *Opisthacanthus*, is found in both Africa and northern South America (Lourenço 1981, 1985). This distribution pattern has sparked considerable interest among scorpion biologists, and it is now generally accepted that the ancestors of *Opisthacanthus* were established in the New World before the separation of Africa and South America in the Mesozoic (Lamoral 1980, Lourenço 1985). The African species are considered to represent a separate subgenus, *Nepabellus*, whereas the New World species make up the nominal subgenus.

Characterization

CHELICERAE. On the cheliceral fixed finger, the medial and basal teeth are fused into a bicusp. The movable finger has one subdistal and one basal tooth on the external margin; the distal external tooth is smaller than the distal internal tooth and is usually apposable to it; the internal margin is smooth. This morphology is consistent with that of Fig. 3.1*G*.

TRICHOBOTHRIAL PATTERN. The trichobothrial pattern is Type C (as in Fig. 3.5). *Cheloctonus, Chiromachus, Iomachus, Liocheles*, and *Opisthacanthus* possess the fundamental number and pattern; *Heteroscorpion* and *Hadogenes* possess accessory trichobothria on the pedipalp patella and chela manus (Vachon 1973).

COXOSTERNAL REGION. The coxapophyses do not possess broadly expanded anterior lobes; the sternum is subpentagonal (as in Fig. 3.7*G*).

LEG SPINATION. Only prolateral pedal spurs are present; tibial

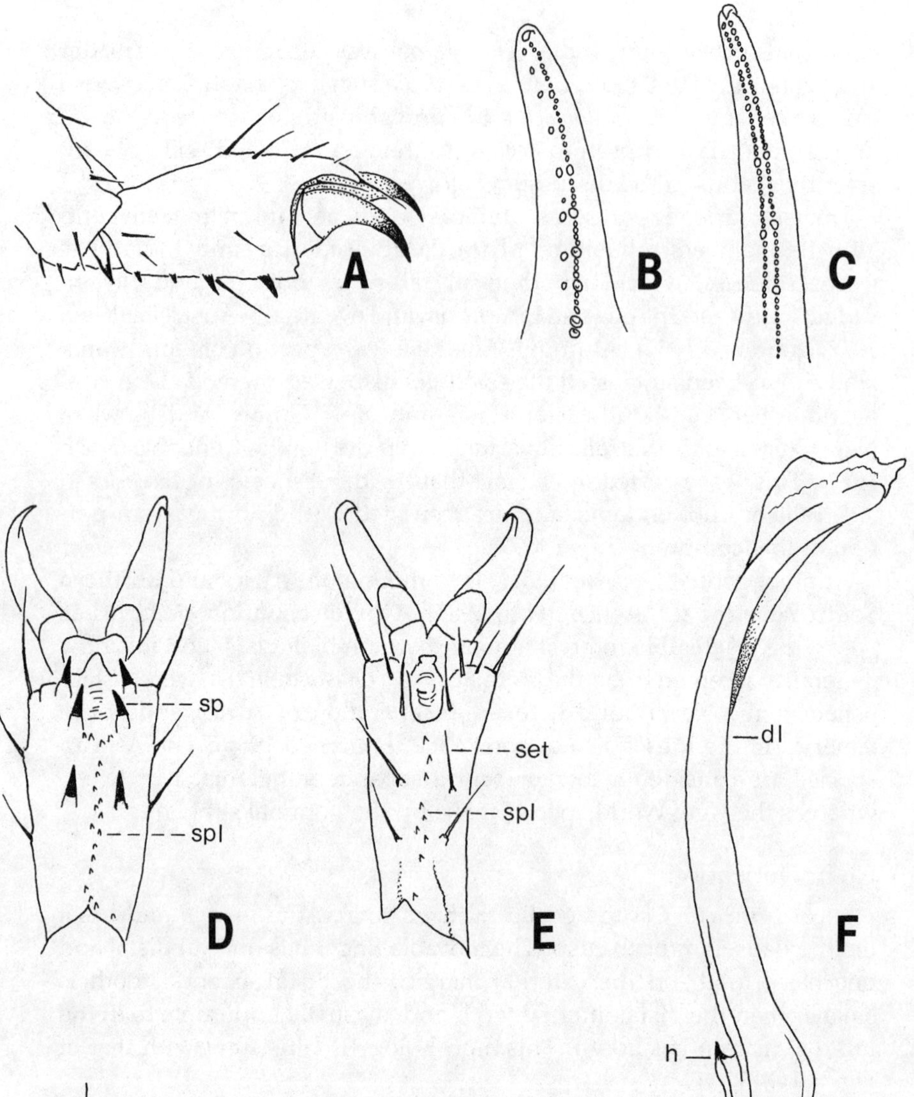
A
B
C
sp
spl
D
set
spl
E
dl
h
F
G
H

spurs and retrolateral pedal spurs are lacking (as in Fig. 3.8C). The genera *Cheloctonus, Hadogenes, Heteroscorpion,* and *Opisthacanthus* possess paired submedian spines on the ventral surface of the tarsi; setae may or may not be present there. *Chiromachus, Iomachus,* and *Liocheles* possess only setae on the ventral surface of the tarsi.

VENOM GLANDS AND TELSON. The epithelium of the venom glands is variable, ranging from smooth to moderately folded (Lourenço 1985). Most of the genera possess smooth venom glands, but some species of *Chiromachus* and *Hadogenes* have moderately folded glands. The venom glands of *Opisthacanthus* may be either smooth or slightly folded. The telson never bears a subaculear protuberance, and the venom is opalescent.

MALE REPRODUCTIVE SYSTEM. The hemispermatophore is lamelliform and bears an elaborate capsule. The paraxial gland lacks accessory glands.

FEMALE REPRODUCTIVE SYSTEM. The ovariuterus forms a reticulate mesh of six cells (as in Fig. 3.13*F, G*). Numerous diverticula arise from the lateral and mesial branches of the ovariuterus. The embryos develop within these diverticula.

DEVELOPMENT. Katoikogenic.

Key to the Genera of the Ischnuridae

The following key is based mainly on the characteristics utilized by Lourenço (1985) in his cladistic study of the family Ischnuridae. The genus *Chiromachetes* is of doubtful validity and therefore does not appear in the key below.

Fig. 3.24. Characteristics of ischnurid genera: *A, Opisthacanthus lepturus,* lateral aspect of tarsomere II of right leg IV, showing straight lateroapical margin; *B, Cheloctonus jonesii,* dentition of pedipalp-chela movable finger; *C, O. madagascariensis,* dentition of pedipalp-chela movable finger; *D, O. madagascariensis,* ventral aspect of tarsomere II of right leg IV, showing ventral submedian spines flanking row of small midventral spinules; *E, Iomachus politus,* ventral aspect of tarsomere II of right leg IV, showing ventral submedian setae flanking row of small midventral spinules; *F, Chiromachus ochropus,* lateral view of right hemispermatophore; *G, Heteroscorpion opisthacanthoides,* lateral view of hook of right hemispermatophore; *H, Hadogenes taeniurus,* lateral view of doubled hook of right hemispermatophore. *dl,* distal lamina of hemispermatophore; *h,* hook of hemispermatophore; *set,* ventral submedian seta of tarsomere II; *sp,* ventral submedian spine of tarsomere II; *spl,* midventral spinule row. (Views *B–G* redrawn from Lourenço 1985; view *H* redrawn from Lamoral 1979)

Pedipalp patella and chela manus with fundamental number and pattern of trichobothria (Fig. 3.5); pectines with less than 10 teeth in both sexes:

Tarsi of legs armed ventrally with stout spines (Fig. 3.24*A, D*):

Dentate margins of pedipalp-chela fingers with a single row of granules (Fig. 3.24*B*) .... ***Cheloctonus***

Dentate margins of pedipalp-chela fingers with 2 rows of granules (these rows are sometimes fused at the base; Fig. 3.24*C*) ... ***Opisthacanthus***

Tarsi of legs with only setae on ventral aspect (Fig. 3.24*E*):

Ventral setae of tarsi numerous (usually in excess of 10 pairs); hemispermatophore with elongated lamella; lamellar hook positioned distinctly in basal half of hemispermatophore (Fig. 3.24*F*) .... ***Chiromachus***

Ventral setae of tarsi fewer in number (usually only 4 or 5 pairs); hemispermatophore with very short lamella; lamellar hook positioned in distal half of hemispermatophore:

Tarsi with a ventromedian series of small spinules between the setae (Fig. 3.24*E*) .... ***Iomachus***

Tarsi lacking a ventromedian series of small spinules between the setae .... ***Liocheles***

External face of pedipalp patella and ventral face of chela manus with numerous accessory trichobothria; pectines with more than 10 teeth (always in males, usually in females):

Dentate margins of pedipalp-chela fingers with 2 parallel series of granules (Fig. 3.24*C*); hemispermatophore with a double lamellar hook (Fig. 3.24*H*); metasomal segments I–IV with paired ventral submedian carinae .... ***Hadogenes***

Dentate margins of pedipalp-chela fingers with numerous granules not arranged into rows; hemispermatophore with a single lamellar hook (Fig. 3.24*G*); metasomal segments I–IV with a single ventromedian carina .... ***Heteroscorpion***

List of Genera and Their Distributions

Cheloctonus Pocock, 1892 (Botswana, Lesotho, Mozambique, South Africa), 5 spp.

Chiromachus Pocock, 1899 (Seychelles Is.), 1 sp.

Hadogenes Kraepelin, 1894 (southern Africa), 12 spp.

Heteroscorpion Birula, 1903 (Malagasy Republic), 1 sp.

Iomachus Pocock, 1893 (eastern Africa, southern India), 5 spp.

Liocheles Sundevall, 1833 (southern and eastern Asia, many Pacific islands, Australia), 4 spp.

Opisthacanthus Peters, 1861 (southern, eastern, and western Africa; Central America, Cocos Is., Hispaniola, northern South America), 18 spp.

Family Iuridae Thorell, 1876

The Iuridae (Figs. 3.25, 3.26) is another small family, comprising five genera and 19 species. Iurids are found in southwestern Eurasia, western North America, and western South America. Although disjunct, this distribution follows a well-known generalized biogeographical track, the Tethys geosyncline (Francke and Soleglad 1981).

The Iuridae includes some genera that were formerly included in the Chactidae and Vaejovidae. That the genera are related is indicated by the presence of a large darkened tooth on the internal margin of the cheliceral movable finger. Francke and Soleglad (1981) suggested that this feature is a synapomorphy for the group, because the condition is unknown in other scorpions.

Characterization

CHELICERAE (Fig. 3.1*F*). The cheliceral fixed finger has the medial and basal teeth fused into a bicusp. The movable finger has one (Iurinae) or two (Caraboctoninae) subdistal teeth and a single basal tooth on the external margin; the distal external tooth is smaller than the distal internal tooth, and is apposable to it; the internal margin possesses a single, large, darkened tooth.

TRICHOBOTHRIAL PATTERN. The trichobothrial pattern is Type C (as in Fig. 3.5). Only *Paraiurus* and *Iurus* have the fundamental number and pattern; the remaining genera have accessory trichobothria, usually in high numbers (Vachon 1973, Francke and Soleglad 1981).

COXOSTERNAL REGION (Fig. 3.7*E*). The coxapophyses do not possess anteriorly expanded lobes. The sternum is subpentagonal.

LEG SPINATION. Tibial spurs are present on legs III and IV in *Paraiurus* but absent in other genera. All taxa possess both prolateral and retrolateral pedal spurs.

VENOM GLANDS AND TELSON. The venom glands are thick-walled and folded; the venom is opalescent. The telson lacks a subaculear protuberance.

MALE REPRODUCTIVE SYSTEM. The hemispermatophore (Fig.

Fig. 3.25. Representative Iuridae: *top, Hadrurus concolorous,* Guerrero Negro, Baja California Sur; *bottom, Hadruroides lunatus,* near Lima, Peru (photo courtesy Edward S. Ross).

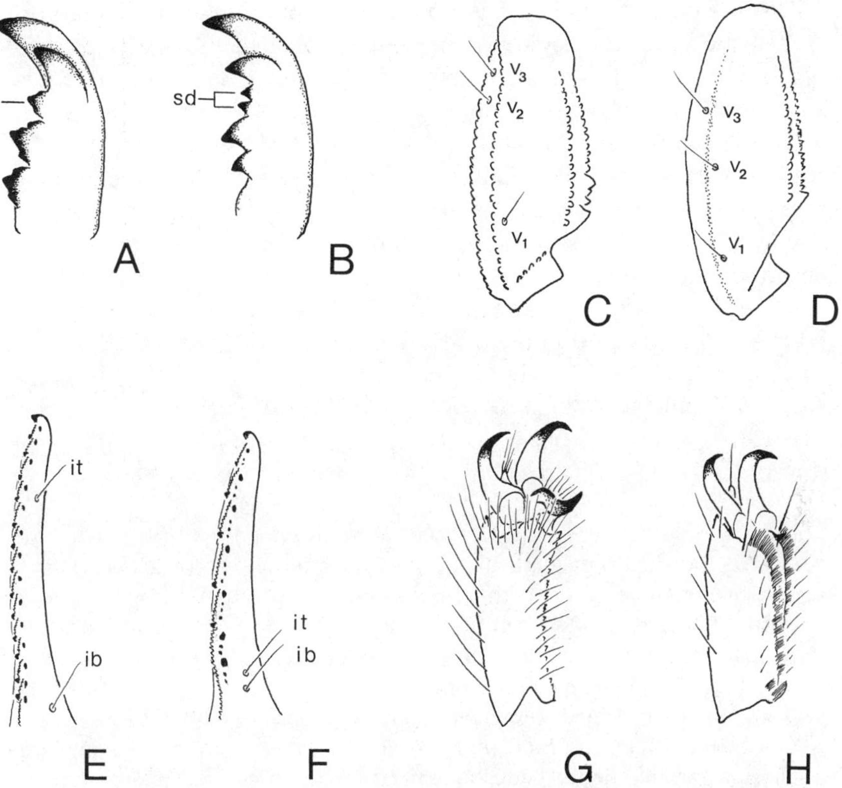

Fig. 3.26. Characteristics of iurid genera: *A*, *Iurus dufoureius*, dorsal aspect of right cheliceral movable finger; *B*, *Hadrurus spadix*, dorsal aspect of right cheliceral movable finger; *C*, *I. dufoureius*, ventral aspect of pedipalp patella, showing trichobothrial pattern; *D*, *Hadruroides lunatus*, ventral aspect of pedipalp patella, showing trichobothrial pattern; *E*, *I. dufoureius*, dentate margin of pedipalp-chela fixed finger, showing trichobothrial pattern and dentition; *F*, *H. lunatus*, dentate margin of pedipalp-chela fixed finger, showing trichobothrial pattern and dentition; *G*, *Hadrurus spadix*, lateral aspect of right tarsomere II of leg III; *H*, *Hadruroides lunatus*, lateral aspect of tarsomere II of leg III. *sd*, subdistal tooth. Trichobothrial designations are as given in text (see "Trichobothria").

3.12*B*) is lamelliform. There is no truncal flexure (except in *Iurus*) or embellished capsular region (Francke and Soleglad 1981). The paraxial organ has one pair of anterior accessory glands in *Paraiurus*, but accessory glands are lacking in other genera.

FEMALE REPRODUCTIVE SYSTEM. The ovariuterus forms a reticulate mesh of six cells (Fig. 3.13*D*). In *Hadrurus*, two large saclike structures (seminal receptacles?) extend from the genital atrium and give rise to the lateral branches of the ovariuterus. Diverticula are not present in any genera.

DEVELOPMENT. Apoikogenic, but the ova are yolkless. The ovarian follicles are connected to the ovariuterus by a short pedicel.

Key to the Subfamilies and Genera of the Iuridae

The following key is slightly modified from that of Francke and Soleglad (1981).

Cheliceral movable finger with 1 subdistal tooth on external margin (Fig. 3.26*A*); pedipalp patella with 1 trichobothrium on ventral aspect (Fig. 3.26*C*); pedipalp chela with trichobothrium *it* situated in distal half of fixed finger, well separated from *ib* (Fig. 3.26*E*) . . . Subfamily IURINAE:

- Dentate margin of pedipalp-chela fingers composed of 6 or 7 rows of granules; tibial spurs present on legs III–IV . . . . ***Paraiurus***
- Dentate margin of pedipalp-chela fingers composed of 14 or 15 rows of granules (Fig. 3.26*E*); tibial spurs absent . . . . ***Iurus***

Cheliceral movable finger with 2 subdistal teeth on external margin (Fig. 3.26*B*); pedipalp patella with more than 1 trichobothrium on ventral aspect (Fig. 3.26*D*); pedipalp-chela fixed finger with both internal trichobothria situated basally (Fig. 3.26*F*) . . Subfamily CARABOCTONINAE:

- Dentate margins of pedipalp-chela fingers composed of 9 or 10 rows of granules; pedipalp patella with about 30 or more trichobothria on ventral aspect; ventral aspect of tarsomere II of legs with setae not arranged in tufts (Fig. 3.26*G*); pedipalp chela with 13 to 27 trichobothria on ventral aspect . . . . ***Hadrurus***
- Dentate margins of pedipalp-chela fingers composed of 6 or 7 rows of granules; ventral aspect of tarsomere II of legs with 2 rows of setal tufts (Fig. 3.26*H*); pedipalp patella with only 2 ventral trichobothria (Fig. 3.26*D*); chela with only 4 ventral trichobothria:
  - Dentate margins of pedipalp-chela fingers in adults with supernumerary granules flanking the rows of granules (Fig. 3.26*F*); sternum as long as wide, with a deep longitudinal furrow . . . . ***Hadruroides***

Dentate margins of pedipalp-chela fingers in adults lacking supernumerary granules; sternum wider than long, without distinct furrow, but with a deep pit posteriorly .... ***Caraboctonus***

List of Genera and Their Distributions

Subfamily Caraboctoninae Pocock, 1905
- *Caraboctonus* Pocock, 1893 (Chile), 1 sp.
- *Hadruroides* Pocock, 1893 (Ecuador, Peru), 7 spp.
- *Hadrurus* Thorell, 1876 (Mexico, western United States), 8 spp.

Subfamily Iurinae Thorell, 1876
- *Paraiurus* Francke, 1985 (Samos I. [Greece], Turkey), 1 sp.
- *Iurus* Thorell, 1876 (Greece, Turkey), 2 spp.

Family Scorpionidae Pocock, 1893

The family Scorpionidae (Figs. 3.27, 3.28) includes seven genera distributed among three subfamilies. The members of the Scorpionidae differ from the Ischnuridae in the possession of moderately to highly folded venom glands and tarsi with rounded lateroapical lobes. These characters actually represent synapomorphies for the two sister groups Scorpionidae and Diplocentridae. There are currently no known synapomorphies for the family Scorpionidae, although most of the genera possess accessory trichobothria on the pedipalp patellae and chelae (obviously a derived feature, compared with that found in most Ischnuridae). The Diplocentridae, in contrast to scorpionids, possess only the fundamental number of trichobothria, a distinct subaculear tubercle, and venom with a reddish pigment. An exception to the scorpionid characterization is the Hemiscorpiinae: they usually possess moderately complex venom glands and have straight lateroapical tarsal lobes; in fact, they cannot be distinguished from the ischnurid genera on the basis of these characters. Lourenço (1985) retained them in the Scorpionidae pending further phylogenetic studies; it is doubtful that they actually belong there.

The Scorpionidae contains some of the world's largest living scorpions. *Heterometrus* (*Gigantometrus*) *swammerdami* reaches lengths up to 16.8 cm (Couzijn 1981), and *Pandinus imperator* 18 to 20 cm (Vachon 1953). A third species, *Opistophthalmus gigas*, has been recorded at 16 cm in length (Lamoral 1979).

Fig. 3.27. Representative Scorpionidae: *top, Opistophthalmus carinatus*, Etosha Pan, South-West Africa; *bottom, Scorpio maurus*, Ain Tuta, Algeria. (Photos courtesy Edward S. Ross)

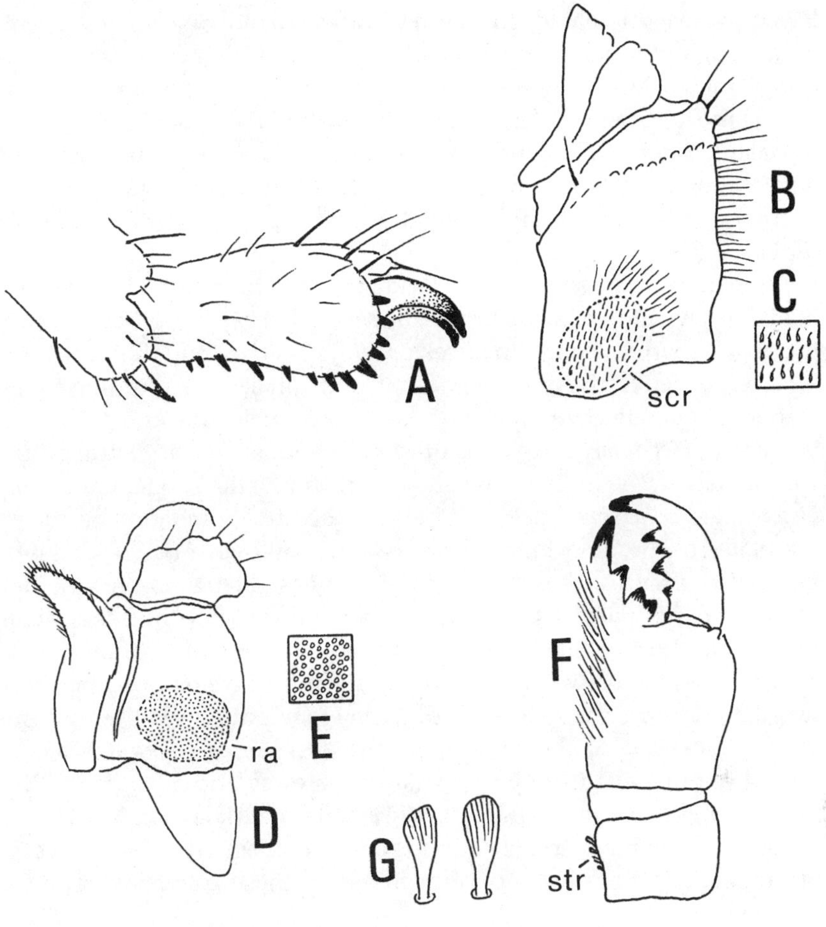

Fig. 3.28. Characteristics of scorpionid genera: *A*, *Opistophthalmus* sp., lateral aspect of right tarsomere II, showing rounded lateroapical lobe; *B–E*, *Heterometrus fulvipes*, details of coxal stridulatory organs (*B*, posterior face of right pedipalpal coxa, showing location of scraper; *C*, close-up of scraper surface, showing minute spines; *D*, anterior face of right coxa of leg I, showing position of rasp; *E*, close-up of rasp surface, showing minute tubercles); *F*, *O. ecristatus*, right chelicera; *G*, stridulatory setae, enlarged. *ra*, rasp; *scr*, scraper; *str*, stridulatory setae.

The family Scorpionidae is exclusively an Old World group and is believed to have originated in eastern Gondwanaland. Several groups are of biogeographical interest. The subfamily Urodacinae (*Urodacus* spp.) is endemic to Australia and evolved there from a scorpionine ancestor during the long period of isolation after Australia separated from Gondwanaland (Koch 1977). This genus is the most diverse in Australia, and its radiation can be explained in terms of allopatric speciation resulting from changing climate and vegetation patterns during the Cenozoic (Koch 1977).

The genus *Heterometrus* is widely distributed from India throughout Southeast Asia to Wallace's Line. Couzijn (1981) has hypothesized that the genus shared a common ancestor with *Pandinus*, and that its prototype originated in the eastern part of Gondwanaland. According to his scenario, separate Gondwanaland fragments carried the ancestral populations of *Heterometrus* to their present locations. One ancestral group was carried on the Indian Plate and diverged into the subgenera *Gigantometrus* and *Chersonesometrus*. A second ancestral population occupied the plates of the Indochinese shelf complex and gave rise to the subgenus *Javanimetrus*; this part of the hypothesis depends on whether Indochina was part of Gondwanaland (Stauffer 1974; but see McElhinny et al. 1974). When these plates collided with Laurasia, migration became possible between the two areas. The subgenus *Srilankametrus* was derived from the spreading of *Javanimetrus* into India through the Assam Gateway. At the same time, the ancestor of *Chersonesometrus* moved out of India through the Assam Gateway and gave rise to the subgenus *Heterometrus*, which now inhabits the Malay Peninsula and Indonesia. The present distributions are hypothesized to be the result of range expansions and contractions since those original vicariance and dispersal events.

Characterization

CHELICERAE. On the cheliceral fixed finger, the medial and basal teeth are fused into a bicusp (Fig. 3.1G). The movable finger has one subdistal and one basal tooth on the external margin; the distal external tooth is smaller than the distal internal tooth and is usually apposable to it; the internal margin is smooth.

TRICHOBOTHRIAL PATTERN. The trichobothrial pattern is Type C (as in Fig. 3.5). *Hemiscorpius*, *Heterometrus*, and *Scorpio* possess the fundamental number; *Habibiella*, *Opistophthalmus*, *Pandinus*, and *Urodacus* possess accessory trichobothria on the pedipalp-chela manus and/or patella (Vachon 1973).

COXOSTERNAL REGION. The coxapophyses do not possess broadly expanded anterior lobes; the sternum is subpentagonal (Fig. 3.7*G*).

LEG SPINATION. Only prolateral pedal spurs are present; tibial spurs and retrolateral pedal spurs are lacking (Fig. 3.8*C*). All genera except *Hemiscorpius* and *Habibiella* are characterized by stout spines on the ventral surface of the tarsi; in these two genera spines are present, but are much finer.

VENOM GLANDS AND TELSON. The venom glands are moderately to highly folded (Lourenço 1985). In *Scorpio, Opistophthalmus, Heterometrus, Pandinus,* and *Urodacus,* they tend to exhibit complex folding. The telson never bears a subaculear protuberance, and the venom is opalescent.

MALE REPRODUCTIVE SYSTEM. The hemispermatophore is lamelliform and bears an elaborate capsule. The paraxial gland lacks accessory glands.

FEMALE REPRODUCTIVE SYSTEM. The ovariuterus forms a reticulate mesh of six cells (Fig. 3.13*F, G*). Numerous diverticula arise from the lateral and mesial branches of the ovariuterus. The embryos develop within these diverticula.

DEVELOPMENT. Katoikogenic.

Key to the Subfamilies and Genera of the Scorpionidae

The following key is derived from several sources, including Werner (1934), Vachon (1973), Lamoral (1979), and Couzijn (1981), as well as from the examination of preserved material.

Metasomal segments I–IV with a single ventral submedian carina:
- Tarsi with lateroapical margin distinctly produced into rounded lobes (Fig. 3.28*A*) .... Subfamily URODACINAE .... ***Urodacus***
- Tarsi with lateroapical margin more or less straight, not produced into rounded lobes (see Fig. 3.24*A*) .... Subfamily HEMISCORPIINAE:
  - Pedipalp patella with 15 trichobothria on external surface .... ***Habibiella***
  - Pedipalp patella with only 13 trichobothria on external surface .... ***Hemiscorpius***

Metasomal segments I–IV with paired ventral submedian carinae .... Subfamily SCORPIONINAE:
- Stridulatory (sound-producing) organ located on opposing surfaces of the coxae of the pedipalps and first pair of legs (Fig. 3.28*B–E*):

Pedipalp chela with 26 trichobothria; patella with 19 trichobothria, 13 of which are on the external surface .... ***Heterometrus***

Pedipalp chela with more than 26 trichobothria and/or patella with more than 13 trichobothria on the external surface........ ***Pandinus***

No stridulatory organ on the coxae of the pedipalps and first pair of legs:

Pedipalp patella with 19 trichobothria, 13 of which are on the external surface; first cheliceral segment (that immediately preceding the manus) lacking stridulatory setae on medial surface .... ***Scorpio***

Pedipalp patella with more than 19 trichobothria, always more than 13 on external surface and usually more than 3 on the ventral surface; first cheliceral segment (that immediately preceding the manus) with stridulatory setae on the medial surface (Fig. 3.28*F, G*) .... ***Opistophthalmus***

List of Genera and Their Distributions

Subfamily Hemiscorpiinae Pocock, 1893

Habibiella Vachon, 1974 (Iran), 1 sp.

Hemiscorpius Peters, 1861 (Eritrea, Arabian Peninsula, Socotra I., Iran, Iraq, Pakistan), 6 spp.

Subfamily Scorpioninae Pocock, 1893

Heterometrus Hemprich and Ehrenberg, 1829 (southern and southeastern Asia, much of Indonesia), 5 subgenera, 21 spp.

Opistophthalmus C. L. Koch, 1837 (southern and eastern Africa), 47 spp.

Pandinus Thorell, 1876 (central Africa to Yemen), 5 subgenera, 23 spp.

Scorpio Linné, 1758 (western Africa and throughout most of the Middle East), 1 sp., over 20 ssp.

Subfamily Urodacinae Pocock, 1893

Urodacus Peters, 1861 (Australia), 19 spp.

Fossil Scorpions: Systematics and Evolution

Only about 90 species of fossil scorpions have been discovered, but some of these date as far back in time as the Silurian Period, over 400 million years ago. Although the fossil record is quite fragmentary, our understanding of these ancient scorpions was greatly enhanced in 1986 by the posthumous publication of Erik N. Kjellesvig-Waering's *A Restudy of the Fossil Scorpionida of the World*. Because of this outstanding contribution, we now have some tenable ideas about how scorpions evolved since their origin as gill-breathing forms inhabiting Silurian estuaries and lagoons.

Kjellesvig-Waering (1986) divided the order Scorpiones into two suborders: the Branchioscorpionina and the Neoscorpionina. Almost all of the fossil species are assigned to the Branchioscorpionina, a group that possessed gills and whose members obtained their oxygen from water. The second group, the Neoscorpionina, developed book lungs for breathing air and were almost certainly terrestrial. The genera *Compsoscorpius* and *Palaeopisthacanthus* are the only Paleozoic scorpions to be included in this group. Obviously, the claim in many paleontology and invertebrate-zoology textbooks that scorpions were the first terrestrial arthropods has been refuted.

In general appearance, a fossil scorpion looks much the same as a modern one (Fig. 3.29). The pedipalps are chelate, the prosoma is provided dorsally with a carapace, and the opisthosoma is divided into a mesosoma of seven segments and a narrow, taillike metasoma of five segments. Following the last segment of the metasoma is the telson, modified into a stinger. All scorpions, whether fossil or living, possess pectines, and no other organism possesses such structures.

In spite of these similarities, fossil scorpions often exhibit some conspicuous differences from their extant relatives. The position of the leg coxae in relation to the sternum is highly variable in fossil taxa (Fig. 3.30), but remarkably consistent in Recent forms. Many Branchioscorpionina possessed compound lateral eyes (Wills 1947; Kjellesvig-Waering 1966a, 1986; Paulus 1979) as well as gills concealed beneath their ventral abdominal plates (Kjellesvig-Waering 1986); both of these features are unknown in modern scorpions. Most Branchioscorpionina lacked trichobothria, although some appear to have possessed sensory setae of various types (Wills 1959, 1960; Størmer 1963; Kjellesvig-Waering 1969, 1986). Finally, a few of the fossil species are noted for their large size: *Brontoscorpio anglicus*, known from the single pedipalp-chela movable finger shown in Figure 3.29, was estimated to be about 94 cm in total length; a more complete specimen of *Praearcturus gigas* (formerly considered a giant isopod) also measures almost a full meter in total length! Certainly, today's largest scorpions seem rather puny in comparison.

Systematics of Fossil Scorpions

Prior to Kjellesvig-Waering's revision, several inadequate attempts to classify the fossil scorpions were made (Pocock 1911; Petrunkevitch 1953, 1955), although certain of the observations and ideas from those

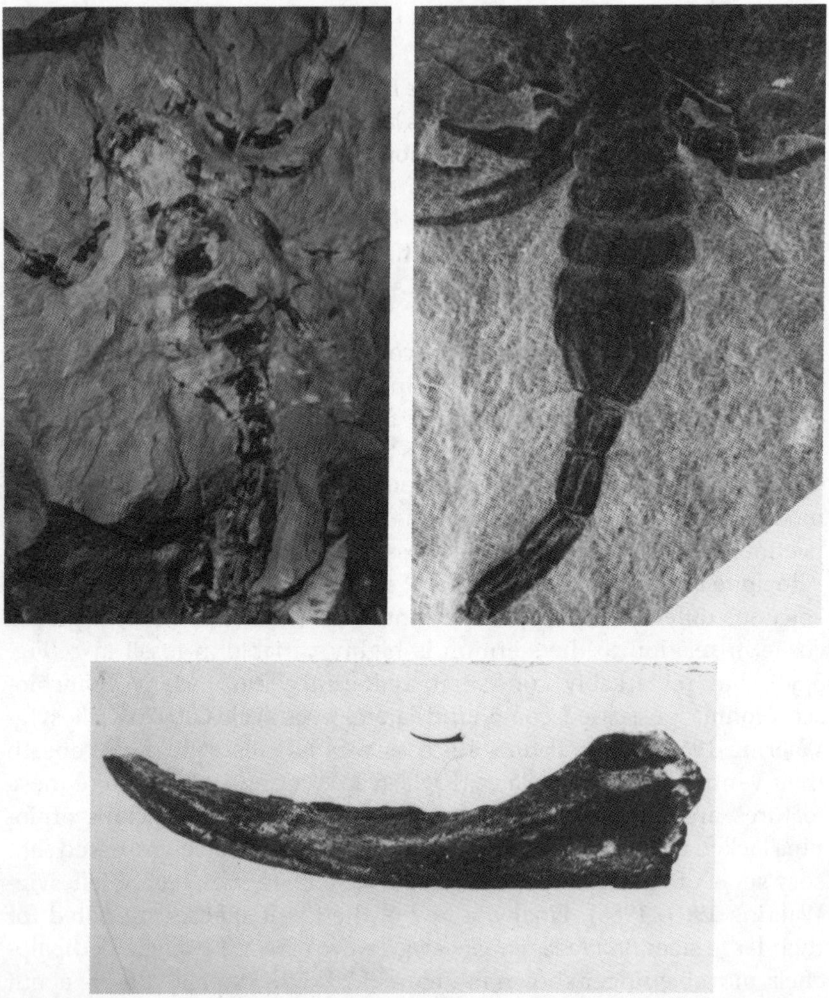

Fig. 3.29. Representative fossil scorpions: *above left, Isobuthus rakovnicensis,* Upper Carboniferous of Czechoslovakia, contained in the British Museum of Natural History (photo courtesy Edward S. Ross); *above right, Proscorpius osborni,* Silurian of New York (reprinted from the *Journal of Paleontology* with permission of the Paleontological Society of America); *below,* photographic comparison of the holotype of the large *Brontoscorpio anglicus* with a Recent species, *Tityus trinitatis,* showing the movable fingers of the pedipalp chelae (reprinted from the *Journal of Paleontology* with permission of the Paleontological Society). The total body length of the adult *T. trinitatis* male from which the finger was taken was 6.95 cm; the finger alone of *B. anglicus* measures 9.75 cm. Kjellesvig-Waering (1972) estimated that *B. anglicus* was probably almost a meter in length!

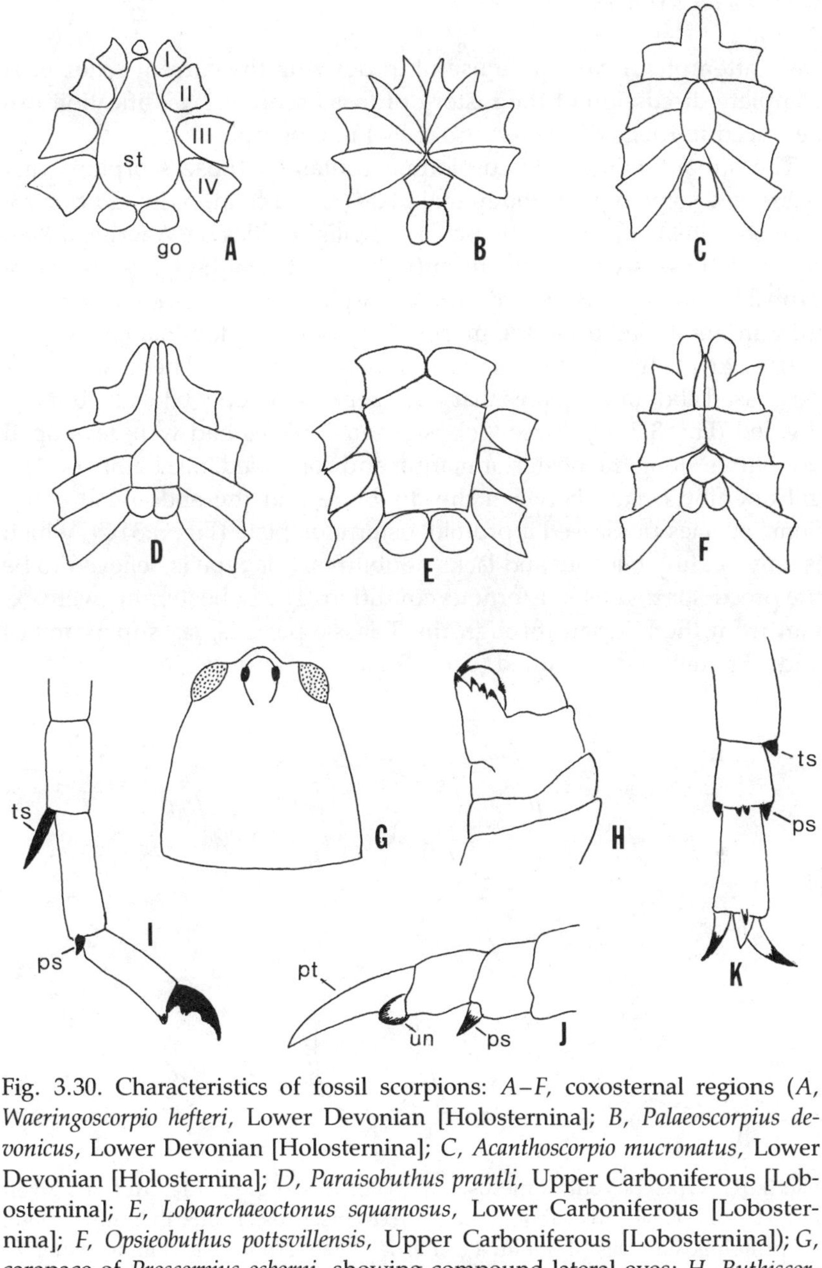

Fig. 3.30. Characteristics of fossil scorpions: *A–F,* coxosternal regions (*A, Waeringoscorpio hefteri,* Lower Devonian [Holosternina]; *B, Palaeoscorpius devonicus,* Lower Devonian [Holosternina]; *C, Acanthoscorpio mucronatus,* Lower Devonian [Holosternina]; *D, Paraisobuthus prantli,* Upper Carboniferous [Lobosternina]; *E, Loboarchaeoctonus squamosus,* Lower Carboniferous [Lobosternina]; *F, Opsieobuthus pottsvillensis,* Upper Carboniferous [Lobosternina]); *G,* carapace of *Proscorpius osborni,* showing compound lateral eyes; *H, Buthiscorpius buthiformis,* four-segmented right chelicera; *I, Boreoscorpio copelandi,* spination of distal segments of leg IV; *J, Allopalaeophonus caledonicus,* spination of distal segments of left leg I; *K, P. osborni,* spination of distal segments of right leg III. *go,* genital operculum; *ps,* pedal spur; *pt,* posttarsus; *ts,* tibial spur; *st,* sternum; *un,* unguis; *I–IV,* coxae of legs I–IV. (After Kjellesvig-Waering 1986)

early attempts proved quite useful in deriving the current scheme. A complete discussion of the history of fossil scorpion classification can be traced in Kjellesvig-Waering's (1986) monograph.

The suborder Branchioscorpionina contained those scorpions that possessed some type of abdominal plate (rather than, or in addition to, true sternites) concealing branchia, or gills (Kjellesvig-Waering 1986). Most of these scorpions were entirely aquatic, although some were probably amphibious. The Branchioscorpionina was divided into four infraorders, based upon the morphology of the abdominal plates:

INFRAORDER LOBOSTERNINA POCOCK, 1911. These scorpions possessed abdominal plates that were gently to deeply bilobate, but undivided (Fig. 3.31*D*); these "lobosternous" plates had well-developed doublures along the posterior margin and concealed gill chambers. The gill opening occurs between the doublure and the abdominal plate. Some species possessed a protolobosternous plate (Fig. 3.31*C*), which is only slightly bilobate and lacks doublures; this type is believed to be the precursor to the lobosternous condition. The Lobosternina were extant from the Silurian through the Triassic periods; six superfamilies and 15 families are included (Table 3.2).

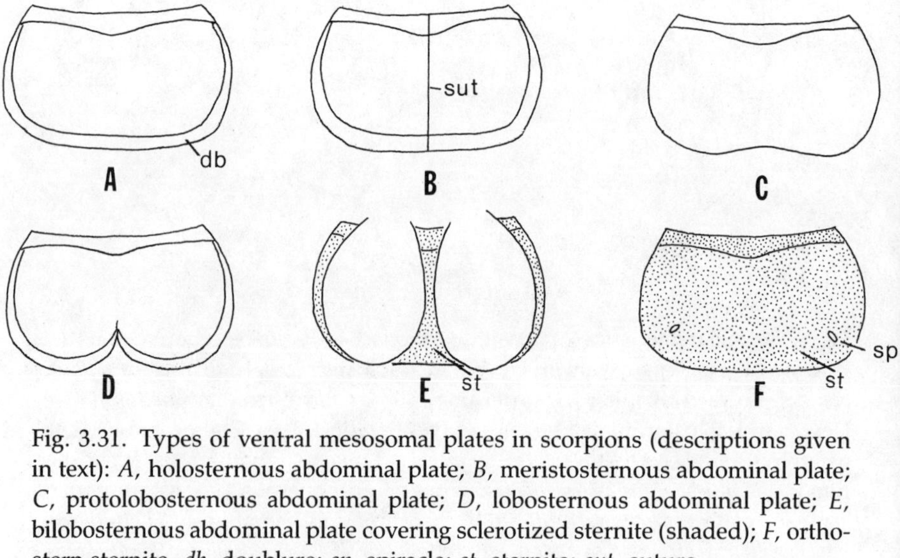

Fig. 3.31. Types of ventral mesosomal plates in scorpions (descriptions given in text): *A*, holosternous abdominal plate; *B*, meristosternous abdominal plate; *C*, protolobosternous abdominal plate; *D*, lobosternous abdominal plate; *E*, bilobosternous abdominal plate covering sclerotized sternite (shaded); *F*, orthostern sternite. *db*, doublure; *sp*, spiracle; *st*, sternite; *sut*, suture.

INFRAORDER HOLOSTERNINA KJELLESVIG-WAERING, 1986. The abdominal plates of these scorpions were rectilinear, undivided, and possessed doublures (Fig. 3.31*A*). The abdominal plates concealed well-developed gill chambers; the gill opening was found either on the doublure or between the doublure and the abdominal plate. The Holosternina are known from the Silurian through the Jurassic periods; ten superfamilies and 25 families are included (Table 3.2).

INFRAORDER MERISTOSTERNINA KJELLESVIG-WAERING, 1986. The abdominal plates in this group are merely divided by a median longitudinal suture (Fig. 3.31*B*). The plates possessed doublures and concealed gill chambers; the gill opening lies between the abdominal plate and the doublure. This was a relatively small group, containing three superfamilies and four families (Table 3.2). It persisted from the late Devonian Period through the Pennsylvanian.

INFRAORDER BILOBOSTERNINA KJELLESVIG-WAERING, 1986. In this group, the abdominal plates were completely divided and lobiform (Fig. 3.31*E*). Doublures are absent, although the Bilobosternina apparently still had gills. Interestingly, true sternites are found dorsal to the abdominal plates, and Kjellesvig-Waering suggested that primitive lungs may have been present. There were one superfamily and two families of bilobostern scorpions, which occurred in the Silurian and Devonian periods (Table 3.2).

The Neoscorpionina includes all terrestrial, pulmonate (air-breathing) scorpions. It consists of a single infraorder, the Orthosternina Pocock, 1911, which is characterized as having sternites rather than abdominal plates with doublures (Fig. 3.31*F*). Paired spiracles, which perforate each sternite, are present and open into book lung chambers. There are only four sternites on the venter of Neoscorpionina, in contrast to the five abdominal plates found in the Branchioscorpionina; hypothetically, the first sternite lying underneath the first abdominal plate was lost when the remaining abdominal plates were lost. The Neoscorpionina includes the Pennsylvanian family Palaeopisthacanthidae (Table 3.2) and all Recent scorpions.

Characterization of Fossil Superfamilies and Families

The fossil superfamilies are based primarily on the arrangement of the coxosternal region (Fig. 3.30*A–F*). This concept was originated by Petrunkevitch (1955), but was utilized most effectively and comprehen-

TABLE 3.2

Classification, age, and distribution of the fossil scorpions of the world

Superfamily	Family	Genus	Age	Locality
SUBORDER BRANCHIOSCORPIONINA				
INFRAORDER BILOBOSTERNINA				
Branchioscorpionoidea	Branchioscorpionidae	*Branchioscorpio*	Lower Devonian	Wyoming
	Dolichophonidae	*Dolichophonus*	Middle Silurian	Scotland
INFRAORDER HOLOSTERNINA				
Proscorpioidea	Proscorpiidae	*Proscorpius*	Upper Silurian	New York
		Archaeophonus	Upper Silurian	New York
	Labriscorpionidae	*Labriscorpio*	Middle Carboniferous	Illinois
	Waeringoscorpionidae	*Waeringoscorpio*	Lower Devonian	Germany
Stoermeroscorpionoidea	Stoermeroscorpionidae	*Stoermeroscorpio*	Silurian	New York
Allopalaeophonoidea	Allopalaeophonidae	*Allopalaeophonus*	Silurian	Scotland
Palaeoscorpioidea	Palaeoscorpiidae	*Palaeoscorpius*	Lower Devonian	Germany
	Hydroscorpiidae	*Hydroscorpius*	Lower Devonian	Wyoming
Archaeoctonoidea	Archaeoctonidae	*Archaeoctonus*	Lower Carboniferous	Scotland
		Pseudoarchaeoctonus	Lower Carboniferous	Scotland
Acanthoscorpionoidea	Acanthoscorpionidae	*Acanthoscorpio*	Lower Devonian	Wyoming
	Stenoscorpionidae	*Stenoscorpio*	Triassic	England
Gigantoscorpionoidea	Gigantoscorpionidae	*Gigantoscorpio*	Lower Carboniferous	Scotland
Mesophonoidea	Mesophonidae	*Mesophonus*	Jurassic	Siberia
			Triassic	England
	Mazoniidae	*Mazonia*	Carboniferous	England, Illinois
	Centromachidae	*Centromachus*	Lower Carboniferous	Scotland
	Heloscorpionidae	*Heloscorpio*	Carboniferous	England
	Liassoscorpionidae	*Liassoscorpionides*	Jurassic	Germany
	Phoxiscorpionidae	*Phoxiscorpio*	Lower Carboniferous	Scotland
	Willsiscorpionidae	*Willsiscorpio*	Triassic	England
Eoctonoidea	Eoctonidae	*Eoctonus*	Upper Pennsylvanian	Illinois
	Buthiscorpiidae	*Buthiscorpius*	Carboniferous	England
			Pennsylvanian	Illinois
	Allobuthiscorpiidae	*Allobuthiscorpius*	Carboniferous	England
		Aspiscorpio	Carboniferous	England

	Anthracoscorpionidae	*Anthracoscorpio*	Upper Carboniferous	Czechoslovakia
			Carboniferous	England
		Lichnoscorpius	Upper Carboniferous	England
		Allobuthus	Carboniferous	England
		Coseleyscorpio	Carboniferous	England
	Garnettiidae	*Garnettius*	Pennsylvanian	Kansas
Spongiophonoidea	Spongiophonidae	*Spongiophonus*	Triassic	England
	Praearcturidae	*Praearcturus*	Lower Devonian	Wyoming
		Brontoscorpio	Siluro-Devonian	England
INFRAORDER LOBOSTERNINA				
Palaeophonoidea	Palaeophonidae	*Palaeophonus*	Silurian	England, Sweden
Anthracochaeriloidea	Anthracochaerilidae	*Anthracochaerilus*	Lower Carboniferous	Scotland
Isobuthoidea	Isobuthidae	*Isobuthus*	Upper Carboniferous	Czechoslovakia
		Boreoscorpio	Upper Carboniferous	Nova Scotia
		Feistmantelia	Permian	Czechoslovakia
		Bromsgroviscorpio	Triassic	England
	Eobuthidae	*Eobuthus*	Upper Carboniferous	Czechoslovakia, England, Netherlands
	Eoscorpiidae	*Eoscorpius*	Upper Carboniferous	England, Illinois
			Lower Pennsylvanian	Nova Scotia
		Trachyscorpio	Lower Carboniferous	Scotland
		Eskiscorpio	Lower Carboniferous	Scotland
	Pareobuthidae	*Pareobuthus*	Carboniferous	England
	Kronoscorpionidae	*Kronoscorpio*	Pennsylvanian	Illinois
Paraisobuthoidea	Paraisobuthidae	*Paraisobuthus*	Upper Carboniferous	Czechoslovakia, England, Illinois
		Leioscorpio	Upper Carboniferous	England
	Telmatoscorpionidae	*Telmatoscorpio*	Pennsylvanian	Illinois
	Scoloposcorpionidae	*Scoloposcorpio*	Lower Carboniferous	Scotland
		Benniescorpio	Carboniferous	Scotland
	Opsieobuthidae	*Opsieobuthus*	Upper Carboniferous	Indiana
Loboarchaeoctonoidea	Loboarchaeoctonidae	*Loboarchaeoctonus*	Lower Carboniferous	Scotland
Pseudobuthiscorpioidea	Pseudobuthiscorpiidae	*Pseudobuthiscorpius*	Upper Carboniferous	England
	Petaloscorpionidae	*Petaloscorpio*	Upper Carboniferous	Quebec
	Waterstoniidae	*Waterstonia*	Carboniferous	Scotland

TABLE 3.2
(*continued*)

Superfamily	Family	Genus	Age	Locality
		SUBORDER BRANCHIOSCORPIONINA		
INFRAORDER MERISTOSTERNINA				
Tiphoscorpionoidea	Tiphoscorpionidae	*Tiphoscorpio*	Upper Devonian	New York
Cyclophthalmoidea	Cyclophthalmidae	*Cyclophthalmus*	Carboniferous	Czechoslovakia, Siberia
			Upper Carboniferous	England
	Microlabiidae	*Microlabis*	Upper Carboniferous	Czechoslovakia
Palaeobuthoidea	Palaeobuthidae	*Palaeobuthus*	Upper Carboniferous	Illinois
		SUBORDER NEOSCORPIONINA		
INFRAORDER ORTHOSTERNINA				
Scorpionoidea	Palaeopisthacanthidae	*Palaeopisthacanthus*	Pennsylvanian	Illinois
		Compsoscorpius	Carboniferous	England
	Scorpionidae	*Mioscorpio*	Miocene	Germany
		GENERA *INCERTAE SEDIS*		
		Titanoscorpio	Upper Carboniferous	Illinois
		Wattisonia	Upper Carboniferous	England
		Palaeomachus	Upper Carboniferous	England

sively by Kjellesvig-Waering (1986). The great variation in the arrangement of the coxae and sternum among fossil scorpions has led to the recognition of 21 fossil superfamilies, which demonstrates how diverse the order was in earlier geological times. Because the fossil record is so fragmentary, these 21 superfamilies must have been only a fraction of the number that actually existed. Kjellesvig-Waering (1986) considers all Recent scorpions to belong to a single superfamily, the Scorpionoidea, claiming that the characters utilized by neontologists are not of value at the superfamilial level.

Families of fossil scorpions are based on four features (Fig. 3.30): the development of the maxillary lobes, the shape of the sternum, the terminations of the tarsi, and the presence or absence of lateral, faceted eyes (Kjellesvig-Waering 1986). The diversity in these features exhibited by the fossils results in the recognition of 47 families. Kjellesvig-Waering (1986) suggested that, on the basis of these characters, there is reason to recognize only three families of Recent scorpions, rather than the nine recognized by neontologists. This is basically a taxonomic problem, and the recognition of an extra suprafamilial category could easily resolve it. The number of categories recognized should be dependent on the cladistic branching of the group, according to the phylogenetic method. Difficulties of this nature are often the result of the paleontologist's working from the higher categories down, versus the neontologist's working from the lower categories up.

To gain a more meaningful appreciation of the diversity of the order in earlier geological times, the fauna of today should be compared to the fauna living during a given period in the past. The Recent scorpions comprise three to four superfamilies (depending on the source consulted) and nine families, which we know have been present at least for 35 to 40 million years. The family Scorpionidae dates to the Miocene Epoch (25 M.Y.B.P.), and the Buthidae to the Oligocene (35–40 M.Y.B.P.), assuming the species from Baltic amber are indeed referable to that family (Schawaller 1979). These must be, of course, the minimum ages of those families rather than the maximum ages, because we have no knowledge of earlier Cenozoic scorpions. In comparison, eight superfamilies and 11 families (that we know of) lived from the Middle Silurian to the Lower Devonian, which probably represents a comparable time period. Scorpions seem to have been most diverse during the Carboniferous Period (a span of about 80 million years), when 13 known superfamilies and 30 families existed. These numbers are more impressive when one considers the fragmentary nature of the fossil record.

Evolution in the Order Scorpiones

EVIDENCE OF AN AQUATIC EXISTENCE IN FOSSIL FORMS. Much of the past controversy surrounding the evolution of scorpions centered on the type of breathing mechanism they employed. Until the publication of Kjellesvig-Waering's monograph there was little direct evidence to demonstrate either a gill-breathing or an air-breathing existence. In many cases the same evidence, which was usually circumstantial, was used to support both sides of the argument. For example, the tarsi of *Palaeophonus nuncius* from the Middle Silurian were thought to terminate in single claws, rather than the double claws seen in more modern forms. Pocock (1901) and Størmer (1963) suggested that such legs were suitable for an aquatic habitat, enabling the scorpion to climb among underwater vegetation or on rocks in the intertidal. Because crabs have single spines at the tips of their tarsi, this seemed to be a reasonable argument. However, others pointed out that single tarsal spines were found in some terrestrial arthropods as well, and that the the tips of the tarsi really did not provide much evidence for determining the type of habitat the early scorpions lived in (Kjellesvig-Waering 1966a, Weygoldt and Paulus 1979a). Kjellesvig-Waering (1986) has now shown that the elongate spine is actually the posttarsus (Fig. 3.30*J*), and that two tarsal spurs (homologous to the tarsal claws of modern forms) are present at its base.

To prove what type of habitat the Paleozoic scorpions lived in requires an analysis of their breathing mechanisms, and this is precisely what Kjellesvig-Waering (1986) carried out. Although the Lower Devonian *Waeringoscorpio* was described and illustrated as having tubelike external gill structures (Størmer 1970), there was still some skepticism among researchers who suggested that the structures in question might not be gills (Weygoldt and Paulus 1979a). Kjellesvig-Waering (1986) believes that the specimen had recently molted before its death and that the gills may have been in internal gill chambers when the animal was alive. Although he could not discount the possibility that the structures were branchial leaves with trachioles, Kjellesvig-Waering thought them to be true gills.

The early arguments about Paleozoic habitat now seem academic in light of Kjellesvig-Waering's recent publication. He has clearly described and illustrated gills and gill chambers in many fossil species. *Tiphoscorpio hueberi* is remarkable among these for its almost-complete preservation of the abdominal plates, the inner aspect of which could be studied in detail (Kjellesvig-Waering 1986). The plates were bounded by a wide doublure, with gill slits located between the doublure and abdominal plate. Three types of gills were concealed by these plates in *Tiphoscorpio*: rounded, subrectangular, and irregular. Each gill was made of a spongy white material with small triangular spines on the surface and larger spines on the posterior margins. The anterior portions of the gills were attached to the gill chambers (body wall) and possessed distinct turned-in folds, or ribs, with some interconnections between them. Water apparently entered at the anterolateral corner of the gill chamber, circulated over the gills, and exited through the posterior gill slit.

The gills of *Paraisobuthus* were quite different, but were composed of thick, white spongy material with surface spines similar to those of eurypterids (Kjellesvig-Waering 1986). These gills apparently formed an oval surface on the outer face of the body wall (not the inner surface of the abdominal plate). The circulation of water over the gill surface was probably much like that described for *Tiphoscorpio*. Similar oval structures on the body wall were described for *Eoctonus miniatus* and on the abdominal plates themselves in other taxa (e.g., *Palaeobuthus* and *Eoscorpius*; Kjellesvig-Waering 1986).

It is interesting to note here that only a single fossil scorpion, *Palaeopisthacanthus schucherti*, clearly shows spiracles on true sternites. Abdominal plates are, of course, lacking, because this scorpion belongs to the Neoscorpionina. There is no question that it was a terrestrial pulmonate form that lived in the Carboniferous Period. This places a minimum age for terrestriality at that time, and Kjellesvig-Waering (1986) speculates that air-breathing mechanisms must have arisen sometime during the Devonian.

MAJOR EVOLUTIONARY TRENDS IN THE ORDER SCORPIONES. Kjellesvig-Waering (1986) lists 13 evolutionary trends in the order that led to the development of modern forms. Most of them are directly involved in making the transition from an aquatic to a terrestrial existence; the seven most interesting and important trends are listed below.

The abdominal plates and associated gills were lost; concurrently, sternites and book lungs developed and replaced them.

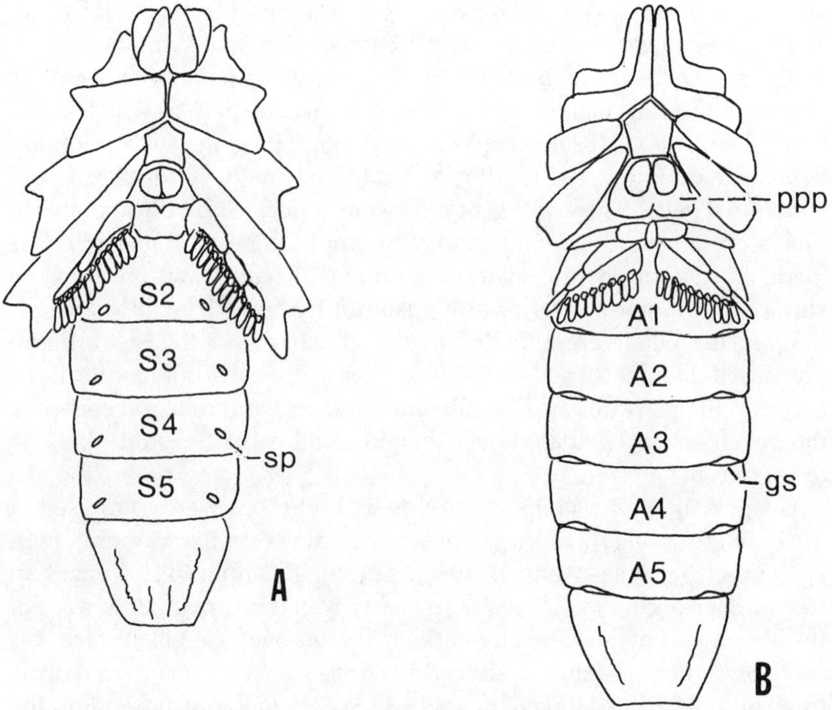

Fig. 3.32. Comparison of ventral segmentation of scorpions: *A*, Neoscorpionina (Orthosternina); *B*, Branchioscorpionina. *A1–5*, abdominal plates 1–5; *gs*, gill slit; *ppp*, prepectinal plate; *S2–5*, sternites 2–5; *sp*, spiracle for book lung.

Certain abdominal segments, specifically, the prepectinal plate and first abdominal plate, were lost. The presence of a ventral prepectinal plate means that the early scorpions possessed eight segments in the mesosoma (Fig. 3.32*B*), rather than the seven typical of modern adult forms (Fig. 3.32*A*)—although males of New World buthids retain this extra plate. This prepectinal plate lies ventral to the embryonic pregenital tergite found in all scorpions (see Chapter 4, "Prenatal Development and Birth"). Interestingly, however, the presence of a pregenital sternite has never been mentioned in any embryological study, and apparently one does not develop. Following the prepectinal plate and the pectines were five pairs of abdominal plates in the Branchioscorpionina; when the abdominal plates and gills were lost, the first sternite was also lost, leaving only the four sternites that persist in modern scorpions. The eighth abdominal segment in the Branchioscorpionina (the seventh segment in Recent scorpions) did not function in respiration; embryologists actually consider this last segment to be the first metasomal segment (Anderson 1973). In this respect, the last ventral mesosomal plate is neither a true abdominal plate in the Branchioscorpionina nor a true sternite in the Neoscorpionina.

The first two leg coxae were modified to form an oral tube. This was accomplished by the development of the coxal endites (maxillary lobes), and was especially necessary to process food in the terrestrial environment. At the same time, the chelicerae (Fig. 3.30*H*) were reduced in size (and also reduced from four segments to three segments).

The coxosternal region was shifted anteriorly so that the fourth pair of coxae abutted the sternum, rather than the genital operculum.

The compound eyes were disaggregated to form five or fewer individual lateral eyes on each side.

The median eyes were displaced posteriorly to the middle of the carapace or beyond.

The tarsi and posttarsi of the legs were shortened, and some of the spines on the legs were lost.

PHYLOGENETIC RELATIONSHIPS OF THE FOSSIL SCORPIONS. Kjellesvig-Waering (1986) also provided some insight about how the various groups of scorpions were related (Fig. 3.33). Sometime after the scorpions and eurypterids diverged from a common ancestor in the Cambrian, the scorpions split into two lineages. One line produced the Holosternina and Meristosternina; the other led to the Lobosternina, Bilobosternina, and Orthosternina. Kjellesvig-Waering suggested that the Bilobosternina split from the Lobosternina by losing the doublures and forming completely separate lobes on each abdominal plate; these modifications exposed the sternites. This condition is intermediate between the lobosterns and orthosterns and is well exemplified by *Branchioscorpio richardsoni*, which was probably amphibious. As the reduction continued, the sternites became sclerotized and developed book lungs as the body wall was exposed to the terrestrial environment. The subsequent loss of the abdominal plates (along with the first sternite) was the key event in the evolution of the Orthosternina, which includes modern scorpions.

Phylogeny of Recent Scorpion Families

The reconstruction of phylogeny depends on the uncovering of genealogical relationships among the taxa considered. The cladistic method,

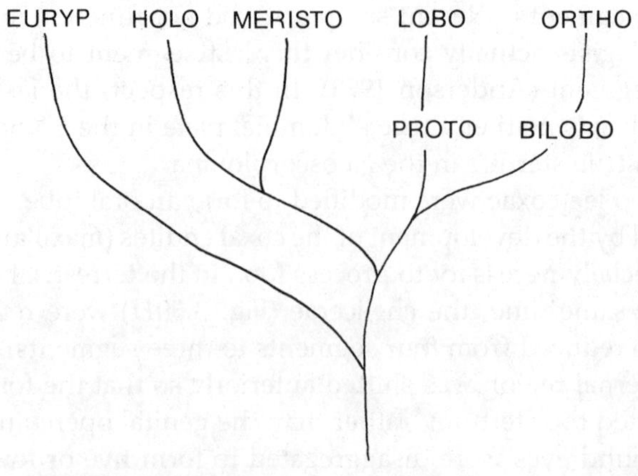

Fig. 3.33. Hypothesized phylogeny of scorpions as the sister group of the Eurypterida, based on morphology of abdominal plates: *BILOBO,* Bilobosternina; *EURYP,* Eurypterida; *HOLO,* Holosternina; *LOBO,* Lobosternina with lobosternous abdominal plates; *MERISTO,* Meristosternina; *ORTHO,* Orthosternina; *PROTO,* Lobosternina with protolobosternous abdominal plates. (After Kjellesvig-Waering 1986)

derived mainly from the works of Hennig (1950, 1966), is specifically designed to uncover these relationships. Hennig's method relies on the recognition of primitive (plesiomorphic) and derived (apomorphic) character states for each character under consideration. The character states must be homologous for the analysis to be valid. Only the possession of shared derived character states (synapomorphies) indicates phylogenetic relationships between taxa. The sharing of primitive character states is uninformative, because such character states are acquired in older, nonrelevant clades. Once the phylogeny is reconstructed, the results are depicted in a branching diagram, or cladogram.

The Recent scorpions can be placed with certainty in the Neoscorpionina (Kjellesvig-Waering 1986), but their relationships to their early Carboniferous relatives cannot be well understood, because of the absolute lack of fossil material between the Carboniferous and Miocene. The single Miocene scorpion, *Mioscorpio zeuneri,* was placed in the family Scorpionidae when it was described in 1931, and its placement there seems appropriate (Kjellesvig-Waering 1986). Several fossils have also been described from amber, but these have all been assigned to modern genera (for a discussion, see Schawaller 1979, 1981).

As for the Recent scorpions themselves, a cladogram has been constructed to demonstrate familial relationships (Lamoral 1980), although it remains largely untested. Primarily on the basis of trichobothrial patterns, male reproductive systems, shape of the sternum, and condition of the coxapophyses, Lamoral recognized three main evolutionary groups in the Scorpiones: the buthoids, chaeriloids, and diplocentroids (Fig. 3.34), each of which is monophyletic. Lamoral's hypothesis states that the diplocentroids and chaeriloids are more closely related to each other than either is to the buthoids (i.e., they are sister groups sharing an immediate common ancestor not shared with the buthoids). The buthoids, then, represent the primitive sister group of all other scorpions.

Within the diplocentroid line, there are several monophyletic subgroups. The Diplocentridae, Ischnuridae, and Scorpionidae form one of these groups. They are characterized by the loss of retrolateral pedal spurs on the tarsi of the legs, by numerous lateral diverticula in the female reproductive system, and by their method of embryonic nutrition (Lamoral 1980, Francke and Soleglad 1981). The Bothriuridae is also monophyletic, as is indicated by the modification of the sternum into a slitlike sclerite, by the possession of a crest on the distal lamina of the hemispermatophore, and by truncated tarsi (Lamoral 1980, Maury 1980, Francke 1982b).

The remaining families (Chactidae, Vaejovidae, and Iuridae) pose quite a problem for the elucidation of phylogeny. The three considered together probably represent a monophyletic group (Francke and Soleglad 1981), but the relationships within this group are obscure. Francke and Soleglad (1981), in recognizing the Iuridae as a valid family (the synapomorphy for the group is the large tooth on the ventral surface of the cheliceral movable finger), took the first step in clarifying relationships between the genera in this assemblage. The problem remains largely unsolved, however, and awaits future research.

Prior to Lamoral's (1980) study, most authors recognized three superfamilies (established by Birula 1917a) in the Scorpiones: the Buthoidea (containing the Buthidae), the Scorpionoidea (containing the Diplocentridae, Ischnuridae, and Scorpionidae), and the Chactoidea (containing all the remaining families). In addition, some authors (e.g., Mello-Leitão 1945) chose to place the Bothriuridae in its own superfamily, the Bothriuroidea. It is interesting to note that, with the exception of the position of *Chaerilus* (which in Birula's time was a problematic taxon

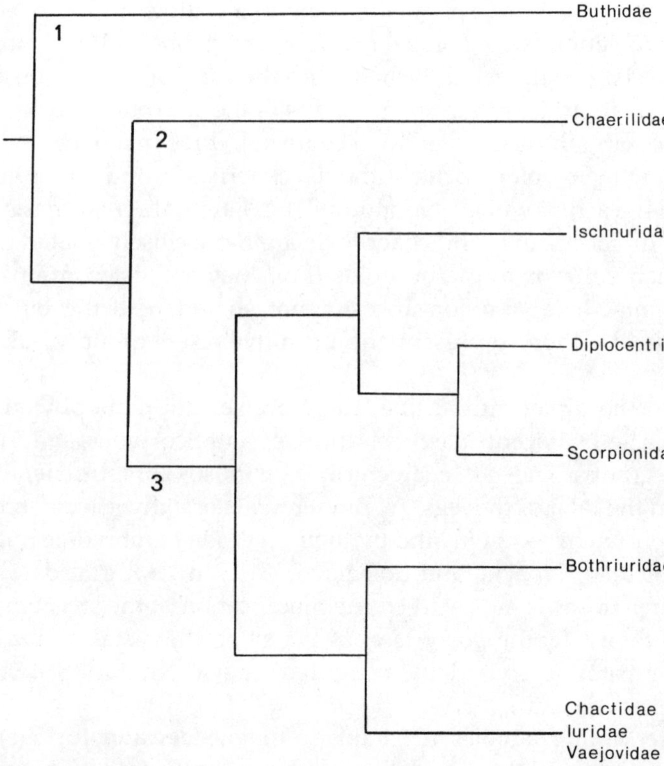

Fig. 3.34. Cladogram showing relationships of Recent scorpion families according to Lamoral (1980). The cladogram has been modified by the inclusion of the Ischnuridae and Iuridae, which were revalidated after Lamoral's analysis. The uncertainty of relationships among the "chactoids" (Chactidae + Iuridae + Vaejovidae) does not permit further analysis of those groups at this time. The three main lines of Recent scorpion evolution (mentioned in the text) are the buthoids (*1*), chaeriloids (*2*), and diplocentroids (*3*).

usually placed in the Chactidae), there is considerable congruence between these older classifications and Lamoral's cladogram, specifically in the positions of the families Buthidae, Diplocentridae, Scorpionidae, and Bothriuridae. Aside from this, the older classifications are mentioned here because they are so prevalent in the literature that they will surely be encountered by the reader.

Scorpions and Chelicerate Phylogeny

Because scorpions are the oldest known arachnids, they always occupy an important place in interpretations of chelicerate phylogeny. Although interest in that subject has surged in the last decade, there still remains room for widely divergent opinions. This section is not intended as a comprehensive summary of all the views of chelicerate phylogeny or as an evaluation of those views. Rather, I hope to provide the reader with a general knowledge of the hypothesized relationship between the various chelicerate taxa.

The subphylum Chelicerata has traditionally included three main groups: the Merostomata, which includes the Xiphosura (horseshoe crabs) and the Eurypterida ("water scorpions"); the Arachnida (including scorpions, spiders, harvestmen, mites, ticks, and other groups); and the Pantopoda (or Pycnogonida, the sea spiders). Of these, the Pantopoda are not important to this discussion, because their origin and phylogenetic affinities are largely obscure. Boudreaux (1979) and Weygoldt and Paulus (1979b) suggested that they diverged early from the main line of chelicerate evolution, and that they may be the sister group of the remaining Chelicerata (i.e., the groups mentioned above).

Scorpions are generally hypothesized to have evolved out of the eurypterid line (Størmer 1963; Bergström 1979; Boudreaux 1979; Weygoldt and Paulus 1979a, b; Kjellesvig-Waering 1986). Most authorities also agree that they subsequently gave rise to or shared a common ancestor with the remaining Arachnida. If both hypotheses are true, they render the class Merostomata an artificial (paraphyletic) group recognized only because of its aquatic grade of existence. Addressing the first hypothesis, there is certainly a considerable body of evidence to indicate close phylogenetic relationship of scorpions and eurypterids, but some characters may be convergent (Størmer 1963, Weygoldt and Paulus 1979a). Paleozoic scorpions in particular shared a number of interesting features with eurypterids (reviewed and evaluated by Størmer 1963 and Kjellesvig-Waering 1986). Some of the more noteworthy features are:

1. Homologies in segmentation and appendages (see text-fig. 2 in Kjellesvig-Waering 1986 for a clear comparison of segmentation in the two groups).

2. Reduction of the prosomal shield to form a carapace, coupled with development of more powerful prosomal appendages. Both of

these represented advances over the primitive xiphosuran and aglaspid pattern, and contributed to a more active predatory lifestyle.

3. The possession of ventral abdominal plates with doublures concealing gills (Kjellesvig-Waering 1986). Lobostern, meristostern, and holostern scorpions had abdominal plates and gills that were virtually identical to the types found in eurypterids (Kjellesvig-Waering 1986).

4. A constricted "metasoma" of five segments, with a terminal telson in some eurypterids (Carcinosomatidae and Mixopteridae) and in scorpions. There is no evidence that the telson contained venom glands in the eurypterids. Størmer (1963) regarded these features to be convergent, rather than indicative of close relationship.

5. Similar sensory macrochaetes (neither the Paleozoic scorpions nor the eurypterids possessed trichobothria, but they possessed similar setae).

6. Compound lateral eyes (among scorpions, known only in fossil taxa; the lateral eyes of modern scorpions represent vestiges of compound eyes).

7. Similar tarsal structure (consisting of a median posttarsus flanked by a pair of subdistal spines in both groups; in scorpions, the posttarsus was later shortened, and the spines became the tarsal claws).

Scorpions diverged from eurypterids in the following ways (Størmer 1963, Kjellesvig-Waering 1986):

1. The development of pectines as specialized appendages from the second mesosomal somite. These structures represent the single uniquely derived character (synapomorphy) shared by all scorpions. Kjellesvig-Waering (1986) speculates that if the earliest scorpions retained the egg-laying habits of their ancestors, the original function of the pectines may have been to scoop out a nest on the muddy aquatic bottom for the deposition of eggs. Størmer (1963) suggested that they could also have been used as swimming organs. Regardless of their function in the primitive scorpions, a suitable homology with any eurypterid structure has not been found.

2. The development of powerful chelate pedipalps. Enlarged pedipalps are found in most arachnids, but only in scorpions and pseudoscorpions are they truly chelate.

3. The development of paired poison glands and a syringelike aculeus on the telson. Although the telsons of some eurypterids resembled those of scorpions, it is thought that they did not possess venom glands.

This evidence overwhelmingly demonstrates phylogenetic relationship between scorpions and eurypterids, and some authors have specified in cladograms or phylogenetic trees that the two groups shared a common ancestor (Boudreaux 1979; Weygoldt and Paulus 1979a, b; Kjellesvig-Waering 1986). An important question to be addressed now is whether the sister group of the eurypterids was purely a scorpion ancestor (i.e., did the remaining Arachnida evolve from a separate merostome ancestor?), or was it an ancestor both to scorpions and to all other arachnids? Opinion is somewhat divided on this question, and some of the key arguments are summarized below.

Since the early 1800's, scorpions have been included in the Arachnida. Adaptations for terrestriality were the key features that linked scorpions and other arachnids (Petrunkevitch 1955; Boudreaux 1979; Weygoldt and Paulus 1979a, b), and the following characters are usually used to define the class Arachnida. The prosoma bears six pairs of appendages (a pair of chelicerae, a pair of pedipalps, and four pairs of legs); the opisthosoma is twelve-segmented and bears such structures as the genital opercula, pectines (scorpions only), book lungs, and spinnerets (spiders only). Arachnids developed book lungs housed in an internal chamber for respiration in air (augmented or replaced by tracheal or cuticular respiration in advanced forms). They also developed a preoral chamber for extraintestinal digestion, reduced the lateral compound eyes to groups of simple lenses, and developed endodermal Malpighian tubules for the excretion of guanine. If the class Arachnida is monophyletic, then one or more of these characters should be synapomorphic for the group (or else new synapomorphies should be identified). As mentioned above, segmentation and tagmatization of the arachnid body (i.e., in scorpions) can be matched readily to those features of the eurypterid body and cannot be considered synapomorphic for arachnids. The remaining characters, however, usually are taken as synapomorphic for arachnids (Boudreaux 1979; Weygoldt and Paulus 1979a, b). But it should be reemphasized here that among scorpions probably only the Neoscorpionina (orthosterns) possessed all these characters. This leads to the interpretation that the nonscorpion arachnids may have shared an immediate common ancestor with the Neoscorpionina, but not with the Branchioscorpionina. The difficulty with this hypothesis is that pulmonate Acari and Trigonotarbi are known from early Devonian deposits, but there are no known Devonian Neoscorpionina. Kjellesvig-Waering (1986) estimates that the development of book lungs in scorpions occurred "at least" by the late Devonian, but he does not speculate on the possible derivation of nonscorpion arachnids. This should provide the impetus for further phylogenetic studies.

Weygoldt and Paulus (1979a, b) presented a detailed cladistic study of chelicerate phylogeny. I have chosen to provide a simplified version of their cladogram showing only details relevant to this discussion (Fig. 3.35), the characters used to construct it (Table 3.3), and the resulting classification scheme they proposed (Table 3.4). According to their cladogram, which follows the earlier description here, the Chelicerata is a monophyletic group. The chelicerate ancestor diverged into the Aglaspida (considered by some to be subordinate to the Xiphosura) and the Euchelicerata; this basal part of the chelicerate cladogram is not shown in Fig. 3.35. The euchelicerate ancestor (shown as the ancestral branch in Fig. 3.35) also produced two groups: the Xiphosurida and a new taxon, the Metastomata. The Metastomata is a monophyletic group that includes the eurypterids, scorpions, and nonscorpion arachnids; its members are characterized by the reduction of the head shield and associated doublures, the possession of an unpaired ventral plate derived from the seventh somite (the metastoma in eurypterids and the sternum in others), and the possession of two simple median eyes.

The metastome ancestor produced two sister groups, the Eurypterida and the Arachnida (the traditional grouping). The Scorpiones (= Ctenophora) and the remaining Arachnida (= Lipoctena) are de-

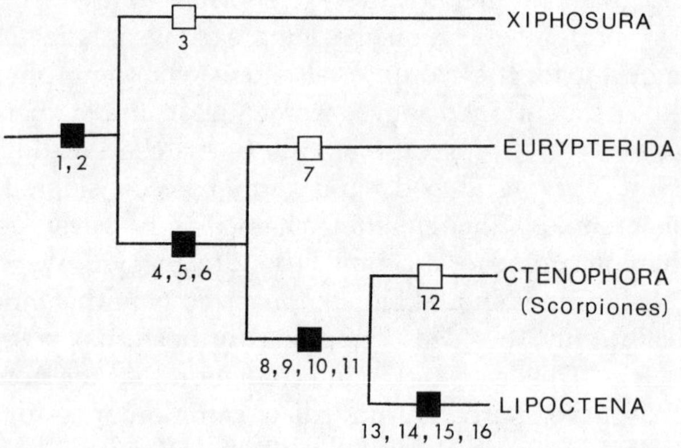

Fig. 3.35. Hypothesized relationships of the Xiphosura, Eurypterida, Ctenophora (scorpions), and Lipoctena (nonscorpion arachnids). This cladogram is a very abbreviated version of Weygoldt and Paulus's (1979b) original cladogram. The numbers refer to the characters presented in Table 3.3. ■, synapomorphy; □, autopomorphy.

TABLE 3.3
Characters used in constructing a cladogram for chelicerates

Character	Derived	Primitive
1	chelicerae three-segmented	chelicerae four-segmented
2	opisthosomal appendages paddle-shaped	opisthosomal appendages leglike
3	?	?
4	cephalothoracic doublures and pleurotergites reduced	cephalothoracic doublures and pleurotergites well developed
5	seventh segment with an unpaired, platelike appendage (i.e., a metastoma or sternum)	seventh segment with a paired appendage (i.e., chilaria)
6	median eyes two	?
7	?	?
8	digestion extraintestinal	digestion intraintestinal
9	Malpighian tubules endodermal	no Malpighian tubules
10	compound eyes reduced to five simple lenses	compound eyes present
11	slit sense organs present	no slit sense organs
12	pectines present	no pectines
13	rhabdomeres in lateral eyes in netlike arrangement	rhabdomeres in lateral eyes in starlike arrangement
14	spermatozoa coiled	spermatozoa filiform (not coiled)
15	slit sense organs lyriform	slit sense organs single
16	first legs sensory	—

SOURCE: Weygoldt & Paulus 1979a, b.

NOTE: The autapomorphies (uniquely derived characters) for the Xiphosura (#3) and Eurypterida (#7) are unknown. Cladogram appears in Figure 3.35; character numbers in this table and in the cladogram are identical.

rived from the arachnid ancestor (Fig. 3.35). The four characteristics listed above (development of book lungs, Malpighian tubules, a preoral chamber for extraintestinal digestion, and loss of compound eyes) were considered the synapomorphies linking the scorpions and other arachnids. Weygoldt and Paulus (1979a, b) accepted the Paleozoic scorpions as terrestrial pulmonates; consequently, they believed that the scorpion–arachnid ancestor began the single chelicerate invasion of land. This position was defended on the basis of available information (see Weygoldt and Paulus 1979a, pp. 88, 89). Unfortunately, the recent

TABLE 3.4
A cladistic classification of the Chelicerata Heymons, 1901

Aglaspida Walcott, 1911
Euchelicerata Weygoldt & Paulus, 1979
Xiphosurida Latreille, 1802
Metastomata Weygoldt & Paulus, 1979
Eurypterida Burmeister, 1843
Arachnida Lamarck, 1801
Ctenophora Pocock, 1893
Scorpiones Latreille, 1817
Lipoctena Pocock, 1893

NOTE: The table is based on the work of Weygoldt and Paulus (1979a, b). As in the cladogram in Figure 3.35, only the relevant aspects of the classification are given; that is, the Lipoctena are not subdivided as in the original.

monograph by Kjellesvig-Waering (1986), in which an aquatic existence for Paleozoic scorpions seems irrevocably established, was unpublished at the time. Other arguments for and against terrestriality in the Paleozoic scorpions had appeared frequently prior to its publication (e.g., Pocock 1901, 1911; Petrunkevitch 1953, 1955; Wills 1959, 1960; Størmer 1963, 1970; Kjellesvig-Waering 1966a, 1969). It will be interesting to see the future effects of Kjellesvig-Waering's monograph on the various cladograms of chelicerate phylogeny and classifications thus far generated.

Boudreaux (1979) also studied chelicerate phylogeny with the cladistic method, although he did not consider scorpions separately from the other arachnids. His cladogram is congruent with that of Weygoldt and Paulus (1979b), but he added a few more characteristics to justify certain branches of his cladogram (e.g., the modification of the fourth pair of legs into swimming paddles was considered synapomorphic for the eurypterids). His monophyletic grouping of the Eurypterida and Arachnida was named the Cryptopneustida (= Metastomata of Weygoldt and Paulus 1979b), but otherwise his groupings were the same. Boudreaux also thought all scorpions were terrestrial, and the synapomorphies he gives for the Arachnida are based on that opinion.

On the other hand, some have suggested that scorpions and nonscorpion arachnids may have had a separate ancestry. Anderson (1973), in his detailed study of arthropod embryology, and Yoshikura (1975) both demonstrated that scorpions are quite different from other arachnids in many aspects of development. The development of viviparity is virtually unique to scorpions, occurring elsewhere only in a few Acari (in which it must represent convergence; Yoshikura 1975). Anderson (1973) pointed out that despite their specialization for viviparity scorpions retain a hypothesized primitive pattern of early development similar to that of the Xiphosura. In scorpions, as in xiphosurans, proliferation of the germ band proceeds in an anterior-to-posterior direction from an embryonic primordium consisting of a cephalic lobe, a pedipalpal segment, and a growth zone; the growth zone produces the four ambulatory segments and the opisthosoma (see Chapter 4 for further details of scorpion development). In other arachnids, the prosoma (including the ambulatory segments) develops directly from the blastoderm, and the growth zone produces only the opisthosoma. Taking this and other embryological differences into consideration, Anderson concluded that either scorpions must have diverged very early from the other arachnids, or perhaps the two groups may both have evolved from merostomes independently. The conclusions of Anderson (1973) have been questioned by Weygoldt (1979) on the grounds that they are based on patterns of overall similarity (including primitive patterns of development), rather than on the utilization of synapomorphies to indicate relationship.

Bergström (1979) agreed with Anderson that scorpions and other arachnids probably had separate origins, and suggested that scorpions were merostomes that exhibited the terrestrial grade of existence. The possession of a "keeled tail spine" (i.e., the telson) was cited as the synapomorphy linking the scorpions to the Merostomata. He hypothesized that the Arachnida (minus the Scorpiones) arose from an unknown nonmerostome ancestor. This, of course, implies that there were several independent colonizations of land by chelicerate arthropods.

Van der Hammen (1977) likewise believed that chelicerate groups were independently derived and colonized land at different times. He studied the leg joints of the various chelicerates and found seven distinct types of legs. Because such diversity in leg joints is unknown in other arthropod lineages, Van der Hammen suggested that the seven types of legs arose independently of one another. Combining his observations on leg structure with other characters (e.g., embryology, methods of digestion, respiration, and sperm transfer), he proposed that seven separate classes of chelicerates should be recognized, instead of the traditional Merostomata and Arachnida. (His analysis did not include the Pantopoda and lumped the Aglaspida with the Xiphosura.) The Scorpionidea represented a monotypic class in this assemblage, separate from both merostomes (horseshoe crabs and eurypterids) and nonscorpion arachnids.

It is obvious that a considerable amount of work remains to be done in this area before a definitive assessment of chelicerate phylogeny can be reached. In particular, the relationship between the nonscorpion Arachnida and the various infraorders of scorpions should be reinvestigated in light of Kjellesvig-Waering's (1986) most recent discoveries.

ACKNOWLEDGMENTS

I wish to express my gratitude to the following individuals who greatly facilitated the writing of this chapter: Drs. Oscar F. Francke, Herbert W. Levi, Wilson R. Lourenço, Gary A. Polis, and an anonymous referee, for reviewing various drafts of the manuscript; Mr. William Carver, Editor of Stanford University Press, for comments on style; Drs. Edward S. Ross (California Academy of Sciences), Robert W. Mitchell (Texas Tech University), and Gary A. Polis, for kindly providing most of the photographs used in this chapter; and the Paleontological Society of America, for granting permission to use several photographs published in the *Journal of Paleontology*.

Drs. L. E. Acosta, Matt E. Braunwalder, Victor Fet, Oscar F. Francke, Wilson R. Lourenço, and Sherman A. Minton kindly provided specimens, through gifts and exchanges, from which many of the drawings were made. The taxonomic keys would hardly have been possible without these specimens. I am grateful for the many suggestions and comments of Dr. Lourenço and of Scott A. Stockwell (University of California, Berkeley), both of whom reviewed the keys to genera at my request.

Dr. Lourenço allowed access to some manuscripts and information that were unpublished at the time much of this chapter was written. Dr. Kenneth E. Caster and his wife, Anne, gave me the hospitality of their home to examine Erik N. Kjellesvig-Waering's manuscript on fossil scorpions while it was still unpublished.

To everyone above, I wish to express a heartfelt thanks. Finally, a special thanks is in order to my wife, Shari, for her patience and encouragement throughout the course of this work.

4

Life History

GARY A. POLIS AND W. DAVID SISSOM

Scorpions are rather unique among terrestrial arthropods in many of their life-history characteristics. Following a ritualized and complex courtship, the male deposits a spermatophore, from which the female receives a sperm packet. Once fertilization is accomplished, embryos undergo a highly specialized viviparous development that lasts from several months to well over a year, depending upon species. Once born, the young climb onto the mother's back to continue development and molt for the first time. The young then disperse to assume an independent existence.

Juveniles feed, increase in weight, and molt an average of six times before reaching maturity. The period before maturity is extraordinarily long, lasting from 6 to 83 months. In total, scorpions may live up to 25 years. During this time they reproduce repeatedly, having broods of from one to over a hundred young.

The details of scorpion life history are highly variable, even at the species level. The purpose of this chapter is to discuss each phase in the life of various species of scorpion. Comparisons between taxa and with other arachnids are given. In general, there is a high degree of specialization in scorpions, a group considered to be among the most primitive arthropods.

Courtship and Mating Behaviors

Courtship and sperm transfer in scorpions is a complex process involving several characteristic behaviors. Basically, the male leads the female in a classical *promenade à deux* (Fig. 4.1) until a suitable location is found for depositing the spermatophore. The male assists the female in positioning her genital aperture over the spermatophore so that the female can take up the sperm. The partners separate after sperm transfer.

The promenade, but not the sperm transfer, was first described by Maccary in 1810 and then again by Fabre (1923) in his famous *The Life of the Scorpion*. It was not until the mid-1950's, however, that a series of researchers independently discovered that a spermatophore was the vehicle involved in sperm transfer (Angermann 1955, 1957; Bücherl 1955/56; Zolessi 1956; Alexander 1956, 1957, 1959a; Shulov 1958; Shulov and Amitai 1958, 1960).

Courtship and mating are described for 29 species in six families, not including the Chaerilidae and Iuridae (Table 4.1). A number of basic behaviors common to all families are listed in the table and described below (descriptions from Alexander 1959a, Rosin and Shulov 1963, Garnier and Stockmann 1972, Polis and Farley 1979a), along with the percentage of all species and the number of families exhibiting each behavior. Speculations on the possible function of these behaviors are also included.

Finally, pheromones may be important in sex and species recognition. Alexander (1959a) cited evidence that the males of *Opistophthalmus latimanus* positively respond to the smell of responsive females. She suggested that the female scent could be an initial releaser of courtship behavior (*O. latimanus* also uses juddering—but this behavior is important only at short distances, whereas pheromones would function at greater distances). Alexander also noted that *Parabuthus planicauda* has the ability to smell and that sex recognition may depend solely on the distinctive odors emitted by females and recognized by males. This may also be the case for *Isometrus maculatus* (Probst 1972). Bücherl (1955/56) argued that a female pheromone in two species of *Tityus* allows conspecific males to discriminate fertilized from unfertilized females. S. Williams (pers. comm. 1982) has noted that the air is thick with a strong, pungent odor during the discrete periods of mass mating by *Didymocentrus* spp. in southern Baja California. He speculated that this odor is a mating pheromone produced by receptive females. *Paruroctonus* males may also be attracted to and/or recognize receptive conspecific females via pheromones (S. McCormick and G. Polis, pers. obs.). For example, receptive females of *P. luteolus* are often surrounded (at a 1–8-m distance) by clusters of two to eight mature males. Because this species is relatively uncommon, such localized aggregations of males are significantly nonrandom, especially considering the preponderance of females. Long-distance recognition and pair formation via pheromones would be particularly important among low-density species. (Very recent unpublished research by P. Brownell [pers. comm. 1987] indicates that pheromones are produced by glands associated with the anus, at the end of the metasoma. The pectines of both sexes, particularly of males, contain numerous chemoreceptors that are extremely sensitive to such pheromones.)

INITIATION. In scorpions, as in other arachnids, there is some variability among species in how courtship is initiated. In 73 percent of the species, courtship began when the male approached the female; the female initiated courtship in 11 percent of the species, and for 16 percent of the species it appears that either sex will initiate courtship. In 8 percent of the species, initiation appears to be accidental. Several cues were suggested to function in sexual recognition (Alexander 1959a, Polis and Farley 1979a). Alexander suggested that initiation for some species (e.g., *Tityus trinitatis*) is stimulated by male recognition of characteristics of the female's anatomy after physical contact is established. She noted that *Tityus trinitatis* is markedly sexually dimorphic in both the size and coloration of the metasoma and that during the initial contact the male preferentially grasps the female's tail. This suggests that sexual differences in the metasoma are important in sex recognition. It is unknown if sexual differences in anatomy are used by other species for mate recognition.

Vibrations that travel through the substrate may serve a similar function. Rosin and Shulov (1963) noted that female *Nebo* definitely sense the pre-promenade juddering of the male from a distance of several centimeters. In *Opistophthalmus latimanus* the primary releaser of courtship behavior appears to be a change in female behavior in response to juddering by the approaching male (Alexander 1957). In *Paruroctonus mesaensis* and probably most other scorpions, animals are able to receive a great deal of information from substrate vibrations (Brownell 1977a). Polis and Farley (1979a) showed that receptive *P. mesaensis* females sense the substrate vibrations of moving males (see also Armas 1986). The female then initiates courtship behavior with a series of "mating attacks," consisting of a series of rapid physical encounters followed by the retreat of both sexes.

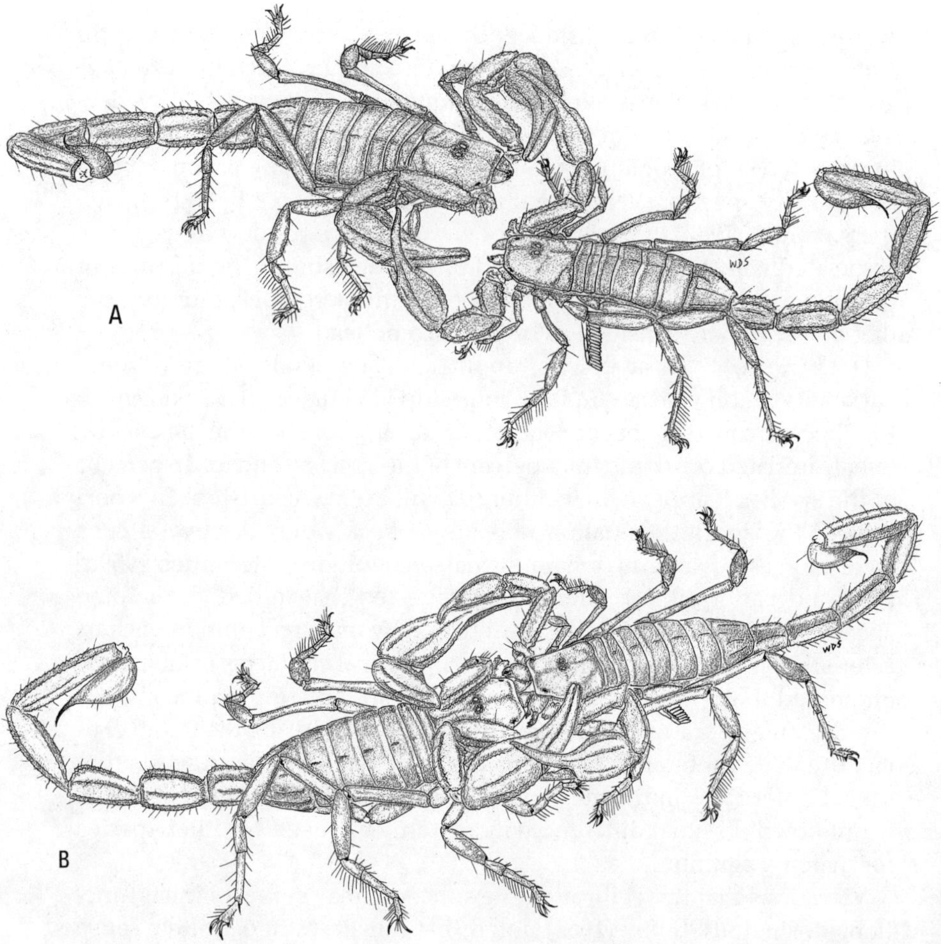

Fig. 4.1. Mating behaviors. *A, Promenade à deux* of *Paruroctonus mesaensis*. The smaller male (right) is grasping the female's pedipalp chelae. The male's pectines wave laterally at this time, and are pictured as extended. *B,* Cheliceral massage in *P. mesaensis*. The male (right) grasps and kneads the female's chelicerae with his own. *C,* close-up of cheliceral massage in *P. luteolus*. Note the sexual dimorphism exhibited by the larger pedipalp of the male (left). *D,* Sperm transfer in *P. mesaensis*. The male (right) has deposited a spermatophore on the stick. The female is arched over the spermatophore; her genital opercula are spread to receive the sperm droplet on the apex of the spermatophore.

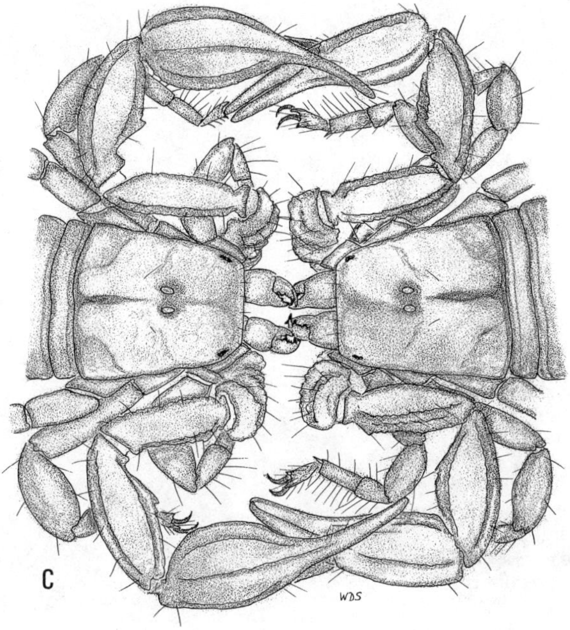

C

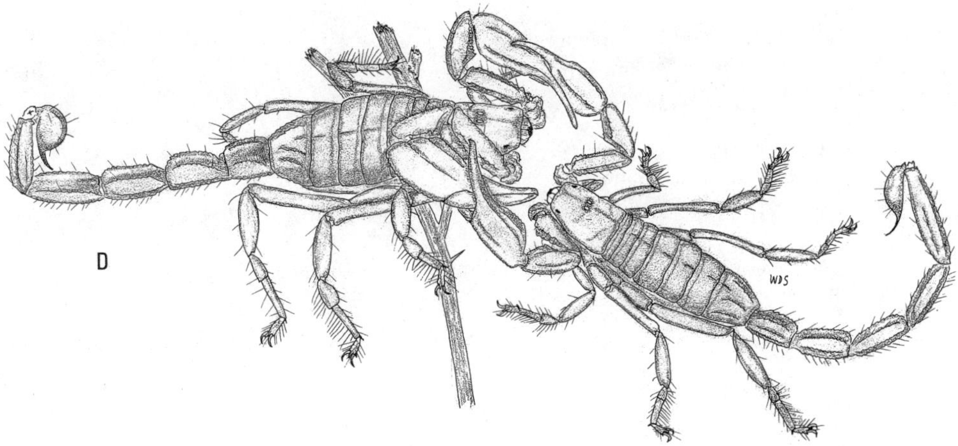

D

TABLE 4.1

Scorpion courtship and mating behaviors

Taxon	Male movement	Initiation	Male juddering	Clubbing	Promenade with: Pedipalp grip	Promenade with: Chelicera grip	Sexual sting	Cheliceral massage
BOTHRIURIDAE								
1. *Bothriurus asper* Pocock		♂	−		+			+
2. *B. bonariensis* (C. L. Koch)		♂	−		+	−		
3. *B. flavidus* Kraepelin					+			
4. *Urophonius brachycentrus* (Thorell)		♂	−		+	+	+	+
BUTHIDAE								
5. *Androctonus australis* (Linné)		♂			+			+
6. *Buthus occitanus* (Amoreux)		♂	+	+	+	−		+
7. *Centruroides insulanus* Thorell		♂			+	−		+
8. *C. vittatus* (Say)		♀ A	+	+	+			+
9. *Hottentotta judaica* (Simon)		♂ ♀	+	+	+	−		+
10. *Isometrus maculatus* (DeGeer)		♀ A	+	+	+	−		+
11. *Leiurus quinquestriatus* (Hemprich & Ehrenberg)		♂ ♀	+	+	+	−		+
12. *Parabuthus planicauda* Pocock		♂	+	+	+	−		+
13. *Tityus bahiensis* (Perty)		♂	+		+	−		
14. *T. fasciolatus* Pessôa		♂	+		+			+
15. *T. trinitatis* Pocock		♂	+	+	+	−		+
16. *T. trivittatus* Kraepelin		♂	+		+			+
CHACTIDAE								
17. *Euscorpius carpathicus* (Linné)		♂		+	+		+	+
18. *E. flavicaudis* (DeGeer)		♂		+	+		+	+
19. *E. italicus* (Herbst)		♂		+	+		+	+
20. *Megacormus gertschi* Diaz		♂	+		+		+	+
DIPLOCENTRIDAE								
21. *Nebo hierichonticus* (Simon)		♂ ♀	+		+	+		+
SCORPIONIDAE								
22. *Heterometrus scaber* (Thorell)		♂			+			+
23. *Opistophthalmus latimanus* (C. L. Koch)		♂	+		+	+		+
24. *Pandinus imperator* (C. L. Koch)		♂ ♀	+		+		+	+
25. *Urodacus manicatus* (Thorell)	+				+			+
VAEJOVIDAE							+	
26. *Paruroctonus luteolus* (Gertsch & Soleglad)	+		+	+	+	−		+
27. *P. mesaensis* Stahnke	+	♀	+	+	+	−		+
28. *Vaejovis carolinianus* (Beauvois)	+	♂		+	+	−	+	+
29. *Vejovoidus longiunguis* (Williams)	+		+		+			+

SOURCES: 1, Matthiesen 1968. 2, Zolessi 1956, Varela 1961. 3, Abalos & Hominal 1974. 4, Maury 1968a. 5, Aube Thomay 1974. 6, Fabre 1923, Auber 1963a. 7, Baerg 1961. 8, McAlister 1965. 9, Shulov & Amitai 1958. 10, Prob 1972. 11, Thornton 1956, Shulov & Amitai 1958, Abushama 1968. 12, 15, Alexander 1959b. 13, 16, Bücherl 1956. 1 Lourenço 1978, 1979b. 17–19, Angermann 1955, 1957. 20, Francke 1979b. 21, Shulov & Amitai 1958, Rosin & Shul

Taxon	Male pecten movement	Male sand scraping	Female headstand	Postmating escape	Female swaying	Spermatophore consumed	Mate cannibalism
BOTHRIURIDAE							
1. *Bothriurus asper* Pocock	+		+	−			
2. *B. bonariensis* (C. L. Koch)							+
3. *B. flavidus* Kraepelin			+	♂		♂	
4. *Urophonius brachycentrus* (Thorell)	+		+	♂		♀	+
BUTHIDAE							
5. *Androctonus australis* (Linné)	+						
6. *Buthus occitanus* (Amoreux)				♂			+
7. *Centruroides insulanus* Thorell	−	+					
8. *C. vittatus* (Say)	+			♀			
9. *Hottentotta judaica* (Simon)	+	+		♀			
10. *Isometrus maculatus* (DeGeer)	+	−				−	+
11. *Leiurus quinquestriatus* (Hemprich & Ehrenberg)	+	+		♂		−	+
12. *Parabuthus planicauda* Pocock	+	+		♀	+	−	
13. *Tityus bahiensis* (Perty)	+	+		♂		♂	−
14. *T. fasciolatus* Pessôa	+					♂	+
15. *T. trinitatis* Pocock	+	+		♀	+	−	
16. *T. trivittatus* Kraepelin	+	+		♂		♂	−
CHACTIDAE							
17. *Euscorpius carpathicus* (Linné)	+			−		♀	
18. *E. flavicaudis* (DeGeer)	+			−		♀	
19. *E. italicus* (Herbst)	+			−		♀	
20. *Megacormus gertschi* Diaz						♀	
DIPLOCENTRIDAE							
21. *Nebo hierichonticus* (Simon)	+	+		♀		♂	
SCORPIONIDAE							
22. *Heterometrus scaber* (Thorell)	+						+
23. *Opistophthalmus latimanus* (C. L. Koch)	+	+		♂ ♀	+	−	−
24. *Pandinus imperator* (C. L. Koch)	+				+	−	+
25. *Urodacus manicatus* (Thorell)	+						
VAEJOVIDAE							
26. *Paruroctonus luteolus* (Gertsch & Soleglad)	+	+	+	♂	+	−	+
27. *P. mesaensis* Stahnke	+	+	+	♂	+	−	+
28. *Vaejovis carolinianus* (Beauvois)	+	+					
29. *Vejovoidus longiunguis* (Williams)	+		+	♂		−	+

1963. 22, Mathew 1957. 23, Alexander 1956, 1957, 1959a. 24, Garnier & Stockmann 1972. 25, Southcott 1955, Smith 1966. 26, Polis & Farley 1979b, unpubl. data. 27, Polis & Farley 1979b. 28, Benton 1973. 29, Polis, unpubl. data. Vaejovidae: Francke 1979a.

NOTE: +, behavior observed; −, behavior reported to be absent; ♂, behavior by male; ♀, behavior by female; *A*, accidental initiation of courtship. *Paruroctonus luteolus* was reported as *P. borregoensis* Williams in Polis & Farley 1979a.

MALE JUDDERING. This behavior consists of a series of rapid rocking or shaking movements of the male's body back and forth. During this quivering the pectines are usually spread out and the legs firmly planted on the substrate. Sporadic juddering occurs throughout the initiation and promenade, and during and after spermatophore deposition. It functions either as a highly ritualized sex or species-recognition behavior, as a releaser of mating behavior so that the female is stimulated to cooperate, or as a simple byproduct of intense male sexual excitation (59 percent, 5 families).

CLUBBING. Either scorpion strikes its partner with the metasoma while the sting is tucked away. Clubbing may function as an inhibited form of aggression between partners (45 percent, 4 families).

PROMENADE À DEUX. The promenade (Fig. 4.1) is the mating dance, during which the male grasps the female and then leads her as the pair moves together. The male may grasp the female in a pedipalp-to-pedipalp grip (100 percent; Fig. 4.1*A*), occasionally also leading her in a chelicera-to-chelicera grip (11 percent, 3 families). The promenade provides the mating pair with the mobility and coordination necessary to find a suitable substrate on which to deposit the spermatophore (Fig. 4.1*D*). The duration of courtship seems to be determined primarily by the length of time it takes to find a solid surface for the spermatophore (Alexander 1957, 1959a; Shulov and Amitai 1958; Rosin and Shulov 1963; Polis and Farley 1979a). The promenade may last from 5 minutes to over 2 days, although long promenades probably reflect a researcher's error in not providing a suitable substrate in laboratory chambers. Most courtships and matings last 30 to 60 minutes. In the field, promenades of *Paruroctonus mesaensis* were observed to range from 5 to 35 minutes; during this time the pair may move from 3 to 25 meters or more.

SEXUAL STING. The male uses its sting to puncture the female's body, usually at the membrane adjacent to the tibial joint of the pedipalp. The sting remains within the female for from 3 to 20 minutes or more. It is difficult to determine if envenomation occurs. This behavior occurs early in the promenade and then sporadically afterward. If envenomation occurs, the sexual sting may drug the female and thus function to subdue her normal aggressive behavior (> 21 percent, 4 families; Francke 1979b reported sexual stinging by several unnamed species of Vaejovidae).

CHELICERAL MASSAGE. The male grasps and kneads with his chelicerae the female's chelicerae (Fig. 4.1*B*, *C*), the prosomal edge, and/or the articulation of the pedipalps ("kissing" of Southcott 1955). Cheliceral massage is usually observed during the promenade when the female ceases to move and during spermatophore deposition and sperm uptake. In these situations the female appears to become more docile and cooperative, suggesting that this behavior may function to suppress the aggressive tendencies of the female (86 percent, 6 families).

MALE PECTEN MOVEMENT. The pectines are spread wide and sporadically swept across the substrate. Pecten movement occurs during the entire promenade but intensifies upon encounter with a substrate that is proper for spermatophore deposition. Thus, the behavior most likely ensures that the spermatophore will be placed on a surface where it will adhere and remain upright during sperm transfer (86 percent, 6 families).

MALE SAND SCRAPING. The male's legs move rapidly and sporadically over the substrate. Dirt is often moved, and the surface is occasionally smoothed. Alexander (1959a) observed that scraping never occurs on hard or smooth surfaces or on soil devoid of loose sand particles. She speculated that scraping clears a space suitable for spermatophore deposition. Occasionally, the male digs into the soil, producing grooves rather than a smooth area. This type of scraping occurs in vaejovids and may not function to prepare the surface for the spermatophore (36 percent, 4 families).

SPERM TRANSFER. Once a suitable substrate is encountered, the male lowers its mesosoma until its genital aperture touches the ground. A spermatophore is then extruded until a sticky basal plate just contacts the surface. The entire spermatophore leaves the aperture in 5 to 6 seconds (*Leiurus* or *Hottentotta*) to 6 minutes (*Nebo*). After extrusion, the spermatophore stands nearly vertical on the surface. The male then repeatedly jerks backward, eventually succeeding in pulling the female into a proper position over the spermatophore. The female's genital opercula spread as she lowers herself onto the spermatophore. The spermatophore bends slightly under the female's weight, thereby opening its valves and allowing sperm to enter the female's reproductive tract. The opercula may close on the spermatophore, apparently exerting a suction that facilitates sperm uptake. The female may be on the spermatophore for a few seconds (*Leiurus, Paruroctonus mesaensis*) to 5 to 6 minutes (*Opistophthalmus*). The method by which the spermatophore opens to release sperm is discussed by Alexander (1957, 1959a) and by Rosin and Shulov (1963).

FEMALE HEADSTAND. This behavior occurs while the female is over the spermatophore. The female stilts her aperture over the apex of the spermatophore, apparently achieving insemination (21 percent, 2 families).

POSTMATING ESCAPE. This behavior consists of a rapid and usually violent disengagement by one mating partner immediately after sperm uptake. The escaping animal may use its metasoma to club its partner or may repeatedly probe its partner with its sting. The escaping animal often runs a distance (5–50+ cm) from its partner. Because scorpions are notoriously cannibalistic (see below), it is likely that escape behavior decreases the incidence of cannibalism by removing one animal (the smaller sex?) from the vicinity of the other. Escape behavior is performed primarily by males (36 percent, 4 families) but also by females (21 percent, 3 families).

FEMALE SWAYING. The female raises her body above the ground while slowly moving side to side or backward and forward on her immobile legs. Swaying occurs immediately after the disengagement following sperm uptake. It probably serves a mechanical function in aiding the sperm's travel farther into the female's reproductive tract (21 percent, 3 families).

SPERMATOPHORE CONSUMPTION. The spermatophore is eaten after sperm uptake. Both females (50 percent of all spermatophores eaten) and males (50 percent) exhibit this behavior. This probably represents a nutritional or energetic gain for one of the partners (34 percent, 4 families).

MATE CANNIBALISM. After sperm uptake the female may capture and eat the smaller male (39 percent, 4 families; Fig. 4.2).

An inspection of Table 4.1 reveals much similarity in courtship and mating behavior among all scorpions. All behaviors are observed repeatedly in species that are only distantly related. This implies that reproductive behavior appeared at a point in scorpion evolution before the various modern families diverged. Garnier and Stockman (1972) suggested that buthid behavior is less complex than that of other families (see also Matthiesen 1968). This suggestion is consistent with the fact that buthids separated very early from all other living scorpions.

After initiation, the male is the active partner in the courtship of all species so far observed. He leads the promenade, chooses the substrate for spermatophore deposition, and initiates all major behaviors involved in sperm transfer. Many of the male's behaviors not directly

Fig. 4.2. Mate cannibalism terminating the mating of *Paruroctonus mesaensis*. The larger female (left) has already envenomated the male and is beginning to feed. The spermatophore was deposited on the stick.

involved with sperm transfer apparently evolved to minimize the probability that the male will be cannibalized by his mate (Alexander 1957, 1959a; Polis and Farley 1979a). These behaviors include those that allow females to recognize males (juddering and initiation), those that decrease female aggression (clubbing, cheliceral massage, sexual sting), and those that enable the male to leave the female successfully after mating (postmating escape). The high incidence of mate cannibalism (39 percent of all reported species) argues that there is substantial selective pressure favoring the evolution of any behavior by the male that would allow him to avoid being eaten before sperm transfer. (Under some conditions, selection favors males that allow themselves to be eaten by their mates after insemination: Polis 1980b.) For *Paruroctonus mesaensis*, males were eaten by their mates after about 10 percent of all matings. The origin and function of male behaviors that inhibit female aggression are discussed at length by Alexander (1959a) and by Polis and Farley (1979a).

Similarities exist in courtship behavior between scorpions and other groups of arachnids (Kästner 1968, Weygoldt 1969, Savory 1977, Foelix 1982). Except for Opiliones and possibly some Acari, insemination is indirect, and not achieved by true copulation. Consequently, courtship is usually complex, with the male as the initiator and active partner. Spermatophore transfer is used by amblypygids, uropygids, schizomids, pseudoscorpions, and some Acari. Further, the male is often at risk from the predaceous female. Mate cannibalism by females is widespread among the arachnids. In general, courtship behavior in arachnids has been interpreted as an adaptation both to produce the cooperation and coordination necessary to ensure successful sperm transfer and to decrease the probability that the male will be cannibalized by the female (Polis 1980b, 1981).

Postmating Biology

In some species, fertilized females can be identified by the presence of a vaginal plug (spermatocleutrum). Pavlovsky (1924b) used the term spermatocleutrum for the whitish, shrunken mass found in the vaginal opening of fertilized females. It is occasionally found in bothriurids (Varela 1961), buthids (Bücherl 1956, Shulov and Amitai 1958, Probst 1972), scorpionids (Smith 1966, Shorthouse 1971), and vaejovids (Fox 1975). The origin of the plug is uncertain (Shulov and Amitai 1958), but it is apparently secreted by the female sometime after mating, and it is often lost prior to birth. There is no consensus on its function; whether its presence interferes with additional inseminations is unclear (Shulov and Amitai 1958).

Males of most if not all species are capable of mating more than once. There is considerable evidence that newly mated males can produce new spermatophores and remate in a very short time (Shulov and Amitai 1958; Matthiesen 1960, 1968; Maury 1969; Probst 1972; Koch 1977; G. Polis, pers. obs.). For example, one *Tityus bahiensis* male mated

11 times in 102 days; 6 days was the minimum time between successive matings (Matthiesen 1968). Some buthid females also mate repeatedly (Baerg 1961, Matthiesen 1968, Probst 1972, Armas and Contreras 1981). Armas (1987) presented evidence to suggest that some species of diplocentrids and buthids (especially *Rhopalurus*) must mate again after each parturition in order to produce another brood. Females of *Centruroides, Tityus,* and *Isometrus* were repeatedly observed to mate while carrying newly born first instars. No reports address the question of multiple matings by females from other families.

Hybridization and Reproductive Barriers

Mistakes occasionally occur in the identification of mates. There are reports of seven pairs of different species engaging in promenade and courtship behavior (Auber 1963a, Matthiesen 1968, Probst 1972, Le Pape and Goyffon 1975), including four pairs of species from different genera. However, spermatophore deposition and sperm uptake occurred only between congeneric species of *Euscorpius* (Auber 1963a) and *Androctonus* (Le Pape and Goyffon 1975). In one case (*A. australis* ♂ × *A. mauretanicus* ♀), 42 hybrid young were successfully produced.

Barriers to interspecific hybridization in scorpions were discussed by Le Pape and Goyffon (1975). The general classification of these barriers (prezygotic vs. postzygotic) is the same as in other groups of animals. Important prezygotic barriers for scorpions probably include species-specific pheromones, anatomical characteristics, courtship behavior, and differences in seasonal phenology, range, and habitat. For example, *A. australis* and *A. mauretanicus* do not hybridize in nature because they are geographically separated. In several of the reported attempts of interspecific courtship, the female became uncooperative and ended the courtship before sperm transfer (Matthiesen 1968, Probst 1972). Postzygotic barriers may include mortality, sterility, or decreased fitness of the hybrid offspring.

Prenatal Development and Birth

In previous chapters it was noted that two types of development occur in scorpions: apoikogenic and katoikogenic. The ova of apoikogenic scorpions have variable amounts of yolk; the ova of katoikogenic scorpions are yolkless, and the embryos are nourished via a specialized oral feeding apparatus. Traditionally, apoikogenic scorpions were regarded as ovoviviparous and katoikogenic scorpions as viviparous. In ovoviviparity, ova develop inside the mother, but without receiving any nutrients from her during development; all nourishment is derived from yolk reserves. Preliminary evidence has demonstrated that apoikogenic embryos receive part of their nourishment from the mother during development, possibly by diffusion of nutrients through the embryonic membranes (Francke 1982d). This observation led Francke to suggest that both types of scorpion development should be regarded as viviparous. Toolson (1985) used radioactive tracers to demonstrate that apoikogenic *Centruroides* embryos take up water and amino acids from the tissue of their mother. These embryos take up leucine rapidly and are adapted to utilize nutrients provided by the mother throughout development. Toolson concluded that *Centruroides* (and probably other apoikogenic scorpions) is not qualitatively different from the katoikogenic scorpions, despite the obvious morphological specialization for embryonic nutrition in the latter.

Details of the embryological development of both types are considered separately below, along with other aspects of early development.

Apoikogenic Scorpions

The oocytes of apoikogenic scorpions develop in ovarian follicles, which are attached directly to the ovariuterus (Fig. 4.3*A*). These oocytes are variable in size and yolk content. In the Iuridae and Vaejovidae, the ova are small and yolkless: the ova of *Iurus dufoureius*, for example, measure 0.15 × 0.10 mm (Laurie 1896a, b). The other apoikogenic families have large ova rich in yolk: the ova of *Buthus* sp. measure 0.40 × 0.20 mm (Werner 1934); of *Centruroides vittatus*, 0.40–0.63 mm (Francke 1982d); of *Lychas tricarinatus* (which, interestingly, has alecithal ova), around 0.03 mm (Mathew 1960); and of *Euscorpius italicus*, between 0.20 × 0.40 and 1.20 × 0.83 mm, depending on the report (Laurie 1890, Brauer 1895, Werner 1934). Ova of apoikogenic scorpions are either alecithal, isolecithal, or telolecithal (Yoshikura 1975).

The following account of the embryology of *Euscorpius* spp., which are typical apoikogenic scorpions, is derived mainly from Laurie (1890) and Anderson (1973). In *E. italicus*, the zygote nucleus lies in the yolk-free polar cap. Cleavage is discoidal, with the first division producing two flattened, equal cells, and subsequent divisions producing a germ disc, or blastoderm (Fig. 4.3*B, C*), which spreads over the yolk (Anderson

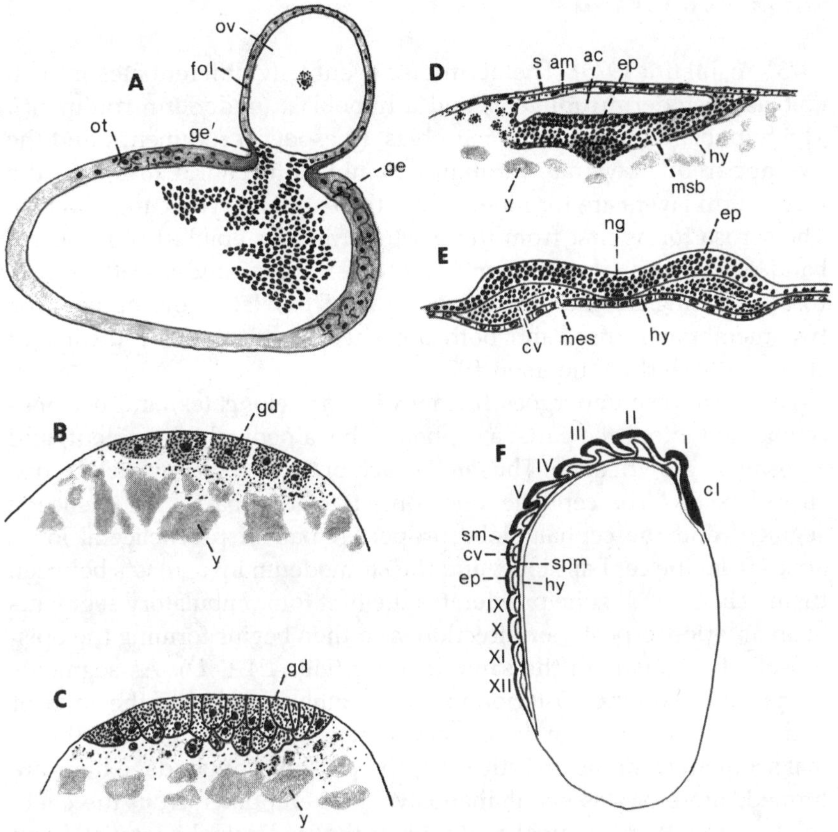

Fig. 4.3. Embryology of *Euscorpius italicus* (Chactidae), a representative apoikogenic scorpion: *A*, longitudinal section through ovum attached directly to ovariuterine tubule; *B*, section through one-layered blastoderm; *C*, section through later stage, showing proliferation of cells at one pole of the ovum; *D*, transverse section through three-somite embryo (posterior end), showing formation of three germ layers (hypoblast, epiblast, mesoblast) and extraembryonic membranes; *E*, transverse section through the anterior somite of a seven-somite embryo, showing initial formation of coelomic sacs inside mesodermal somite; *F*, longitudinal section of embryo at the twelve-somite stage. *ac*, amniotic cavity; *am*, amnion; *cl*, cephalic lobe; *cv*, coelomic vesicle (sac); *ep*, epiblast; *fol*, follicle; *gd*, germ disc (blastoderm); *ge*, germinative epithelium; *ge′*, yolk-producing cells of germinative epithelium; *hy*, hypoblast; *mes*, mesoderm; *msb*, mesoblast; *ng*, neural groove; *ot*, ovariuterine tubule; *ov*, ovum; *s*, serosa; *sm*, somatic mesoderm; *spm*, splanchnic mesoderm; *y*, yolk; *I–XII*, somites, numbered anterior to posterior (*I*, cheliceral somite; *II*, pedipalpal somite; etc.). (After Laurie 1890)

1973, Yoshikura 1975). The germ disc eventually differentiates into an epiblast (ectoderm rudiment) and a hypoblast (endoderm rudiment). The hypoblast produces the mesoblast (mesoderm rudiment), and the two apparently separate through delamination (Laurie 1890). As the three germ layers are forming, so are the extraembryonic membranes. The serosa forms first from the blastoderm. The epiblast of the germ band produces a second membrane, the amnion, underneath the serosa, enclosing an amniotic cavity (Fig. 4.3*D*). In later development the two membranes fuse, and both are shed simultaneously during or shortly after birth (Anderson 1973).

The germ disc undergoes further division, elongates, and becomes segmented into three parts: a cephalic lobe, a pedipalpal segment, and a posterior growth zone. The fate of each of these segments will be discussed below. The cephalic lobe consists of two parts, the cheliceral segment and the cephalic lobe proper. A pair of precheliceral lobes arises from the cephalic lobe, and the stomodeum invaginates between them. The growth zone proliferates the first four ambulatory segments in an anterior-to-posterior direction, and then begins forming the opisthosomal segments in the same manner (Fig. 4.4*A–D*). As segments are produced on the posterior end of the embryo, those of the anterior end become more elaborate. Upon proliferation of the third opisthosomal segment from the growth zone, the prosomal appendage buds are formed, more or less simultaneously: cheliceral buds (from the cheliceral segment), pedipalpal buds (from the pedipalpal segment), and the first four ambulatory limb buds (from somites III–VI).

Proliferation of the growth zone produces the remaining segments (note the progression in Fig. 4.4). In all, seven mesosomal segments and seven pairs of mesosomal limb buds are formed. The first pair is resorbed, the second pair forms the genital opercula, the third forms the pectines, and the fourth to seventh pairs merge into the body surface. The merging of the four posterior limb buds into the body is accompanied by the formation of four pairs of book lungs from ectodermal invaginations developing behind the disappearing limb buds. The last mesosomal segment of adults (somite XIV) was considered by Anderson (1973) to represent the first metasomal segment, which, like the following five metasomal segments, does not produce limb buds or book lungs. The metasomal segments and telson are proliferated from a rudimentary tail segment, and curl along the ventral surface of the abdomen as they are produced (Fig. 4.4*E, F*).

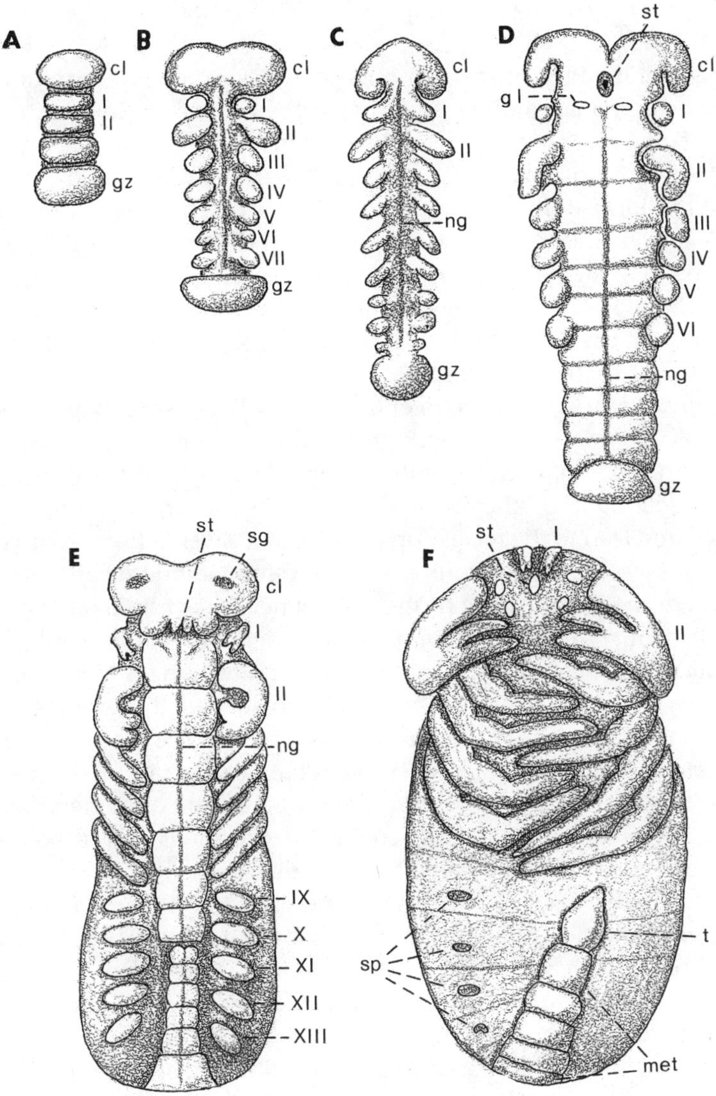

Fig. 4.4. Embryology of *Euscorpius italicus* (Chactidae), as shown by ventral surface views of representative stages of development: *A*, 3-somite embryo; *B*, 7 somites; *C*, 9 somites; *D*, 10 somites; *E*, early 19-somite embryo, showing limb buds to be replaced by pectines (somite IX) and (X–XIII) book lungs (last pair of legs hides those the genital opercula will replace, on somite VIII; only 18 somites are seen, since the pregenital somite [VII] has disappeared); *F*, later (fully formed) 19-somite embryo; *g I*, ganglion of cheliceral segment; *gz*, growth zone; *met*, metasomal segments; *sg*, semilunar groove; *sp*, spiracles of book lungs; *st*, stomodeum; *t*, telson; other symbols as in Figure 4.3. (After Laurie 1890)

The lateral edges of the prosoma and mesosoma grow up and out toward the dorsal midline underneath the serosa, enclosing the yolk mass. This event is known as dorsal closure.

As the opisthosoma is formed, a neural groove arises on the epiblast along the longitudinal axis of the embryo, and the epiblast in the cephalic lobe begins to thicken, forming the rudimentary central nervous system. At a later stage (corresponding approximately to the time the tail segment is formed), the ventral nerve cord can be seen as a pair of longitudinal bands of epiblast, one on each side of the neural groove. Still later (after the six metasomal segments have been formed), a semilunar groove arises on each cephalic lobe. These extend posteriorly, and their lumina meet medially inside the prosoma. The cells around the lumina and on the sides of the cephalic lobe become the future cerebral ganglia. Cells on the dorsal side of the invagination become tightly packed and lead to the formation of the retinal layer of the median eyes.

The final stages of nervous system development occur after the formation of the telson. Cells of the ventral nervous system have formed ganglia, from which a number of nerves originate. The cerebral ganglionic mass has differentiated and has lost connection with the epiblast. The lateral eyes are formed by small cup-shaped thickenings of the prosomal ectoderm in the anterolateral corners of the carapace. The invagination over the median eyes eventually closes, and the ectoderm above them becomes the vitreous layer. Later, the optic nerve grows out of the cerebral ganglion and connects to the median eyes. Nerves to the lateral eyes are presumably formed still later.

The digestive system comprises three parts of different embryological origin. Anteriorly, the stomodeum is formed very early as an invagination of the epiblast between the precheliceral lobes (Fig. 4.4*D*). During its development, it grows as far back as the cerebral ganglion, but remains closed off until it unites with the midgut. The proctodeum arises as an invagination on the underside of the telson, with the intestine in the metasoma forming from hypoblast cells shortly thereafter. Apparently, a midgut rudiment forms early from the hypoblast, separating from the mesoderm rudiment by delamination. This rudiment proliferates and spreads around the yolk mass, eventually penetrating the metasomal rudiment. It does not connect with the stomodeum until shortly before birth. The caecae of the hepatopancreas are formed during dorsal closure by the upgrowth of mesoderm into the midgut sac. Malpighian tubules of excretory function arise as outgrowths of the midgut in the first metasomal (actually the last mesosomal) segment.

The mesoblast of each segment rudiment separates into paired somite rudiments. There are six prosomal pairs and seven mesosomal pairs, each of which is trilobed and contains coelomic spaces (Fig. 4.3*E*, *F*). The six pairs of metasomal somites lack the third (appendicular) lobe, which in the preceding pairs penetrates the limb buds. During dorsal upgrowth and closure, the dorsolateral lobes of the somites give rise to blood cells. The dorsal edges of the somites combine to form the heart. Somatic mesoderm produces the cells that differentiate into the dorsolongitudinal muscles. Splanchnic muscles are produced from the splanchnic mesoderm, and somatic muscles are produced from the appendicular lobes. The details of their formation are unclear (Anderson 1973).

Gonads are formed from the coelomic sacs of the genital (second opisthosomal) segment. The coelomic sacs of each side unite, and, with cells from the somites, produce the genital coelom. The gonoducts are outgrowths from the coelomic sacs that go around the gut and meet midventrally, joining an ectodermal invagination from the genital aperture. The coxal glands, which in scorpions are limited to the third ambulatory segment, form in a similar manner from coelomic sacs, joining an ectodermal invagination at the base of the coxa.

Katoikogenic Scorpions

The oocytes of katoikogenic scorpions develop in specialized diverticula that branch from the female ovariuterus, rather than in contact with the ovariuterus itself. Laurie (1891) described the diverticulum as consisting of four distinct parts: a stalk (the pedicel), a thickened collar, a conical portion that contains the ovum, and an appendix. Variations in this design exist (Fig. 4.5*A*). The specializations of the diverticulum with regard to the oral feeding apparatus will be described below.

The ova of katoikogenic scorpions are very small and alecithal. The size of the oocytes of only two species are known: in *Heterometrus fulvipes*, ova measure 0.12 mm (Laurie 1891) or 0.06 × 0.04 mm (Werner 1934); and in *H. scaber*, 0.04 mm (Mathew 1956). The oocytes of *Liocheles australasiae*, though no measurements are available, must be equally small—a four-celled embryo was estimated to be 0.08 mm in length (Francke 1982d).

The embryological development of katoikogenic species is considerably different from that of apoikogenic scorpions. The following account of the major differences between the two groups is based on the works of Laurie (1891), Mathew (1956), and Anderson (1973).

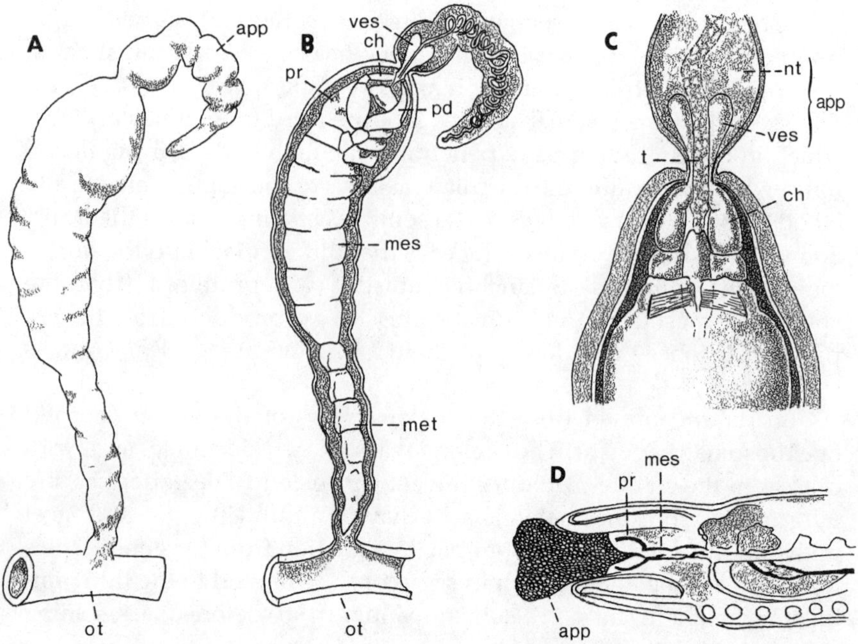

Fig. 4.5. Morphology of the katoikogenic diverticulum and oral feeding apparatus: *A*, external structure of the diverticulum of *Chiromachus ochropus* (the diverticulum of this species has a very short pedicel and lacks the collar); *B*, diverticulum of *C. ochropus*, containing an embryo; *C*, horizontal section of anterior part of diverticulum, showing the relationship between the chelicerae and the appendix; *D*, longitudinal section through anterior half of embryo of *Urodacus manicatus*. *app*, appendix; *ch*, chelicerae; *mes*, mesosoma; *met*, metasoma; *nt*, nutritive fluid; *ot*, ovariuterine tubule; *pd*, pedipalp; *pr*, prosoma; *t*, teat; *ves*, vesicular organ of chelicera. (Views *A–C* after Vachon 1950, 1953; view *D* after Mathew 1968)

The zygote nucleus of *Heterometrus* undergoes holoblastic, equal cleavage through the second division. Cleavage is then unequal, and the result is a coeloblastula (Mathew 1956). In *L. australasiae*, the blastula is a stereoblastula (Pflugfelder 1930). An extraembryonic membrane, the embryonal capsule, derived from follicle cells, forms to enclose the embryo. Soon, a second membrane, the trophamnion, appears around the first. It is derived from the polar-body cells that attach the embryo to the follicle wall. A third extraembryonic membrane, the embryonal envelope (also derived from follicle cells), forms around the trophamnion, and this event is accompanied by the degeneration of the first embryonal envelope. Later, as the oral feeding mechanism develops (see below), the trophamnion disintegrates. It appears that the trophamnion aids the embryo in absorbing maternal nutrients, and this function is no longer needed. The second embryonal envelope stays with the embryo through the remainder of development, and crumbles when the embryo passes out of the diverticulum during parturition (Mathew 1956, Francke 1982d). There is apparently no correspondence between these membranes and those of apoikogenic scorpions, at least in terms of homology.

The basic pattern of anterior-to-posterior development that occurs in apoikogenic scorpions is also found in the katoikogenic species (Fig. 4.6*A–D*), although it is considerably modified. The embryo elongates with its anterior end toward the free end of the diverticulum (see Fig. 4.5*B*), and develops a pronounced flexure in the mesosomal region (Fig. 4.6*B–D*). The body is divided into three regions, which differ

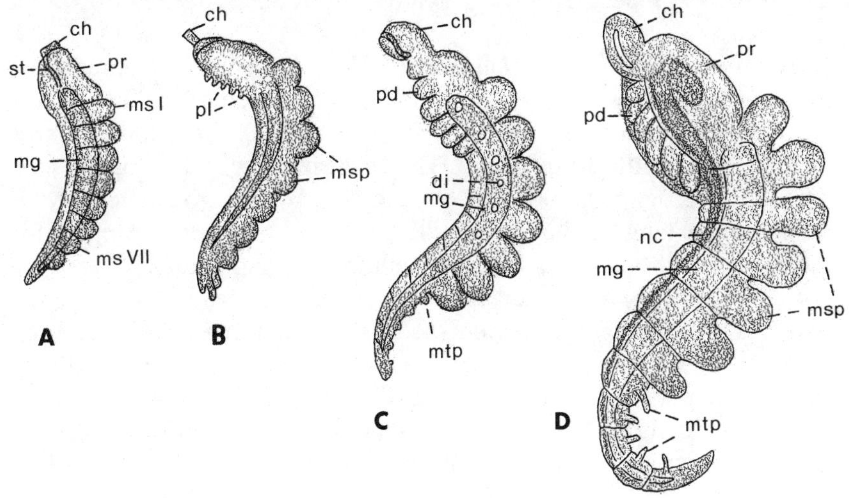

Fig. 4.6. Representative stages in the development of *Heterometrus: A*, 3-mm embryo; *B*, 4-mm embryo; *C*, 6-mm embryo; *D*, 9-mm embryo. *ch*, chelicera; *di*, diverticulum of midgut; *mg*, midgut; *msp*, mesosomal dorsolateral protrusion; *ms I*, mesosomal segment 1; *ms VII*, mesosomal segment 7; *mtp*, metasomal dorsolateral protrusion; *nc*, nerve cord; *pd*, pedipalp; *pl*, prosomal limb bud; *pr*, prosoma; *st*, stomodeum. (After Mathew 1956, Anderson 1973)

from those of apoikogenic scorpions: a prosomal rudiment, a mesosomal rudiment, and a growth zone. The mesosoma develops more rapidly than the prosoma, producing its seven segments by the time the embryo is 2 mm long. The mesosomal segments possess dorsolateral protrusions (Fig. 4.6*B–D*), which persist throughout development and apparently serve as exchange surfaces between the embryo and mother. The dorsolateral protrusions are resorbed shortly after birth.

As the metasomal segments are formed by the growth zone, the dorsal flexure of the mesosoma becomes strong, and the prosoma begins developing pedipalpal and ambulatory limb buds. The metasomal segments also display dorsolateral protrusions, but they are much smaller than those of the mesosoma (Fig. 4.6*C, D*). When the embryo becomes 6–6.5 mm in length, the metasoma takes on an upward flexure of its own (Fig. 4.6*D*). Simultaneously, the mesosoma develops limb buds on the second to sixth segments (no limb buds are formed on mesosoma 1 and 7); the second and third limb buds are rudiments of the genital opercula and pectines, respectively, and the fourth to sixth buds are resorbed. Because the embryo is already tubular, there is no dorsal closure.

The most important specializations of katoikogenic development are those associated with the oral feeding apparatus. The digestive system develops very early. By the time the embryo is 0.2 mm long, a large mouth forms at the anterior end. The chelicerae begin developing soon after as outgrowths flanking the mouth, and by the time the embryo is 0.75 mm in length, they are well developed and have gripped the "teat" of the appendix of the diverticulum (as in Fig. 4.5*C*). The appendix is very closely associated with the hepatopancreas of the mother; it receives and transports nutrients from the hepatopancreas to the embryo (Mathew 1956). These nutrients are pumped into the mouth of the embryo by pharyngeal musculature. The midgut also arises somewhat precociously and connects to the stomodeum early in development; the proctodeum, however, invaginates and connects to the midgut much later.

The feeding mechanism of *Urodacus* differs morphologically from that of *Heterometrus* (Mathew 1968). The appendix is shortened and is a tiny bun-shaped structure at the free end of the diverticulum. The periphery of the bun has large gland cells of secretory function; inner to this layer is a mass of capillaries. The central portion of the appendix, into which the capillaries converge, produces a conical projection that extends toward the mouth of the embryo (see Fig. 4.5*D*). This conical projection is gripped by the chelicerae, in a manner analogous to that found in *Heterometrus*.

Largely neglected in recent times, embryology is one aspect of scorpion biology that would greatly benefit from future study. For additional information on what is known, see Laurie 1890, 1891; Brauer 1894, 1895; Pavlovsky 1925; Pflugfelder 1930; Werner 1934; Abd-el-Wahab 1952, 1954; Mathew 1956, 1960, 1968; Anderson 1973; Yoshikura 1975; and Weygoldt 1979.

Gestation and Birth

Gestation periods are rather variable in scorpions, ranging from only a couple of months to a year and a half (Table 4.2). It appears that buthids have much shorter gestation periods than the other families (average gestation period for buthids is 5.2 ± 2.3 months, for non-buthids 11.4 ± 3.56 months; data expressed as mean ± standard deviation). This conclusion is tentative, however, because adequate sample sizes are available only for buthids. It is notable that some scorpions (particularly scorpionids) exhibit rather long gestation periods (10–18 months). Such long periods of gestation appear to have no parallel among other arachnids (Savory 1977) and are longer than those of many vertebrates.

Scorpion species may give birth year round or seasonally (Table 4.2). In temperate regions, species usually give birth in the spring and summer months (May to October appear to be the preferred months in the Northern Hemisphere, November to April in the Southern Hemisphere). Tropical species, as a group, seem to have no preferred time for birth, although many of them restrict parturition to a couple of months per year. For example, *Isometrus maculatus* in Tanzania gives birth in January and February (Probst 1972); *Centruroides arctimanus* from Cuba gives birth in August and September (Armas and Contreras 1981); and *Tityus bahiensis* from Brazil apparently gives birth year round (Matthiesen 1961, 1969–70). Year-round parturition is not unique to tropical species, and several temperate forms exhibit this characteristic as well (e.g., *Euscorpius italicus* and *Belisarius xambeui*). Geographical trends, derived from the data in Table 4.2, are depicted in Figure 4.7.

It appears that synchrony in parturition times is common among scorpions. Varela (1961) reported that all females of *Bothriurus bonarien-*

TABLE 4.2

Parameters of scorpion parturition

Taxon	Gestation period (months)	Duration of parturition (hrs.)	Birth dates (month)	Litter size	Reference
			APOIKOGENIC SCORPIONS		
BOTHRIURIDAE					
Bothriurus bonariensis (C. L. Koch)	12	several	12	41 (35–48)	Varela 1961
Urophonius brachycentrus (Thorell)	6.5		11–1	33 (21–46)	Maury 1969
U. granulatus Pocock	9–10		12–1		Maury 1978
U. iheringi Pocock	10–11		8–2	47 (31–60)	San Martín & Gambardella 1974
BUTHIDAE					
Androctonus amoreuxi (Audouin)				23	Birula 1917a
A. australis (Linné)				46	Auber-Thomay 1974
A. bicolor Hemprich & Ehrenberg			8	30–40	Levy & Amitai 1980
Buthus occitanus (Amoreux)	10 (9–11.5)	6	9–10, 2–3	12–50	Auber 1963a, LePape 1974, Levy & Amitai 1980
Centruroides anchorellus Armas	4 (2–9)		3–6, 10		Armas & Contreras 1981
C. arctimanus Armas			8–9	14	Armas & Contreras 1981
C. exilicauda (Wood)		<7	5–10	20 (7–42)	Stahnke 1966; Williams 1969; Polis, unpubl. data
C. gracilis (Latreille)		<12	5, 6, 10	47 (26–91)	Armas & Contreras 1981, Francke 1982d, Francke & Jones 1982
C. guanensis (Franganillo)	4.5 (2–9.5)				Armas & Contreras 1981
C. insulanus Thorell	5 (4.5–7.5)		12–6	50 (6–105)	Baerg 1961
C. margaritatus (Gervais)			1–2	39 (26–70)	O. Francke, pers. comm. 1978
C. robertoi Armas	4 (3–6)		5, 8		Armas & Contreras 1981
C. vittatus (Say)	8		5–9	31 (13–47)	O. Francke, pers. comm. 1981
Compsobuthus werneri (Birula)			5–6	5–14	Levy & Amitai 1980
Hottentotta judaica (Simon)			7–8	15–18	Levy & Amitai 1980
Isometrus maculatus (DeGeer)	2.5	2–3	1–2	17	Probst 1972, Shulov & Levy 1968

TABLE 4.2
(*continued*)

Taxon	Gestation period (months)	Duration of parturition (hrs.)	Birth dates (month)	Litter size	Reference
		APOIKOGENIC SCORPIONS			
Leiurus quinquestriatus (Hemprich & Ehrenberg)	5		6–8, 1–12	12–17	Thornton 1956, Shulov & Amitai 1958, Abushama 1968, Levy & Amitai 1980
Liobuthus kessleri Birula				20–25	Birula 1917a
Mesobuthus caucasicus (Fischer)				12–14	Birula 1917a
M. eupeus (C. L. Koch)			7–8	23	Birula 1917a
Microtityus fundorai Armas				7	Armas 1974
Orthochirus innesi Simon	1.5–3.5		2–9	9 (5–12)	Shulov & Amitai 1960
O. scrobiculosus (Grube)			5–9	5–12	Levy & Amitai 1980
Rhopalurus garridoi Armas	2–7	2	5, 9	19 (14–27)	Armas 1986
R. rochae Borelli			2–3	28–49	Matthiesen 1968; O. Francke, pers. comm. 1978
Tityus bahiensis (Perty)	2.5–4.2		1–12	35–82	Matthiesen 1961, 1969–70; Bücherl 1971
T. fasciolatus Pessôa	3–10	3.5–9	8–4	12.2 (1–26)	Lourenço 1978, 1979b
T. mattogrossensis Borelli			4	12 (7–20)	Lourenço 1979a
T. serrulatus Lutz & Mello-Campos	4.5		1–12	17 (2–21)	Matthiesen 1961, 1962, 1971; San Martín & Gambardella 1966
T. trinitatis Pocock			1–12	28	Waterman 1950
Uroplectes insignis Pocock			1–2	12.5	Eastwood 1976
U. lineatus (C. L. Koch)			1–2	8	Eastwood 1976
CHACTIDAE					
Belisarius xambeui Simon		~24	1–12	5–24	Auber 1959
Euscorpius carpathicus (Linné)			9	15	Angermann 1957, Vannini et al. 1978
E. flavicaudis (DeGeer)				65	Angermann 1957, Vannini et al. 1978
E. italicus (Herbst)		2–5	1–12	23–62	Birula 1917a, Angermann 1957, Vannini et al. 1978

TABLE 4.2
(continued)

Taxon	Gestation period (months)	Duration of parturition (hrs.)	Birth dates (month)	Litter size	Reference
			APOIKOGENIC SCORPIONS		
CHACTIDAE (*continued*)					
E. tauricus (C. L. Koch)				25–32	Birula 1917a
Megacormus gertschi Diaz		<12	5	46 (19–75)	Francke 1979b
IURIDAE					
Hadrurus arizonensis Ewing		1	8	10	Williams 1969
H. pinteri Stahnke			6	12–34	Polis, unpubl. data
VAEJOVIDAE					
Anuroctonus phaiodactylus (Wood)			8	13	Williams 1966
Nullibrotheas allenii (Wood)				8–23	Williams 1974
Paruroctonus baergi (Williams & Hadley)			10		Fox 1975
P. boreus (Girard)			8	34–52	Tourtlotte 1974
P. mesaensis Stahnke	10–14		7–9	33 (9–53)	Polis & Farley 1979b
P. utahensis (Williams)	11.5		7–8	11 (2–32)	Bradley 1984
Serradigitus gertschi (Williams)			7	12	Williams 1968
S. minutus (Williams)			6	10	K. Sculteure, pers. comm. 1982
S. wupatkiensis (Stahnke)			7–9	13.4 (7–17)	O. Francke, pers. comm. 1978
Syntropis macrura Kraepelin		37	10	33	Hjelle 1974
Uroctonus mordax Thorell		8–12	8–9	32	Bacon 1972, Haradon 1972, Toren 1973, Francke 1977
Vaejovis bilineatus Pocock			8–9	22 (17–26)	Sissom & Francke 1983
V. carolinianus (Beauvois)	12–13		8–9	23 (9–36)	Taylor 1971, Benton 1973
V. coahuilae Williams			6–8	20–41	Francke & Sissom 1984
V. confusus Stahnke			4, 6, 9	17–25	Polis, unpubl. data
V. jonesi Stahnke			8–9	35	O. Francke, pers. comm. 1978
V. littoralis Williams			6	5 (2–8)	D. Due, pers. comm. 1983
V. spinigerus (Wood)		7.5		35.3	McAlister 1960, Stahnke 1966, Williams 1969

TABLE 4.2
(*continued*)

Taxon	Gestation period (months)	Duration of parturition (hrs.)	Birth dates (month)	Litter size	Reference
			APOIKOGENIC SCORPIONS		
V. vorhiesi Stahnke			7–8	14 (8–19)	Williams 1969; O. Francke, pers. comm. 1978
			KATOIKOGENIC SCORPIONS		
DIPLOCENTRIDAE					
Diplocentrus spitzeri Stahnke		56.5	9	7–12	Francke 1981b
D. trinitarius (Franganillo)	13.5		5–6	46	Armas 1982b
D. whitei (Gervais)		36–72	9–11	36 (33–38)	Francke 1982c
Heteronebo bermudezi (Moreno)				11	Armas 1974
Nebo hierichonticus (Simon)		72	7–9	30–45	Shulov et al. 1960, Rosin & Shulov 1963
ISCHNURIDAE					
Hadogenes sp.		240	1		Williams 1971b
Liocheles waigiensis (Gervais)				20 (18–26)	Koch 1977
Opisthacanthus africanus Simon				16	Lourenço 1985
O. asper (Peters)	15–18		3	22	Lourenço 1985
O. cayaporum Vellard	16–18		5–7	18.2 (15–23)	Lourenço 1985
O. madagascariensis Kraepelin				26	Lourenço 1985
SCORPIONIDAE					
Heterometrus longimanus (Herbst)	12	24+	7	34	Schultze 1927
H. scaber (Thorell)				30–35	Mathew 1956
Pandinus gambiensis Pocock		72		17	Vachon et al. 1970
P. imperator (C. L. Koch)	7	24+	3	32	Larrouy et al. 1973
Scorpio maurus L.	14–15		8–9	8–43	Birula 1917a, Levy & Amitai 1980, Shachak & Brand 1983
Urodacus manicatus (Thorell)	16		2–3	11–18	Smith 1966, Koch 1977
U. novaehollandiae Peters				20–33	Koch 1977
U. planimanus Pocock			4	22	Koch 1977
U. yaschenkoi (Birula)	18		2–3	11.8 (8–31)	Shorthouse 1971

NOTE: There are no published data on the parturition parameters of the Chaerilidae. Where possible, data are presented as means followed by the range in parentheses. Otherwise, reports are of single observations or ranges as obtained from the original sources.

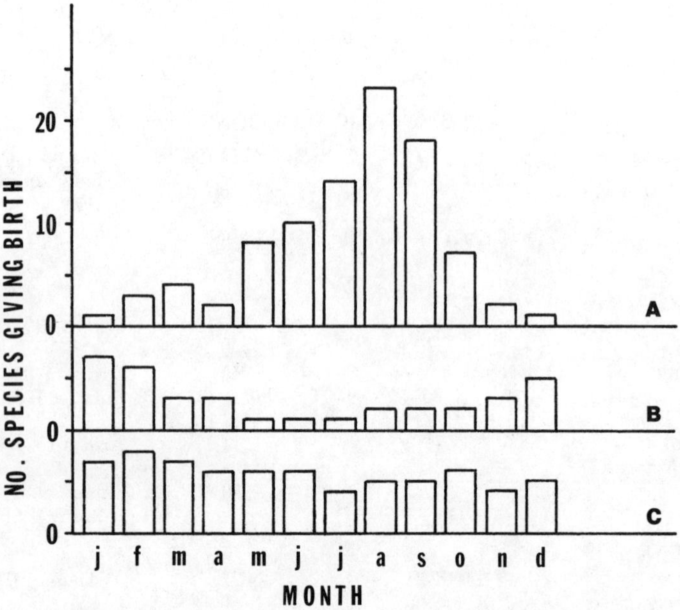

Fig. 4.7. Date of birth as a function of latitude: *A*, north-temperate species; *B*, south-temperate species; *C*, tropical species (Tropic of Cancer to Tropic of Capricorn). Birth by temperate-latitude scorpions occurs primarily during the warmer months, whereas birth by tropical species is more evenly distributed throughout the year. Note that in the Southern Hemisphere winter is from June to September.

sis gave birth in the month of December. Williams (1969) reported that 50 females of *Vaejovis vorhiesi* from Arizona gave birth in a two-week period in midsummer. Likewise, parturition is synchronized in *Hadrurus arizonensis* (Williams 1969; Polis and McCormick, unpubl. data). Polis and Farley (1979b) reported that all second-instar young of *Paruroctonus mesaensis* appeared in early August for five consecutive years at Palm Springs, California. Francke (1979b) observed four females of *Megacormus gertschi*, collected in March, giving birth during a two-day period in May.

Several factors may promote synchrony. Francke (1979b) suggested that a highly synchronized mating season could produce a synchronous birth period, but an alternative explanation is needed for species lacking such a synchronized mating season (e.g., *P. mesaensis*, which mates from May to October). In *P. mesaensis*, substantial embryonic development does not appear to occur while the mature females are inactive during winter months. Likewise, embryonic growth of *P. utahensis* is arrested at a very early stage of development during the winter; the majority of growth occurs after the onset of surface activity in spring (Bradley 1983, 1984). Polis and Farley (1979b) proposed that a pulse in prey availability in the spring serves to accelerate greatly embryonic growth and thus to synchronize development. Just how this synchrony is achieved, however, is still unclear, since embryos of females mating in May should undergo considerable development before the onset of winter, relative to the amount of development in the embryos of females mating in late summer. In general, developmental times appear to be quite variable because of differences in the environment and the feeding history of the mother (see "Environmental Influences on Life History").

Birth Behavior

As discussed above, all scorpions are truly viviparous, distinguishing them from the remaining Arachnida, which, with the exception of one family of mites, deposit eggs. These eggs may be carried with the female (usually attached to her venter in an egg sac, as in Amblypygi, Uropygi, Schizomida, Pseudoscorpiones, and some Araneae), deposited in the soil with an ovipositor (as in the Opiliones), or placed in a specially made egg case in some retreat (most Araneae and Solifugae). Many Acari deposit their eggs in vegetation or on their hosts, although the Pyemotidae are viviparous (Yoshikura 1975). Ricinuleid females lay a single egg, which the female carries in her pedipalps and cucullus until it hatches.

Birth behavior is recorded for species representing six scorpion families (excluding Iuridae and Chaerilidae, for which no published data are available). Because birth usually occurs in protected places, such as in burrows or under objects (Maury 1969, Williams 1969), most observations on birth behavior are derived from laboratory studies. One interesting fact that has emerged from these studies is the remarkable similarity of stereotyped behaviors across species and even families.

A general description of parturition will suffice because of the similarities between taxa. The following account is derived mainly from Shulov et al. (1960), Williams (1969), Haradon (1972), Hjelle (1974), Francke (1982c), and Armas (1986).

Immediately prior to parturition, the female assumes a posture

Fig. 4.8. Birth behavior: stilting posture of female *Vaejovis spinigerus*. Note the elevation of the prosoma above the substrate and the flexure of the first pairs of legs underneath the body. (Photo courtesy Stanley C. Williams)

known as stilting (Fig. 4.8). This posture is characterized by the elevation of the anterior portion of the body above the substrate (often the posterior segments rest on the substrate); the arching of the metasoma over the mesosoma; the flexing of the pedipalps; and the flexing of the first two pairs (sometimes only the first pair) of legs underneath the mesosoma near the genital opercula, forming a "birth basket" (Figs. 4.9, 4.10). Stilting is maintained throughout the entire parturition process.

As the female stilts, the genital opercula open, and the young begin to emerge, one at a time. Emergence is either continuous or sporadic, with each young dropping into the birth basket. Shortly after falling into the birth basket, the young become active. Those born enclosed in membranes (see below) begin writhing free of them at this time. Once free, the young climb up the female's legs until they reach her back, where they eventually settle (Figs. 4.11, 4.12). After parturition is complete and all the young have ascended and settled, the female resumes normal behavior. The young remain with the female through their first molt, and then disperse in 3 to 14 days, depending on species.

Fig. 4.9. Birth behavior: formation of the birth basket in *Vaejovis spinigerus*. (Photo courtesy Williams)

Fig. 4.10. Birth behavior: formation of the birth basket in *Tityus fasciolatus*. (Photo courtesy Wilson R. Lourenço)

Fig. 4.11. First instars ascending to the female's back in *Tityus fasciolatus*. (Photo courtesy Lourenço)

Fig. 4.12. *Centruroides exilicauda* female carrying first-instar young.

Apparently, some of the Old World species (*Compsobuthus acutecarinatus, Leiurus quinquestriatus, Nebo hierichonticus,* and *Orthochirus innesi*) do not form a birth basket during parturition. Instead, the young are deposited directly on the substrate, but are aided in their ascent by the placement of the female's legs on the ground underneath the genital aperture and (in *O. innesi*) by the lowering of the female's body (Shulov and Amitai 1960, Williams 1969).

As previously stated, some scorpions are born enclosed in membranes, and others are not. Scorpions born with membranes are apoikogenic; those without are katoikogenic (Francke 1982d). The extraembryonic membranes of apoikogenic scorpions are the fused serosa and amnion, shed during or shortly after parturition. Katoikogenic scorpions also possess membranes as embryos (see "Katoikogenic Scorpions," above), but these disappear before parturition.

There are some interesting differences between these two groups (Francke 1982d). First, the young of katoikogenic scorpions are always born tail first; the young of apoikogenic scorpions are randomly oriented at birth, some head first, others tail first. The anatomy of embryological development seems to account for the difference. Embryos of katoikogenic scorpions develop in diverticula with the head directed toward the distal end of the diverticulum; as parturition occurs, embryos simply back out of the diverticulum and move toward the genital aperture. Embryos of apoikogenic scorpions develop in the ovariuterus itself, and are not oriented in any particular manner within that structure. When the young reach the genital aperture, they consequently appear either head first or tail first, depending upon their orientation within the ovariuterus.

Second, parturition times are shorter in apoikogenic scorpions (Table 4.2). For example, the apoikogenic scorpions *Vaejovis* and *Centruroides* are characterized by individual parturition times of about 1 minute and total parturition times of less than 12 hours (Williams 1969, Francke 1982d). In katoikogenic scorpions, these times are much longer: in *Diplocentrus*, for example, individual times average from 6.9 to 61.6 minutes; the time for complete birth ranges from 36 to 72 hours (Francke 1982d). In *Hadogenes* the total time is 10 days (Williams 1971b).

The shorter individual parturition times in apoikogenic scorpions may be explained by the moist membranes that facilitate the passage of the young through the genital aperture (Williams 1969, Francke 1982d). The differences in total parturition times again reflect the differences in embryological development. Embryos of apoikogenic scorpions, which develop in the ovariuterus, move toward the genital aperture simultaneously. Therefore, there are short intervals between births, and a shorter total parturition time. Katoikogenic scorpions must move out of the diverticula before they enter the ovariuterus. Apparently, only five or six young may occupy the ovariuterus at one time, and the ovariuterus must be cleared before the next group enters from the diverticula (Mathew 1956, Francke 1982d). Young are born in groups or sometimes pairs, with long intervals between groups or pairs (Francke 1982d). This, coupled with the long individual parturition times, acts to increase the total parturition time.

Litter Size and Sex Ratio

Litter size is quite variable, ranging from 1 to 105 young per parturition (Table 4.2). The average litter size for the entire order, based on all available data, is about 26. Two families appear to have substantially larger average litters than the other families studied: Bothriuridae ($\bar{X}$ = 40.5) and Chactidae ($\bar{X}$ = 35.1). There may be sampling error, however, as the data for these families include only a few species.

These brood sizes are small compared to those for spiders, which lay from 20 to 2,500 eggs per oviposition (usually <100: Foelix 1982). However, the brood size of scorpions is comparable to those for other arachnid groups: Amblypygi, 15–50; Pseudoscorpiones, 3–40; Schizomida, 6–30; Uropygi, 12–40; Solifugae, 20–164; and Opiliones, 6–100+ (Kästner 1968). The Ricinulei are exceptional, usually laying a single egg (Kästner 1968).

In attempting to identify the factors responsible for this variation in litter size among scorpions, Francke (1981b) demonstrated that litter size among diplocentrids was directly proportional to the size of the female and inversely proportional to the size of the young. The size of the mother and the size of the young accounted for 81 percent of the variation in litter size between species. But this is rather intuitive, and preliminary analyses of buthids and vaejovids have not revealed any significant trends (Sissom, unpubl. data). In fact, Bradley (1984) found no correlation between the size of *Paruroctonus utahensis* females and the clutch or body size of their offspring.

The sex ratio at birth is 1:1 for all non-parthenogenetic scorpions for which such data were reported (see Table 6.3). An equal proportion of males and females at birth is widespread among arachnids (Savory 1977) and, indeed, throughout the animal kingdom.

Iteroparity

Some species of scorpion are reported to be iteroparous, that is, capable of repeatedly giving birth after one or more matings. Iteroparity may be advantageous in variable environments, where the probability of successful reproduction during any given year may vary. The ability to produce young repeatedly over several years would then constitute a hedge against reproductive failure. Females from at least three genera in the family Buthidae (Table 4.3) require only one insemination to produce multiple (1–5) broods. For example, female *Isometrus maculatus* produce four or five litters regularly spaced by 66–84 days (Probst 1972). This phenomenon may have evolved because, in these species, mature males are relatively less frequent than females. In *Isometrus* and many species of *Tityus*, males are outnumbered by females in ratios of 1:2–1:4 (see Table 6.3). Multiple births from a single mating requires that females store sperm, often for extended periods. Kovoor et al.

TABLE 4.3

Scorpion species producing multiple broods from a single insemination

Species[a]	Number of females	Litters: Total	Litters: Per female	Reference
Centruroides anchorellus Armas	7	27	5/17 mos.	Armas & Contreras 1981
C. exilicauda (Wood)			2/season	Williams 1969,[b] Kovoor et al. 1987
C. gracilis (Latreille)			2/season	Williams 1969[b]
C. guanensis (Franganillo)	7	24		Armas & Contreras 1981
C. robertoi Armas	1	4	4	Armas & Contreras 1981
Isometrus maculatus (DeGeer)			4 or 5	Probst 1972
Tityus bahiensis (Perty)	53		1	Bücherl 1956; Matthiesen 1968, 1969–70, 1971
	8		2	
	8		3	
T. cambridgei Pocock	3	7		Matthiesen 1969–70, 1971
T. fasciolatus Pessôa	84		1	Lourenço 1978, 1979b
	15		2	
	2		3	
T. serrulatus Lutz & Mello-Campos[c]	2		3	Matthiesen 1969, 1971
	1		4	
T. stigmurus (Thorell)	1	3	3	Matthiesen 1969, 1971

[a] All species are in the family Buthidae.
[b] The number of inseminations was not reported.
[c] *Tityus serrulatus* is parthenogenetic.

(1987) demonstrated that females from at least four genera from two families (Buthidae and Diplocentridae) store spermatozoa in their genital tract. This strategy allows maximum reproduction by the individual female even in the prolonged absence of males. Evidently, these females will readily mate again if they encounter another male. Females of some species will even mate when gravid or when carrying young on their backs (Matthiesen 1968, 1971; Williams 1969): for example, *Centruroides exilicauda* females with young readily court with males (G. Polis, pers. obs.).

Other species are iteroparous but produce only one litter per year (often synchronously throughout the population) for several years. This appears to be the case for most North American vaejovids (e.g., *Anuroctonus, Paruroctonus, Vaejovis*) and iurids (*Hadrurus*; Williams 1969; Polis and Farley 1979a, b, unpubl. data; Bradley 1983, 1984). Marked females of *Paruroctonus mesaensis* were observed to produce at least two consecutive litters 12 months apart (Polis and Farley 1979b), and individual females of Australian *Urodacus* are known to bear three to five litters in their lifetime (Smith 1966, Shorthouse 1971).

Since mature scorpions live for many years (Table 4.4), it is likely that iteroparity is widespread, if not universal, throughout the entire order. This contrasts markedly with the majority of terrestrial arthropods, which reproduce once (semelparity) and live for less than a year. Of course, some long-lived arthropods live for several years and reproduce repeatedly (e.g., mygalomorph spiders, uropygids, amblypygids, some large myriapods: Savory 1977).

Parthenogenesis

Parthenogenesis, or reproduction from unfertilized eggs, is known to occur in *Tityus serrulatus* (Matthiesen 1962, 1971; San Martín and Gambardella 1966) and in *Liocheles australasiae* (Makioka and Koike 1984, 1985). It was thought to occur in *Tityopsis inexpectatus inexpectatus* (Armas 1980), but males were recently discovered and described (Armas 1984). No valid reports of males exist for *Tityus serrulatus*, and previous reports of such (e.g., Mello-Leitão 1945) are considered spurious.

Liocheles australasiae is widespread throughout Asia and Australia. Females from two Asian populations isolated in the laboratory became pregnant three times (Makioka and Koike 1984). Analysis of the morphology and germ cells of 569 specimens from the Asian populations indicated that males were absent in nature (Makioka and Koike 1985; in

TABLE 4.4
Parameters of scorpion post-birth life history

Taxon	Duration of First instar (days)	Number of Molts	Age to Maturity (months)	Longevity (months)	Mating (month)	Reference
BOTHRIURIDAE						
Bothriurus bonariensis (C. L. Koch)	6–10				12	Varela 1961
Urophonius brachycentrus (Thorell)					5–7	Maury 1969
U. iheringi Pocock					12–2	San Martín & Gambardella 1974
BUTHIDAE						
Androctonus australis (Linné)	10	7	21.7–26			Auber-Thomay 1974
Buthus occitanus (Amoreux)	4–8	6	6		4–5	Auber 1963a, Le Pape 1974, Levy & Amitai 1980
Centruroides anchorellus Armas	6 (4–8)	♂ 4, 5; ♀ 4, 5		24–36		Armas & Contreras 1981
C. arctimanus Armas	4					Armas & Contreras 1981
C. exilicauda (Wood)	6–15			60		Stahnke 1966; Williams 1969; C. Myers, pers. comm. 1988
C. gracilis (Latreille)	5–9	♂ 5, 6; ♀ 6	8–10	31–52		Armas & Contreras 1981, Francke & Jones 1982
C. guanensis (Franganillo)	6 (5–9)	♂ 4, 5; ♀ 5		24–36		Armas & Contreras 1981
C. insulanus Thorell	7–9				10–12	Baerg 1961
C. robertoi Armas	6–9					Armas & Contreras 1981
C. vittatus (Say)	3–7	♂ 5, 6; ♀ 5, 6	36–48		4–10	Sissom & Francke, unpubl. data
Hottentotta minax occidentalis Vachon & Stockmann		♂ 5, 6; ♀ 5, 6, 7				Stockmann 1979
Isometrus maculatus (DeGeer)	4 (4–5)	♂ 6; ♀ 5, 6	7–10	32	10–11	Shulov & Levy 1968, Probst 1972
Leiurus quinquestriatus (Hemprich & Ehrenberg)	1.5		8		1–12	Thornton 1956, Shulov & Amitai 1958, Abushama 1968, Levy & Amitai 1980

Table 4.4
(*continued*)

Taxon	Duration of First instar (days)	Number of Molts	Age to Maturity (months)	Longevity (months)	Mating (month)	Reference
Orthochirus innesi Simon		♂ 4, ♀ 5				Shulov & Amitai 1960
Rhopalurus garridoi Armas	5	♂ 5, 6; ♀ 6, 7	13–25		8–10	Armas 1986
R. rochae Borelli		♂ 4, ♀ 5			1–12	Matthiesen 1968; O. Francke, pers. comm. 1978
Tityus bahiensis (Perty)	6	♂ 4; ♀ 4, 5	16–36	21–37	1–12	Matthiesen 1961, 1969–70; Bücherl 1971
T. fasciolatus Pessôa	6.5 (4–9)	♂ 4, 5; ♀ 5	25–27	48	1–12	Lourenço 1978, 1979b
T. mattogrossensis Borelli	7	5				Lourenço 1979a
T. serrulatus Lutz & Mello-Campos	6	♀ 5	17–25	52	partheno-genetic	Matthiesen 1961, 1962, 1971; San Martín & Gambardella 1966
T. trinitatis Pocock	7					Waterman 1950
Uroplectes insignis Pocock	9–12					Eastwood 1976
U. lineatus (C. L. Koch)	9–12					Eastwood 1976
CHACTIDAE						
Belisarius xambeui Simon	1	6, 7			1–12	Auber 1959, Francke 1977
Euscorpius carpathicus (Linné)	12	6				Angermann 1957, Vannini et al. 1978
E. flavicaudis (DeGeer)	15					Angermann 1957, Vannini et al. 1978
E. italicus (Herbst)	6–14	♂ 5; ♀ 5, 6	6.2		4–5	Angermann 1957, Vannini et al. 1978
Megacormus gertschi Diaz	10–12					Francke 1979b
DIPLOCENTRIDAE						
Diplocentrus spitzeri Stahnke	12–17	5	33			Francke 1981b
D. trinitarius (Franganillo)	6–8	♂ 8, ♀ 9	47		4	Armas 1982b
D. whitei (Gervais)	18–28	♂ 7–8, ♀ 8				Francke 1982c
Nebo hierichonticus (Simon)	10–16	6, 7				Shulov et al. 1960, Rosin & Shulov 1963

TABLE 4.4
(*continued*)

Taxon	Duration of First instar (days)	Number of Molts	Age to Maturity (months)	Longevity (months)	Mating (month)	Reference
ISCHNURIDAE						
Hadogenes sp.	37–51					Williams 1971b
Opisthacanthus asper (Peters)	12	♂ 5, ♀ 6	♂ 13.1–17.0, ♀ 16.7–21.5	15.7–27.3+[a]		Lourenço 1985
O. capensis Thorell		5				Lourenço 1985
O. cayaporum Vellard	10–11	6	22.3–26.5	33.2–56.2+[a]		Lourenço 1985
O. lepturus (Beauvois)		6				Lourenço 1985
SCORPIONIDAE						
Heterometrus longimanus (Herbst)	8	7	11.5–14.5		7	Schultze 1927
Pandinus gambiensis Pocock	10–13	6, 7	39–83	96	8	Vachon et al. 1970
P. imperator (C. L. Koch)	17					Larrouy et al. 1973
Urodacus manicatus (Thorell)	10–16	5	18–25		10–11	Smith 1966, Koch 1977
U. planimanus Pocock		5				Koch 1977
U. yaschenkoi (Birula)	14–28	5	54		9–11	Shorthouse 1971
VAEJOVIDAE						
Anuroctonus phaiodactylus (Wood)	17					Williams 1966
Paruroctonus mesaensis Stahnke		6, 7	19–24	> 60	5–10	Fox 1975, Polis & Farley 1979b
P. utahensis (Williams)				72–84	6–9	Bradley 1983
Syntropis macrura Kraepelin	18–20					Hjelle 1974
Uroctonus mordax Thorell	14–15	6			5	Francke 1977
Vaejovis bilineatus Pocock	9.8 (9–10)	5				Sissom & Francke 1983
V. carolinianus (Beauvois)	7–13		22–34	72	8–9	Taylor 1971, Benton 1973
V. coahuilae Williams	9–12	♂ 5, 6; ♀ 6, 7				Francke & Sissom 1984
V. spinigerus (Wood)	7–8					McAlister 1960, Stahnke 1966
V. vorhiesi Stahnke	9.5					O. Francke, pers. comm. 1978

NOTE: There are no published data on the parameters of post-birth life history for the Chaerilidae or the Iuridae. Where possible, data are presented as means followed by the range in parentheses. Otherwise, reports are of single observations or ranges as obtained from the original sources.

[a] Plus-sign (+) indicates that some specimens were still alive when the report was published, and, hence, that the maximum longevity will be greater than is shown in the range.

Australia, males are present, but rare: Koch 1977). Makioka and Koike concluded that some populations of *L. australasiae* reproduce normally by thelytokous parthenogenesis (females produce only females without any participation of males).

Parthenogenesis is rare in other arachnids as well. In the Araneae, only *Theotima* (Ochyroceratidae) is known to be parthenogenetic, and *Steatoda triangulosa* (Theridiidae) may be (Kästner 1968). There are also reports of the phenomenon among the Acari and Opiliones (Matthiesen 1962, Kästner 1968).

Parthenogenesis in *Tityus serrulatus* was discovered by Matthiesen (1962), who collected pregnant females of that species for laboratory studies. The young were reared separately after the second molt. Three individuals reached maturity at the sixth instar, and each of these gave birth seven to eight months later. The same three females gave birth to other litters: one female gave birth a total of four times, and the two others gave birth three times (Matthiesen 1971). Since females average 17 young per litter, one female can thus produce up to 70 young in her lifetime without fertilization (Matthiesen 1971).

Postnatal Development

Many characteristics of scorpion growth and development do not generally occur in other terrestrial arthropods. In particular, several stages of postnatal development are quite long, compared with other arachnids: maternal care of young, the juvenile period, and adult life. In this section, the biology of growth and development after birth is presented and analyzed.

Postnatal Association of Mother and Young

Newborn scorpions stay with their mother through the first instar (or larval stage) and, in most cases, begin dispersing in three days to two weeks after molting to the second instar. During the first instar, the young do not feed; instead they absorb stored nutrient reserves (McAlister 1960, Williams 1969). In this characteristic, the first instars of scorpions are similar to the larvae of Uropygi, Amblypygi, and spiders, the prelarva of a species of actinedid mite (Actinotrichida: Actinedida), and the deutonymph of many acaridid mites (Actinotrichida: Acaridida; Van der Hammen 1978).

This early mother-young association is obligatory for newborns; without it, first instars do not molt successfully, and usually die (Williams 1969, Vannini and Ugolini 1980, Ugolini et al. 1986). Possible functions of the postbirth association include protection from predators or harsh environmental conditions and protection during the teneral period after the first molt, when the new cuticle is still soft. Finally, Vannini et al. (1985a, b) used tritiated water to show that first instars gain water from their mother during their tenure on her dorsum. Water intake by first instars is important, because this stage does not possess a waterproof cuticle, and accordingly is subject to quite high rates of water loss.

The mother–offspring bond among scorpions appears to be maintained by chemoreception. Torres and Heatwole (1967a) found that the young of *Centruroides nitidus* and *Tityus obtusus* settle and remain only on conspecific females. If placed on females of other species, they do not remain there (on congenerics, they exhibit brief exploratory movements before leaving). When body fluids of a *C. nitidus* female were placed on the back of a female of a different species, the latter female became attractive to the young. Vannini et al. (1978), in a series of controlled experiments, found chemical stimuli to be important in maintaining the association in *Euscorpius carpathicus*, including the mother's tolerance of the young. In *Euscorpius*, females pick up fallen young to aid them in regaining their position on the back. The researchers disguised fallen young with strange scents; these young were either rejected or eaten, suggesting that females recognize their young by scent. In *Euscorpius*, however, the chemical stimulus is not species specific: the young of *E. italicus* were accepted by an *E. carpathicus* female and remained with her "indefinitely."

Maternal behavior is restricted to females with young: those with first instars exhibit strong maternal behavior, whereas those with second instars exhibit weaker maternal behavior. Predatory behavior toward the young was observed in males, in unfertilized females, and in mothers 10–12 days after the first molt of their young, that is, after the association with them had terminated (Vannini et al. 1978).

In spiders there is apparently no recognition between mother and young (Foelix 1982). Spiderlings are known to climb onto any adult spider, even those of a different species.

In a few species the mother–young relationship is continued for an extended period (Polis and Lourenço 1986). From reports of the biology of *Scorpio maurus palmatus* (Shachak and Brand 1983), it appears that newborn of this species also reside in the maternal burrow for an extended period. They are born in August but do not disperse until three months later, in November, at the beginning of the rainy season. In *Urophonius brachycentrus* from Argentina, the young are born in December and January in an underground "gestation chamber" (Maury 1969). After birth the young may climb to the female's back or simply rest on the walls and floor of the chamber. This is the only reported exception to the observation that first instars must remain on the mother's back in order to complete the stage. These young bear an extraordinary amount of nutritive reserves, and molt to the second instar at 25 days of age, after an unusually long first instar. For the duration of time spent in the chamber, the female and young remain in a semilethargic, nonfeeding state. In May, four months after birth, the female digs out of the chamber, and the young begin an independent existence (Maury 1969). A similar association within chambers occurs between the mother and newborns of *Opisthacanthus cayaporum* (Lourenço 1981b, pers. comm. 1983; see also Chapter 6).

Preliminary observations of *Didymocentrus comondae*, a diplocentrid from the cape region of Baja California Sur, show that second instars share burrows and prey with females (D. Lightfoot, pers. comm. 1979). It is quite possible that such phenomena occur in other burrowing species but are rarely observed. For example, Schultze (1927) reported that female *Heterometrus longimanus* provided food for the young evidently until they dispersed, in their third instar. If such associations do in fact occur more commonly than has yet been observed, future studies will be of interest to those studying the evolution of social behavior, since scorpions were among the earliest terrestrial animals.

Growth Characteristics

In arthropods, growth occurs in two ways: by a more or less continuous increase in body weight due to feeding and a stepwise increase in body and exoskeleton length that occurs during ecdysis, or molting (see "Characteristics of Ecdysis," below). Exoskeletal growth occurs only at ecdysis because arthropods possess a rigid, nonliving cuticle made of chitin. Scorpions also show some increase in total body length (but not in cuticular structures) during intermolt periods because intersegmental membranes stretch, thus allowing the body to telescope slightly with increases in weight.

Morphometric analysis of growth is often conducted to determine growth rates and the number of molts and instars. The basic growth equation used in morphometry, $y = bx^k$, was derived by Huxley (1932). It relates the size (usually length) of the dependent structure, y, to that of an independent structure, x, where b is the initial growth difference or index (the value of y when x equals zero) and k is the constant differential growth ratio. When x and y are allometrically transformed into logarithms, the equation becomes linear: $\log y = \log b + k \log x$. The ratio of the logarithmic growths of the two structures, k, gives the slope of the line. Teissier (1960) argued that Huxley's growth equation does not provide a complete description of growth and that a progression factor (Dyar 1890, Przibram and Megusâr 1912) should also be calculated. The progression (or growth) factor is the ratio of the size of a mensural characteristic from one instar to the next. Morphometric analysis of the growth statistics of scorpions suggests that exoskeletal growth is more or less constant during the developmental period and on to maturity (Smith 1966, Shorthouse 1971, Probst 1972, Polis and Farley 1979b). Further, in *Isometrus maculatus* and *Paruroctonus mesaensis* there were no significant differences between males and females in the growth equation for the prosoma; for at least these two species, then, the growth in body length (excluding the metasoma) is not sexually dimorphic.

Growth may also be described by changes in weight and total body length. The only data for such growth were collected for a field population of *P. mesaensis* (Polis and Farley 1979b). The growth of an average cohort is depicted in Figure 4.13. Growth is gradual, with 0.03-g newborn animals averaging 0.35 g at 1 year of age and 1.85 g at 2 years.

The newborn brood of *P. utahensis* varies in weight from 11 to 36 percent of the weight of the gravid female (Bradley 1983). The average *P. mesaensis* female loses 43 percent of her body weight when she gives birth. *Urodacus manicatus* and *U. yaschenkoi* show similar weight changes associated with parturition and birth (Smith 1966, Shorthouse 1971); for example, *U. manicatus* gravid females weigh 2.65 g, postpartum females 1.45 g (weight decrease = 45 percent). In contrast, males show a continuous increase in size even after their last molt (Fig. 4.13). These general characteristics of growth were also observed for other species (G. Polis, pers. obs.).

Scorpions reach maturity in as few as 6 months (*Centruroides guanensis*, Armas and Contreras 1981) to as many as 39–83 months (*Pandinus gambiensis*, Vachon et al. 1970; see Table 4.4, and, for a summary of the life history data for each family, Table 4.5). As a group, buthids appar-

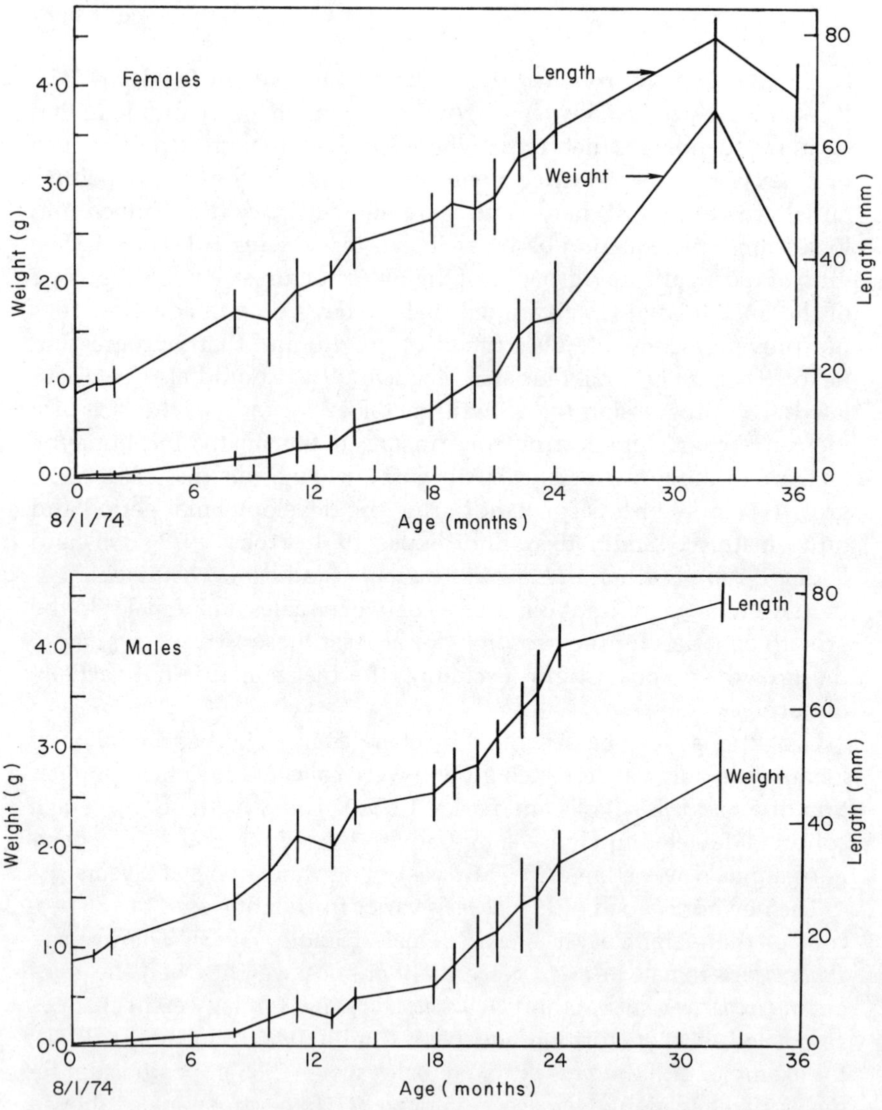

Fig. 4.13. Growth of *Paruroctonus mesaensis* in the field. The mean and standard deviation of total length and weight are graphed as a function of age. Data from birth to maturity (age 24 months) are taken from a cohort born in August 1974. For females (above), the data at 32 months represent the average of gravid females of all ages, and the data at 36 months represent the average of postpartum females of all ages. For males (below), the data past 24 months represent the average of mature males of all ages. (From Polis & Farley 1979b)

TABLE 4.5
Summarized life history data for the families of the order Scorpiones

Taxon	Gestation period (months)	Litter size	Duration of first instar (days)	Number of molts	Age to maturity (months)	Longevity (months)
Bothriuridae	9.6 ± 2.3 (9–12)	40.3 ± 7.0 (21–60)	8.0 ± 2.0 (6–10)			
Buthidae	5.0 ± 2.2 (2–11.5)	23.7 ± 14.5 (1–105)	6.5 ± 2.1 (1–15)	5.1 ± 0.7 (4–7)	18.9 ± 12.0 (6–48)	37.9 ± 11.8 (24–60)
Chactidae		35.3 ± 19.7 (5–75)	9.8 ± 5.3 (1–15)	5.9 ± 0.6 (5–7)	6.2	
Diplocentridae	13.5	28.5 ± 15 (7–46)	14.4 ± 6.6 (6–28)	6.8 ± 1.5 (5–9)	39.2 ± 7.15 (33–47)	
Ischnuridae	16.8 ± 0.4 (15–18)	20.4 ± 3.8 (16–26)	22.2 ± 18.9 (10–51)	5.6 ± 0.5 (5–6)	20.8 ± 5.2 (13–26.5)	33.1 ± 16.4+[a] (15.7–56.2+)[a]
Scorpionidae	12	33.25 ± 1.1 (30–35)	14.1 ± 5.03 (8–28)	5.7 ± 1.0 (5–7)	37.4 ± 23.7 (11.5–83)	96
Vaejovidae	12 ± 2 (10–14)	23.0 ± 11.4 (2–53)	12.6 ± 4.0 (7–20)	5.8 ± 0.7 (5–7)	21.5 ± 2.5 (19–24)	70 ± 9.1 (60–84)
ORDER SCORPIONES	7.6 ± 4.3 (2–18)	25.0 ± 13.6 (1–105)	10.7 ± 7.2 (1–51)	5.4 ± 0.8 (4–7)	27.7 ± 20.1 (6–83)	49.0 ± 21.8 (24–96)

NOTE: There are no data for the Chaerilidae or the Iuridae. Where possible, means are given with their standard deviations, followed by ranges in parentheses; otherwise, reports are for single species.

[a] Plus-sign (+) indicates that some specimens were still alive when the report was published, and, hence, that the maximum longevity will be greater than is shown in the range.

ently mature faster (18.9 ± 12.0 months) than other families of scorpions (all nonbuthids = 36.6 ± 22.9 months). Lifespans are reported to range from 1 year (*C. insulanus*, Baerg 1961) to 25 years or longer (*Hadrurus arizonensis*, Stahnke 1966, pers. comm. 1975). Longevity for most species studied ranges from 2 to 5 years.

Scorpions are exceptional among terrestrial arthropods both in their potential longevity and in the length of time spent as juveniles. Most insects, spiders, solpugids, and mites are annual species, which complete their entire life cycle within a year. A few arachnids have longer cycles. Some of these, including some species of spiders, Ricinulei, Opiliones, and even mites (e.g., *Arrenurus*), reach maturity in 1 to 3 years and may live 2 to 5 years (Kästner 1968, Weygoldt 1969, Savory 1977, Foelix 1982). Opiliones, for example, may live 3 years (*Trogulus*) to 9 years (*Siro*). Some mygalomorph spiders, uropygids, amblypygids, and Cyphopthalmi take 4 or more years to mature and may live 14 to 25 years or more (Savory 1977, Foelix 1982).

Characteristics of Ecdysis

Specific events associated with molting have been poorly studied for scorpions. It is known that scorpions increase in weight until the exoskeleton becomes too small to allow further growth. As the new exoskeleton is secreted by the epidermis underneath the old exoskeleton, some materials are resorbed from the old cuticle. Before the molt, individuals become more reclusive and inactive; some species become totally motionless during the 24 hours immediately before the cuticle is shed (Rosin and Shulov 1963). The blood pressure of scorpions, like that of other arachnids, probably increases just before the molt. The cuticle ruptures at the side and front margins of the carapace. The chelicerae, pedipalps, and legs are all withdrawn from the exuviae. The body emerges slowly during short periods of vigorous movement, which alternate with long periods of relaxation. The entire process of emergence lasts about 12 hours. After emergence, scorpions apparently expand body volume temporarily while the new cuticle hardens. Such volumetric expansion in other arthropods is caused by changes in blood pressure or by taking in air. However, pneumatic expansion in scorpions cannot be large because of the limited capacity of the book lungs, as opposed to the highly elastic capacity of the insect tracheal system (Francke 1976b). As the cuticle tans it hardens, darkens, and gradually acquires the ability to fluoresce under ultraviolet light. Armas (1986) details the events that occur during the molt of the buthid *Rhopalurus garridoi*.

During most molts the growth of cuticular structures occurs with little change in morphology. Extensive changes do occur, however, at the first and last (adult) molt. (Changes at the adult molt are discussed under "Secondary Sexual Characteristics," below.) Much differentiation occurs at the molt between the first and second instars (McAlister 1960, Williams 1969, Vachon and Condamin 1970, Shorthouse 1971, Francke 1976b, Polis and Farley 1979b, Sissom and Francke 1983). Setae and trichobothria, previously sparse or nonexistent, first appear in normal numbers over the entire body. Tarsal claws and the dentate margins of the pedipalp chelae also first appear and harden. The chelicerae and the telson likewise harden and darken as they become fully differentiated. Finally, the second instar marks the beginning of carinal development and the appearance of granulation on the cuticular surfaces. The cuticle of the second instar fluoresces brightly under ultraviolet light, whereas the cuticle of the first instar fluoresces only weakly.

Development of the newborn scorpion on the dorsum of the mother continues during the first instar until the young scorpion is able to assume a fully independent existence after its first molt. There is a decrease in length and weight during the first instar, since these animals do not feed but instead draw on their internal yolk reserves (McAlister 1960; G. Polis, pers. obs.). However, Vannini et al. (1986a, b) have discovered that first instars receive at least water from their mother during the time that they stay with her. Evidently the water is transferred from the mother's cuticle and absorbed by the first instars. The first instar lasts from 1 to 51 days, depending on the species (Table 4.4). It passes rather quickly in the buthids, lasting an average of only 6 days. By contrast, the average duration for all other families ranges from 10 to 20 days.

The biochemistry of the chemical messengers that produce ecdysis in scorpions is unknown. So, too, are the precise cues that stimulate the ecdysis cycle. For many arthropods, a change in photoperiod or seasonal temperatures or an increase in weight beyond some threshold triggers ecdysis. For some species of scorpion it appears that the initiation of ecdysis is dependent proximally on weight gain and ultimately on prey availability. Thus, many species of buthids exhibit a wide range in instar duration (e.g., Matthiesen 1969, Probst 1972, Auber-Thomay 1974, Lourenço 1978, Armas and Contreras 1981). The instar duration of the scorpionid *Pandinus gambiensis* is also quite variable, with up to 11 months' difference in the length of a particular instar (Vachon and Condamin 1970). But other species molt more regularly. For example, immature *Urodacus yaschenkoi* molt once a year between December and February (Shorthouse 1971). R. Bradley (pers. comm. 1982) observed that equally aged animals of *Paruroctonus utahensis* all molt within a month of each other.

Currently, there is information concerning the number of molts to maturity for about 30 species and subspecies representing five scorpion families (Tables 4.4, 4.5). Scorpions achieve maturity in as few as four molts (some buthid males). In *Didymocentrus trinitarius*, females require nine molts to reach maturity (Armas 1982b). But these are the extremes: most species molt 5 or 6 times. On the average, buthids molt fewer times (5.1 ± 0.7) than all other families (5.8 ± 0.8).

In some species, different sexes or individuals of the same sex may undergo a different number of molts before maturity (Table 4.4). Such indeterminacy is particularly evident in the buthids, although it has been reported from four other families. In these species, males will often molt fewer times than females (Armas and Contreras 1981, Francke and Jones 1982, Armas 1986). The exact causes of this phenomenon are unknown; males may mature earlier because of the higher mortality rate that they experience as compared with females (see Chapter 6). In general, differences in feeding history between individuals of the same population or differences in the physical environment (e.g., temperature) or resource availability between populations of the same species may be important (see below).

Other arachnids molt both fewer and more times than scorpions (Kästner 1968, Weygoldt 1969, Savory 1977, Foelix 1982). The lowest number of molts (2) occurs in the spider *Mastophora*; the most molts (22) occur in the mygalomorph spider *Eurypelma*. Uropygids (6), Opiliones (6 or 7), and Schizomida (5) molt about the same number of times as scorpions. Pseudoscorpions and Ricinulei molt 4 times, and some Solifugae molt 8 or 9 times.

Reports of postreproductive molts do not appear in the literature. Most researchers indicate that their species do not molt as adults, although some are uncertain (Rosin and Shulov 1963; Vachon et al. 1970; R. Bradley, pers. comm. 1982). In *P. mesaensis* (Polis and Farley 1979b) and *U. yaschenkoi* (Shorthouse 1971) adults do not appear to molt; if they do, they do not show the same growth ratio that occurs between earlier molts. *A priori*, postreproductive molts would not be expected for males. At ecdysis all chitinous structures are replaced. The hemispermatophore of mature male scorpions is a complex sclerotized structure that is situated well inside the reproductive tract. Hemispermatophores would be difficult to shed and have never been associated with cast exuviae (G. Polis, pers. obs.; O. Francke, pers. comm. 1979). Postreproductive molts are uncommon among some groups of terrestrial arthropods (insects), but they occur regularly in myriapods and some arachnids, particularly in long-lived spiders (Savory 1977).

Environmental Influences on Life History

The exact life-history traits of any species are subject to modification by various environmental factors. In particular, the food supply may greatly influence growth and reproduction. Further, in poikilothermic ectotherms (such as scorpions), temperature directly affects metabolism, and indirectly affects growth.

Evidence that these factors influence growth comes from an analysis of changes in growth parameters and developmental times. For example, there were significant differences in both seasonal and year-to-year growth rates in a field population of *Paruroctonus mesaensis* (Polis and Farley 1979b). The growth rates of four cohorts (each spaced by one year) were significantly different from each other. Such annual differences are likely due to year-to-year differences in prey availability (Polis and McCormick, unpubl. data). Maximum growth occurred seasonally in the spring and fall; growth was lowest in the winter and summer, and actually negative in July (Fig. 4.14). These differences correspond to seasonal differences in prey availability and in metabolic rates: prey is much more abundant in the spring and fall compared with winter and summer, and metabolism is a function of seasonal differences in burrow temperature (Polis and Farley 1979b) and in ambient temperatures.

Equally aged individuals of the same population may differ greatly in

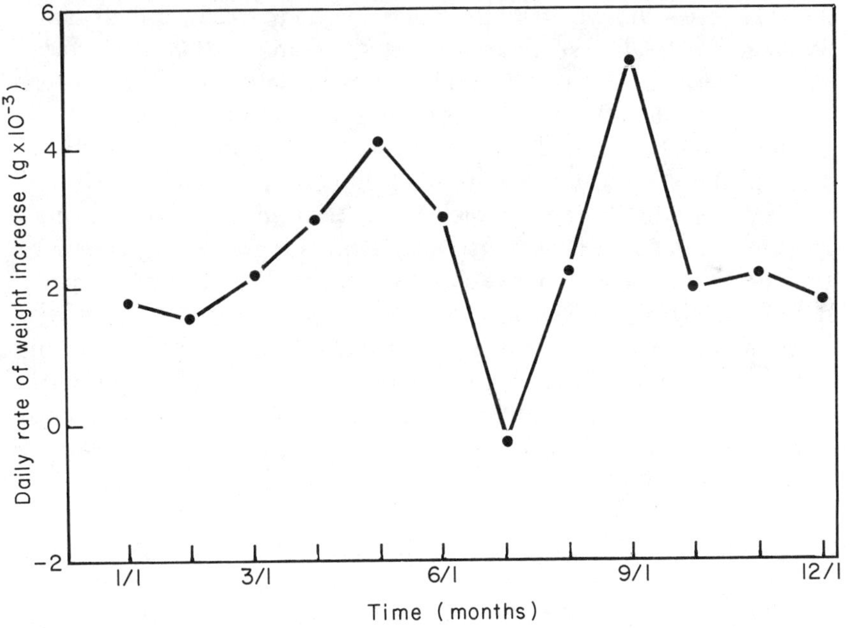

Fig. 4.14. Monthly average growth rate of *Paruroctonus mesaensis* in southern California. Maximum growth occurred in the spring and fall. Growth rates were uniformly low in the winter and actually became negative in summer (June). The average growth rate as calculated from data pooled from all months and ages was 0.0025 ± 0.0021 g/day. (From Polis & Farley 1979b)

size. This occurs within field populations (Polis and Farley 1979b; Polis, unpubl. data) and for many species in the laboratory (e.g., Armas and Contreras 1981, Sissom and Francke 1983). For example, prior to maturation, the largest *P. mesaensis* individuals in one age cohort are frequently more than twice as large as the smallest. Adults of other species, too, show marked differences in body size within the same population; thus, adults may be "small," "normal," or "giant." This phenomenon was observed in many species of *Centruroides* (Williams 1980; Armas and Contreras 1981; G. Polis, pers. obs.) and in *Tityus* (W. Lourenço, pers. comm. 1982), *Rhopalurus* (Armas 1986), and *Vaejovis punctipalpi* (S. Williams, pers. comm. 1980). Sizes may also vary significantly between different populations of the same species. Koch (1977, 1981) found that scorpions from the arid central regions of Australia were consistently larger than conspecifics from wetter areas. Adults from populations of *R. laticauda* increase in size by 70 percent with a 525-m increase in elevation (Scorza 1954). In North and South America several species show marked differences in body size over their range (Williams 1974, 1980; W. Lourenço, pers. comm. 1982; G. Polis, pers. obs.). For example, *V. confusus* individuals in Idaho are approximately two-thirds the length and width of those in Nevada (Anderson 1975); southern California populations of *V. confusus* are about half the size of those to the east, south, and north (comparisons from Gertsch and Allred 1965; Williams and Hadley 1967; Anderson 1975; S. Williams, pers. comm. 1980; G. Polis, pers. obs.). Populations of *P. mesaensis* separated by as little as 15 km may show 20 percent differences in the body size of adults (McCormick and Polis, 1986).

Differences in body size may result from feeding history, ambient temperature, differences in the number of molts to maturity, and/or genetic differences. Nonquantitative observation suggests that difference in food is the key determinant (Scorza 1954; Williams 1980, pers. comm. 1980; G. Polis, pers. obs.).

Developmental times may also be a function of prey availability and temperature. For example, embryonic size and developmental stage were significantly greater for embryos within *P. mesaensis* females fed *ad libitum* than for embryos in females whose food intake was limited (Polis and McCormick 1987). Decreased food and temperatures were noted to increase instar duration in the laboratory (Smith 1966; Lourenço 1978, 1979b; Armas and Contreras 1981). These observations may explain the frequent reports of large variation in instar duration and age to maturity (e.g., Shorthouse 1971; Probst 1972; Auber-Thomay 1974; Lourenço 1978, 1979b; Armas and Contreras 1981; Sissom and Francke 1983; see also Table 4.4).

Food supply and temperature also influence litter size and gestation period. Litter size is known to vary greatly in some species of scorpion (Table 4.2), and it may be that the feeding history of individual females determines this variation. It is known, for example, that females deprived of food resorb their embryos. Resorption decreases potential litter size by 16 to 22 percent in *Tityus bahiensis* (Matthiesen 1961), 27 percent in *Urodacus manicatus* (Smith 1966), 32 percent in *P. mesaensis* (Polis and Farley 1979b), and 35 percent in *P. boreus* (Tourtlotte 1974). However, in an experimental analysis of the relationship between food intake and litter size, Bradley (1983, 1984) found no significant differences in brood size or weight of newborn between *P. utahensis* females fed *ad libitum* and those on a low-food regime. Likewise, there were no differences in brood size between *P. mesaensis* females on high versus low feeding regimes (Polis and McCormick 1987). These findings stand in contrast to the widespread relationship between female size and brood size throughout the animal kingdom (Bradley 1984).

The period of gestation is likewise variable (see Table 4.2) and appears to be influenced by food and temperature. For example, Armas and Contreras (1981) noted that the gestation period of *Centruroides robertoi* increased from 60–88 days to 115–215 days when ambient temperatures were lowered. In *T. fasciolatus*, gestation averaged 97.6 days in the wet season; during the dry season, a diapause in embryonic growth occurred, and gestation lasted 7 to 10 months (Lourenço 1978, 1979b). The widespread occurrence of synchronous birth in species with mating periods that last several months indicates that the length of the gestation period is quite variable (see "Birth Dates," above).

In summary, it is clear that environmental factors can modify scorpion life history. This is true of other arthropods as well (e.g., Weygoldt 1969, Labeyrie 1978, Scriber and Slansky 1981, Foelix 1982, Huffaker and Rabb 1984). In scorpions, the data indicate that differences in food and temperature are particularly important, but another factor that may play a significant role is photoperiod. One of the primary goals of life-history research should be to describe the way that changes in these three environmental factors affect development and reproduction.

Because of the observed interaction between environment and life history, research conducted under laboratory conditions should be carefully evaluated. Specifically, since growth, ecdysis, developmental times, reproduction, and survival are all functions of food intake and temperature, laboratory estimates of these parameters probably only approximate what really occurs in nature. It is difficult, if not impossible, to maintain a feeding schedule, thermal regime, and photoperiod that represent actual field conditions.

Laboratory-reared individuals often suffer extremely high mortality (90–100 percent), in spite of the fact that they are raised in the absence of predation and potentially harsh environmental conditions. High mortality rates in the laboratory are due to stillborns, failure of newborns either to eclose from the birth membranes or to ascend to the female's dorsum, maternal cannibalism, or failure to feed or molt successfully (Stahnke 1966; Francke 1976b, 1979b, 1982c; Polis and Farley 1979b). Francke (1976b) noted that about one-third of the scorpions entering each ecdysis died in the process under laboratory conditions. Unfortunately, because of the secretive nature of birth and molting in the field, the natural frequencies of these mortality factors are essentially unknown.

The life history of those individuals that do survive may be modified under laboratory conditions. Thus, many researchers have noted that differences in laboratory conditions inhibit ecdysis, decrease growth, increase instar duration, and lengthen gestation period (Smith 1966; Stahnke 1966; Auber-Thomay 1974; Lourenço 1978, 1979b; Polis and Farley 1979b; Armas and Contreras 1981). This can result in inflated estimates of instar duration, time to maturity, and generation time.

Phenological differences in birth between laboratory and field have been observed (Williams 1966; Francke 1976b; Polis and Farley 1979b; G. Polis, pers. obs.). For example, birth for *P. mesaensis* occurred in the laboratory between September and December, but in the field occurred in late July and August (Polis and Farley 1979b). Estimates of clutch size from laboratory-reared animals may also be unreliable. Varela (1961) found that for *Bothriurus bonariensis* the number of offspring born in the laboratory was only 18 percent of the number born in the field.

Field estimates of life-history parameters are not perfect either (Polis and Farley 1979b; Francke and Sissom 1984; O. Francke, pers. comm. 1979, 1983). Growth, maturity, and ecdyses of individuals are not actually observed, but are estimates of population averages. Further, average size parameters from the population may represent a mean lower than the true population mean because hungry (thinner) animals are more likely to emerge to the surface to feed than satiated (heavier) individuals.

Field estimates of the number of molts and instars are obtained by measuring individuals of all sizes and plotting the data in a bivariate morphometric plot (Huxley 1932, Teissier 1960, Polis and Farley 1979b). The plot is then more or less arbitrarily inspected for gaps and clusters, which are equated to molts and instars, respectively. When large samples are plotted, gaps and clusters become quite obscure, and the potential for error increases. Small samples, on the other hand, introduce the possibility of not detecting one or more instars in the plot and do not permit adequate analysis of the size range of instars.

Field estimates of litter size must be carefully analyzed as well. Estimates are often obtained by collecting females with offspring on their backs and counting the young. Because offspring may be lost if they fall off the female's back, and because second instars may have dispersed already, these counts may be underestimates of the actual number born. With this in mind, counts involving first-instar young are probably more accurate. A second method is to dissect field-collected pregnant females and count the number of embryos. Caution is also advised here, because the female may contain embryos at different stages of development (to be born at different parturitions). Some species of scorpions (primarily buthids) are known to give birth to more than one litter per year (Table 4.3) and more than one litter per fertilization event. In these cases dissections would yield overestimates of litter size.

Obviously, measuring parameters of life history is a difficult task, and problems exist with both field and laboratory methods. Perhaps the best way to conduct a life-history analysis is to use field cages to enclose individual scorpions. In this manner, the growth of individuals can be monitored under nearly natural conditions of photoperiod, temperature, and prey availability. Further, manipulation of food abundance allows an experimental analysis of the interaction between food intake and various life-history parameters. Preliminary evidence (Bradley 1982, 1983, 1984) suggests that this method may be of great value for future workers.

Secondary Sexual Characteristics

Sexual dimorphism in scorpions was reviewed by Kraepelin (1907), Birula (1917a), Werner (1934), Koch (1977), and Farzanpay and Vachon

(1979). Secondary sexual characteristics may be placed in the following categories (modified from Koch 1977):

1. Differences in body size
2. Differences in the shape of body structures
3. The presence of a feature in one sex but not in the other
4. Stronger development of features in one sex than in the other
5. Differences in the texture of the body surface
6. Higher meristics in one sex than in the other.

Each of these is discussed below, as well as some nonmorphological differences.

In general, females are larger than males (Fig. 4.15). Differences in body length may be quantified by some absolute measurement, such as carapace length, which is not strongly affected by allometry. Females are also characterized as having more robust bodies, as indicated by morphometric ratios such as the length of the carapace to the width of sternite VII. Sexual dimorphism in body size appears to be extremely common among scorpions, although in many taxa the differences are not significant. Among Australo-Papuan scorpions, differences in carapace length may average as much as 40 percent, but they are sometimes less than 1 percent (calculated from measurements in Koch 1977). It is interesting to note, however, that females are not always the larger sex: the male of *Liocheles australasiae* averages 12.8 percent larger, as indicated by carapace lengths (Koch 1977); and Kjellesvig-Waering (1966b) noted that males of *Tityus trinitatis* on the islands of Trinidad and Tobago are 33 to 58 percent larger than females in total length.

The shape of structures may often be quite variable between the sexes. Kraepelin (1907) presented numerous examples of sexual dimorphism in shape, and a few of those will serve to demonstrate. Dimorphism is commonly expressed by elongation of the pedipalps and/or metasoma in males. Males from many taxa exhibit such lengthening to some degree. Extreme cases of pedipalp elongation are found in a few taxa, notably *Isometrus* and a few species of *Chactas* and *Tityus* (Kraepelin 1907, Mello-Leitão 1945, González-Sponga 1978c). Extreme elongation of the male metasomal segments (Fig. 4.16) is more common, for example, in *Centruroides*, some *Hadogenes, Hemiscorpius, Isometrus,* and some *Urodacus* (Kraepelin 1907, Koch 1977). Elongation of the metasomal segments in the male of *Hadogenes troglodytes* makes that species one of the longest scorpions in the world at 21.0 cm (Newlands 1974b).

Fig. 4.15. Sexual dimorphism in body size of *Vaejovis gravicaudus:* adult male (left) and adult female (right) from Puerto Escondido, Baja California Sur.

Sexual dimorphism in shape affects other structures as well. In many cases (e.g., *Centruroides, Heterometrus, Isometrus,* and *Tityus*) the male pedipalp chela is longer and narrower than in the female, but in other cases (e.g., *Buthus,* some *Tityus,* and *Scorpio*) the situation is reversed. The telson of females is generally larger and more rounded than in males (Fig. 4.17), but there are cases (e.g., *Anuroctonus* and *Euscorpius*) in which the reverse is true. In *Anuroctonus phaiodactylus* the male telson bears a swelling at the base of the aculeus (Fig. 4.17); in *Hemiscorpius* the male telson is also modified, being elongate and bilobed distally.

Perhaps some of the best-known examples of sexual dimorphism in

Fig. 4.16. Sexual dimorphism in metasoma length of *Hadogenes troglodytes: above,* adult male; *below,* adult female. (Photo courtesy Gerald Newlands)

Fig. 4.17. Sexual dimorphism in telson shape: *above,* telson of male and female *Anuroctonus phaiodactylus; below,* telson of male and female *Centruroides vittatus.*

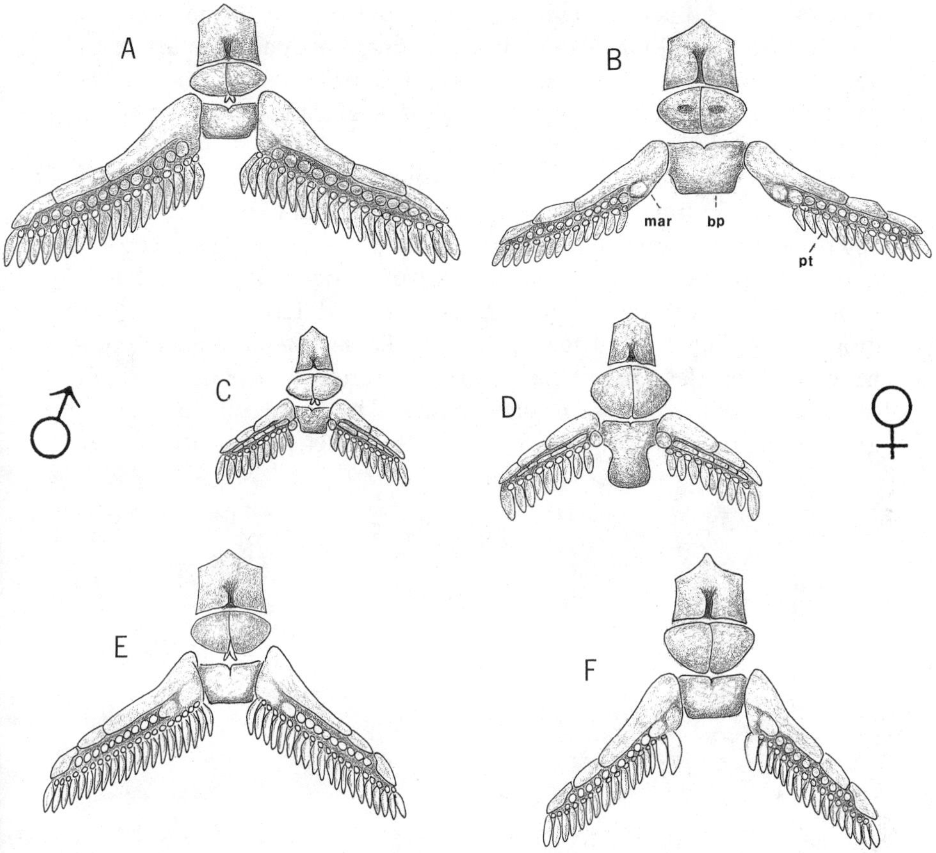

Fig. 4.18. Sexual dimorphism in pectinal structure. *A, B,* male and female *Paruroctonus luteolus;* note the length of the pectinal teeth (*pt*) and the angle formed by the medial margin of the pecten (*mar*) and the basal piece (*bp*). *C, D,* male and female *Microtityus waeringi;* note the expanded basal piece of the female. *E, F,* male and female *Serradigitus wupatkiensis;* note the enlarged proximal pectinal teeth of the female.

shape occur in the pectines (Fig. 4.18). Females are characterized by having smaller pectines, with shorter and straighter teeth. The medial margins of the pectines of females form an obtuse angle with the posterior edge of the basal piece; the medial margins in the male form more of a right angle with the basal piece. Sexual dimorphism affects the pectines in other ways as well. In *Microtityus* females, the basal piece of the pectines is broadly expanded posteriorly. Sexual modifications in the medial lamellae and/or proximal pectinal teeth are exhibited in other taxa, including *Grosphus*, *Opistophthalmus opinatus*, *Parabuthus liosoma*, many *Tityus*, and *Serradigitus* (see Kraepelin 1907; Stahnke 1974; W. Lourenço, pers. comm. 1983).

Polis and Farley (1979b) examined growth in the pectines of *Paruroctonus mesaensis*, and their analysis probably applies to most other sexually dimorphic characters. In females, the pectines grow at a constant linear rate throughout life; in males, however, pectinal growth demonstrates an early-life linear growth rate and a different late-life linear growth rate (Fig. 4.19). The slopes of the linear growth rates of young males and females are not significantly different, but both are significantly different from that of older males. The change in the pectinal growth rate in males probably occurs at the onset of maturity.

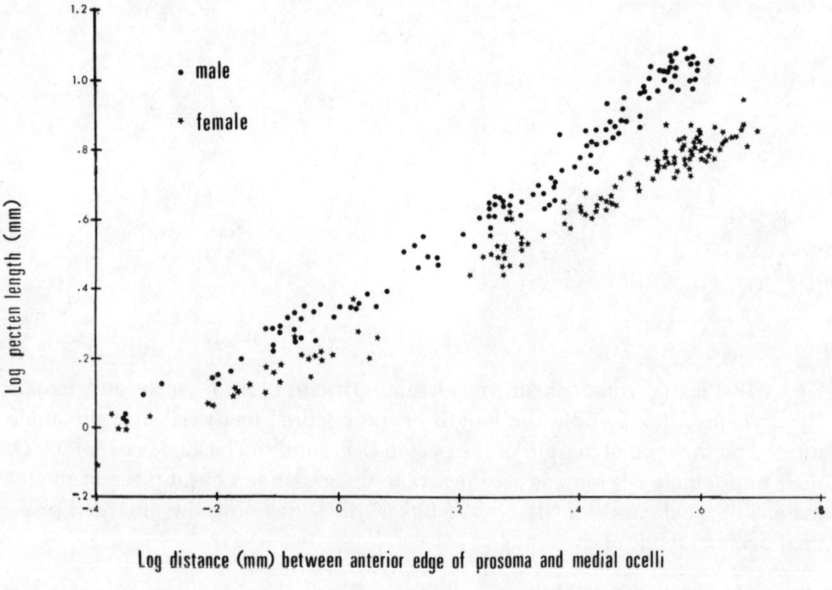

Fig. 4.19. Sexual dimorphism in the pectinal growth rate of *Paruroctonus mesaensis*. The logarithms for the lengths of these structures for males and females are plotted against each other. Pectines are shown to be a secondary sexual characteristic, because the growth rate of the pectines in older males is significantly greater than that either in young males or in all females. (From Polis & Farley 1979b)

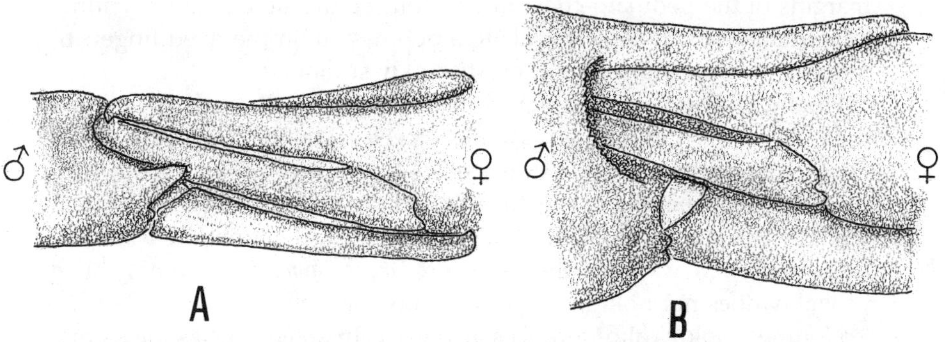

Fig. 4.20. Structure and function of sexually dimorphic features of the pedipalp chelae in male bothriurids: *A*, *Brachistosternus alienus,* male chela grasping female with the aid of an apophysis on the inner surface of the chela; *B*, *Timogenes mapuchi,* male chela grasping female with the aid of a depression on the inner surface of the chela. (After Maury 1975a)

The presence of extra features in one sex occasionally occurs among scorpions. In bothriurids, the male pedipalp chela is modified in one of two ways (Maury 1975a), so that the inner face displays either an apophysis or a semicircular depression at the base of the fixed finger. The apophyses and depressions in the male appear to aid him in securing the female's chelae during mating (Fig. 4.20; see Maury 1975a).

Two other characters are unique to males. In most taxa mature males may be easily identified by the presence of a pair of genital papillae (crochets) protruding from the posterior margin of the genital operculum. Genital papillae are notably absent in some taxa (e.g., *Hadrurus*). The second character is a pair of caudal exocrine glands. Such glands are present on the dorsal aspect of the fifth metasomal segment in males of *Brachistosternus* and *Timogenes* (Bothriuridae) and on the dorsal aspect of the telson in *Timogenes, Bothriurus* (Bothriuridae), and *Hadrurus* (Iuridae) (Williams 1970a; Cekalovic 1973; Maury 1975b, 1977).

There are also sexual differences in carinal development and the texture of the body surfaces. Males typically possess stronger carinae on the pedipalps and metasoma. However, in some Buthidae (e.g., *Centruroides, Rhopalurus, Tityus*) females have stronger carinae. The carapace, mesosoma, and metasoma are usually more granular in males. Males of many scorpion species possess distinct scalloping on the inner margins of the pedipalp-chela fingers: on the movable finger a distinct lobe occurs, which corresponds to a depression on the fixed finger. In females, these margins are at most weakly scalloped.

Finally, the sexes may differ in meristic characters, such as pectinal tooth counts. Males tend to have more pectinal teeth than females, and in some cases (e.g., many *Paruroctonus*) the counts do not overlap. Although it is generally taken for granted that the number of pectinal teeth is sexually dimorphic, Lourenço (1981c) demonstrated statistically that such was not the case for *Tityus cambridgei*. Dimorphism probably does not hold in many other taxa as well.

Nonmorphological differences also exist. In some species males mature in fewer days or fewer molts than females (Table 4.4). Conversely, females have never been observed to mature earlier or with fewer molts than males. For example, males of *T. bahiensis* mature in 16 months, whereas females take almost two years to mature (Matthiesen 1969). These same sexual differences occur in spiders (Savory 1977).

Behavioral and ecological differences also exist. In general the survivorship and average longevity of males appear to be lower than for females (see Chapter 6). The mortality rate for adult males is much higher than for adult females; this difference produces the heavily skewed sex ratio (♂♂:♀♀ = 1:2–1:4) observed for many species of scorpion (see Table 6.3). Differential mortality is attributed to several behavioral differences between males and females. Mature males of many species are often vagrant and mobile during the mating season (see below); this movement predisposes them to a disproportionately higher incidence of cannibalism, predation by other scorpion species or by vertebrate and invertebrate predators, starvation, and thermal death (Chapter 6; Polis and Farley 1979a). This mobility also makes it likely that mature males feed less and construct more temporary burrows than females or young males. In *P. mesaensis*, mature males feed proportionately less often than mature females (Polis and Farley 1979a), and the burrows of adult males are significantly shallower than those of adult females (Polis, unpubl. data).

Sexual differences in behavior and ecology pose problems for the analysis of scorpion populations when data are collected with can traps. Analysis of such data almost always indicates a severe trap bias for mature males (Gertsch and Allred 1965; Hibner 1971; Roig-Alsina 1978; Polis, unpubl. data). The bias is almost certainly the result of male movement during the breeding season (Allred 1973, Polis and Farley 1979a).

Similar sexual differences in mortality and behavior also occur in spiders and may occur generally among arachnids.

Reproductive Behavior

There is a regular geographical variation in the mating season. Mating in temperate areas occurs during the warm months. Thus, in the Northern Hemisphere, matings occur from April through October, with peaks in May and September. In the Southern Hemisphere, matings occur from September through February, with insufficient data to detect trends. (The one Southern Hemisphere exception is *Urophonius brachycentrus*, which mates in the winter during the peak activity of its "inverted life cycle"; see Chapter 6 and Maury 1968.) No monthly trends in the frequency of mating are discernible for tropical species. Four species were observed to mate in all months of the year; other species were observed to mate during more restricted periods, but it is likely that these species have a wider (but unobserved) mating period.

Mature males of many (most?) species of scorpion abandon their burrows during the mating season and travel on the surface, apparently in search of mature females. During this period, the only time when males of obligate fossorial species are found on the surface (Williams 1966; Shorthouse 1971; Anderson 1975; Koch 1978; Lamoral 1979; S. Williams, pers. comm. 1980), males are frequently observed moving, and may travel great distances. The behavior of *Paruroctonus mesaensis* will serve as an example (Polis and Farley 1979a). Mature males during the breeding season are more frequently observed to be moving at the time of first observation than all other components of the population: 48.3 percent, versus 5.2 percent for mature males out of the breeding season and 1.8 percent for mature females and all immature scorpions. Further, on the average, mature males move farther than mature females between successive sightings on different nights (34.7 ± 24.9 m vs. 4.0 ± 8.7 m). Some marked individuals were observed to travel more than 100 m in a night and over 23 m in half an hour. This movement is seasonal and corresponds to the mating period (Fig. 4.21). Similar vagrant behavior was observed in the field for many other species (Williams 1970a; Tourtlotte 1974; Roig-Alsina 1978; Lamoral 1979;

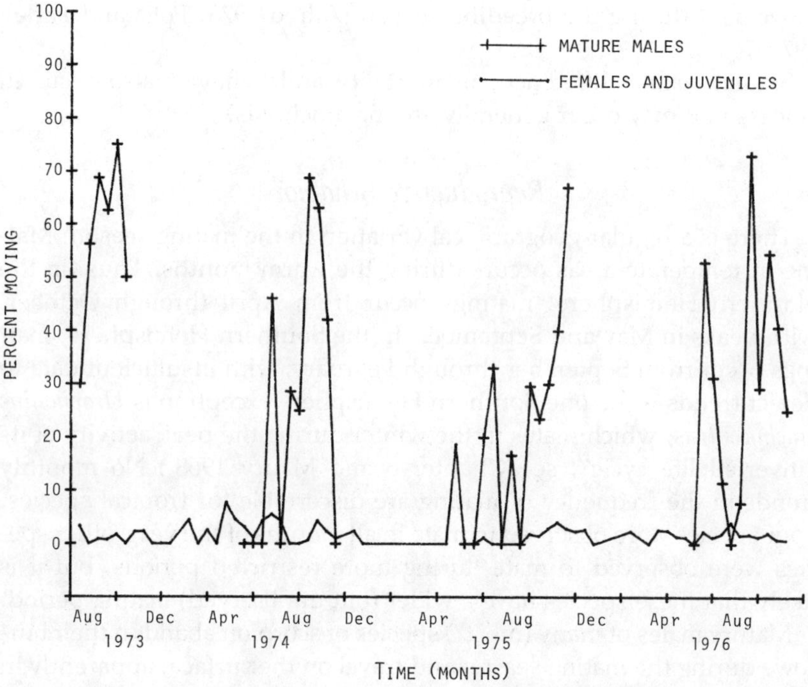

Fig. 4.21. Seasonal movement of mature *Paruroctonus mesaensis* males during their breeding season, from late spring through early fall. Such behavior appears to be widespread among mature male scorpions during the breeding season. (From Polis & Farley 1979a)

Polis and Farley 1979a; Polis, unpubl. data). Similar movement by mature males during the mating season is probably widespread among spiders and other arachnids. This movement by mature males represents the main avenue of gene flow among populations.

ACKNOWLEDGMENTS

We wish to thank Herbert W. Levi, Wilson R. Lourenço, and Stanley C. Williams for reviewing the manuscript and providing many valuable suggestions. Likewise, we wish to thank Stanford Press, and in particular William W. Carver and Jean McIntosh, for comments that improved the chapter. Wilson Lourenço, Gerald Newlands, and Stanley Williams kindly provided most of the photographs used in this chapter. The California Academy of Sciences granted permission to use Dr. Williams' photographs, which had appeared in the *Proceedings*.

A number of colleagues have allowed us to use their unpublished observations: Richard Bradley, Denise Due, Oscar F. Francke, David Lightfoot, Wilson Lourenço, Sharon McCormick, and Stanley Williams. David Shorthouse has allowed access to his doctoral dissertation, much of which is yet unpublished. Finally, Oscar Francke has given us years of stimulating discussion about scorpion life history, and his contribution is sincerely appreciated.

5

Behavioral Responses, Rhythms, and Activity Patterns

MICHAEL R. WARBURG AND

GARY A. POLIS

Scorpions perceive the environment through a diversity of sensory channels. Various stimuli produce a number of behavioral responses that adapt the individual to its environment. Biological rhythms are an example of a physiological-behavioral adaptation to regular changes in the environment. Such rhythms are modified by less predictable physical factors (e.g., precipitation, temperature fluctuations) and by biological factors (e.g., presence of predators, prey, and mates) to produce daily and seasonal patterns of surface activity.

Behavioral Responses

There is evidence to suggest that scorpions possess special receptors for chemical stimuli, located on their chelicerae (in *Opistophthalmus latimanus*: Alexander and Ewer 1957). Furthermore, they can locate ambient air currents by means of the trichobothria located on the pedipalps (Linsenmair 1968; D. Krapf, pers. comm. 1985). Brownell (1977) found that the sand scorpion *Paruroctonus mesaensis* responds to substrate vibrations by detecting surface waves of low velocities. The compound slit sensillae responsible for this perception are located on the basitarsal leg segment. Brownell was able to show that this scorpion could precisely locate the direction and distance of prey (e.g., the sand cockroach *Arenivaga investigata*) up to 50 cm away. At greater distances only the direction of the prey is sensed (Brownell and Farley 1979a). In addition, sensory hair sensillae on the tarsi of *P. mesaensis* respond to high-frequency compressional waves that travel through the substrate (Brownell and Farley 1979b).

Hygroreaction

One species of scorpion, *Leiurus quinquestriatus*, exhibited a positive hygrokinetic response to water vapor when the scorpion was in either the hydrated or the dehydrated state (Abushama 1964). Hygroreceptors are apparently located on the distal tarsal segments of the legs. In these experiments (which lasted up to 5 hrs) the humidity reaction of the scorpion could be altered by its strong negative protoreaction. Abushama also showed that scorpions were incapable of detecting small differences in relative humidities (58, 63, and 67 percent).

Thermoreaction

Although scorpions in general are thermophilic arthropods, there are interspecific differences in their thermal preferences. Thus, *Opistophthalmus latimanus* "preferred" a thermal zone of 32–38° C (Alexander and Ewer 1958), whereas *Heterometrus petersii petersii* limited its preference to a thermal zone of 34–35° C (Nemenz and Gruber 1967). The thermoreceptors in *O. latimanus* are located on the pedipalps and the poison bulb. In experiments lasting up to a day, Abushama (1964) found that *Leiurus quinquestriatus* stayed for the longest period in a broad thermal zone of 15–39° C. In experiments lasting for only 30 minutes (Warburg and Ben-Horin 1981), this scorpion has shown a preference for the narrower zone of 24–27° C (Fig. 5.1 and Table 5.1). Other scorpions exhibit either a pattern similar to *L. quinquestriatus* (e.g., *Hottentotta judaica*) or a slightly lower thermal preference zone, 23–25° C (e.g., *Nebo hierichontichus* and *Scorpio maurus fuscus*: see Figs. 5.1 and 5.2, and Table 5.1; Warburg and Ben-Horin 1981). In the field, different species and different age groups of the same species are maximally active at different temperatures (see below).

Photoreaction

Scorpion behavior generally follows a nocturnal pattern, which is often associated with marked photonegative behavior. The response to photic stimuli was studied in *Euscorpius italicus* by Angermann (1957), who showed that a photonegative response is already present in juve-

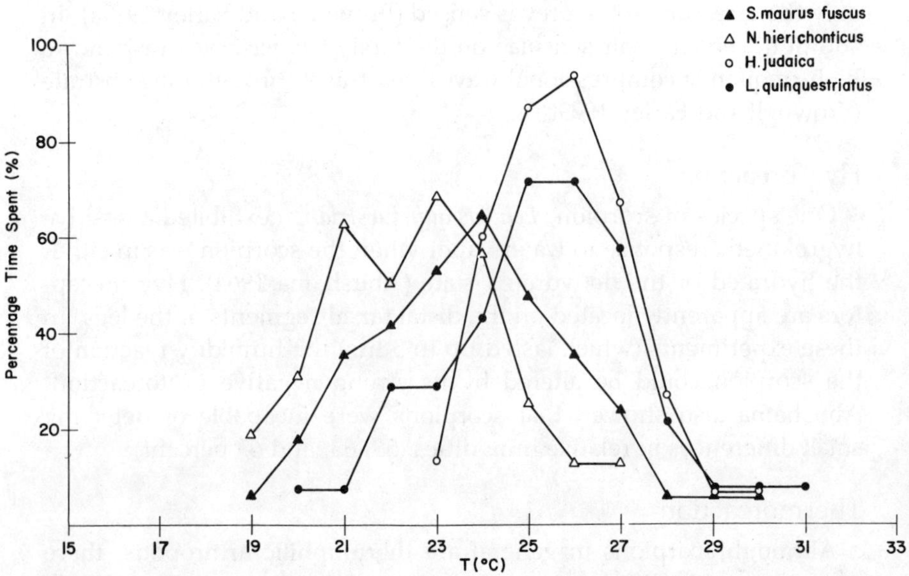

Fig. 5.1. Thermal preference by four species of scorpion. Scorpions were placed in a thermal gradient and allowed to choose their location freely. The data indicate that different species prefer different temperature ranges. (For details, see Warburg & Ben-Horin 1981.)

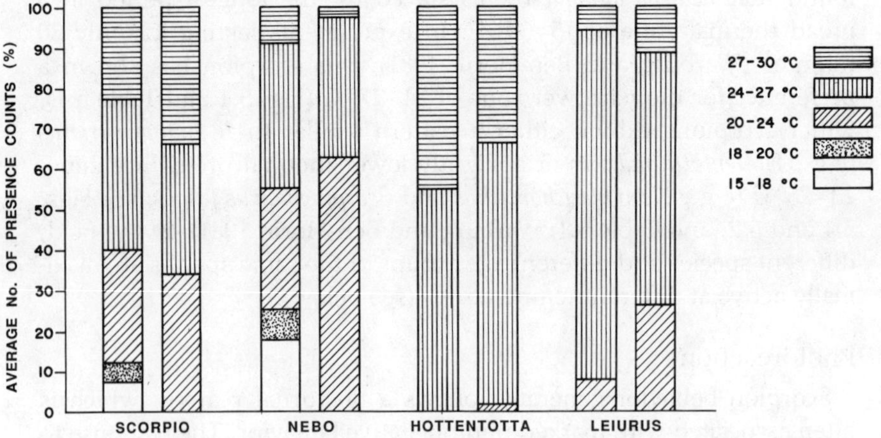

Fig. 5.2. Thermal preference by four species of scorpion. Scorpions were exposed to two different temperature gradients (15–30° C and 20–30° C) and allowed to choose freely their "preferred" locations.

TABLE 5.1
Scorpion temperature preference

Temperature (°C)	Number of scorpions: *Hottentotta judaica* (Simon)	*Leiurus quinquestriatus* (H. & E.)	*Nebo hierichonticus* E. Simon	*Scorpio maurus fuscus* L.
15				
16				
17				
18				
19			3	1
20		1	5	3
21		1	10	6
22		4	8	7
23	2	4	11	9
24	9	6	9	11
25	13	10	4	8
26	14	10	2	6
27	10	8	2	4
28	4	3		1
29	1	1		1
30	1	1		1
31		1		
TOTAL	54	50	54	58
Mean temperature ± S.D.	25.8 ± 1.5	25.3 ± 2.2	22.6 ± 2.0	23.8 ± 2.3

NOTE: Scorpions were placed in a thermal gradient, 15–31° C, and individuals were allowed to move freely within the chamber. The number of individuals resting at each temperature was recorded. Total frequency indicates the sum of individuals in all trials. (For details, see Warburg & Ben-Horin 1981.)

niles three to five days old; this pattern lasts throughout life. This scorpion also manifests a positive response to a black surface (skototaxis). In *Leiurus quinquestriatus*, the negative photokinetic response is predominantly a reaction to the light intensity rather than to its direction (Abushama 1964). Scorpions in general have long been known to "prefer" the lowest light intensity. A similar pattern of behavior was observed by Torres and Heatwole (1967) for several scorpion species: *Opisthacanthus lepturus*, *Cazierius scaber*, *Centruroides nitidus*, *C. vittatus*, *Tityus obtusus*, and *T. cambridgei*.

Photoreceptors

Photoreception occurs primarily in the lateral eyes of most scorpion species; the median eyes are generally less responsive to light (Machan 1968). There is also some evidence for extraocular photoreception at the tip of the tail or at the sting (see Chapter 9; Torres and Heatwole 1967b, Zwicky 1970b).

In an extensive series of studies, Fleissner and his coworkers were able to demonstrate a difference in function between the lateral and median eyes of *Androctonus australis* (Fleissner 1974). The lateral eyes are better suited as receptor organs for phase-shifting stimuli (Fleissner 1975); thus, they do not exhibit marked circadian rhythms (Fleissner 1977b). But they are also more sensitive to night vision than the median eyes, and they apparently possess a lower-sensitivity oscillator (Fleissner 1977d). This would mean that they function as light detectors for *Zeitgeber* stimuli (Fleissner 1977c, Schliwa and Fleissner 1980).

In contrast, the median eye shows an electroretinogram pattern that coincides with the circadian rhythm (Fleissner 1977b). This signifies that the median eyes apparently change sensitivity endogenously under the control of a circadian oscillator (Fleissner 1977d).

Because of the great sensitivity of their eyes to light, scorpions are apparently able to orient toward light reflecting from stones on a clear moonless night and may actually be able to orient by starlight (Fleissner 1977a). Some scorpions are much less active during moonlit nights (see below), and Fleissner (1977d) has suggested that such inhibition may be caused by the relatively intense luminescence of the moonlight. Light on bright moonlit nights is nine logarithmic units above the absolute threshold of the median eye and ten log units above the night phase of the lateral eyes (Fleissner 1977a). Scorpions in the laboratory, however, are capable of orienting at night even under a bright light (K. Linsenmair, pers. comm. 1984).

Biological Rhythms

The biological rhythms observed in scorpions so far belong to two types (see Table 5.2). The first type consists of activity patterns that are entrained by various environmental factors. Such a factor is called a *Zeitgeber*, or reference point, and may be a light impulse or a thermal fluctuation (Cloudsley-Thompson 1961, 1978). Light is known to be the main entraining factor for the cyclic pattern of scorpion activity. The temperature effect on rhythmic activity was also studied (see Figs. 5.3–5.7; Warburg and Ben-Horin 1979). The second type is characterized by self-sustained endogenous oscillations that persist under either constant illumination or constant darkness.

Table 5.2

Studies of scorpion locomotory and physiological rhythms

Taxon	Rhythm	Reference
SCORPIONIDAE		
Hadogenes bicolor (Purcell)	L	Constantinou 1980
Heterometrus fulvipes (Koch)	E	Venkateswara & Govindappa 1967; Venkatachari & Muralikrishna Dass 1968; Chengal Raju et al. 1973; Chandra Sekhara Reddy & Padmanabha Naidu 1977; Chandra Sekhara Reddy et al. 1978; Mastanaiah et al. 1977, 1978, 1979; Jayaram et al. 1978
	H	Devarajulu Naidu & Padmanabha Naidu 1976
	NC	Pampathi Rao & Gropalakrishnareddy 1967
	NSC	Pampathi Rao & Habibulla 1973, Uthaman & Srinivasa Reddy 1985
	R	Uthaman & Srinivasa Reddy 1979
H. swammerdami Simon	L	Cloudsley-Thompson 1981
Pandinus exitialis (Pocock)	L	Cloudsley-Thompson 1963
P. gregoryi (Pocock)	L	Constantinou 1980
P. imperator (C. L. Koch)	L	Toye 1970
	NC	Goyffon et al. 1975
Scorpio maurus L.	L	Cloudsley-Thompson 1956
S. m. fuscus H. &. E.	L	Warburg & Ben-Horin 1979
Urodacus sp.	H	Zwicky 1970b
BUTHIDAE		
Androctonus australis (L.)	NC	Goyffon et al. 1975
	NSC	Fleissner & Schliwa 1977
	L	Constantinou & Cloudsley-Thompson 1980
Babycurus buttneri Karsch	L	Cloudsley-Thompson 1975
Buthus occitanus E. Simon	L	Cloudsley-Thompson 1956, Constantinou 1980, Constantinou & Cloudsley-Thompson 1980
Centruroides exilicauda (Wood)	L	Crawford & Krehoff 1975
Hottentotta hottentotta (Fabricius)	L	Toye 1970
H. judaica (Simon)	L	Warburg & Ben-Horin 1979
H. minax (C. L. Koch)	L	Cloudsley-Thompson 1963, 1973a
Leiurus quinquestriatus H. & E.	L	Cloudsley-Thompson 1961, Abushama 1962, Warburg & Ben-Horin 1979
DIPLOCENTRIDAE		
Didymocentrus lesueurii (Gervais)	NC	Carricaburu & Muñoz-Cuevas 1986b
Diplocentrus peloncillensis[a] Francke	L	Crawford & Krehoff 1975
Nebo hierichonticus E. Simon	L	Warburg & Ben-Horin 1979
CHACTIDAE		
Euscorpius carpathicus L.	R	Dresco-Derouet 1961
E. flavicaudis (DeGeer)	NC	Carricaburu & Muñoz-Cuevas 1987
E. italicus (Herbst)	R	Dresco-Derouet 1961

NOTE: E, enzyme; H, heart; L, locomotor; NC, nerve cord; NSC, neurosecretory cells; R, respiration.
[a]Population reported as *D. spitzeri* Stahnke but redescribed by Francke (1975).

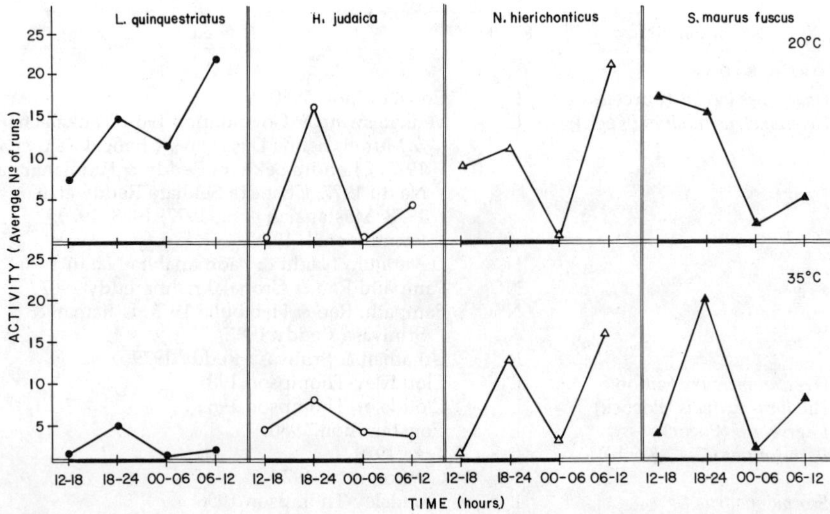

Fig. 5.3. Diel activity of four scorpion species at low (20° C) and high (35° C) temperatures. Activity is expressed in "runs," or movement in an actograph. Each species exhibits marked changes in daily activity as a function of temperature. (For details, see Warburg & Ben-Horin 1979.)

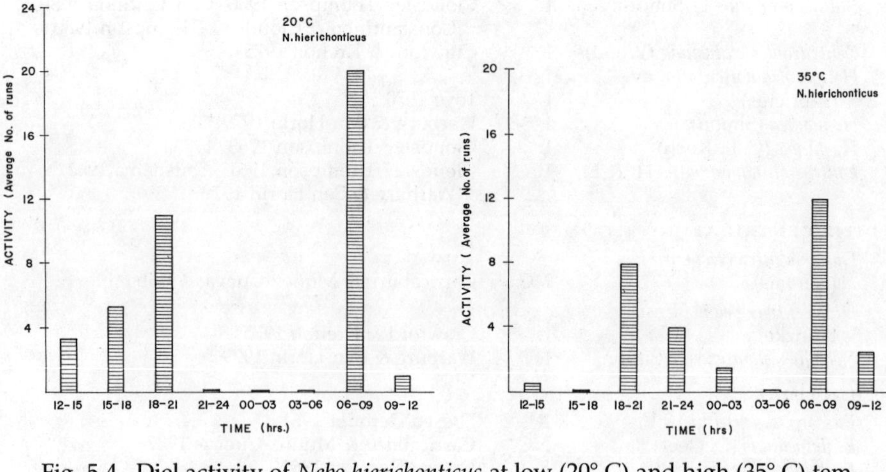

Fig. 5.4. Diel activity of *Nebo hierichonticus* at low (20° C) and high (35° C) temperatures. Activity is expressed in "runs," or movement in an actograph. (For details, see Warburg & Ben-Horin 1979.)

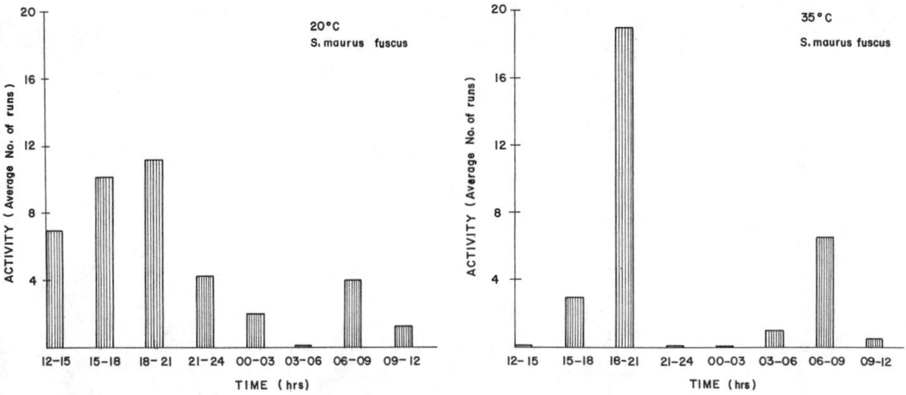

Fig. 5.5. Diel activity of *Scorpio maurus fuscus* at low (20° C) and high (35° C) temperatures.

Rhythmic Locomotory Activity

Laboratory research using actographs indicates that scorpions generally show an activity pattern that is highest during the first few hours of the night (Cloudsley-Thompson 1978, 1981). Thus, Cloudsley-Thompson (1956) found a peak of activity during the hours 1800–2000 in *Scorpio maurus, Buthus occitanus,* and *Androctonus australis*. This last species was found to be strongly nocturnal (Constantinou 1980). A similar pattern was observed in *Hottentotta minax, Pandinus exitialis,* and *Leiurus quinquestriatus* (Cloudsley-Thompson 1963). *Leiurus quinquestriatus* was found to have an additional peak of activity just before midnight (Abushama 1962). These studies were conducted at 30° C and 15 percent relative humidity in natural daylight and darkness (April–June). Similar small peaks of activity were observed at various temperatures in other species as well (*H. judaica, S. m. fuscus* and *Nebo hierichonticus*: Warburg and Ben-Horin 1979). Abushama observed that when the rhythm consisted of 6-hour light intervals and 12 to 18 hours of darkness, the response was less marked. Peaks of activity also occurred at the highest temperatures (32–38° C). He concluded that *L. quinquestriatus* was a nocturnal animal with an endogenous rhythm. A similar observation was noted by Crawford and Krehoff (1975) for *Centruroides exilicauda*.

The European scorpion *Euscorpius carpathicus* exhibited peaks of activity similar to those of desert species—namely, during the first hours

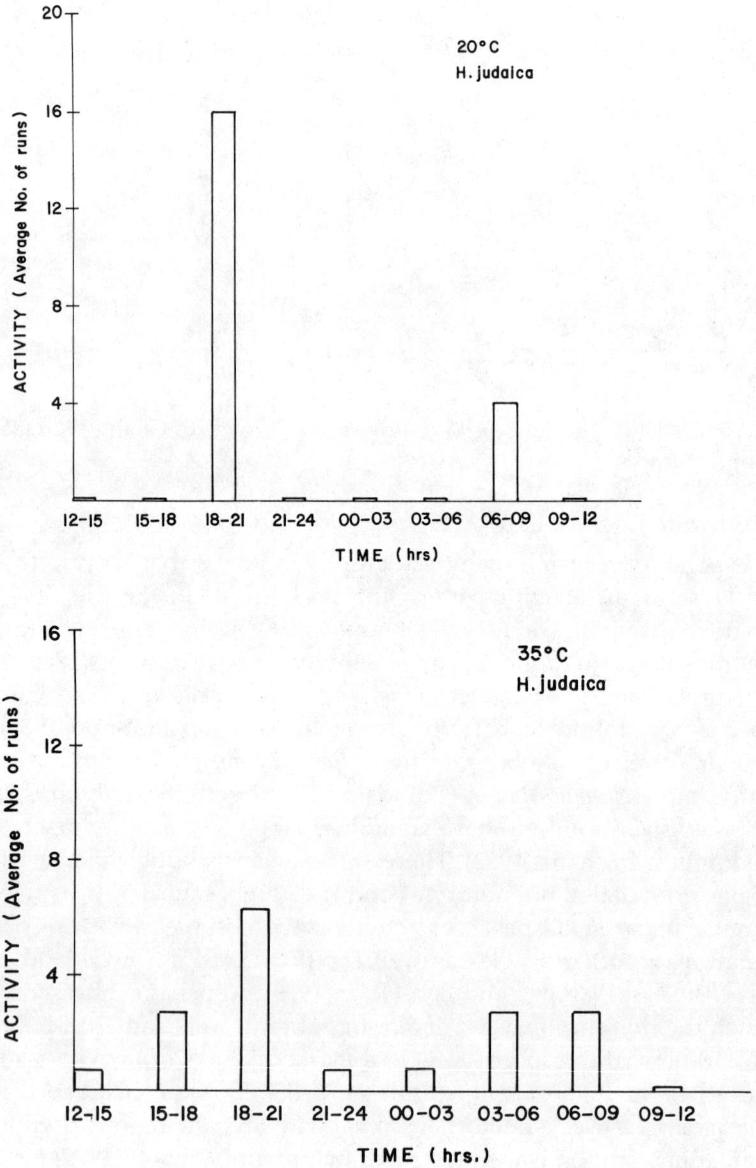

Fig. 5.6. Diel activity of *Hottentotta judaica* at low (20° C) and high (35° C) temperatures.

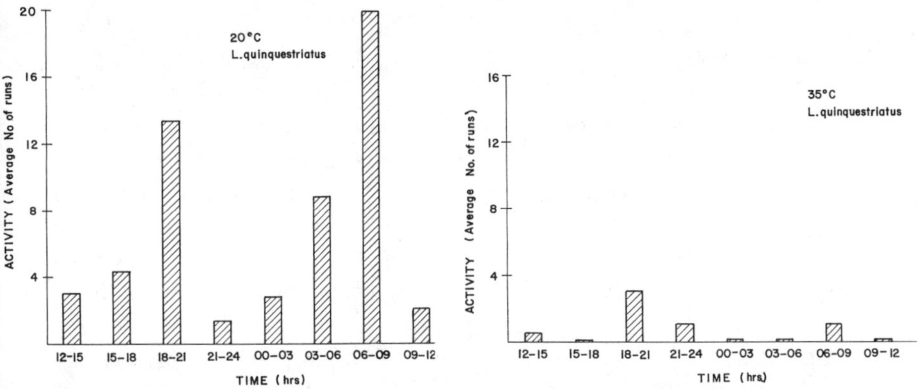

Fig. 5.7. Diel activity of *Leiurus quinquestriatus* at low (20° C) and high (35° C) temperatures.

of darkness, 1900–2100 (Wuttke 1966). Activity dropped considerably after midnight. Wuttke further observed a different type of hunting activity ("doorkeeping") that does not involve locomotory activity (see also Chapter 7). Field observations show that *S. m. fuscus* females also remain at the entrance of their burrows during the first hours of the night. Nemenz and Gruber (1967) observed similar doorkeeping behavior in *Heterometrus longimanus* females, and Shorthouse (1971) observed this pattern in *Urodacus yaschenkoi*.

Toye (1970) studied the diurnal activity of two tropical species. *Pandinus imperator*, under seminatural conditions, was active during the day and night, especially at 1500–1800 and 1800–2100. The activity dropped toward morning (0600). This pattern of behavior persisted under constant darkness, and to a certain extent under constant light as well. In the other case, *Hottentotta hottentotta* was active primarily during the early morning (0600–0900) and early night (1800–2100). Under constant light no definite pattern emerged, whereas under constant darkness intense activity took place at 0600–1200 rather than at the normal hours (1800–2100). Similar activity was exhibited by a semidiurnal species, *Hadogenes bicolor* (Constantinou 1980). In another tropical scorpion, *Opisthacanthus* sp., no clear-cut rhythm could be observed (Cloudsley-Thompson and Constantinou 1985a). So far Wuttke (1966) is the only one to describe a 3-to-4-day rhythm superimposed on the 24-hour daily rhythm. The purpose of such double rhythms is uncertain. Thus desert scorpions seem to be largely nocturnal, whereas tropical species are more diurnal (Constantinou and Cloudsley-Thompson 1980).

Cloudsley-Thompson (1973a, 1975) found that rising temperatures are an important synchronizing factor in the entrainment of circadian rhythms. In this respect, they are similar to the transition from light to dark. These studies were conducted on *Hottentotta minax* and *Babycurus centrurimorphus*.

Increased temperature influences different scorpion species in one of two ways (Warburg and Ben-Horin 1979). The first, an increase in activity during normally nonactive periods, is illustrated by *Nebo hierichonticus* (Figs. 5.3 and 5.4). The second, a drop in overall activity at high temperature, is manifested by *Scorpio maurus fuscus, Hottentotta judaica*, and especially *Leiurus quinquestriatus* (Figs. 5.5–5.7). Cloudsley-Thompson (1981) found that the locomotory activity of *Heterometrus swammerdami* is affected by temperature. The literature suggests that circadian rhythms may depend to a certain extent upon the animal's state of water balance (Cloudsley-Thompson 1963), but so far there is no direct support of this contention (Toye 1970), and direct tests are still needed.

Rhythmic Behavior in Other Arachnids

Rhythmic behavior in spiders has been studied by a few authors (Cloudsley-Thompson 1957, Tongiorgi 1959, Krafft 1970, Le Guelte and Ramousse 1979, Seyfarth 1980; and see recent review in Cloudsley-Thompson 1987). Both locomotory activity (Seyfarth 1980) and web-building activity (Le Guelte and Ramousse 1979) apparently peak at night for the spiders studied (see also Williams 1962, Edgar and Yuan 1968). In some species the pattern is similar to that found in scorpions: activity peaks during the first three hours of the night (Seyfarth 1980). Furthermore, the nature of this cyclic activity was endogenous and continued even in constant darkness. In amblypygids a similar peak of activity occurs during the early hours of the night, with another peak before sunrise (Beck and Görke 1974). The arboreal tarantula *Avicularia* also shows a weak bimodal rhythm of locomotory activity (Cloudsley-Thompson and Constantinou 1985b).

Rhythmic Patterns in Physiological Phenomena

The observed cyclic patterns in locomotory behavior are reflected in various physiological phenomena, such as enzymatic activity and respiratory functions. Rhythmic activity in enzyme level was studied most intensively in *Heterometrus fulvipes* (Table 5.2). In these studies enzyme activity peaked at 2000 hr and reached a low at 0800. The pattern is related to this scorpion's highest peak and lowest levels of locomotory activity. Thus, Venkateswara Rao and Govindappa (1967) found a peak in the action of muscle dehydrogenase at night. The need for more energy during the peak periods of locomotor activity is also indicated both by the increased activity of acid and alkaline phosphatases during the night (Chandra Sekhara Reddy and Padmanabha Naidu 1977) and by the increased lipid activity in the afternoon (1600–2000: Mastanaiah et al. 1978). Similarly, the muscle-activity pattern is reflected in the maximum level of acetylcholine esterase during the night (Chandra Sekhara Reddy et al. 1978). This increase in enzyme activity is accompanied by an increase in the electrical activity of the nerve cord during the afternoon (1600); the lowest activity occurs at 0400 (Venkatachari and Muralikrishna Dass 1968). The heart muscle shows a similar pattern of activity (Devarajulu Naidu and Padmanabha Naidu 1976). The glycogen content of heart muscle was lowest at 2000 and highest at 0800, whereas the phosphorylase (A and AB) was highest at night (Jayaram et al. 1978). A similar peak in isocitrate dehydrogenase in the heart muscle was observed at that time by Mastanaiah et al. (1979). Such a pattern was also observed previously in pedipalpal muscles and in the hepatopancreas (Mastanaiah et al. 1977). Blood glucose peaked at 2000, whereas liver glycogen was lowest at that time and highest in the morning (Chengal Raju et al. 1973).

In other scorpions (*Euscorpius italicus* and *E. carpathicus*), respiratory activity was highest at 2000 (Dresco-Derouet 1961). This agrees well with previous observations of peak nocturnal activity. The oxygen consumption of isolated hepatopancreas and pedipalp muscles was studied in *Heterometrus* (Uthaman and Srinivasa Reddy 1979). These tissues have two respiratory peaks: one at 0800 and another at 2000. Such respiratory peaks are not necessarily related to the rhythm of locomotor activity in this scorpion. They may indicate the existence of an innate bimodal oscillation in these tissues; such rhythmic activity was observed in the ventral nerve cord and subesophageal ganglion in *Androctonus mauretanicus, A. australis, Buthus occitanus,* and *Pandinus imperator* (Goyffon et al. 1975). Recently Fleissner (1986) came to the conclusion that bilaterally symmetrical oscillators are located close to the optic ganglion and control the locomotor-activity rhythm (see review in Fleissner and Fleissner 1986).

Endocrine Control of Circadian Rhythms

There is some evidence that the endogenous cycles are under endocrine control. Pampathi Rao and Gropalakrishnareddy (1967) found that a neurohormone produced by the neurosecretory cells (NSC) of the subesophageal ganglion is released into the blood. Blood obtained from active scorpions induced spike activity in an isolated nerve cord of nonactive scorpions. Conversely, blood taken from nonactive scorpions (at 0200) caused a depression of activity in the treated scorpions. Apparently the Type A cells in the subesophageal ganglion show two peaks in their neurosecretory activity (Uthaman and Srinivasa Reddy 1985). These peaks precede the locomotory peaks by six hours.

Pampathi Rao and Habibulla (1973) have shown that continuous light and continuous darkness caused changes in the neurosecretory cells (NSC groups 3, 4, and 5). Neurosecretory cells exhibited similar changes with the circadian rhythms (peaking at 1800). When extracts of these cell groups taken at the peak of locomotory activity were injected into animals in their resting state at midday, the injected animals became active. Indirect support for the role of neurosecretory substances was provided by Fleissner and Schliwa (1977), who described neurosecretory fibers in the median eyes (see details in Heinrichs and Fleissner 1987). These fibers may signify that NSC do control the movement of circadian pigment in the retinula cells of the eyes. The whole process is mediated through the optic nerve (Fleissner and Fleissner 1978), and the NSC are arranged in two bilateral groups (Fleissner and Heinrichs 1982). Whenever the NSC fibers were severed from the supraesophageal ganglion, the circadian electroretinogram rhythm of the median eyes was disrupted (Fleissner 1983). The visual circadian rhythm was later confirmed also in the diplocentrid *Didymocentrus* and the scorpionid *Euscorpius* (Carricaburu and Muñoz-Cuevas 1986b, 1987).

Recently Carricaburu and Muñoz-Cuevas (1986a) described four kinds of electrical waves from the subesophageal ganglion of *Euscorpius flavicaudis*. One of them, a periodic wave, recorded only at night, possibly marks the activity of a subesophageal pacemaker.

Temporal Patterns of Field Activity

Scorpions spend the vast majority of their time in their burrows or other hiding places. What little time they spend on the surface is to feed and, less commonly, to mate. Most scorpion populations adhere to predictable daily and seasonal patterns of surface activity. The determinants and general pattern of these activities are presented in this section.

Daily Patterns

It is likely that scorpions living on mountaintops and in caves, tropical forests, and the intertidal zone may commonly exhibit activity during both the day and the night. Other, normally nocturnal species are occasionally active during the day. Diurnal activity in the field has been reported in many species (Toye 1970; Wanless 1977; Lamoral 1979; Fet 1980; Harington 1982; K. Linsenmair, pers. comm. 1984; G. Polis, pers. obs.). For example, diurnal surface activity occurs sporadically in *Leiurus quinquestriatus* and *Scorpio maurus palmatus* during the early spring or after the first rains in autumn (Shulov and Levy 1978, Amitai 1980). In fact, *S. m. palmatus* regularly feeds during the day (Krapf 1986a, pers. comm. 1986). Daytime activity after the rains appears to be in response to the sudden mass appearance of great numbers of arthropod prey, particularly alate termites.

But by far, most scorpion activity in the field occurs during the night (Hadley and Williams 1968; Shorthouse 1971; Bacon 1972; Tourtlotte 1974; Fet 1980; Polis 1980a, pers. obs.). Several factors are hypothesized to explain this (Cloudsley-Thompson 1978, 1981), among them the suggestion that nocturnal activity may allow scorpions to avoid predation and/or climatic extremes. Cloudsley-Thompson (1978, 1981) proposed that nocturnal activity may have evolved partially because scorpions are somewhat vulnerable to the harmful effects of solar ultraviolet radiation.

Most species first begin their surface activity at dusk (Hadley and Williams 1968; Shorthouse 1971; Bacon 1972; Toren 1973; Tourtlotte 1974; Fox 1975; Fet 1980; Polis 1980a, pers. obs.). For example, *Vaejovis confusus* and *Paruroctonus mesaensis* were active between 2100 and 0300 (Hadley and Williams 1968, Polis 1980a), with maximum activity occurring in the first hour. Almost all *Urodacus yaschenkoi* individuals synchronously begin foraging activity (doorkeeping in the burrow) at dusk (Shorthouse 1971). The initiation of activity is an inverse function of light intensity for *P. mesaensis* (Polis 1980a) and probably for most other nocturnal scorpions. The surface density of *P. mesaensis* increases rapidly during the early evening, as more animals emerge from their burrows and occupy the surface. Although new individuals emerge throughout the night, there is a significant decrease in net surface density in the late evening (after 2200), and there are usually no scorpions on the surface by dawn. During the breeding season, however, mature males are occasionally observed at dawn on the surface either moving rapidly or burrowing. Similar patterns of surface activity were observed for other species in the field (Hadley and Williams 1968, Shorthouse 1971, Toren 1973, Tourtlotte 1974, Fox 1975, Fet 1980).

The decreased surface activity during late evenings is independent of the light level. However, increasing light in the hours before dawn serves as a cue for *P. mesaensis*, either to stimulate a return to the burrow or to initiate digging behavior if the burrow is not located (Polis et al. 1986). This response ensures that animals will be below ground before the occurrence of extreme surface temperatures or possible diurnal predators.

Seasonal Patterns

Two general patterns describe the seasonal phenology of scorpions (Maury 1973b, 1978). For the majority of species, populations are most active in warmer months, and animals are absent or present only in low numbers during the colder months (Gertsch and Allred 1965; Zinner and Amitai 1969; Hibner 1971; Shorthouse 1971; Bacon 1972; Maury 1973b, 1978; Fox 1975; Tourtlotte 1975; Fet 1980; Levy and Amitai 1980; Polis 1980a, unpubl. data; Polis and McCormick 1986a). This is especially true for scorpions that live at high latitudes (e.g., *Paruroctonus boreus*: Tourtlotte 1975, Anderson 1975) or high elevations (e.g., *Urophonius granulatus* or *Bothriurus prospicuus*: Maury 1973b, 1978), where activity is impossible during the freezing winter months. One variation of this pattern is that in desert areas the surface activity significantly

---

Fig. 5.8. Seasonal activity of year groups of *Paruroctonus mesaensis*. The activity of each age group is significantly different from that of other age groups. Note particularly that the 1–2-year-old animals are active during periods when larger, cannibalistic adults (> 2 years old) are inactive. Further, 0–1-year-olds are active during some periods when adult surface density is relatively high, especially late summer after their young are born and they must forage before winter. During this time newborns suffer high rates of cannibalism from adults.

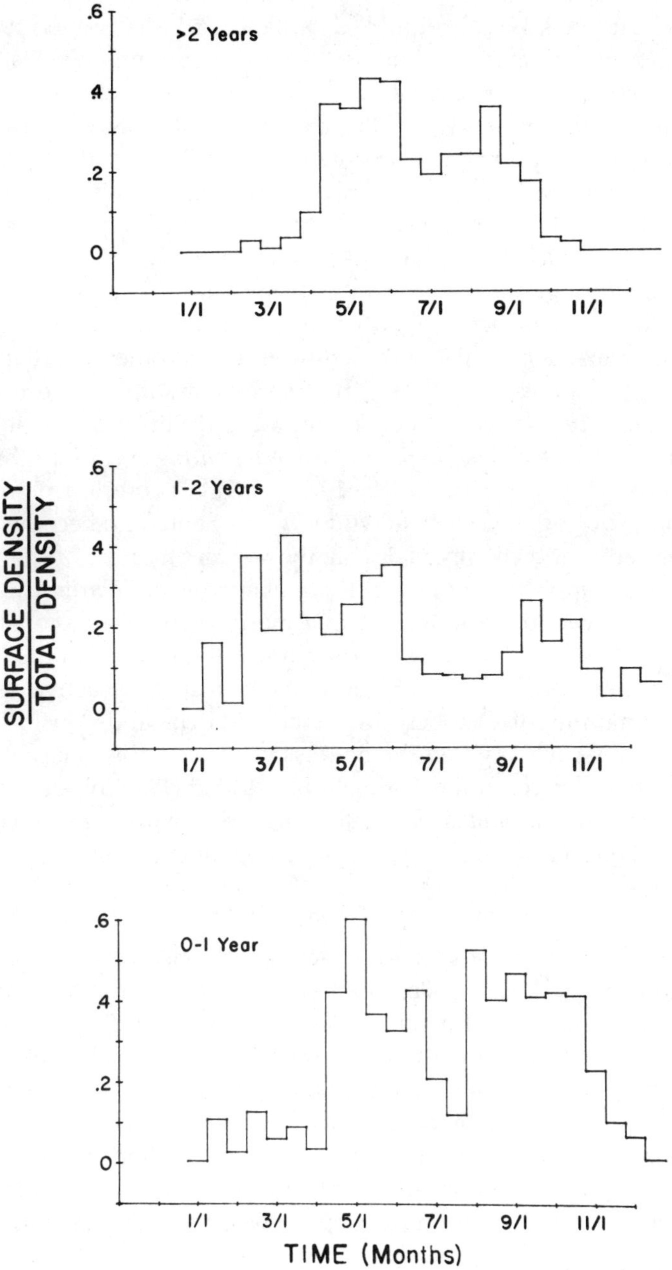

>2 Years
1-2 Years
0-1 Year
SURFACE DENSITY / TOTAL DENSITY
.6
.4
.2
0
1/1
3/1
5/1
7/1
9/1
11/1
TIME (Months)

decreases for all animals, including scorpions, during times of extreme summer heat (Polis 1980a, unpubl. data; Polis and McCormick 1986a). *Paruroctonus mesaensis* is a typical desert species: maximum surface activity occurs in the spring and fall, with minimum activity in midwinter and midsummer (Fig. 5.8; Fox 1975, Polis 1980a). Although there are few data on tropical scorpions, it is likely that they provide a different variation, remaining active during all months, with only slightly greater activity in the summer months. This conjecture is supported by the observation that life history events (e.g., mating and birth) occur throughout the year in the tropics (see Chapter 4).

The second general activity pattern is inverted from the first. A few species are maximally active in cooler months and undergo a diapause in feeding and movement during the warmer months. Accordingly, these species are relatively inactive when the majority of species are most active. *Urophonius iheringi* and *U. brachycentrus* exemplify this inverted activity cycle (Maury 1969, 1973b, 1978). These scorpions actively forage during the Brazilian winter (May–Sept.) and become inactive in underground chambers during the warmer months of the year. This is exactly opposite from the activity of species of *Bothriurus* found in the same area. At least two North American species show similar patterns: *Superstitionia donensis* (Gertsch and Allred 1965, Hibner 1971) and *Paruroctonus baergi* (Fox 1975), both of which are most active during the cooler months of October, November, March, and April. Thus, *P. baergi* is most active when the aggressive *P. mesaensis* is largely absent from the surface. In the Turkmen S.S.R., Fet (1980) observed that one species (*Anomalobuthus rickmersi*) is active primarily in the colder months, when four sympatric species are largely inactive.

Environmental Determinants of Seasonal Activity

Many different biotic and abiotic factors influence activity patterns (MacArthur 1972, Polis 1980a, Bradley 1982, Polis and McCormick 1986a). For the majority of poikilothermic species, including scorpions, most activity occurs during the warm months. It is probable that scorpions respond to thermal cues rather than seasonal ones. Evidence to support this hypothesis comes from activity patterns during periods of unseasonably high or low temperatures. Periods when temperatures are warm during the winter are characterized by unusually high surface densities, and cold periods during the warm months are characterized by relatively low surface densities (Zinner and Amitai 1969, Levy and Amitai 1980, Polis 1980a). Further, there is often a positive correlation between surface temperature and surface density (Shorthouse 1971, Bacon 1972, Toren 1973, Tourtlotte 1974, Fox 1975, Fet 1980, Polis 1980a).

Scorpions, like other poikilotherms, exhibit a limited temperature range within which field activity on the surface is maximized (Shulov and Levy 1978; Polis 1980a, unpubl. data). There are threshold temperatures: a minimum (4–10° C), below which no activity occurs (Bacon 1972, Toren 1973, Tourtlotte 1974, Polis 1980a); and a maximum, above which activity decreases because of thermally induced physiological stress (Chapter 8; Polis 1980a).

It is likely that a fluctuation in prey abundance is the ultimate causal factor underlying the direct temperature-dependent activity pattern (Chapter 7). Temperature acts as a proximal cue indicating prey availability, and increased activity at higher temperature thus allows scorpions to be on the surface during favorable periods of high prey abundance. In fact, for *Paruroctonus mesaensis* both the proportion of animals active on the surface and the proportion observed feeding are significantly correlated with both prey abundance and temperature (Polis, unpubl. data).

Scorpions with inverted phenologies exhibit peak activity during the cooler months, when prey are much less abundant. The most likely explanation for such a cycle is that it evolved to lessen the overlap in surface activity either with other scorpion species or with predators (or both). Decreased temporal overlap with other species of scorpion may reflect an avoidance of either exploitation competition or interference (see Chapter 6; Fox 1975; Maury 1978; Polis 1980a, b; Polis and McCormick 1986a, 1987). Exploitation competition may be for a specific microhabitat, burrow site, or food. Interference can be manifested either as aggression or as outright predation by other scorpion species on their potential competitors (intraguild predation). Since all species that exhibit the inverted cycle are small, it may be that interference by larger species is the ultimate causal factor favoring the evolution of the inverted activity pattern. Intraguild predation is widespread among scorpions (Chapter 7; Polis et al. 1981) and may be the paramount interaction in some scorpion guilds (see Chapter 6; Polis and McCormick 1987).

Nonscorpion predators may also influence activity (Polis et al. 1981).

The low level of surface activity observed in scorpions (see below) may ultimately be attributed to strong predation (Chapter 7; Polis 1980a).

It appears that abiotic factors such as precipitation and moonlight also influence surface activity (Hadley and Williams 1968; Polis 1980a, unpubl. data). Some species increase surface activity after rain and are able to feed on the pulsed abundance of arthropod prey (see above and Chapter 7). But many species exhibit little or no surface activity during or immediately after rainstorms (Tourtlotte 1974; Polis 1980a, unpubl. data; Bradley 1984): burrows tend to collapse in wet sand; further, the movement of wet soil during burrowing is probably difficult and may effectively prohibit emergence to the surface until the substrate dries.

In general, it appears that moonlight suppresses scorpion activity. Hadley and Williams (1968) analyzed the surface activity of five species of scorpion and determined that more scorpions were active on nights with little or no moon, in comparison with full-moon nights. Similar behavior was observed for some scorpions in the Negev Desert (Shulov and Amitai 1978, Amitai 1980, Levy and Amitai 1980). An exception once again, *P. mesaensis* showed no significant difference in surface activity between nights with no moon and ones with a full moon (Polis 1980a). However, particular surface behaviors of *P. mesaensis* were significantly inhibited by moonlight: 95 percent of all matings occurred on nights with little or no moonlight; and the proportion of scorpions observed feeding on these nights was more than double that on nights with a full moon.

It is not surprising that scorpion activity is lowered with increasing moonlight. This pattern is a general phenomenon among most animals and is particularly marked in desert dwellers; it most likely evolved to reduce predation by visual hunters on moonlit nights.

Age- and Sex-Specific Surface Activity

In many species, different age groups are active at different times (Fig. 5.8). The adults of these species are usually most active during warmer months, whereas younger age groups or instars are active during cooler periods (Shorthouse 1971; Anderson 1975; Fox 1975; Koch 1977; Polis 1980a, 1984; W. Riddle, pers. comm. 1976; Polis and McCormick, unpubl. data). Age groups may also show different periods of activity within the same night. *Paruroctonus mesaensis* adults were more active in the early evening, and immature animals were more active later in the evening (Polis 1980a).

These differences in activity probably evolved in response to selective forces similar to those that favor interspecific differences in activity (inverted versus direct temperature-dependent activity patterns) (see above and Polis 1984). Adult seasonal activity is synchronized with prey abundance, whereas the activity of younger age groups reflects an avoidance of adults and/or competition for food (Polis 1980a, Polis and McCormick 1986a). For example, the surface activity of *P. mesaensis* adults occurs primarily in the late spring, early summer, late summer, and early fall. This activity is significantly correlated with insect abundance. The two younger age groups (especially 1–2-year-olds) show more activity during cooler months. Consequently, there are no significant correlations between the surface densities of these younger age classes and insect abundance. For these species, then, the overall pattern is maximum adult activity during the best times, with the activity of younger animals primarily restricted to less favorable periods not used by adults.

Age-specific seasonal patterns of *P. mesaensis* are maintained by differential responses to surface temperature by each group (Polis 1980a). In *P. mesaensis* (and probably other species) cannibalism by older scorpions is the apparent ultimate selective force favoring the temporal separation of age groups (Polis 1980a, b).

A different pattern exists among age groups of obligate fossorial species. Immature animals are never on the surface except for dispersal away from their maternal burrow; and adults come to the surface only during a discrete mating season (Williams 1966, pers. comm. 1982; Shorthouse 1971; Anderson 1975; Koch 1978; Lamoral 1979). The great majority of individuals on the surface are mature males apparently in search of mature females residing in their burrows. For example, of the 1,552 *Didymocentrus comondae* scorpions observed by Williams (31 July–3 Aug. 1968: Williams 1980, pers. comm. 1982) 100 percent were adults, and only 8.3 percent were females. It is likely that most of these females had recently been involved in courtship. In periods other than the discrete mating season, fewer than ten individuals of this fossorial species would normally be observed on any night. These patterns closely resemble those of burrowing spiders (e.g., mygalomorphs).

The seasonal surface activity of mature males differs markedly from that of females in many species. For many North American species, the sex ratio of adults on the surface changes from month to month: females are more active in the spring (until June) and again in the fall (after September), whereas males are more frequent on the surface during midsummer. Sex-specific differences in surface activity are illustrated by *P. mesaensis* (Fig. 5.8; Polis 1980a). This general trend was also observed for other species studied using ultraviolet light (Anderson 1975; Polis and McCormick, unpubl. data). For example, the sex ratio (males:females) of *P. boreus* adults active on the surface changes monthly: 0.4 in June, 1.8 in July, 15.7 in August, and 1.8 in September (Tourtlotte 1974). Although sex-ratio data from can traps are severely biased in many species to favor mature males (see Chapters 4 and 11), similar sex-specific patterns were found in many species sampled by can traps (e.g., Gertsch and Allred 1965, Allred 1973).

The biological basis of these patterns derives from life-history phenomena. In *P. mesaensis*, the maximum surface density and the greatest proportion of males occur in July and August of each year (Polis and Farley 1979a, b; Polis 1980a). This peak corresponds to the large number of virgin males that recently reached maturity. The conspicuous absence of mature males in early spring and late summer is caused by the high mortality of male scorpions after they mature.

For *P. mesaensis*, maximum adult-female surface density occurs in the spring and then again in the middle of August. The spring peak of mature females occurs during the maximum insect abundance and correlates with high rates of embryonic growth. In contrast, the second peak occurs during a period of low insect abundance but is synchronized with maximum adult-male surface density. It is during this period, in August and September, that 90 percent of the matings occur (Polis and Farley 1979a). The inactivity of adult females in July and early August corresponds well with the period of birth and subsequent maternal brooding of the young through their first molt (Polis and Farley 1979b). The subsequent reappearance of postpartum females on the surface in the late summer, combined with the arrival of newly emerged virgin females, produces the second peak in mature-female surface density.

Levels of Surface Activity

Low levels of surface activity are characteristic of all scorpion populations that were analyzed (Bacon 1972; Toren 1973; Tourtlotte 1974; Polis 1980a; Bradley 1982, 1983; Polis and McCormick, unpubl. data). For example, an average of only 15 percent of all *Uroctonus mordax* and 5 percent of *Paruroctonus utahensis* emerged to the surface on any night (Bacon 1972, Bradley 1982). In *P. mesaensis*, neither individuals nor entire populations of each of three year-classes emerged to the surface more than 60 percent of the nights during any particular time period (see the ordinate in Fig. 5.8; Polis 1980a). The average yearly surface activity (scorpion nights on the surface divided by the number of possible scorpion nights) was 48.6 percent for scorpions under age 1, 44.8 percent for ages 1–2, and 20.4 percent for those older than 2. The total subsurface population of *P. mesaensis* is always quite large, ranging in size from approximately equal to the surface population to more than ten times as great. Shorthouse (1971; Shorthouse and Marples 1982) observed that only 56 percent of all *Urodacus yaschenkoi* were active during the several nights that composed each of his field trips. The average individual is obviously less active on a per-night basis.

Such activity patterns may reflect highly efficient foraging, the very low metabolism of scorpions, and/or the avoidance of predation and cannibalism. There is support for the hypothesis that intraspecific and interspecific predation are the major factors contributing to the evolution of such low levels of surface activity.

Many scorpions fit Schoener's (1971) classification of a "time minimizer." This foraging strategy is defined by short feeding periods, which last only until a set amount of energy is assimilated. The contrasting strategy is pursued by an "energy maximizer," in which the amount of energy assimilated is maximized during feeding periods bounded only by the physical constraints of the environment. A time minimizer, which is classified as a K-strategist, is characterized by determinate growth, relatively constant brood size from year to year, and a wide breadth of diet (because prey are selected largely on the basis of their ease of capture). An energy maximizer is characterized by the converse features.

Scorpions exhibit many traits of time minimizers (Polis 1980a). For example, in *P. mesaensis*, growth is determinate, and the production of offspring per gravid female is remarkably constant (Polis and Farley 1979b). Diet is quite varied (> 125 prey species were recorded: Polis 1979, unpubl. data). Finally, many life-history and demographic characteristics (see Chapter 4) indicate that *P. mesaensis* is an equilibrium species, or K-strategist (Polis and Farley 1980).

Schoener (1971) argued that time minimizers are forced into this strategy by the occurrence of strong predation pressure: the increased probability of predation that accompanies extended foraging periods favors those animals that minimize foraging time. High rates of cannibalism and predation can establish a selective regime that favors the evolution of a low-risk, low-reproductive-gain foraging strategy, as in *P. mesaensis.* Life-history characteristics (see Chapter 4) and low levels of activity suggest that many species of scorpion may likewise be time minimizers. If this hypothesis is true, the implication is that predation is a paramount force in the evolution and ecology of the surface activity of scorpions.

ACKNOWLEDGMENTS

It is a pleasure to acknowledge our indebtedness to K. E. Linsenmair and C. S. Crawford for the many useful discussions we have had about scorpions, and to Shoshana Goldenberg for her skillful technical assistance throughout the period of these studies.

6

Ecology

GARY A. POLIS

Ecology is the least-known aspect of scorpion biology. The reason is historical: before the use of the ultraviolet-light technique (Lawrence 1954, Pavan and Vachon 1954) in field research (Stahnke 1972a), it was very difficult to conduct field research on this group of nocturnal predators. Ecological studies were restricted to data collected during the day by rolling over rocks or searching under surface debris.

With the discovery that scorpions fluoresce under UV light (Fig. 6.1), scorpions have come to represent a near-ideal organism for all types of ecological and behavioral investigations. In some locations they are both very dense and diverse. Since they are so easily detected at night, it is relatively simple to manipulate the densities of entire populations. Diet is readily quantified, because scorpions digest their prey externally in a process that may last hours. Individuals of many species are long-lived and may remain in single burrows for long periods, and their behavior can be monitored by uniquely marking them with oil-based fluorescent paint in different combinations of colors and dots. Finally, because behaviors are not greatly modified by UV light, courtship, homing, orientation, prey detection, and prey capture are easily investigated in the field.

Unfortunately, scorpions are not perfect research organisms. Individuals lose their marks at each molt. Surface activity may be so sporadic that some individuals will appear to have left the study area or

Fig. 6.1. Fluorescence of the scorpion cuticle under ultraviolet light. The scorpion is *Paruroctonus mesaensis,* a desert species from the southwestern United States. (Photo courtesy Philip Brownell)

died. There is also some personal danger involved, particularly with the study of potentially lethal species. Further, although scorpions fluoresce under UV light, poisonous snakes do not, and they also pose a serious threat to field workers.

Spatial Patterns

Predictable patterns characterize the spatial distribution of most taxa, including scorpions (MacArthur 1972). Such patterns are detected at several levels of organization: worldwide, between and within habitats, and between and within species.

Geographical Patterns

Scorpions live on all major land masses except Antarctica. Although New Zealand, England, and some of the islands in Oceania have no native species, scorpions are now present there owing to their introduction by man in historical times (Koch 1977, Wanless 1977). Scorpion diversity (here, the number of sympatric species at one site) is maximal in subtropical areas (23–38° latitude), decreasing toward the poles and the equator (Fig. 6.2; this figure and the following discussion are bi-

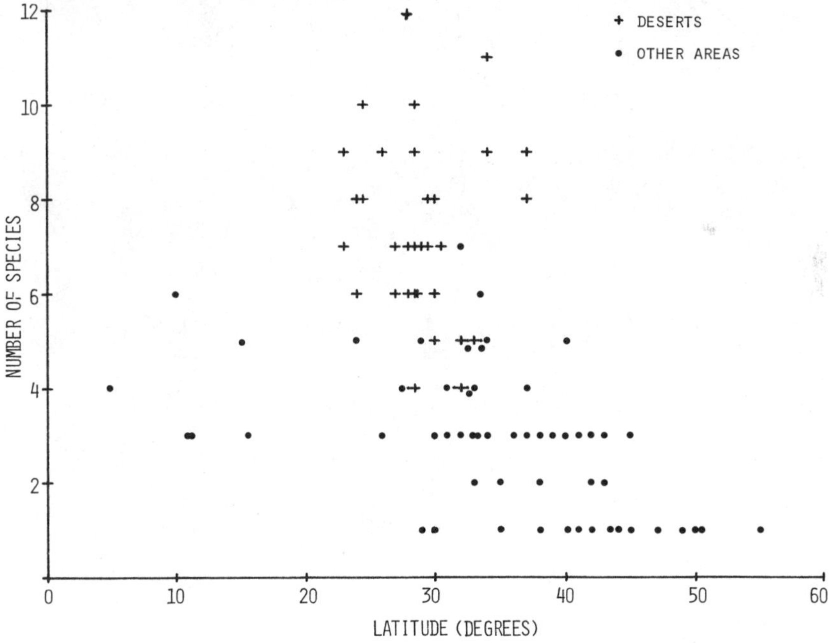

Fig. 6.2. Scorpion diversity as a function of latitude (Northern Hemisphere species only).

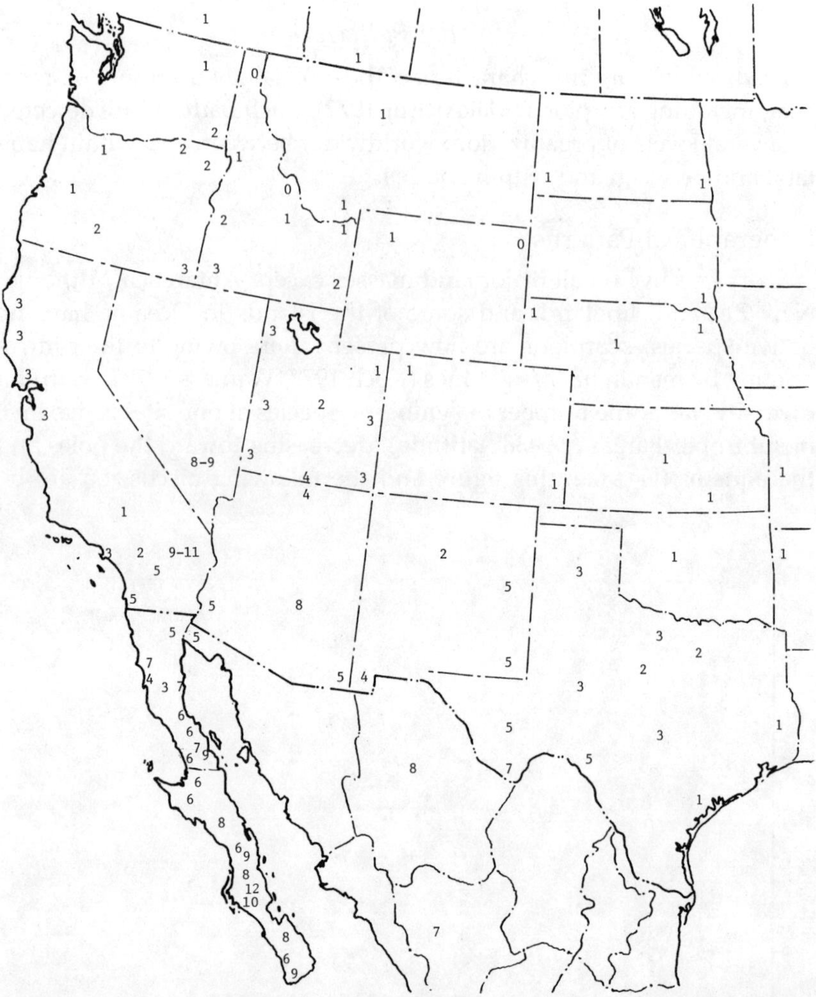

Fig. 6.3. Scorpion diversity in western North America, shown by the number of different species recorded at each site. Data are from various published studies and unpublished field notes of S. C. Williams (pers. comm. 1979) and G. A. Polis.

ased, because a large number of the studies reported data on North American scorpions). In contrast to many (but not all) groups of animals that are most diverse in the tropics (MacArthur 1972), the most diverse communities of scorpions (4–13 species) occur in desert areas (Gertsch and Allred 1965; Hibner 1971; S. Williams, pers. comm. 1981; Polis, unpubl. data): 24 of the 28 communities with at least 6 species occur in subtropical deserts. Point, or local, diversity in tropical communities ranges from 5 to 7 species (Kjellesvig-Waering 1966b [Trinidad]; González-Sponga 1970 [Venezuela]; W. Lourenço, pers. comm. 1982 [French Guiana]) to 3 species (Lourenço 1976 [Brazil], Francke 1978b [Costa Rica]). At latitudes greater than 40°, diversity is substantially reduced, and at high latitudes, no scorpion species is to be found. In the New World, there are at least 61 species described in Baja California (Williams 1980), and 37 in California (Williams 1976); but only 1 species as far north as southern Canada (*Paruroctonus boreus*, to 50° N: Sissom and Francke 1981), and 1 as far south as Tierra del Fuego (*Bothriurus burmeisteri*, to 55° S: Maury 1968b). This is also true in Europe, where scorpions occur as far as 47–49° north latitude, though not in Scandinavia (Kästner 1940, Gertsch 1974), and only 14 species have been described for all of the U.S.S.R. (Fet 1980; there are undoubtedly more species undescribed).

Scorpion diversity in the western United States and parts of Mexico is illustrated in Figure 6.3 (S. Williams kindly allowed me access to many of his unpublished field notes, especially for Baja California). The highest diversity in the world occurs in the southern third of Baja California: from 10 to 13 species can be found within 1 km^2 between Loreto and La Paz (Due and Polis 1986). It is clear that from the desert areas in the southwestern United States and Baja California, diversity decreases dramatically toward the mountains, forest, or plains (Fig. 6.3). On a larger scale, there are no species of scorpion in the high Rocky Mountains or in the north-central and northeastern United States. One species, *Vaejovis carolinianus*, occurs in most of the southeast. Florida has 1 (north) to 3 (south) species (Due and Polis 1986).

Habitats

Although scorpions are most diverse in deserts, they also occur in all other terrestrial habitats with the exception of tundra, high-latitude taiga, and some high-elevation mountaintops. There are no scorpions in boreal areas, but some species are found in mountain and alpine habitats. They occur up to 2,000 m in the Alps, 3,000 m in the southwestern United States, 5,500 m in the Andes, and 4,000 in the Himalayas (Birula 1917a, Kästner 1940, Millot and Vachon 1949, Mani 1968, Gertsch 1974, Roig-Alsina and Maury 1981). *Scorpiops rohtangensis* is found at 4,300 m under snow-covered stones in the northwestern Himalayas, and scorpions in the genera *Hottentotta* and *Chaerilus* also occur at about 4,000 m in the Himalayas (Mani 1959). *Scorpiops hardwickei* occurs in areas of frozen snow at 3,500 m in Kashmir (Birula 1917a, Kästner 1940). The elevation record is held by *Orobothriurus crassimanus*: W. Lourenço (pers. comm. 1982) found this species at 5,500 m in the Andes.

Such high-elevation species are all small. They feed on a diverse array of arthropods that are also found at these heights (Mani 1968), and their small size may be due to the short period during which they are able to forage (Maury 1978). "Cold hardiness" allows at least some species to survive freezing temperatures (Crawford and Riddle 1974). Surprisingly, high-altitude scorpions live under rocks, in scrapes, and in relatively short burrows (Mani 1968, Abalos and Hominal 1974, Crawford and Riddle 1974, Maury 1978, Roig-Alsina and Maury 1981), rather than in deep burrows with terminal chambers below the frost line.

Several species are found exclusively in caves (troglobites); and others may be found both in caves and in other favorable habitats (troglophiles; Lourenço and Francke 1985). Troglophiles often occur near the entrance to the cave, whereas troglobites live throughout the cave systems. Some troglobites occur at great depths; for example, *Alacran tartarus* was collected at 812 m below the surface (Francke 1982a). Remarkably, troglobites have been collected only from caves in the New World (Ecuador and Mexico). There are 12 troglobitic species (10 in Mexico) in four families: Diplocentridae (Francke 1977b, 1978c), Chactidae (Mitchell 1968, 1971; Francke 1981a; Lourenço 1981a), Chaerilidae (Vachon and Lourenço 1985), and Vaejovidae (Gertsch and Soleglad 1972). In addition, 9 other species are reported to be troglophiles (see Millot and Vachon 1949, Gertsch and Soleglad 1972, Armas 1973, González-Sponga 1974, Lamoral 1979). It is likely that many additional species of cave-dwelling scorpions throughout the world remain undiscovered.

Cave-dwelling scorpions show various degrees of adaptation to their subterranean existence (Fig. 6.4). Adaptations may include greatly attenuated appendages, pigmentation reduced or absent, lateral and me-

Fig. 6.4. *Sotanochactas elliotti*, a troglobitic scorpion from Sotano de Yerbani Cave, San Luis Potosí, Mexico. (Photo courtesy Robert Mitchell)

dian eyes reduced or absent, and carinae reduced or absent on the metasoma and pedipalps (Francke 1978c, Lourenço and Francke 1985). Species of *Typhlochactas* and *Sotanochactas*, the most highly adapted cave dwellers, lack eyes and pigment and possess long narrow legs and pedipalps. But some species (troglophiles) exhibit no external modifications and resemble epigean (surface) species (see González-Sponga 1974). Most species show an intermediate degree of adaptation. Such a range of adaptations occurs in other troglobitic arthropods.

Scorpions also live along the seashore and in the intertidal (littoral) zone. At least ten species representing four families are described from four continents and several offshore islands. Supralittoral species live just above the high-tide line under debris (Vachon 1951, Kinzelbach 1970, Williams 1971c, Armas 1976, Lamoral 1979, Roth and Brown 1980, Due and Polis 1985). *Opistophthalmus litoralis* lives in scrapes, under driftwood, or in decaying carcasses of marine mammals along the Skeleton Coast of Africa (Lamoral 1979). Scorpions (primarily *Centruroides exilicauda*) are very dense on the beaches of several offshore islands in the Gulf of California (Williams 1971c; Due and Polis 1985, unpubl. data); Williams (1971c) collected over 200 animals in two hours. Some species live under rocks or in sandy burrows along the high-tide line (Millot and Vachon 1949, Kinzelbach 1970). Other species live in areas that are submerged daily. Millot and Vachon (1949) reported that over 100 *Mesobuthus martensi* were collected under wet rocks. *Euscorpius carpathicus* lives in the same zone as such marine organisms as eelgrass (*Zostera*), crabs (*Pachygrapsus*), periwinkles (*Littorina*), and isopods (*Ligia*).

Vaejovis littoralis is very dense (8–12/m^2) along the drift line formed by masses of *Sargassum* algae in the Gulf of California (Williams 1980; Due and Polis 1985, unpubl. data; see also Fig. 6.5). This species co-

Fig. 6.5. *Vaejovis littoralis*, an intertidal scorpion from the Gulf of California. The scorpion is feeding on *Ligia*, a common intertidal isopod. Note the alga *Sargassum* accumulated at the drift line.

occurs with several mollusks, crabs, isopods, and marine worms, and with a diverse "terrestrial" arthropod fauna in the littoral zone (Roth and Brown 1980). It follows the waterline as the tide changes, and it is active both day and night. Observed prey include isopods (*Ligia*), ants, spiders, pseudoscorpions, and other scorpions.

Littoral and supralittoral scorpions have not been well characterized. In general, they are relatively small: over half the described species range from 8 to 15 mm in total body (prosoma and mesosoma) length, and no species is longer than 35 mm. With the possible exception of a tidal rhythm, they apparently exhibit little external adaptation to the intertidal habitat (Kinzelbach 1970). It is likely that the relative abundance of potential prey in the littoral zone is the primary selective force favoring the evolution of intertidal habits (Remy and Leroy 1933, Due and Polis 1985). Potential prey are abundant under debris at the drift line and throughout much of the intertidal zone. In the Gulf of California, the abundance and biomass of potential prey in the littoral zone are more than two orders of magnitude greater than in adjacent terrestrial habitats. The relatively high humidity of the littoral zone is another factor that may favor intertidal existence (Remy and Leroy 1933, Vachon 1951). However, the intertidal habitat is not totally favorable to invading terrestrial arthropods. Storms and spring tides may produce high, unpredictable mortality.

Microhabitats

Scorpions are not distributed randomly within a habitat. Rather, particular species are normally found in specific microhabitats. Scorpions can be roughly divided into a group that lives on or in the ground and a group that lives in vegetation. Scorpions that live on plants may hide under bark, in tree holes, at the base of large leaves and branches, or in epiphytes that grow on trees. Some of these species are the "bark scorpions" (Stahnke 1966), which may also be associated with dead vegetation, fallen logs, and human dwellings. Many genera of buthids are almost exclusively bark scorpions (Kjellesvig-Waering 1966b), such as *Centruroides* (Stahnke 1966), *Compsobuthus* (Zinner and Amitai 1969), and *Uroplectes* (Newlands 1974b).

Some arboreal scorpions live at great heights (González-Sponga 1978b; W. Lourenço, pers. comm. 1982). For example, in Australia, *Liocheles australasiae* lives on branches of the pine *Araucaria huntsteinii* at heights of up to 40 m (Koch 1977). Some species live in the petioles of palm trees or in epiphytic bromeliads (Kjellesvig-Waering 1966b, González-Sponga 1978b), where they hunt at night around the water that collects in the base of bromeliads. In Baja California, individual scorpions feed and molt within the epiphytic *Tillansia* bromeliads growing on desert shrubs and cactus (G. Polis, pers. obs.). The epiphytic and arboreal habitats of scorpions may have been favored by three factors (González-Sponga 1978b): such habitats allow escape from seasonal floods that inundate lower levels (e.g., on the Amazon floodplain); they provide a more mesic habitat in arid zones; and prey may be more abundant in plants than on the ground.

Some species live in burrows during the day and climb onto shrubs and herbs at night (Williams 1970b; Maury 1975b; Polis 1979, pers. obs.). These animals forage on both the plants and the ground and also carry prey from the ground into vegetation before they begin ingestion (Polis 1979). Such behavior is most common among smaller species (e.g., *Paruroctonus luteolus, Vaejovis confusus*) and younger age classes of larger species (e.g., *Hadrurus* spp., *P. mesaensis, P. utahensis*: W. Riddle, pers. comm. 1976; Polis 1979; Polis and McCormick 1986a). Smaller scorpions may avoid predation by foraging or feeding in vegetation. It is also possible that prey is more abundant in plants than on the ground, and foraging therefore more profitable.

Ground-dwelling scorpions often build burrows (see "Homing Behavior and Burrows," below). They also live in crevices and under rocks, logs, and other surface debris, including vegetation litter. They often improve the quality of these natural retreats by digging scrapes or small chambers beneath them. Some may live in burrows made by other animals.

Many factors influence the spatial distribution of scorpions. Important physical factors include temperature, precipitation, soil or rock characteristics, stone or litter cover, and environmental physiognomy. (Biotic factors will be discussed under "Community Structure." Historical and biogeographical factors are discussed in Chapter 3; in Koch 1977, 1981; and in Due and Polis 1986.) Temperature and precipitation are probably the most important determinants of general geographical range (MacArthur 1972; Koch 1977, 1981; Newlands 1978a). For example, temperature appears to be the primary factor limiting the southward expansion of species on the east coast of Australia (Koch 1977); the south is simply too cold during the winter. On a world scale, the fact that scorpions are not found at higher latitudes may be attributed to the cold.

Some species reach their highest densities only in areas with extensive ground cover from rocks, logs, or other vegetation litter (Smith 1966, Koch 1978, Warburg and Ben-Horin 1978). Evidently, trees and the resulting litter often occur in areas characterized by a lower probability of flooding than in adjacent areas (Koch 1978).

It is unclear whether vegetation influences scorpion distribution. Although a species may occur with a specific plant association for part of its range, this co-occurring distribution may simply reflect similar requirements for water temperature or edaphic (soil) conditions. For example, Bradley (1986) used principal-components analysis to determine that the density of *P. utahensis* was significantly correlated with both "vegetation characteristics" and soil hardness (they occurred in softer, sandy soils). Bradley's analysis could not discriminate to which of these two factors scorpions responded. Koch (1977, 1981) concluded that the distribution of Australian scorpions does not correlate strongly with vegetation patterns. Studies of scorpion-vegetation associations in North America (Gertsch and Allred 1965; Williams 1970c; Polis and McCormick, unpubl. data) suggested that most species are generalists living in several vegetation types.

By contrast, scorpion distribution does seem to be influenced by edaphic factors (San Martín 1961; Smith 1966; Lamoral 1978, 1979; Bradley 1983, 1986; Bradley and Brody 1984; Polis and McCormick 1986a; but see Koch 1977, 1981). Several taxonomic reports characterized the substrate on which a particular species is found (e.g., Lamoral 1979, Williams 1980). The distribution of *Opistophthalmus* species in southern Africa is determined primarily by soil hardness and, to a lesser degree, by soil texture. Lamoral found that each species is restricted to soils within a certain range of hardness (as measured by a penetrometer) rather than a particular soil type. Scorpion response to soil hardness is so defined that as burrows descend they actually track changes in hardness and stay within a preferred range.

North American species restricted to certain soil types consist primarily of animals that live only in sand (psammophiles) or only on rock (lithophiles; Polis and McCormick 1986a, unpubl. data). As noted by Lamoral (1978) and Lawrence (1969), species that burrow in soft soils tend to be restricted to a smaller range of soil hardness than those burrowing in other soils. This pattern is most likely explained by the highly

TABLE 6.1

Ecomorphological adaptations of lithophilic, psammophilic, and fossorial scorpions

	Scorpion type		
Character	Lithophilic	Psammophilic	Fossorial
Body	dorsal-ventral compression	varied, color of substrate	heavy, stout
Pedipalps	long, slender	varied	stout, large, powerful
Chelicerae	robust, powerful	long, slender	robust, powerful
Legs	long; dorsal-ventral compression	long; sometimes dorsal-ventral compression	short, heavy, stout
Leg setae	stout, spinelike	brushlike rows of long setae on outer edges of legs I and II (sometimes III) on tarsi and tibia	short, spinelike
Tarsi & ungues	greatly curved	very long, straight; spine very reduced	short
Metasoma	very long, slender, laterally compressed	often streamlined	very short (≤50% of total length); sting often reduced
Examples	*Hadogenes* spp.	*Opistophthalmus holmi* Lawrence	most Diplocentridae
	Syntropis macrura Kraepelin	*Paruroctonus mesaensis* Stahnke	*Anuroctonus phaiodactylus* (Wood)
	Vaejovis harbisoni Williams	*Vejovoidus longiunguis* (Williams)	*Urodacus yaschenkoi* (Birula)
	Uroplectes planimanus (Karsch)	*Parabuthus stridulus* Hewitt (see Lawrence 1969, Newlands 1972b)	*Opistophthalmus* Pocock (some spp.)
Reference	Newlands 1972a, b, 1985; Eastwood 1978a; Lamoral 1978, 1979; Williams 1980	Lawrence 1969; Lamoral 1969, 1978, 1979; Williams 1969; Newlands 1972b, c, 1974b	Williams 1966; Shorthouse 1971; Hadley 1974; Lamoral 1978, 1979

specialized morphological and behavioral adaptations that characterize psammophilic and lithophilic scorpions (see below and Table 6.1). Such adaptations facilitate burrowing and locomotion in sand or rock but make these animals inefficient in or on other substrates. For example, many psammophiles are unable to burrow in harder or more coarse soil (e.g., *O. holmi*, Lawrence 1969; *P. mesaensis*, *Vejovoidus longiunguis*, Polis and McCormick 1986a, unpubl. data).

Ecomorphotypes

A suite of behavioral and ecomorphological adaptations characterize lithophilic and psammophilic scorpions (Table 6.1). Lithophiles are adapted to cracks and crevices in rocks. They are found in volcanic habitats, in rock slides, and on rocky slopes and cliff faces. Most activity is confined to nonexposed spaces away from the surface (Fig. 6.6). Greatly elongated, flattened bodies and appendages allow them to use such areas. The longest scorpion in the world, *Hadogenes troglodytes* (male to 21 cm), is a southern African lithophile (Newlands 1972a, 1985). Stout, spinelike setae operate in conjunction with highly curved claws to provide the legs with a strong grip on the rough surfaces of rocks. Large unguicular spines are useful in climbing over rough surfaces such as rock and wood (Fig. 6.7). Thus, lithophiles can move rapidly on any plane, even upside down. Some species of *Opistophthalmus* that live in scrapes under rocks show similar ecomorphological adaptations (Eastwood 1978b): they, too, have strongly compressed bodies and elongate chelae.

Fig. 6.6. *Syntropis macrura*, a lithophilic scorpion found on volcanic rock in Baja California Sur. Note the elongated body and slender appendages characteristic of rock-dwelling scorpions.

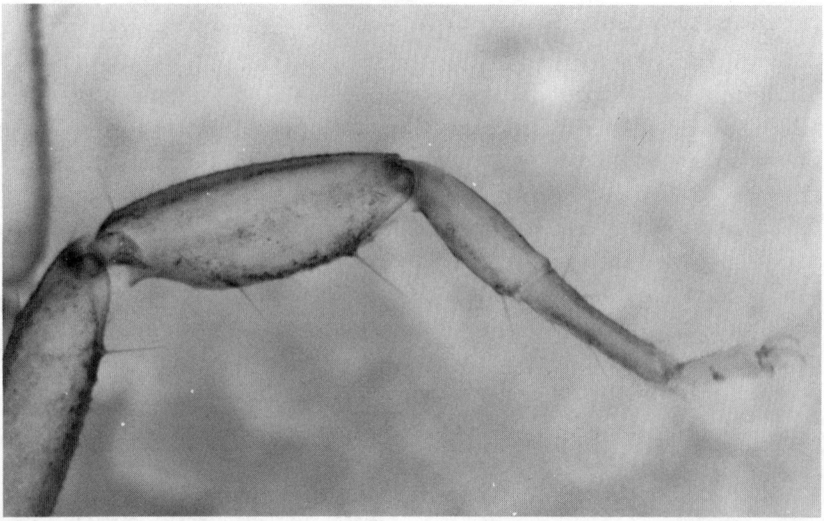

Fig. 6.7. Tibia and tarsus of the lithophilic scorpion *Serradigitus joshuaensis* from rocky areas of southern California deserts. Note the slenderness of the legs and the lengthening of the tarsal claws. (Photo courtesy W. D. Sissom)

Psammophilic scorpions are adapted to loose sand. Long tarsal claws and enlarged macrochaete setae arranged into "sand combs" (Fig. 6.8) increase the effective surface area in contact with the ground (e.g., *O. holmi* has the longest tarsal claws of any known scorpion: Lawrence 1969). This allows scorpions to walk on loose sand without sinking or loss of traction. Such leg morphology also allows efficient movement of sand during burrow construction. Psammophilic scorpions generally burrow to a depth of 0.3–1.0 m below the surface (Polis et al. 1986). Some psammophiles (e.g., *Vejovoidus longiunguis*) have streamlined metasomas and telsons, which may aid in escape when animals become buried in the sand. Scorpions that encounter only sandy substrates (ultrapsammophiles) may have greatly reduced pectines (Newlands 1972b, c). Since pectines function in substrate discrimination during spermatophore deposition (see Chapter 4), such hypotrophy may reflect their relative uselessness for scorpions that are in extremely homogeneous environments (Newlands 1972b, c).

Fossorial scorpions constitute a third distinct ecomorphotype. These species spend almost their entire existence in a burrow of their own construction. They appear on the surface for only short periods: during dispersal of the newborn away from maternal burrows, courtship and sperm transfer, and burrow maintenance (Williams 1966, pers. comm. 1979; Shorthouse 1971; Lamoral 1979). They ambush prey from the entrance of their burrows. Their pedipalps are the primary structure used in both prey capture and defense. Although some fossorial species use venom in prey capture (e.g., *Cheloctonus jonesii*: Harington 1978), other species never or rarely sting their prey (e.g., *Anuroctonus phaiodactylus*: see Chapter 7). Thus, fossorial scorpions are often characterized by reduced telsons and large, crablike pedipalps capable of great crushing power. Their chelicerae, legs, setae, and tarsal claws are short and robust (Fig. 6.9); these structures are clearly adapted for burrowing.

A fourth ecomorphotype is occasionally recognized (Stahnke 1966, McDaniels 1968, Shorthouse 1971). These "errant" scorpions are animals that actively move during foraging. They have long, slender bodies and pedipalps. This ecomorphotype may not be as universal as the

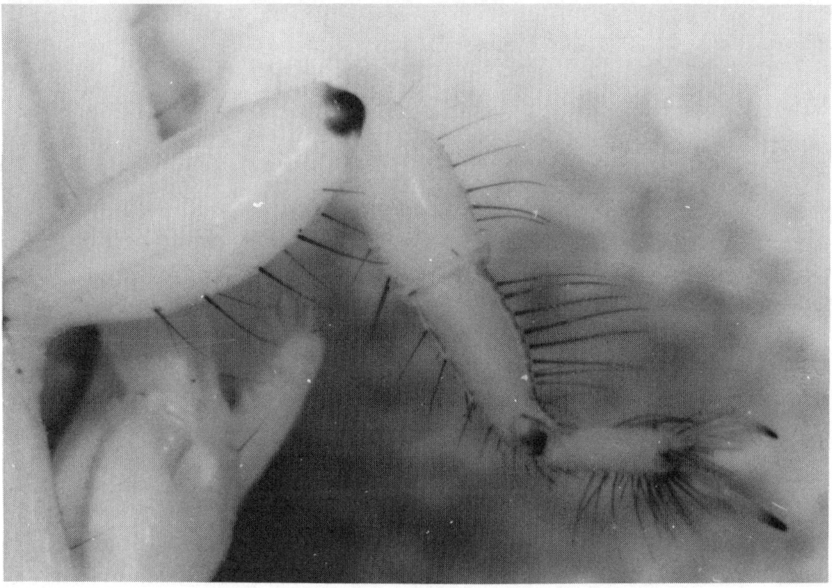

Fig. 6.8. Tibia and tarsus of *Vejovoidus longiunguis*, a scorpion that lives on sand dunes in central Baja California. Setae are elongated to form "sand combs," which provide the animal with traction on loose sand. (Photo courtesy Sissom)

Fig. 6.9. Tarsus of the third leg of *Didymocentrus caboensis* from Baja California Sur. Note the short, robust tarsal claws and setae, which characterize obligate fossorial scorpions. (Photo courtesy Sissom)

other three. In fact, errant scorpions include primarily buthid bark scorpions and only a few other species (e.g., possibly *Paruroctonus sylvestrii*: McDaniels 1968).

Koch (1977, 1981) noted that scorpions (*Urodacus* and *Lychas*) exhibit ecomorphological adaptations to the variation in aridity between the coasts and the central desert areas in Australia. There exists an intraspecific and intrageneric cline from mesic to arid regions: scorpions in more arid regions show increases in size, length of tail segments, metasomal spines, granulation, number of pectinal teeth or trichobothria, and secondary serration of the chelicerae. Koch (1981) discussed how each of these changes is apparently adaptive.

Dispersal Patterns and Sociality

Scorpions are generally nonsocial, solitary animals that interact with conspecifics only at birth, during courtship, or in cannibalism. But exceptional species show varying degrees of sociality, and some even live in groups (Polis and Lourenço 1986). In general, two evolutionary routes have been associated with the occurrence of eusociality (Wilson 1975): the familial route, whereby relatives preferentially aggregate;

and the communal (parasocial) route, whereby nonrelatives associate. Scorpions manifest the early stages of both evolutionary pathways. Because scorpions were among the first terrestrial arthropods, a knowledge of their sociobiology is particularly important for analyzing the evolution of sociality in arthropods.

All newborn scorpions stay with their mothers until just after the first molt. This nonfeeding period, which usually lasts a week to a month, represents the subsocial stage along the familial route (Chapter 4; see also Wilson 1975). Some species retain their young for even longer periods. For example, newborn *Scorpio maurus* remain in the maternal burrow for 3 to 4 months before dispersing (Shachak and Brand 1983). Both *Urophonius brachycentrus* and *U. iheringi* build underground "gestation chambers" (Maury 1969, 1973b). "Lethargic" females remain in this chamber with newborns for 5½ to 6½ months. During this period, there is no feeding and little interaction. The young of the Brazilian *Opisthacanthus cayaporum* also aggregate around their mothers, but young and mothers cooperate to build communal chambers in the centers of termite mounds (Lourenço 1981b, pers. comm. 1982). These chambers also shelter males and other females, and the total group may number 15. The social ecology of *Pandinus imperator* from the Ivory Coast appears to be very similar to that of *O. cayaporum* (D. Krapf, pers. comm. 1986). A different situation exists for *Didymocentrus caboensis*, a fossorial diplocentrid from Baja California. Second instars of this species also remain in the maternal burrow for an extended period (D. Lightfoot, pers. comm. 1978). In this case, the mother and her offspring forage together, and they have been observed to capture and drag prey communally into the burrow. Such active cooperation between mother and offspring characterizes the intermediate subsocial stage of sociality along the familial route (Wilson 1975). Scorpions do not show more advanced stages of familial sociality. Intermediate subsociality is also the most advanced stage of sociality shown by spiders (Wilson 1975).

In some species, unrelated individuals regularly aggregate in the field. *Centruroides exilicauda* individuals are found in "piles" of 20 to 30 in the winter months (Stahnke 1966). Other buthid scorpions also aggregate in groups of 5 to 25 (San Martín 1961, McAlister 1966, Zinner and Amitai 1969, Eastwood 1976): for example, groups of 10 to 17 *Lychas marmoreus* aggregate under *Eucalyptus* bark in southern Australia (A. Lockett, pers. comm. 1987); *Mesobuthus martensi* adults are found in groups of 5 to 10 under wet rocks in the intertidal zone (Remy and Leroy 1933). Groups of *M. martensi* are unique in that all individuals are the same age and all are oriented with the head toward a central point. There are fewer reports for nonbuthids (Birula 1917a, Toye 1970, Benton 1973, Toren 1973). Scorpion aggregations may be due to shortages of suitable overwintering sites or to an active attraction between conspecifics (McAlister 1966, Zinner and Amitai 1969, Toren 1973). Such aggregation of unrelated individuals corresponds to the communal stage of sociality along the parasocial route (Wilson 1975). Apparently this is the highest degree of sociality in scorpions. The communal stage also occurs among pseudoscorpions, Opiliones, and mites. Aside from insects, spiders are the only arthropods to exhibit a higher degree of parasocial sociality (Wilson 1975).

There are other exceptions to the solitary nature of scorpions. During the mating season mature male and female scorpions may live together in burrows or under the same ground cover. This was observed both in the laboratory (Birula 1917a, Fabre 1923) and in the field (McDaniels 1968, Levy and Amitai 1980, Shachak and Brand 1983).

Scorpion populations can aggregate without individuals being in immediate contact with each other. For example, several *Tityus trivittatus* adults live independent existences inside termite mounds (Lourenço, 1978), but individuals are not common outside the mounds. Obligate fossorial scorpions often aggregate in distinct "colonies," with few individuals between colonies (Williams 1966; McDaniels 1968; Shorthouse 1971; Lamoral 1978; Levy and Amitai 1980; Shachak and Brand 1983; Polis, unpubl. data). These colonies most likely form because of the limited dispersal ability of obligate fossorial scorpions. It is likely that colonies are one to a few family groups of mothers and satellite offspring (e.g., *Urodacus yaschenkoi*). Even within such aggregations, however, individuals may be spatially separate from one another. Among some colonial fossorial species, burrow morphology (particularly vertical spiraling) may act to decrease spatial overlap and the probability of encounter (Shorthouse and Marples 1980). Burrowing mygalomorph spiders also frequently aggregate into distinct colonies.

Some species change spatial patterns as a function of season or age. For example, populations of *Compsobuthus* and of *Serradigitus gertschi* seasonally expand or contract their habitats (Zinner and Amitai 1969; S. Williams, pers. comm. 1979). In the summer they occur in a number of habitats, but in the winter they move out of less desirable habitats, such as riverbeds or low meadows that may flood. The spatial distribution of cohorts of *Paruroctonus mesaensis* changes as the animals age (Polis et al. 1985, 1986). The young are significantly aggregated in the first days of their lives around their maternal burrow and then in the first months around large shrubs. Aggregations benefit the young by reducing the probability of mortality from predation by other scorpions. Intermediately aged animals and adults are randomly distributed, although adults tend toward a regular distribution, probably as a consequence of cannibalism (Polis 1980b). These age-specific patterns may occur in other species that live in dense populations (Shorthouse 1971; G. Polis, pers. obs.). For example, newborn and juvenile *Scorpio maurus* are highly aggregated, whereas the distribution of adults is random (Shachak and Brand 1983).

Homing Behavior and Burrows

Many scorpions forage on the surface in the area immediately around their burrows. Field observations in North America suggest that the majority of individuals of many species stay within a meter of their burrows, although individuals occasionally forage at greater distances. The only study to describe home ranges is for *Paruroctonus mesaensis* of the deserts of southern California (Polis et al. 1985); here, too, most surface activity occurred within a meter of the burrow (Table 6.2). It is obvious that the burrow is the center of activity and that individual foraging area expands with age. Remarkably, some individuals were found a considerable distance from their burrows (up to 8 m), and some of these individuals were observed using their original burrows on subse-

TABLE 6.2

Age-specific spatial patterns of Paruroctonus mesaensis

Scorpion age (years)	Distance from burrow (meters)		N	Individuals within 10 cm of burrow (percent)
	$\bar{X} \pm SD$	Range		
< 1	0.46 ± 0.73	0–3	93	67%
1–2	1.25 ± 0.75	0–8	158	59
>2 females	1.18 ± 1.00	0–5	234	44
>2 males	2.13 ± 2.36	0–7	15	33

NOTE: Distance was compiled from the records of individually marked scorpions and from marked burrows. $\bar{X} \pm SD$ = mean ± standard deviation; N = sample size.

quent nights. This suggests that some scorpions (at least *P. mesaensis*) are capable of homing over relatively great distances.

It is likely that scorpions use several mechanisms to return to their burrows, but none are known. Linsenmair (1968, 1972) suggested that scorpions use prevailing winds to orient to their burrows. However, this cue would not work in the habitat of many species (e.g., *P. mesaensis*) where the wind swirls from many directions. Scorpions may also orient using vision (Fleissner 1977c). Fleissner stated that at least some species can readily use their eyes on a clear moonless night to orient by stones and shadows and perhaps even stars. Linsenmair (1968) demonstrated that *Androctonus* individuals orient themselves by the stars and moon (astromenotaxis).

Considerable selective pressures should favor the evolution of homing behavior (Polis et al. 1986). Burrow construction is energetically costly to animals with such low metabolisms as scorpions. This is especially true for sit-and-wait foragers, such as *P. mesaensis*, which expend very little energy in normal activities. Construction of a burrow 3 cm in diameter and 40 cm long necessarily consumes a large proportion of their energy budget. For example, Shorthouse and Marples (1980) estimated that the weight of soil excavated by *Urodacus yaschenkoi* is 200 to 400 times the scorpion's body weight. But scorpions that burrow have developed homing behavior, and several selective pressures must have favored homing scorpions over nonhoming scorpions. Because they expend less energy in burrowing activities, their energetic needs are lower, and they are thus less subject to predation because they need to forage less on the surface. When the threat of predation does arise, homing allows a quick escape. Accordingly, homing scorpions are likely to survive longer and produce more young. Finally, homing ensures that individuals will locate a refuge from the lethal thermal conditions that can occur on the surface during summer days in the desert.

A knowledge of scorpion burrows is central to an understanding of the ecology and evolution of many scorpion species. Burrows provide a refuge not only from predators (see Chapter 7) but also from some potentially harmful physical factors. They represent a major adaptation of many desert arthropods, including scorpions, to the hot, arid environment: climatic extremes are lessened, and animals are sheltered from unfavorable surface conditions (Edney et al. 1974, Hadley 1974, Koch 1978, Polis and Farley 1980, Shorthouse and Marples 1980, Polis et al. 1986; see also Chapter 8). The extreme temperatures that occur at the surface are moderated just a few centimeters below the surface. Further, the relative humidity is much higher than on the surface, and evapotranspiration is correspondingly lower. Some physical factors are also avoided by burrowing animals. For example, fire destroys most life on the surface. However, in both southern Africa (Eastwood 1978a) and southern California (G. Polis, pers. obs.) burrowing scorpions were observed to escape the immediate effects of fires that swept through the area. Thus, the burrowing *Opistophthalmus capensis* survives fire, whereas the surface-dwelling *Uroplectes lineatus* is eliminated (Eastwood 1978a). Burrows may also influence the evolution of life-history strategy for those scorpions that live in harsh environments such as the desert (Polis and Farley 1980; see also "Population Biology," below).

Although species of burrowing scorpion live the great majority of their existence in their subterranean retreats, all scorpion populations so far analyzed are characterized by low levels of surface activity (Shorthouse 1971; Bacon 1972; Toren 1973; Tourtlotte 1974; Polis 1980a; Bradley 1982, 1983; see also Chapter 5). At one extreme, obligate fossorial scorpions spend all their lives in their burrows, except for the short periods of courtship, mating, and dispersal of newborn (Williams 1966; Shorthouse 1971; Anderson 1975; Koch 1978; Lamoral 1979; S. Williams, pers. comm. 1979; Shachak and Brand 1983). Even scorpions that forage on the surface spend most of their time within their burrows. For example, *P. mesaensis* individuals emerge to the surface on only one-fifth to one-half of all possible nights, and even on these nights they are on the surface for a maximum of 8 hours (usually less than 4 hours: Polis 1980a). That is, individuals of this species remain in their burrows for 92 to 97 percent of their existence. Other scorpions may remain in their burrows an even greater proportion of their time (see Chapter 5). Cloudsley-Thompson (1981) compared the activity of *P. mesaensis* with that of *Heterometrus swammerdami* and concluded that *P. mesaensis* is much more active than *Heterometrus*.

The burrow is the location for almost all normal activities of burrowing scorpions: birth, maternal care, molting, feeding, even mating. Females usually give birth in the burrows, and some species even construct special burrows for parturition (Maury 1969, 1973b). Although *Parabuthus planicauda* normally lives in scrapes and not burrows, gravid females dig deep burrows for birth. The newborn scorpions then remain in the maternal burrow until after the first molt, when they disperse to assume an independent existence in their own burrows.

However, even during the first few nights of foraging on the surface, newborns will occasionally return to the maternal burrow (at least in *Paruroctonus mesaensis*). The females of some species remain with their young in the burrows for extended periods (e.g., *Didymocentrus caboensis*) lasting as long as four months (e.g., *Scorpio maurus, Urophonius brachycentrus*: Maury 1969, Shachak and Brand 1983).

The burrow provides a protective environment for the time immediately after molting, a critical stage in the life of any arthropod. The animal is soft and without effective defense against predators, including cannibalistic conspecifics (Probst 1972, Polis 1981). Further, the potential for water loss is high because the waxy layer on the cuticle is not yet established. It appears that burrowing scorpions molt only in the relative safety of the burrow. In fact, exuviae are frequently found in burrows but are only rarely seen on the surface.

Most mating occurs on the surface. However, Williams (1966) found that *Anuroctonus phaiodactylus* may mate within the burrows; in fact, pairs of mature males and females occasionally occupy the same burrow (McDaniels 1968). Likewise male and female *Scorpio maurus* are often found in the same burrow during the breeding season (Shachak and Brand 1984). Many scorpions always or at least occasionally forage from the entrance of their burrows (Chapter 7). The remains of prey in the burrow suggest that even scorpions that forage on the surface occasionally carry prey into burrows.

Scorpion burrow construction has been described by numerous researchers (Birula 1917a; Stahnke 1966; Williams 1966; Newlands 1969a, 1972b; Eastwood 1977, 1978a, b; Koch 1977, 1978; Hadley 1978; Lamoral 1978, 1979; Shorthouse and Marples 1980; Polis et al. 1986). It appears that burrows can be completed in one day. Different species use a combination of chelae, chelicerae, legs, and even the tail to loosen and remove soil and to compact burrow walls. These anatomical structures are adapted to specific substrate characteristics, particularly hardness and particle size. Thus, scorpions in hard soils possess powerful, short legs, setae, and chelicerae; those in looser soils have thinner, longer structures (Lamoral 1978, 1979). Newlands (1972b) noted that the position of the median ocelli is a function of soil hardness and particle size. He reasoned that harder soils require larger prosomal muscles (for cheliceral digging), thus causing an anterior extension of the carapace and thereby locating the median ocelli farther back on the prosoma.

Scorpions use an array of burrow types (Fig. 6.10). Some species consistently use the burrows of other animals such as spiders, isopods, or rodents (see Koch 1977, Levy and Amitai 1980); for example, *Hadrurus* is often found in abandoned rodent or lizard burrows. *Isometroides vescus* is most commonly found in the burrows of trapdoor spiders, which are the specialized prey of this scorpion (Main 1956, Koch 1977). Some Brazilian scorpions are usually found in termite mounds, primarily of *Armitermes* (Lourenço 1976, 1978).

The simplest burrows constructed by scorpions are nothing more than natural openings beneath ground objects that are modified to form short (2–10 cm) "scrapes" or "runs." Such burrows are used by many species (Cekalovic 1965–66; Maury 1968b, 1978; Abalos and Hominal 1974; Armas 1976; Lourenço 1976; Koch 1977; Eastwood 1978a, b; Lamoral 1979). Longer runs are often built on a nearly horizontal plane into slopes (e.g., *Leiurus quinquestriatus*: Levy and Amitai 1980), and some are built nearly perpendicular to the surface (e.g., *Liocheles nigriceps*, Tilak 1970; *Cheloctonus jonesii*, Harington 1978). However, it appears that most burrows are constructed at a 20–40° angle to the surface and are often relatively straight or slightly curved (Rosin and Shulov 1963, Williams 1966, Polis et al. 1986). Burrows of many species form loose spirals as they descend at a 20–40° angle (Koch 1977, 1978; Lamoral 1978, 1979; Shorthouse and Marples 1980; Bradley 1982; Polis, unpubl. data). It is hypothesized that spiraling evolved to reduce the probability of contact between neighboring scorpions (Shorthouse and Marples 1980). Spiraling also facilitates ascent within the burrow.

A number of complexities may be added to these basic burrow types. For example, scorpions sometimes construct enlargements, chambers either near the entrance or at the terminus of the tunnel (Williams 1966; Tilak 1970; Koch 1978; Eastwood 1978a, b; Lamoral 1979). Such chambers are used to turn around in and for feeding. The terminal chamber is often raised above the tunnel, possibly to decrease the effects of water that enters the burrow (Williams 1966). Burrows occasionally have two or more entrances or tunnels (San Martín 1961, Lourenço 1976, Eastwood 1978a, Harington 1978, Shorthouse and Marples 1980).

In general, the entryways of scorpion burrows are characteristic (H. Stahnke, pers. comm. 1975; Eastwood 1978a; Harington 1978; Lamoral 1979; G. Polis, pers. obs.): they tend to be oval in cross section, being wider than high; moreover, they are usually flat on the bottom and crescent-shaped on top, conforming to the cross section of a scorpion's body (ventrally flattened and dorsally rounded). This unique entrance

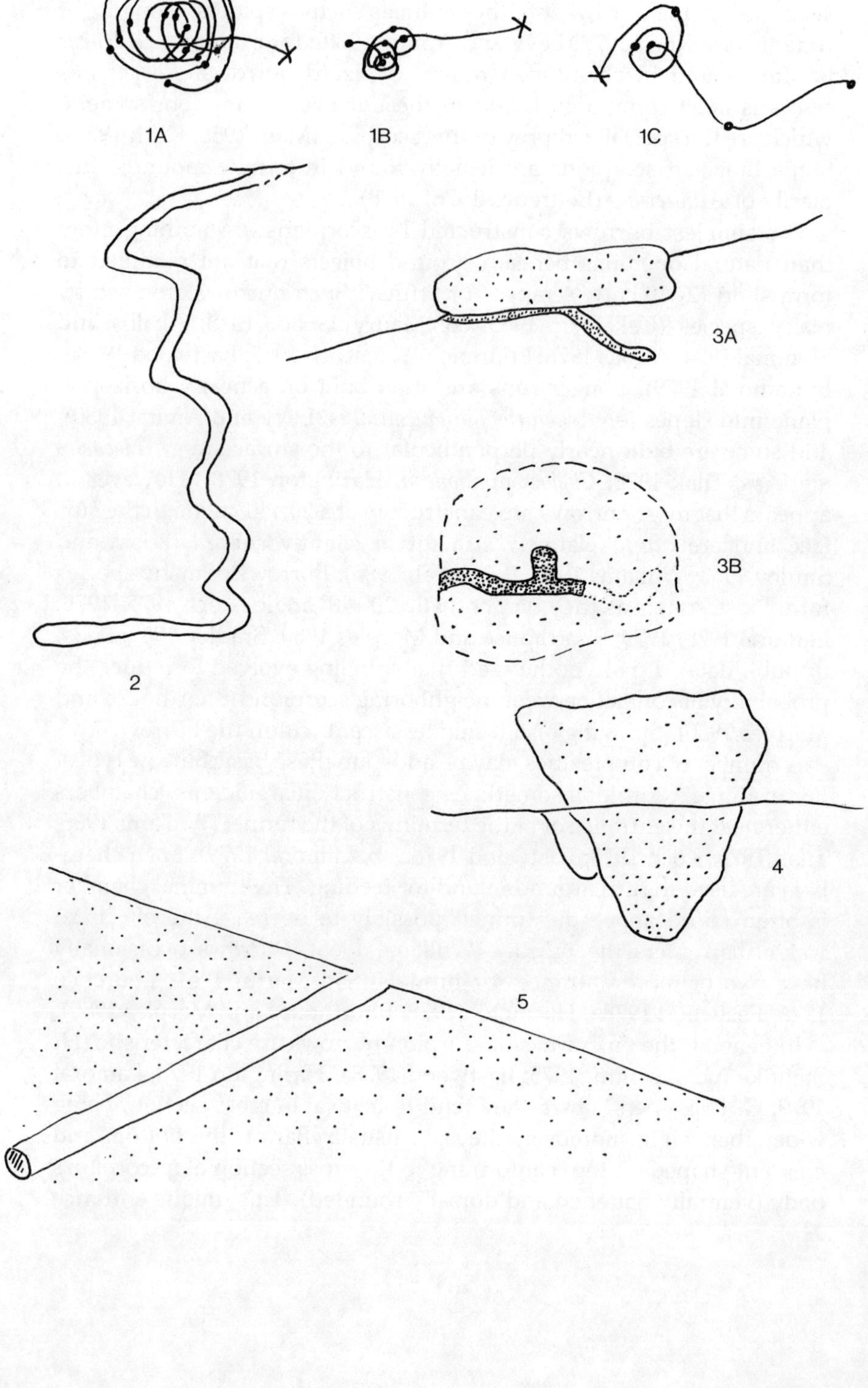
1A
1B
1C
2
3A
3B
4
5

shape allows one to distinguish scorpion burrows from the frequently round burrows of other fossorial arthropods. (However, a few scorpion species have round entrances: e.g., *Didymocentrus caboensis*.) Burrows may also be recognized by the accumulation of a tumulus of excavated soil outside the entrance; this mound may function to inhibit rainwater from entering the burrow (Shulov and Levy 1978).

The majority of burrows appear to be between 15 and 50 cm deep, but they vary in depth from a few centimeters to more than a meter (Millot and Vachon 1949, Koch 1977, Shorthouse and Marples 1980). In fact, *Hadrurus* individuals have been found at soil depths of more than 2 m (Stahnke 1966, Anderson 1975). The depth appears to be a function of scorpion age and sex. Younger age groups and males construct significantly shallower burrows than adult females (Eastwood 1978a, b; Harington 1978; Shorthouse and Marples 1980; Polis et al. 1986). Young scorpions may even build different types of burrows than adults (Eastwood 1978a, b). Differences in (micro)habitat also influence burrow depth and morphology. Koch (1977) found that the burrows of many species of Australian scorpions varied in relation to average rainfall. Burrows are deeper, more spiraled, and under less cover in arid central

---

Fig. 6.10. Burrow types. *1A, B, C,* Schematic representations of the burrow paths of *Urodacus hoplurus,* from above. Dots represent successive equally spaced depths along the burrow paths. Total depths are 68 cm in *A,* 15 cm in *B,* and 25 cm in *C:* most burrows of *U. hoplurus* and other *Urodacus* species also spiral, to maximum depths of 100 cm. Burrow entrances (at the cross) are elongate, elliptical, and slitlike. Burrows often terminate in a horizontal "living chamber." (After Koch 1978) *2,* Diagram of the burrow of *U. yaschenkoi,* from the side. These burrows also spiral, to depths of 16–28 cm for second instars and 30–70 cm for adults. The burrow terminates in a horizontal chamber. (After Shorthouse & Marples 1980) *3A,* Cross section of the burrow of *Opistophthalmus capensis.* Burrows consist of a "run" under a rock or other ground cover; the section shown is excavated through soil and the terminal chamber. (After Eastwood 1976) *3B,* Plan of a typical burrow of *O. capensis* under a rock, showing the upper run, a lateral antechamber, and the actual burrow (stippled border). *4,* Diagram of a nearly vertical burrow of *Bothriurus bonariensis* dug against a rock. Although a common plan includes nearly horizontal burrows dug under rocks or other ground cover, this species also builds burrows at other orientations. Burrows are 10–40 cm long. (After San Martín 1961) *5,* Diagram of the burrow of *Paruroctonus mesaensis.* Burrows are built into sand at 30°–45° angles. Most burrows are straight tubes, 10–75 cm or more in length. Burrow length increases with age. (G. A. Polis, unpubl. data) areas than in wetter coastal areas. Eastwood (1978b) found that burrow depth for *Opistophthalmus* scorpions in southern Africa varied with soil type.

Some species have specific requirements for burrow sites. Hardness and other physical characteristics of the soil are important for many species (Lamoral 1978, 1979; Polis and McCormick 1986a). Additionally, some scorpions construct burrows only on slopes or in other areas where rainwater will not accumulate (Williams 1966; Zinner and Amitai 1969; Koch 1977; G. Polis, pers. obs.). Other species construct burrows predominantly among the roots of trees, shrubs, or grasses (Hadley 1974; Harington 1978; Koch 1978; Lamoral 1979; Couzijn 1981; Polis, unpubl. data). Areas under vegetation offer many advantages not found elsewhere: prey is more abundant, predators may be deterred physically from capturing scorpions, roots stabilize the soil, and some shrubs (e.g., *Larrea divaricata*, creosote bush) produce hydrophobic chemicals, which inhibit the wetting of the soil beneath their canopies (Polis 1980b).

Population Biology

We have information on the population biology of only three species: *Urodacus manicatus* (= *abruptus*), *U. yaschenkoi*, and *Paruroctonus mesaensis*. These three species are similar in that each is a relatively large animal that spends much time within burrows. Unfortunately, then, some trends may not be representative of smaller or nonburrowing species.

Survivorship and Mortality

The survivorship schedule (curve) for each of these three species is presented in Figure 6.11. These curves show the decrease in the population of an average cohort from birth until all individual members have died. Analysis of the survivorship of *Paruroctonus mesaensis* (Polis and Farley 1980) indicates that the rate of mortality can be divided into three general phases:

1. A period of very high early mortality in the weeks directly after birth (14.5 percent per month).

2. A period of constant, relatively low mortality during the months of immaturity (4.4 percent per month).

3. A period of constant, relatively high mortality throughout the years of maturity (for females, 6.8 percent per month, 81.1 percent per year; for males, 7.9 percent per month, 94.2 percent per year). A survivorship curve of this shape represents an intermediate between type 1

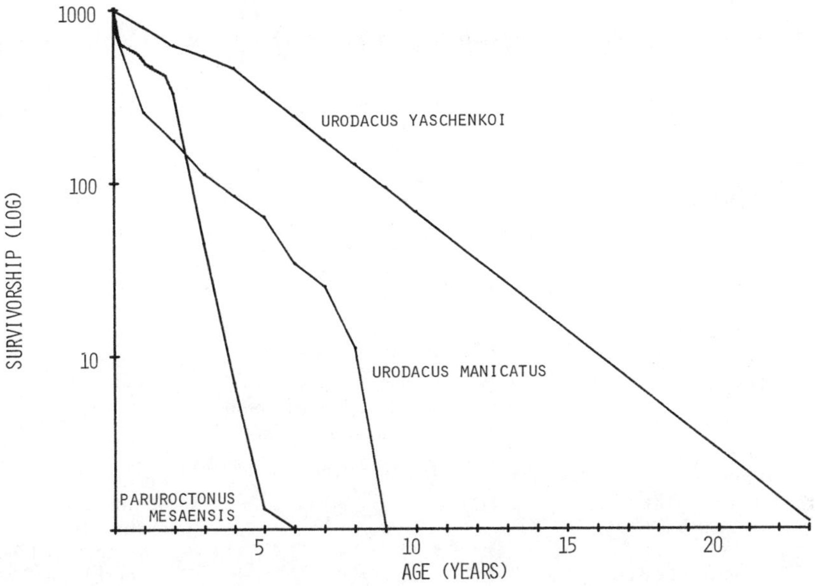

Fig. 6.11. Survivorship curves for three scorpion species. Data for *Paruroctonus mesaensis* from Polis and Farley (1980), for *Urodacus manicatus* from Smith (1966), and for *U. yaschenkoi* from Shorthouse and Marples (1982).

(rectangular: low early mortality, then high mortality after some age) and type 2 (diagonal: constant mortality at all ages).

For the Australian scorpion *Urodacus manicatus*, Smith (1966) reported an intermediate-type surivorship curve, which was similarly divided into three periods of mortality. Young suffer a high mortality of 65 percent, which Smith attributed to predation and cannibalism. There is a relatively low mortality among intermediately aged animals, about 30 percent per year. Old adults suffer a high mortality rate of about 60 percent per year.

Analysis of the life history of another Australian scorpion, *Urodacus yaschenkoi* (Shorthouse 1971, Shorthouse and Marples 1982), indicates that the survivorship curve is similar to that of *U. manicatus* and *P. mesaensis*. There was relatively high early mortality (instar 2, 20.4 percent; instar 3, 22.0 percent), relatively low mortality at intermediate ages (instar 4, 13.6 percent; instar 5, 15.0 percent), and the highest mortality for adults (27.3 percent).

For *P. mesaensis* the age-specific mortality rate of males was greater

TABLE 6.3
Sex ratios of natural populations of scorpions

Taxon	Sex ratio (♂:♀)	Comments	Reference
BOTHRIURIDAE			
Urophonius brachycentrus (Thorell)	1:1 → 1:3	males decrease during season owing to natural mortality and cannibalism by females	Maury 1969
BUTHIDAE			
Isometrus maculatus (DeGeer)	1:3	skew may be caused by cannibalism on males by females	Probst 1972
Isometroides vescus (Karsch)	1:1	UV	Koch 1977
Rhopalurus rochae Borelli	1:2.2		Matthiesen 1968
Tityus bahiensis (Perty)	1:3		Bücherl 1955/56, Matthiesen 1966
T. fasciolatus Pessôa	1:3, 1:4[a]		Bücherl 1955/56; Lourenço 1976, 1978
T. serrulatus Lutz & Mello	0:1	parthenogenetic; all females	Matthiesen 1962
T. trinitatis Pocock	5.6:1	skew may be caused by cannibalism on females by larger males	Kjellesvig-Waering 1966b
ISCHNURIDAE			
Liocheles australasiae (Fabricius)	1:29	skew may be caused by cannibalism on males by females; UV	Koch 1977
	0:1	parthenogenetic; all females	Makioka & Koike 1984, 1985
SCORPIONIDAE			
Urodacus yaschenkoi (Birula)	1:1(b) → 1:2		Shorthouse 1971
U. manicatus (Thorell)	1:1(j) → 1:3		Smith 1966
U. planimanus Pocock	0.8:1	UV	Koch 1977

TABLE 6.3
(*continued*)

Taxon	Sex ratio (♂:♀)	Comments	Reference
VAEJOVIDAE			
Paruroctonus baergi (Williams & Hadley)	1:1	all ages; UV	Fox 1975
P. mesaensis Stahnke	1:1(b) → 1:2	skew may be caused by cannibalism on males; UV	Polis 1980b
Uroctonus mordax Thorell	1:1(b) → 1:2	UV	Bacon 1972
Vaejovis confusus Stahnke	3:1	UV	Anderson 1975; Polis, unpubl. data
V. littoralis Williams	1:2.1	UV	Due & Polis 1985

NOTE: Ratios are for adult animals except where noted by *b* (ratio at birth) or *j* (ratio for juveniles). *UV* indicates that scorpions were collected with ultraviolet light rather than with can traps or by hunting during the day. Arrows indicate change in sex ratio from newborn animals to adults.
[a]The two authors reported different ratios.

than that of females (Polis and Farley 1980). The proportion of males declines significantly from birth (51 percent) to the onset of maturity (42 percent) to full maturity (35 percent). Because of the low annual survivorship of mature males (6 percent), about 80 percent of the reproductive males at the beginning of the breeding season are newly matured virgins. By the end of each breeding season, almost all the mature males (more than 95 percent) had reached maturity that season. By contrast, almost two-thirds of the adult females had matured in an earlier season.

The mortality rate of males is higher than that of females for almost all scorpions for which such data were reported. Smith (1966) noted that mature males of *U. manicatus* rarely last more than two mating seasons, whereas females may live for as long as eight years after maturity. In many species, a sex ratio (male:female) at birth of 1:1 gradually changes until at maturity it is often 1:2 or 1:3 (Table 6.3). In several other species the sex ratio is unknown at birth, but the adult sex ratio is skewed, with females preponderant.

Differential sexual mortality of males was attributed to cannibalism by females (Maury 1969; Probst 1972; Koch 1977; *Liocheles australasiae*, Polis and Farley 1980) and/or to "reproductive stress effects" (Smith

1966, Maury 1969, Polis and Farley 1980). For *P. mesaensis*, at least, there is a significant sexual bias in cannibalistic encounters. Only 23 percent of the cannibalizing scorpions were male, versus 63 percent of the prey scorpions. In *P. mesaensis* and at least 12 other species, intraspecific predation on mature males occurs during mating and/or as a result of behavior associated with mating (Chapter 4; Polis and Farley 1979a). Further, mature males of many species are often vagrant and mobile during the breeding season; this movement predisposes them to a disproportionately high incidence of cannibalism, interspecific predation, and thermal death as compared with mature females (Polis and Farley 1979a). Such movement is also metabolically expensive. Higher metabolic rates combined with lowered feeding rates for mature males (at least for *P. mesaensis*: Polis and Farley 1979a) may subject males to food stress and a higher probability of death by starvation. It is interesting to note the one case of reversed sex ratio (5.6:1 for *Tityus trinitatis*: see Table 6.3), which has been attributed to cannibalism on females by larger males (Kjellesvig-Waering 1966b).

Other biotic factors that contribute to scorpion mortality include predation (see below) and death at birth or during molting (Waterman 1950; Smith 1966; Probst 1972; W. D. Sissom and O. Francke, pers. comm. 1981; G. Polis, pers. obs.). Predation is by far the most important mortality factor. Although interspecific scorpion predators are numerous (Chapter 7; Polis et al. 1981), most mortality to large scorpions and adults is caused by vertebrate predators. The majority of mortality among smaller species and immature animals is caused by predation by various invertebrates and by scorpions of the same or different species. Most mortality experienced by *U. manicatus* (Smith 1966) and *P. mesaensis* (Polis and Farley 1980, Polis 1980b) was attributed to intra- and interspecific predation. In *P. mesaensis* the rate of cannibalism is density-dependent and can thus contribute to the regulation of population size (Polis 1980b). In some scorpion communities, a large proportion of the mortality to smaller species is caused by intraguild predation by larger species (see "Community Structure," below).

Abiotic factors occasionally may also produce considerable scorpion mortality. Such physical factors include freezing (Toren 1973, Crawford and Riddle 1974), lethal high temperatures (Chapter 8; Polis and Farley 1979a), wet soils (Zinner and Amitai 1969), fires (Eastwood 1978a), flash floods (Bradley 1983, 1986), and severe windstorms and sandstorms (Polis and Farley 1980). Crawford and Riddle (1974) estimated that at least 10 percent of the population of *Diplocentrus peloncillensis** may be killed annually by cold temperatures. Entire populations of *Uroplectes* in southern Africa were reported to be exterminated by fire (Eastwood 1978a). Heavy rains and flooding produced a marked decline of localized populations of *P. utahensis* (Bradley 1983, 1986). Severe sandstorms in the deserts of southern California are known to kill large numbers of scorpions (Polis and Farley 1980).

Reproductive Parameters

The ability of a population to grow in a particular environment is determined by its set of reproductive statistics. Reproductive parameters include the generation time *T*, the maximum capacity for daily rate increase r_{max}, and the maximum net reproductive rate R_0, representing the exact amount by which a population is able to increase in each generation (for the formulas for these statistics, see Polis and Farley 1980). These statistics are presented for *P. mesaensis* (Polis and Farley 1980), *U. manicatus* (Smith 1966), and *U. yaschenkoi* (Shorthouse 1971, Shorthouse and Marples 1982) in Table 6.4. The maximum rates of increase for these three species are among the lowest reported in the entire animal kingdom (Pianka 1970, Polis and Farley 1980). Such low values reflect both the long generation time and the low survivorship of females to the age of first reproduction (3.62 percent in *P. mesaensis*). The maximum R_0 indicates that populations of *P. mesaensis* can little more than double in each generation, and *U. manicatus* can increase its population by only 50 percent in its five-year generation time. By contrast, *U. yaschenkoi* can multiply almost seven times in each 8.6-year generation.

For these three species and the many others with similar population dynamics (see below), these statistics suggest that their populations are not resilient to disturbance or to large decreases in size. Thus, if a large proportion of the population dies or is removed, it takes a long time for the population to reestablish itself to its equilibrium size. This phenomenon was observed after the artificial removal of scorpions for reasons of health (Shulov and Levy 1978) and for research (Polis and McCormick, unpubl. data). For example, 4,500 *Leiurus quinquestriatus* were removed from an area 4,000 m^2 in the Gilboa Hills of Israel during

*Crawford and Riddle (1974) reported working with *D. spitzeri* Stahnke, but the population they studied, from the Peloncillo Mts. of New Mexico, was later described as *D. peloncillensis* Francke. See Francke (1975).

TABLE 6.4
Reproductive parameters for three scorpion species

Parameter[a]	Species: *Paruroctonus mesaensis*	*Urodacus manicatus*	*Urodacus yaschenkoi*
T (years)	3.22	5.13	8.62
R_0	2.176	1.505	6.783
r_{max}	6.61×10^{-4}	2.20×10^{-4}	6.08×10^{-4}
Maximum longevity (years)	6	10	24
S_x, adult females	0.19/year	0.40/year	0.73/year
Adult length (mm)	72	54	102

SOURCES: The raw data are from Polis & Farley 1980 (*P. mesaensis*), Smith 1966 (*U. manicatus*), and Shorthouse & Marples 1982 (*U. yaschenkoi*). I calculated the parameters for both *Urodacus* species.
[a] *T*, generation time; R_0, maximum net reproductive rate; r_{max}, maximum capacity for daily rate increase; S_x, annual survivorship. See the text for explanation of these parameters.

a two-month period (Shulov and Levy 1978); the area was very slow to repopulate, and only few scorpions needed to be removed during the next 30 years.

Scorpions as Equilibrium Species

Species can be divided into two general groups according to their life-history strategy: equilibrium, stable, or K-selected; and opportunistic, fugitive, or r-selected (MacArthur and Wilson 1967, Pianka 1970). An equilibrium species develops slowly to a relatively large size and produces (for a series of years) relatively few offspring, each of which has proportionately more parental energy allocated to it; the survivorship curve is type 1 or 2; mortality is predominantly density-dependent, and the magnitude of population fluctuation is relatively small; it inhabits a predictable environment in which species interactions are a paramount force; and it occupies a narrow, specialized niche. An opportunistic species exhibits the inverse of these characteristics: relatively high reproductive rates; less parental investment per offspring; type 3 survivorship curve; high levels of density-independent mortality; large and frequent changes in population sizes; "weedy" existence in temporary, disturbed, or unpredictable habitats where abiotic factors are of prime importance; and a more generalized niche.

Many scorpion species (except some buthids: see below) exhibit several of the characteristics of equilibrium species (Polis and Farley 1980). *Paruroctonus mesaensis* will serve as an example. Mortality is primarily density-dependent, although density-independent episodes (e.g., sandstorms) occasionally occur. The survivorship curve is intermediate between type 1 and type 2. Most significant, the population fluctuates about an equilibrium point: in 4 years, the maximum population of scorpions 22 months of age was only 1.15 times the minimum population; the maximum adult population in June was only twice the minimum June population.

Development and growth are relatively slow in this species (Polis and Farley 1979b). The period from fertilization to parturition is about a year, maturity occurs at 19 to 24 months, and individuals can reach 5 or 6 years of age. Reproduction is postponed until females are about 3 years old, and they exhibit iteroparity (Polis and Farley 1979b). The maximum rate of increase is one of the lowest ever reported. The mother invests a great amount of time and energy in her young. All scorpions are viviparous to at least some extent, in that the developing embryos receive nutrients from the mother (see Chapter 4). At birth, the young remain on the dorsal surface of the female for up to a month before they assume an independent existence. The size of *P. mesaensis* newborns is greater than the adult size of a large proportion of North American insects. At first surface appearance, the average wet weight of newborn animals is 0.035 g and the average total length is 13 mm (Polis and Farley 1979b). Adults (2.0 g, 72 mm) are larger than almost all adult spiders (except some mygalomorphs), most adult insects (Pianka 1970), and more than 14 percent of the total vertebrate fauna (McCormick and Polis 1982).

Although *P. mesaensis* inhabits the desert, its microclimate is actually stable and predictable. Individuals spend from 16 to 24 hours per day and 92 to 97 percent of their entire lives in their burrows. Desert scorpions are not on the surface when the environment is unfavorable, for instance, at extreme temperatures or in desiccating conditions of low humidity (Chapter 8; Hadley 1974). At the depth of the average burrow (25 cm) there is little daily fluctuation in sand temperature or humidity. Seasonal changes in burrow temperature appear to follow predictable cyclic patterns (Polis and Farley 1979b).

A high degree of biotic interaction is characteristic of equilibrium species. For *P. mesaensis,* cannibalism and intraguild predation occur at such high rates that they represent major ecological forces in the evolution of this species (see "Community Structure," below; Polis 1980b; Polis and McCormick 1986a, 1987). Equilibrium species characteristically exhibit high habitat specificity; in this case, *P. mesaensis* is restricted to sandy substrate.

Other scorpions that exhibit many traits of equilibrium species include *U. manicatus* (Smith 1966), *U. yaschenkoi* (Shorthouse 1971, Shorthouse and Marples 1982), and *P. utahensis* (Bradley 1982, 1983, 1984). Although the biologies of other species are less well known, many show at least some of the life-history traits characteristic of equilibrium species. Some scorpions live for 25 years or more (e.g., *Hadrurus*: Stahnke 1966), and others (*Opisthacanthus* spp., *Pandinus gambiensis*, *Leiurus quinquestriatus*) do not reproduce for the first time until they are 4 to 7 years old (Chapter 4). Several species are much larger than *P. mesaensis* (e.g., *Hadrurus, Pandinus, Hadogenes*); Newlands (1972b) reported that nongravid females of *Hadogenes troglodytes* weigh as much as 32 g, and some *Pandinus imperator* exceed 60 g (D. Krapf, pers. comm. 1988)! Like most scorpions, these species live in a subterranean retreat (in a burrow or under a rock), sheltered from adverse climatic conditions.

The data indicate that *Paruroctonus mesaensis* and possibly several other scorpion species resemble long-lived vertebrates in several aspects of their life history. In this resemblance they are similar to mygalomorph spiders and some large myriapods. These animals are among the exceptions to Pianka's (1970) generalization about the r-selection of terrestrial invertebrates, probably because of the stability and predictability of their subterranean habitat. Burrowing simulates some aspects of the selective regime found in stable, predictable environments: less energy need be expended on adapting to a fluctuating abiotic environment, and more energy can be apportioned to other purposes. As the selective regime shifts to increase the importance of the biotic environment, there is a subsequent selection for the characteristics of equilibrium species.

Scorpions as Opportunistic Species

Many scorpions in the family Buthidae exhibit a life history that is in marked contrast to the life history of equilibrium species. Although data are lacking on population dynamics, the life-history characteristics of buthids (Chapter 4) suggest that many of them are opportunistic species. First, their life history is greatly accelerated as compared with that for species from other families (see Tables 4.2, 4.4, 4.5; Armas and Contreras 1981): the average gestation period is 5.0 ± 2.2 months (all other species, 11.6 ± 3.6 months); the duration of the first instar is 6.5 ± 2.1 days (all other species, 14.3 ± 8.1 days); the time to maturity is 18.9 ± 12.0 months (range, 6–48 months; all other species, 36.6 ± 22.9 months, with a range of 6.2–83 months); longevity is 37.9 ± 11.8 months (all other species, 76.5 ± 15.0 months). (All parameters are significantly different, $p < 0.01$.)

Second, buthids are often smaller than other scorpions in their community. Third, they are quite plastic in several developmental features (Chapter 4). For example, the gestation period of *Tityus fasciolatus* is about 3 months in the Brazilian wet season and 7 to 10 months through the dry season (Lourenço 1978). The average birth season of buthids (4.4 ± 3.4 months) is significantly longer than that of all other species (2.5 ± 2.8 months; $p < 0.05$). Clutch sizes vary from 1 to over 100 offspring per litter. Individual buthids may produce several litters per season; in one study, females in five species of *Centruroides* produced three or four litters per year (Armas and Contreras 1981). In addition, they can give birth to multiple litters per insemination; females of *C. robertoi*, for example, can produce four litters without a new mating (see Chapter 4). Some buthids are even parthenogenetic (e.g., *T. serrulatus*: Matthiesen 1962). Furthermore, many buthid species mature at different instars, presumably in relation to individual feeding history (Chapter 4). They may also mature at different sizes, producing a population composed of "little," "normal," and/or "giant" adult males (Armas and Contreras 1981; Armas 1986; G. Polis, pers. obs.). Finally, the time to maturity may be highly variable in some buthids; an extreme case is *C. anchorellus*, which may mature in as few as 171 days or as many as 468 days (Chapter 4; Armas and Contreras 1981).

Thus, it appears that the life-history strategy of many (most?) buthids is more typical of short-lived insects than of longer-lived arthropods or vertebrates. On the average, they are smaller than other scorpions, exhibit shorter generation times, produce offspring more rapidly, and presumably have comparatively high maximum net reproductive rates (R_0) and capacities for increase (r_{max}). It is significant that most buthids differ from almost all other species in being "bark scorpions" (see "Ecomorphotypes," above); that is, they characteristically do not burrow, but live on or in vegetation and under surface debris. Buthids are less buffered from climatic extremes and are thus exposed to a more variable and harsh environment than scorpions that live in burrows.

The markedly different life-history strategies are, arguably, evolved responses to the presence or absence of a burrow-dwelling existence.

Community Structure

The role that scorpions play in natural communities is largely unknown. Patterns emerging from recent studies suggest that in at least some arid areas of the world, scorpions are one of the most important taxa of predators in terms of their density, biomass, and diversity. This section describes community-wide patterns among scorpions and assesses the nature and magnitude of scorpion-scorpion interactions.

Diversity, Density, and Biomass

From 1 to 13 species of scorpion appear sympatrically in localized areas (see Figs. 6.2 and 6.3). Data from 100 different sites were compiled to show the number of species found in each (Table 6.5). The majority of locales (68 percent) are characterized by scorpion communities consisting of 3 to 7 species. The most diverse scorpion communities were reported from the deserts of Nevada (9 species: Gertsch and Allred 1965), California (11 species: Hibner 1971), and southern Baja Cali-

TABLE 6.5
Patterns of scorpion-community composition

Number of sympatric species	Number of sites[a]	Distributions of single most dominant species[b] (% of all species, by sites)			
		21–40%	41–60%	61–80%	81–100%
2	2				2
3	10	1	1	3	5
4	11	1	6	1	3
5	8	1	2	4	1
6	10	5	2	2	1
7	10	1	1	6	2
8	8	1	7		
9	6	5	1		
≥ 10	3	2	1		

[a] 100 populations were studied worldwide; 68 of these yielded sufficient data for determining the prevalence of the different species. For the 100 sites studied, numbers of species were as follows: at 8 sites, only 1 species was found; 7 sites were inhabited by 2 sympatric species, 21 sites had 3 spp., 13 had 4, 12 had 5, another 12 had 6, 10 had 7, 8 had 8, 6 had 9, and 3 had 10 or more.

[b] The 68 communities are characterized in terms of the relative numerical dominance of the one species most abundant in each: e.g., of the 11 communities with 4 sympatric species, the single most abundant species accounted for > 80% of all scorpions found at 3 sites, and for 41–60% at 6.

fornia (10–13 species: Due and Polis 1986; S. Williams, pers. comm. 1979). As noted earlier, the least diverse communities occur at higher latitudes.

Two extremes illustrate the range in distribution of the number of individuals of each species within any community: in one case, 96 percent of all individuals belong to species A, and species B, C, D, and E each account for 1 percent; in another, species A, B, C, D, and E each account for 20 percent of the population. Naturally, most communities fall between these two extremes. Data were sufficiently detailed from 68 of the 100 localities to determine the distribution of individuals among species (Table 6.5). It is important to note that there is an inherent and artificial bias for species-poor communities to be characterized by one dominant species (defined here as at least 80 percent of all individuals) and for species-rich communities to show evenness. Nevertheless, dominance and evenness were found in both species-poor and species-rich communities. In three-species communities generally, the numerical distribution ranges from being dramatically lopsided (96.8/2.7/0.5 percent: Lourenço 1975) to very even (40/35/25 percent: Gertsch and Allred 1965, for scorpions in the *Atriplex–Kochia* plant association). The same is true for more diverse communities. For example, in seven-species scorpion communities west and north of Cabo San Lucas (Baja California Sur), S. Williams (pers. comm. 1979) found one area dominated by *Centruroides exilicauda* (84/4/4/3/3/1/1 percent) and a second area only a few kilometers away that exhibited much greater evenness (35/33/14/10/4/3/1 percent).

As demonstrated by *C. exilicauda*, some species may be dominant in one habitat but much less frequent in another area. At 35 sites in Baja California, *C. exilicauda* accounted for from less than 1 percent to 92 percent of all scorpions ($\bar{X} \pm SD = 39.8 \pm 27.6$: S. Williams, pers. comm. 1979). This pattern also occurs for *Paruroctonus mesaensis*, which constitutes at least 90 percent of all individuals in sandy habitats in California, Arizona, and northern Mexico (G. Polis, pers. obs.), 56 percent in Sonora, Mexico (Hadley and Williams 1968), but only 13 percent in sand dunes near Yuma, Arizona (Fox 1975).

An effort was made to determine which factors are associated with numerical dominance. The only regular (though not universal) trend is for one species to dominate in sandy habitats (see Table 6.6; other species also dominate in sandy habitats throughout the deserts of North

TABLE 6.6

Some dominant scorpion species

Taxon	Percent of individuals	Location	References
BUTHIDAE			
Centruroides exilicauda (Wood)	80–92	17 sites, Baja Calif.	S. Williams, pers. comm. 1979; Polis, unpubl. data
C. vittatus (Say)	>80	many sites, C/W Tex.	W. D. Sissom, pers. comm. 1982
Leiurus quinquestriatus (H. & E.)	86	Jordan Valley, Israel	Warburg et al. 1980
Liobuthus kessleri[a] Birula	65	Turkmen S.S.R.	Fet 1980
Tityus fasciolatus Pessôa	96.8	Brasilia, Brazil	Lourenço 1976
SCORPIONIDAE			
Scorpio maurus fuscus (H. & E.)	75	Lower Galil, Israel	Warburg et al. 1980
VAEJOVIDAE			
Paruroctonus baergi[a] (Williams & Hadley)	83	Yuma, Ariz.	Fox 1975
P. boreus (Girard)	>95	C Nev., N Ariz.	Gertsch & Allred 1965; Polis, unpubl. data
P. grandis[a] (Williams)	50–69	3 sites, N Baja Calif.	S. Williams, pers. comm. 1979
P. mesaensis[a] Stahnke	>90	many sites, SW U.S., N Mexico	Polis, unpubl. data
P. utahensis[a] (Williams)	>95	W of El Paso, Tex.	Polis, unpubl. data
Vaejovis diazi Williams	75	S Baja Calif.	S. Williams, pers. comm. 1979
V. hoffmanni Williams	80	N Baja Calif.	S. Williams, pers. comm. 1979
Vejovoidus longiunguis[a] (Williams)	88–96	5 sites, C Baja Calif.	Polis, unpubl. data

[a] Lives in sand.

America: S. Williams, pers. comm. 1979; G. Polis, pers. obs.). Such dominance may result from the highly specialized suite of adaptations necessary to live in sand (see "Ecomorphotypes," above); or it may be that the prevalent, relatively large and aggressive species prevent the smaller species from occupying sandy habitats (see below). However, large size is not a requisite for dominance. Some very large species (e.g., *Hadrurus* spp.) are consistently uncommon, and some smaller species (e.g., *Vaejovis hoffmanni, V. diazi*) may be numerically dominant.

Scorpions are remarkably dense in some habitats (Table 6.7). Several species occur locally in densities above 0.40/m^2 (4,000/ha). Species with high densities are found in deserts, chaparral, grassy plains, tropical forests, and the intertidal zone. Of course not all species occur in dense populations. Even species that are dense in some habitats are much less common in (apparently similar) adjacent areas (e.g., *P. mesaensis, Vejovoidus longiunguis, C. exilicauda*). For example, the density of *Anuroctonus phaiodactylus* varied by approximately 20 times at two similar study sites near San Diego, California (0.58/m^2, Williams 1963; 0.03/m^2, McDaniels 1968).

The biomass of scorpion populations can be crudely estimated by multiplying the density by the average wet weight (population biomass is a somewhat difficult value to come by because of the obvious differences between ages and sizes within a population). This method was employed by Shorthouse (1971) and Marples and Shorthouse (1982) for *Urodacus yaschenkoi* (weight based on age groups) and by Bacon (1972) for *Uroctonus mordax*. The estimated biomass was 1.23–1.85 kg/ha for *Urodacus yaschenkoi* and 8.3 kg/ha for *Uroctonus mordax*. I estimated the biomass of several species: *Hadrurus* (estimated average weight, 5 g), 20 kg/ha; *P. mesaensis* (1.0 g), 1.5–4 kg/ha; *C. margaritatus* and *C. exilicauda* (1.0 g), 2–5 kg/ha; *Leiurus quinquestriatus* (1.5 g), 16.5 kg/ha. These estimates of population biomass are higher than those for most other arthropods (except termites and possibly ants in some desert habitats) and for almost all vertebrates (Hadley and Szarek 1981; N. Hadley, pers. comm. 1981; Polis, unpubl. data).

Guild Structure

A major topic in community ecology is the assemblage of species that use the environment in similar ways (a guild and the interrelationships among guild members). Four general working hypotheses attempt to explain guild structure. First, exploitation competition for limited re-

TABLE 6.7
Scorpion density

Taxon	Density (per m^2)	Location	Habitat	Reference
BUTHIDAE				
Centruroides margaritatus (Gervais)	0.40	Costa Rica	lowland forest	Polis, unpubl. data
C. exilicauda (Wood)	0.20–0.50	s Baja Calif.	thorn forest	Polis, unpubl. data
Compsobuthus werneri judaicus (Birula)	0.0017	Israel	Mediterranean	Zinner & Amitai 1969
Leiurus quinquestriatus (Hemprich & Ehrenberg)	1.12	Israel	desert	Shulov & Levy 1978
Tityus fasciolatus Pessôa	0.0002	Brasilia, Brazil	in termite mounds	Lourenço 1978
IURIDAE				
Hadrurus spp.	0.50	c Baja Calif.	desert	Williams 1980, pers. comm. 1979
H. hirsutus (Wood)	0.50	s Baja Calif.	thorn forest	Williams 1980, pers. comm. 1979
SCORPIONIDAE				
Cheloctonus jonesii Pocock	0.67	Zululand, Natal, South Africa	grassy plains	Harington 1978
Opistophthalmus carinatus (Peters)	0.07	S.-W. Africa (Namibia)		Lamoral 1978
O. wahlbergi (Thorell)	0.07	S.-W. Africa (Namibia)		Lamoral 1978
Scorpio maurus palmatus (H. & E.)	0.03–0.13	Negev, Israel	desert	Shachak 1980
Urodacus hoplurus Pocock	0.12–0.14	w Australia	open ground	Koch 1978
U. manicatus (Thorell)	0.04–0.22	New S. Wales, Australia	sclerophyll forest	Smith 1966
U. yaschenkoi (Birula)	0.09–0.19	New S. Wales, Australia	sand	Shorthouse 1971

VAEJOVIDAE				
Anuroctonus phaiodactylus (Wood)	0.03–0.58	s Calif.	chaparral	Williams 1963, McDaniels 1968
Paruroctonus baergi (Williams & Hadley)	0.32	sw U.S.	desert, sand	Fox 1975
P. mesaensis Stahnke	0.15–0.40	s Calif.	desert, sand	Polis & McCormick 1986a
P. utahensis (Williams)	0.06–0.27	New Mexico	desert, sand	Bradley 1986
Serradigitus gertschi (Williams) (?)	1.0	c Baja Calif.	intertidal	Due & Polis 1985
S. g. striatus (Hjelle)	0.0007–0.0125	c Calif.	chaparral	Toren 1973
Uroctonus mordax Thorell	0.44	c Calif.	chaparral (adults only)	Bacon 1972
Vaejovis confusus Stahnke	0.02–0.06	s Calif.	desert	Polis, unpubl. data
V. littoralis Williams	8.0–12.0	c Baja Calif.	intertidal	Due & Polis 1985
Vejovoidus longiunguis (Williams)	0.30–0.45	c Baja Calif.	desert, sand	Polis, unpubl. data
		COMMUNITIES		
5 species	0.39	sw U.S.	desert, sand	Fox 1975
3–5 species	0.011–0.10	Israel	Mediterranean	Warburg et al. 1980
4 species	0.25–0.45	s Calif.	desert, sand	Polis & McCormick 1986a

sources (usually food, occasionally space) selects for the ecological divergence of species from one another in the use of resources; such divergence decreases overlap and allows coexistence. Second, interference competition (e.g., aggression, territoriality) selects for subordinate species' avoidance of more dominant species; such avoidance also decreases overlap and may allow coexistence. Third, other factors, particularly predation and strong physical disturbance (e.g., periodic fires, storms, or droughts), restrict the distribution and/or abundance of certain species to a greater degree than for other, more adapted species; for example, physical disturbance may allow coexistence by not allowing sufficient time for biotic processes (competition and predation) to proceed to the point where species are lost or excluded from an assemblage. Predators may differentially exploit competitively superior species and therefore allow coexistence by preventing the exclusion of competitively inferior species. The fourth hypothesis proposes that species assemblages include a random draw of available species, with the specific combination of species in any one area being a product of chance (e.g., climatic and biogeographic history, dispersal, random colonization, or extinction).

It is likely that all four classes of factors contribute both to the distribution and abundance of individual species and to the specific structure of any guild, but data are insufficient to evaluate fully the relative importance of these factors. This is especially true of the predation/physical-disturbance hypothesis and the random-community hypothesis (but see Koch 1977, 1981), and neither hypothesis will be discussed further in this chapter.

Both competitive hypotheses involve scorpion-scorpion interaction and similar predictions with regard to the manner in which species use resources. Both processes theoretically produce a set of characteristics usually associated with "resource partitioning": decreased overlap in the use of critical resources, temporal and spatial patterns that separate species ecologically, and mensural or morphological differences in trophic organs and/or body size.

Do scorpions divide resources by showing significant differences in use of time, space, or food? The existence of such differences would imply that scorpion guilds are not random assemblages of species and that scorpion-scorpion interaction is a major agent shaping guild structure. Evidence suggests that there are niche differences in time and space but not in the use of food.

Different species of scorpion probably do not use significantly different food resources. With one exception (*Isometroides vescus*: Koch 1977), all species of scorpions studied are generalists, eating whatever prey they are able to capture (Chapter 7; Polis 1979). Body-size differences may divide the prey eaten by different-sized adults. However, because adults of big species must grow from a small size at birth, all species overlap in size (and associated size-related prey use) during some parts of their lives. Such "developmental overlap" is generally unrecognized but greatly limits the effectiveness of differences in body size to divide resources among those species that show a wide range in size during development (Polis 1984). (Size is nonetheless of paramount importance in interference among scorpions; see below.)

Temporal differences have been found to exist among all scorpion guilds so far examined. The maximum surface activity of some species occurs in the winter, early spring, or late fall, when other species are relatively inactive (see Chapter 5, "Seasonal Patterns"). Such differences in activity are thought to have evolved to allow certain species to avoid other scorpions (Maury 1973b, 1978; Fox 1975; Polis 1980a, b; Polis and McCormick 1986a).

Other workers have suggested that competition affects the spatial distribution of scorpions (Williams 1970c, 1980; Koch 1977, 1981; Lamoral 1978; Shorthouse and Marples 1980; Bradley and Brody 1984; Polis and McCormick 1986a, 1987). Koch (1977, 1981) argued that the distribution of several species of Australian scorpions is determined by competition for homesites. Scorpion interactions cause a decrease in habitat overlap and may even produce competitive exclusion of some species in certain areas (e.g., *Urodacus* species are almost totally absent where *Liocheles waigiensis* is abundant). Lamoral (1978, 1979) maintained that specific soil-hardness preferences decrease competition for burrow sites and allow coexistence among southern African *Opistophthalmus*. In Williams's (1970c) study, the densities of *Hadrurus arizonensis* and *Vaejovis spinigerus* were related; after Williams removed large numbers of *Hadrurus*, the *V. spinigerus* population increased by over 50 percent. Such reciprocal densities are often interpreted as evidence for competition. Finally, the pattern of habitat use by *V. janssi* (Williams 1980) also implies that interaction may strongly influence spatial distribution. This species, the only one from Socorro Island (on the Pacific side of Baja California), exhibits a marked "ecological release" and is found in every habitat on the island: jungle, heavy brush, rocky terrain, sand, ground surface, vegetation (up to 7.6 m high), and coastline to within 3 m of the surf. The usual explanation for such niche expansion is that the lack of competition from other species allows existing species to expand their niche breadth into areas normally occupied by competitors.

On a finer scale, several scorpion species often live in the same general area separated only by differences in microhabitat. Species may occur primarily in different vegetation types, different soil types, and/or microhabitats differing in ground cover or other factors (e.g., floods, fires, or other disturbances). Further, scorpions in the same area may differ in their foraging station: some hunt in vegetation, others forage on the ground, and some "doorkeep" at the entrance of their burrows; some species constantly move during foraging, and others are sit-and-wait predators (see "Microhabitats," above, and Chapter 7).

Although temporal and spatial patterns thus suggest that scorpion-scorpion interaction is important in structuring at least some guilds, it is difficult to determine the exact nature of the interaction, because of the paucity of data on such processes. For the hypothesis that exploitation competition occurs among scorpions there is little evidence. (In all fairness, exploitation competition is notoriously difficult to observe; it may occur if food is limiting and prey capture by one scorpion decreases the probability of prey capture by another scorpion.) In all likelihood interference is the primary process.

Interference is most commonly manifested as aggression or intraguild predation among species of potentially competing scorpions. Intraguild predation is known to occur in the field among at least 30 pairs of scorpion species at six North American study sites (Polis et al. 1981; Bradley and Brody 1984; Polis and McCormick 1987; Polis, unpubl. data), and it greatly influences the guild structure of four sympatric species in the deserts near Palm Springs, California (Polis and McCormick 1986a, 1987). Considerable intraguild predation occurs among all species, and it appears to be an important cause of mortality. The two largest species, *H. arizonensis* and *Paruroctonus mesaensis*, both eat all other species; heterospecific scorpions represent a large part of their diet (21.9 percent for *Hadrurus*; 9.2 percent for *P. mesaensis*). Heterospecific scorpions form 8 and 9 percent, respectively, of the diet of *P. luteolus* and *V. confusus*.

Intraguild predation at Palm Springs may produce significant mortality to smaller species such as *P. luteolus* (Polis and McCormick 1987). Males of this species normally avoid predation by foraging from the entrance of their burrows. During a two-week period in the fall, however, the males emerge to the surface for courtship and mating; many are eaten by other scorpions at this time. In 1975 and 1976, 27 percent of all

P. luteolus males observed on the surface became prey to other scorpions. Overall, 8 percent of all *P. luteolus* and 6 percent of all *V. confusus* observed on the surface during 12 years of research were being eaten by *P. mesaensis*. Such mortality must represent a significant selective force in shaping the ecology and behavior of these smaller species.

In experiments lasting 30 months, more than 6,000 *P. mesaensis* were removed from 300 100-m^2 quadrats (= 30,000 m^2) of desert near Palm Springs. The density of both smaller species increased significantly in these areas (*P. luteolus* = 600 percent; *V. confusus* = 150 percent), compared with control areas. Since neither the body size nor the feeding rates of these smaller species increased in the experimental areas (relative to the controls), exploitation competition was not responsible for the population increase. These results indicate that intraguild predation is a major factor suppressing the abundance of smaller species, at least at this study site.

The impact of intraguild predation is reflected in spatial and temporal niche shifts by the two smaller species (*P. luteolus* and *V. confusus*). The maximum surface activity of these two species occurs during periods when the surface density of *P. mesaensis* adults is relatively moderate or low (Polis and McCormick 1986a, 1987). Although these smaller species are potentially habitat generalists, they are relatively uncommon in the habitats primarily occupied by *P. mesaensis*. Finally, the subordinate species occupy microspatial refuges. *Paruroctonus luteolus* is largely sympatric with *P. mesaensis* but avoids predation by foraging primarily either from the entrance of its burrow (*P. luteolus* males and females) or from the branches of various plants (*P. luteolus* females). *Vaejovis confusus* both forages from and travels on low grasses and herbs instead of on the ground. These microspatial refuges effectively inhibit predation by ground-dwelling populations of *P. mesaensis* and *H. arizonensis*.

In summary, intraguild predation appears to shape the major spatial, temporal, and demographic patterns of this guild. *P. mesaensis* adults are active primarily during the warm months, when prey abundance is maximal (Polis 1980a), and in sand, where prey abundances are significantly higher than in adjacent areas (Polis and McCormick 1986a; sand holds water much better than other desert soils and is thus characterized by higher vegetation and insect densities than nonsandy habitats). However, *P. mesaensis* is restricted to sand because its psammophilic adaptations preclude burrowing into harder soils (Polis and McCormick 1986a; see also Lawrence 1969, Lamoral 1978). This restriction partially explains the coexistence of smaller species. A very similar situation occurs in the sandy areas dominated by *Vejovoidus longiunguis* in central Baja California.

These general patterns and processes may also occur in other guilds of scorpions. The largest and most aggressive species occupy the best times and places, whereas the smaller, subordinate species are forced into less-than-optimal conditions. Dominant species may be restricted to favorable habitats and times because of specific ecomorphological adaptations or because their large body size and/or energetically costly aggression requires relatively high-productivity areas. Overall, then, it seems likely that interference in the form of intraguild predation is the paramount factor in at least some guilds. Aggression is apparently the structuring force in guilds of other animals (Polis et al. 1989). Specifically, interference competition and intraguild predation play large roles in the community structure of spiders; many researchers have noted that other spiders are a spider's worst enemies. The same appears to be true for scorpions.

Future Research

Ecological research on scorpions is at an early embryological stage of development, and fruitful research areas wait for the picking. Throughout this chapter, I have referred to the dearth of knowledge on scorpion ecology and field behavior. In general, more fieldwork with ultraviolet-light detection should be focused on tropical species, small species, and the many potentially lethal species. Species in each of these groups are relatively dense in their habitats and thus easy to study.

What details of scorpions' life history should be recorded? What facts would be most helpful to know? Interesting and important areas for field research abound.

POPULATION BIOLOGY. What are the time to maturity, clutch size, longevity, shape of the survivorship curve, and mortality factors of scorpions under field conditions? What is the basis of the observation that some species exhibit significantly different densities and/or body sizes in areas separated by only a few kilometers: food abundance, diversity of other scorpion species, microclimatic factors? How do these factors influence the life-history and reproductive parameters of these species? Are populations regulated? If so, what are the mechanisms? Are intraspecific interactions important? Why does cannibalism appear to be tremendously important in the ecology of some species but absent in other species?

COMMUNITY STRUCTURE. What factors determine both the local and geographical diversity of scorpions? What types of interaction exist among different species? Is intraguild predation widespread and important among other scorpion assemblages? How does the age structure of larger species affect interspecific resource partitioning and species coexistence? Why are some scorpion assemblages characterized by a numerically dominant species, whereas assemblages in adjacent areas are characterized by an even distribution of species? What impact do scorpions exert on prey communities? Can scorpions regulate prey populations or act as "keystone predators" to increase community-wide diversity? Have ecological shifts or changes in body size occurred on islands, as compared with the mainland? Are island ecosystems saturated with scorpion species?

ACTIVITY PATTERNS. What factors determine patterns of seasonal activity? What is the significance of age- or species-specific patterns of seasonal activity: resource partitioning, or the avoidance of older age groups and of more aggressive species? What determines the activity of individual scorpions? What is the evolutionary, ecological, and physiological basis for the observation that scorpions are largely inactive during most of their lives? What factors favor or allow diurnal activity in a very few species of scorpion?

BEHAVIORAL ECOLOGY. What degree of sociality is exhibited by different species of scorpion? Do scorpions live in family groups? Do relatives cooperate? What mechanisms do scorpions use to orient themselves and to home to their burrows? What adaptations are exhibited by scorpions that live in atypical habitats such as mountaintops or the intertidal zone?

7

Prey, Predators, and Parasites

SHARON J. MCCORMICK AND
GARY A. POLIS

All animals must both feed and avoid being eaten before engaging in any other activities such as reproduction or competition. Most arachnids, including scorpions, exist at an intermediate level in food chains. They are usually generalist predators on various insects, spiders, and other small animals; at the same time, they are prey to a variety of larger predators. In this chapter, the array of prey, predators, and parasites of scorpions are presented systematically. We also describe foraging tactics, feeding behavior, and digestive physiology. Scorpions manifest many interesting adaptations that allow them to be efficient predators while concurrently minimizing the risk of being eaten themselves. Finally, the importance of cannibalism and intraguild predation among scorpions is assessed; these processes are significant factors in both diet and mortality.

Prey

Foraging Behavior

Rather than using vision or audition to locate prey, scorpions seem to use other specialized neurosensory systems (see Chapters 5 and 9). Although Fleissner's (1977c) study on the eyes of *Androctonus australis* suggested that the scorpion eye should be ranked with the more sensitive arthropod eyes, vision is apparently not important in prey detection. The precision in prey detection that was demonstrated by Brownell's (1977a, b, 1984; Brownell and Farley 1979a, b) extensive work with the sand-dune scorpion *Paruroctonus mesaensis* is probably achieved by many species (Chapter 9). Prey moving within 15 cm of the scorpion are sensed by tarsal sense organs and then located and captured in a single motion; prey more than 30 cm away are located by a series of orientation responses (see Chapter 9). Prey often betray their position by producing air movements, which are then detected by the scorpion's trichobothria, the long and very thin sensory hairs scattered on the pedipalps and sensitive to changes in air pressure (Le Berre 1979). D. Krapf (pers. comm. 1985) has analyzed the role of trichobothria in the capture of aerial prey. Many species use trichobothria to orient accurately toward insects flying within 10 cm of the scorpion. Prey are then captured in midair with a very rapid grab by the pedipalps.

After contacting prey with the pedipalps, a scorpion may or may not sting, depending on the ratio of prey size to predator pedipalp size. Generally, scorpions do not sting relatively small prey, especially if the prey are secured adequately by the pedipalps. However, unwieldy or comparatively large prey are stung. Thus, larger species of scorpion or older animals would not envenomate the same prey that smaller species or younger animals would necessarily sting. Interestingly, *Pandinus imperator* exhibits an ontogenetic change in prey capture: young normally sting prey, whereas adults just use their pedipalps (Casper 1985). Characteristically, scorpion species with slender pedipalps sting, whereas those with robust pedipalps (e.g., fossorial species: see Chapter 6) crush prey. Alexander (1972) experimentally determined that the scorpion actively orients the body of the prey before feeding; for example, many scorpions manipulate their prey such that the head is consumed first (Waterman 1950, Polis 1979).

With the use of ultraviolet light, scorpions are readily observed capturing prey and feeding in their natural habitats. Foraging scorpions may either sit and wait or actively hunt prey. These foraging patterns are species-specific.

Many species are "sit and wait" or "ambush" predators. *Paruroctonus mesaensis* is a typical sit-and-wait predator (Polis 1979). It emerges from its burrow and remains motionless within a meter of the opening. Insects and other prey are detected as they pass close to the waiting scorpion. Although most prey are eaten outside the burrow, usually in the area where they are caught, over a fourth of all prey are carried up into nearby vegetation before being consumed. Generally, North American vaejovids do not return to the burrow with prey but digest it at the site of capture (Hadley and Williams 1968). Not all sit-and-wait predators feed in exactly the same manner. Le Berre (1979) reported that *Buthus occitanus* returned to its burrow with prey caught elsewhere. Several species do not leave the burrow but wait at the entrance ("doorkeeping") and capture prey that move near the burrow (Williams 1966; Shorthouse 1971; Eastwood 1978a, b; Shachak and Brand 1983).

Active search for prey, involving a comparatively greater expenditure of energy, is a less common foraging pattern. Several scorpions of the buthid and scorpionid families hunt for prey in this manner. Newlands (1978a) reported that the arboreal scorpions *Uroplectes otjimbinguensis*, *U. vittatus*, and *Opisthacanthus chrysopus* move about the branches of young acacia trees while foraging at night and retreat diurnally under the bark of older or dead trees. Similarly, Hadley and Williams (1968) observed *Centruroides* crawling over rocks and vegetation and actually chasing insects. Morphologically and behaviorally, these animals are classified as "bark" or "errant" scorpions (Chapter 6). These scorpions handle prey and defend themselves differently than the majority of scorpions.

Evidence indicates that scorpions may eat prey within the burrow. Cekalovic (1965–66) observed *Centromachetes pococki* feeding in shallow burrows just under logs, cow dung, and small pieces of wood. The remains of prey in the burrow indicate that other scorpions also feed there. For example, Newlands (1978a) found burrows of *Opistophthalmus flavescens* littered with tenebrionid beetle remains; burrows of *Scorpio maurus* contain the debris of various arthropod prey (Shachak and Brand 1983). Krapf (1986a) analyzed the diet of *S. maurus* by collecting over 1,000 prey items deposited outside the burrow. Discrete feeding chambers adjacent to the burrow's entrance allow some species to manipulate prey within the burrow (Smith 1966, Eastwood 1978a). Many scorpions are frequently not active on the surface and remain underground for long periods of time (Chapter 5); such periods of apparent inactivity may actually include some feeding below the surface. The difficulty of observing behavior within the burrow is an obstacle to studying many aspects of scorpion ecology.

Digestion

Like most other arachnids, scorpions use the chelicerae to initiate digestion by tearing the prey into smaller particles, which are then collected in a preoral cavity. Dismemberment of heavily sclerotized prey is begun in a relatively exposed area between the head and the thorax

(Eastwood 1978a). Digestion may take from one to several hours (Eastwood 1978a; Hadley and Williams 1968; S. McCormick and G. Polis, pers. obs.). The time required for a prey item to be completely entered into the preoral cavity from the initial contact with the pedipalps is the handling-time component of predation. Temperature, satiation, size of the predator, and the size and cuticle type of the prey probably all determine the duration of handling time.

Digestion by all arachnids, including scorpions, actually begins outside the mouth. The preoral cavity that lies below the chelicerae and carapace is supplied with digestive juices from the gut. Alexander (1972) described a system of channels formed by the coxae of the pedipalps and the coxapophyses of the first two walking legs, through which the enzymes are transported. This enables digestion to begin externally. Consequently, semidigested food is passed directly from the mouth to the gut.

Kästner (1968) described and illustrated the internal digestive anatomy of the scorpion (see also Chapter 2). Setae in the preoral cavity trap any undigestible material, such as the prey exoskeleton. These particles are matted together and expelled from the preoral cavity. The anterior region of the pharynx is a sclerotized pumping organ that draws food into the body from the preoral cavity. The pharynx then connects with the midgut, where digestion also occurs. Cecal glands, which arise from this area of the digestive tract, produce some of the amylases, proteases, and lipases that are transported to the preoral cavity. After passing through a short, sclerotized hindgut, waste passes through the anus, which opens immediately before the bulbous portion of the telson.

An ability to consume a large quantity of food at one time, a highly efficient food-storage organ, and an extremely low metabolic rate may conjunctly enable scorpions to survive food deprivation over a long period. The mesosoma may appear quite swollen after feeding. This occurs because the membranes connecting the segments of the exoskeleton are extensible, allowing substantial increases in body weight: *Tityus bahiensis* can increase its weight by a third (Matthiesen 1961), and *Uroctonus mordax* and *Serradigitus gertschi striatus* increased an average of 17 percent and 16 percent in body weight, respectively, after laboratory feeding (Bacon 1972, Toren 1973).

The hepatopancreas—a large liverlike organ making up about 20 percent of the total body mass—is the only identified food-storage organ in scorpions. High concentrations of glycogen in the hepatopancreas and secondarily in the muscles contribute to survival during food deprivation. Sinha and Kanungo (1967) analyzed the physiological effects of starvation. They found that the concentration of glycogen in the hepatopancreas of *Heterometrus bengalensis* gradually decreased during starvation. After 4 days of starvation the amino acid concentration in the liver dropped to 42 percent of normal; in the muscles, however, there was scarcely any decrease. Remarkably, they reported no change in the water content of the liver and muscle after 12 days of starvation.

Scorpions have extremely low metabolic rates. Keister and Buck (1973) reviewed numerous studies of insect metabolic rates and found the average value to be 1,000 mm of oxygen per gram per hour at 25° C; in contrast, scorpion rates rarely exceed a tenth that level. For example, Dresco-Derouet (1964a, b) determined the metabolic rate of *Euscorpius carpathicus* to be 122.9 mm O_2/g/hr and that of *Nebo hierichonticus* to be 35.1 mm O_2/g/hr, both at 25° C. A low metabolic rate indicates minimal energetic requirements and is thus likely to be a factor contributing to survival under food deprivation.

Some scorpions may survive starvation for 6 to 12 months or more (Stahnke 1945, 1966; Cloudsley-Thompson 1968; Kästner 1968; Savory 1977; Polis 1988). Interestingly, Newlands (1978a) found that *Opisthacanthus validus* from the humid eastern parts of South Africa rarely survived a month without food and water; however, *Parabuthus villosus*, an inhabitant of the more arid Namib Desert, survived for a year. This suggests that the ability to survive starvation may be related to the water stress sustained in the environment to which a species is adapted. Survival over long periods of starvation is characteristic of many arachnids; some spiders can live over two years without feeding (Savory 1977).

Foraging Activity

Only a small proportion of a population feeds on any one night. The feeding rate describes the percentage of all scorpions on the surface that are observed feeding. Hadley and Williams (1968) reported feeding rates of 1–5 percent; Tourtlotte (1974) determined a feeding rate averaged over a season of 2.7 percent; and Polis (1979) calculated bimonthly feeding rates that ranged from zero to 8.1 percent. Polis discovered seasonal variation in the feeding rates of *Paruroctonus mesaensis*: the highest rates occurred in the late spring and in summer, and minimal feeding took place during the late fall and early winter (Fig. 7.1). Seasonal

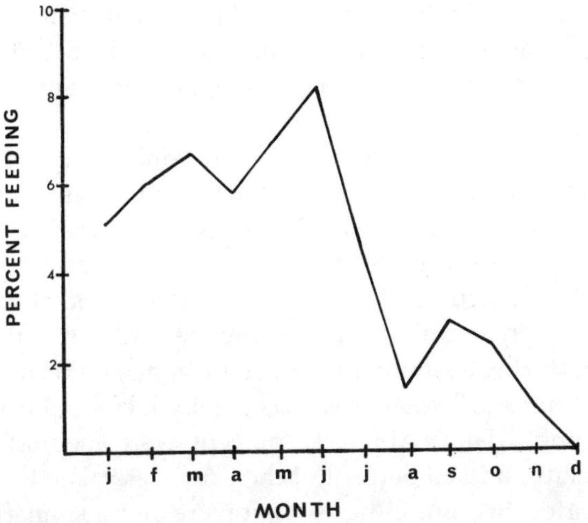

Fig. 7.1. Percentage of all *Paruroctonus mesaensis* individuals observed feeding as a function of season. Data from 5 years were summarized into an average year. The highest feeding rate occurred in June and lowest rate in December. (From Polis 1979)

variation in feeding rates, growth rates, and diet will be discussed further in relation to prey availability. There are other factors that influence scorpion feeding. In a detailed behavioral model, Le Berre (1979) noted that molting, courtship, and maternal behavior were inhibitory to feeding. This is also true for North American vaejovids (Chapter 4).

Prey availability and abundance undergo seasonal changes. In general, prey are classified according to their availability as pulsed, seasonal, or annual (Polis 1979). Pulsed prey—termites, the winged reproductive ants, and other ephemeral arthropods—appear in the diet for relatively short periods each year. Some beetles, moths, crickets, and spiders are seasonal, and they are included in the diet for a few months each year. Other prey are captured evenly throughout the year, including other scorpions and some Orthoptera. Predictable changes in the diet are a consequence of the availability of these classes of prey.

Correlated seasonal changes were observed in the growth and feeding rates of *P. mesaensis* (Polis and Farley 1979b). Feeding rates were highest from January to June. In June the relative abundance of prey was highest, and there was a significant correlation between high growth rate and prey abundance. Prey abundance and the feeding rate dropped precipitously in the summer, and the growth rate correspondingly became negative in July (Polis and Farley 1979b). There was a small increase in feeding rates, growth rate, and prey abundance during the fall.

Scorpion activity is not strictly seasonal, since unusually cool or warm temperatures may alter seasonal activity patterns (Chapter 5). Prey abundance is also correlated with temperature (Maury 1978, Polis 1980a), and consequently the feeding rate may be affected by temperature changes. Sudden fluctuations in other abiotic factors that influence prey abundance may dramatically alter the feeding rate. After 2 cm of rain in an hour during a summer thunderstorm in southern California, the surface densities of *Vaejovis confusus* greatly increased (McCormick and Polis, unpubl. data). About 12 percent were feeding (versus a norm of 2 percent), and 60 percent of the prey were alate termites that had emerged after the rain. Other scorpions respond similarly to rain. It is likely that many scorpions increase surface densities and feeding rates after opportune changes in prey abundance.

Diet

Information about the diets of scorpions may not be reliable. Some reports listed items that captive scorpions would or would not eat (Scorza 1954; Eastwood 1978a; Lourenço 1978, 1979a), and the few reports that have been assembled of prey consumed in the natural environment have not been intensive studies. The most detailed and significant study of scorpion diet was conducted by Polis (1979), who observed the prey of a *Paruroctonus mesaensis* population regularly for five years (see also McCormick and Polis 1986). Table 7.1 summarizes the available descriptions of the natural diets of several species of scorpion.

When viewed systematically and in natural situations, scorpion diet items ranged from annelids to vertebrates. The diet of *P. mesaensis* is practically restricted to the Arthropoda, with only one non-arthropod among 1,600 prey items. Many classes of arthropods are eaten by different scorpions. Arachnids and insects are common prey (Fig. 7.2). Koch (1977) reported that millipedes and centipedes (Fig. 7.3) are also occasionally eaten. Terrestrial isopods are suitable prey (Warburg and Ben-Horin 1979, Shachak 1980). Three species were reported to eat invertebrates other than arthropods: *Urodacus novaehollandiae* ate earth-

TABLE 7.1

Scorpion invertebrate prey in natural habitats

Predator	Prey	Reference
BOTHRIURIDAE		
Bothriurus araguayae Vellard	*Armitermes* sp. (Isoptera)	Lourenço 1976
Centromachetes pococki (Kraepelin)	Gryllidae, *Dalaca noctuides* larvae (Lepidoptera), Geometridae larvae	Cekalovic 1965–66
BUTHIDAE		
Centruroides exilicauda (Wood)	Araneae, Scorpiones, Solifuges, Chilopoda, Orthoptera, Thysanura, Coleoptera, Hymenoptera, Lepidoptera	Polis, unpubl. data (see Table 7.2)
Isometroides vescus (Karsch)	*Arbanitis hoggi*, *Aganippe occidentalis*, *A. latior*, *A. raphiduca*, *Dekana* sp., *Lycosa* sp. (all Araneae)	Main 1956
	Dekana diversicola (Dipluridae), *Lycosa* sp.	Koch 1977
Mesobuthus gibbosus Brullé	*Procurstes banoni*	Birula 1917a
Tityus fasciolatus Pessôa	Dipluridae, Lycosidae, *Pamphobetus* (mygalomorph), Blattidae, Isoptera	Lourenço 1978
T. mattogrossensis Borelli	Araneae, Blattidae, Isoptera	Lourenço 1979a
Uroplectes lineatus (C. L. Koch), *U. insignis* Pocock	*Temnopteryx phalerata* (Blattariae)	Eastwood 1978a
ISCHNURIDAE		
Hadogenes troglodytes (Peters)	Mollusca	Newlands 1978a
IURIDAE		
Hadrurus arizonensis Ewing	Araneae, Scorpiones, Solifuges, Orthoptera, Coleoptera, Hymenoptera, Lepidoptera	Polis & McCormick 1986a (see Table 7.2)
SCORPIONIDAE		
Cheloctonus jonesii Pocock	dung beetles, small Tenebrionidae	Harington 1978
Heterometrus longimanus (Herbst)	Blattidae, Dermaptera, Coleoptera larvae	Schultze 1927
Opistophthalmus capensis (Herbst)	*Dorylinus helvolus* (Formicidae), *Temnopteryx phalerata* (Blattariae), small Tenebrionidae	Eastwood 1978a
O. carinatus (Peters)	*Xerocerastus burchelli* (Mollusca)	Lamoral 1971
Scorpio maurus palmatus (L.)	*Hemilepistus reaumuri* (Isopoda)	Shachak 1980
	Mesor sp. (Formicidae), Tenebrionidae	Shachak & Brand 1983
Urodacus spp.	Araneae, Chilopoda, Diplopoda, Blattidae, Coleoptera	Koch 1977
U. novaehollandiae Peters	Annelida	Koch 1977

TABLE 7.1
(*continued*)

Predator	Prey	Reference
VAEJOVIDAE		
North American Vaejovidae	Araneae, Orthoptera, Coleoptera, Lepidoptera	Hadley & Williams 1968
Paruroctonus boreus (Girard)	Araneae, Orthoptera, Pentatomidae	Tourtlotte 1974
P. mesaensis Stahnke	Isopoda, Araneae, Scorpiones, Solifuges, Hemiptera, Homoptera, Isoptera, Orthoptera, Coleoptera, Diptera, Hymenoptera, Lepidoptera, Neuroptera	Polis 1979 (table II), McCormick & Polis 1986, Polis & McCormick 1986a
Serradigitus gertschi striatus (Hjelle)	Geometridae larvae, Ichneumonidae	Toren 1973
Uroctonus mordax Thorell	Dioptidae (Araneae), Geometridae, Vespidae	Bacon 1972
Vaejovis littoralis Williams	Isopoda, Araneae, Pseudoscorpiones, Scorpiones, Chilopoda, Thysanura, Coleoptera, Hymenoptera	Due & Polis 1985
Vejovoidus longiunguis (Williams)	Araneae, Scorpiones, Solifuges, Orthoptera, Thysanura, Coleoptera, Diptera, Hymenoptera, Lepidoptera, Neuroptera	Polis & McCormick, unpubl. data

worms (Koch 1977), and *Opistophthalmus carinatus* (Lamoral 1971) and *Hadogenes troglodytes* (Newlands 1978a) ate mollusks. Many scorpions are capable of eating vertebrates, both mammals and reptiles. Table 7.2 shows that larger scorpions are able to consume smaller and juvenile vertebrates.

Although some information suggests that the diet of the scorpions studied consists of few types of prey, this is probably not the case. Information from laboratory feedings (Scorza 1954, Stahnke 1966) showed that scorpions generally accept most items of suitable size and palatability. Polis (1979) reported 95 species of prey in the diet of *P. mesaensis*; however, additional unpublished data have increased the number of known prey species to over 125. So far, there has been a direct relationship between the duration of the study and the number of species determined to be in the diet. This increase in prey breadth with increased observation time is also seen for many other species of North American scorpions (Polis, unpubl. data).

Fig. 7.2. Some examples of scorpions with their prey: *A, Paruroctonus mesaensis* feeding on a moth (Sphingidae) at Windy Point in the Colorado Desert of southern California; *B,* also in the Colorado Desert, *P. mesaensis* consuming a solpugid (Eremobatidae); *C, Vaejovoidus longiunguis* eating an adult ant lion (Myrmeleonidae) in Baja California Sur.

Fig. 7.3. *Hadrurus concolorous* eating a centipede of the genus *Scolopendra* in the Loreto area of Baja California Sur.

Prey Selection

A number of factors may determine the suitability of prey for each scorpion species and for each age class of a species. Prey preference is dependent on size, type of exoskeleton, and possibly odor (see, e.g., Stahnke 1966). The exoskeleton and size of the prey determines the efficiency with which the predator manipulates prey in the pedipalps. Hadley and Williams (1968) concluded from their laboratory observations that scorpions accept as prey most animals that can be physically immobilized by the pedipalps.

Some rarely eaten insects are unsuitable because of thick exoskeletons. Hadley and Williams (1968) and Polis (1979, 1988) reported that the thickly cuticled darkling beetles *Eleodes* are generally ignored. Cloudsley-Thompson (1959) found that in captivity *Andoctronus australis* and *Buthus occitanus* would eat the hard beetles *Blaps* and *Akis spinosa* only if starved. But some scorpions do handle heavily sclerotized prey. For example, *Mesobuthus gibbosus* readily sucks out the contents of the large ground beetle *Procurstes banoni*, leaving the thick exoskeleton in-

TABLE 7.2

Scorpion vertebrate prey in natural habitats

Predator	Prey	References
BUTHIDAE		
Centruroides exilicauda (Wood)	*Phyllodactylus* sp. (leaf-toed gecko)	Polis, unpubl. data
Parabuthus villosus (Peters)	*Palmatogecko rangei* (palmate desert gecko)	Lamoral 1971
IURIDAE		
Hadrurus sp. (reported as *Centruroides*)	*Leptotyphlops humilis* (western blind snake)	Anderson 1956
H. arizonensis Ewing	*L. humilis*	Polis & McCormick 1986a
	Cnemidophorus sp. (whiptail lizard)	Stahnke 1966
SCORPIONIDAE		
Opistophthalmus carinatus (Peters)	*Pachydactylus capensis* (common cape gecko)	Lamoral 1971
	Mabuya striata (lizard)	A. Harington, pers. comm. 1980
O. wahlbergi (Thorell)	small rodents	Newlands 1978a
Pandinus pallidus (Kraepelin)	*Leggada minutoides* (harvest mouse)	Koch 1969
VAEJOVIDAE		
Paruroctonus mesaensis Stahnke	*Leptotyphlops humilis*	Polis 1979

SOURCE: McCormick & Polis 1982 (with additions).

tact (Birula 1917a); predation is so intense that the scorpion may cause tremendous mortality in the population of this beetle. *Opistophthalmus flavescens*, a psammophilic scorpion with highly developed and powerful cheliceral muscles, preys on a similar beetle, *Onymacris* (Newlands 1972b). Other arthropods relatively ignored by some scorpions include the ants *Pogonomyrmex* (Polis 1979, 1988) and *Camponotus* (Lourenço 1978) and isopods (Stahnke 1966; Hadley and Williams 1968; Lourenço 1978; S. McCormick and G. Polis, pers. obs.). The exclusion of isopods from the diets of some scorpions is interesting, because at least one scorpion, *Scorpio maurus palmatus*, concentrates its feeding on *Hemilepistus* isopods (Shachak 1980, Krapf 1986a). *Paruroctonus mesaensis* often orients toward isopods (*Venezillo*) but cannot hold one in its chelae when the isopod rolls itself into a ball (conglobation). Conglobating behavior is family-specific (Warburg 1968). Isopods that do not conglobate (e.g., *Ligia* in the littoral zone of the Gulf of California; *Hemilepistus*) are scorpion prey. *Venezillo* and perhaps other isopods of the families Armadillidae and Tylidae escape severe scorpion predation by conglobating.

Toren (1973) suggested that *Serradigitus gertschi striatus* rejects mirid bugs because of their noxious odor. Alexander and Ewer (1957) reported that *Opistophthalmus* rejected prey on the basis of information gained by chemoreceptors located on the chelicerae; they arrived at their conclusion after experimentally removing some receptors and then offering these scorpions quinine-soaked prey. Although scorpions generally reject dead prey, Armas (1975), Bücherl (1971), Vachon (1953), and Krapf (1986b) observed some scorpions accept dead prey under laboratory conditions.

Smaller species of scorpions consistently use smaller prey items than larger scorpion species (McCormick and Polis 1986). There is both an upper and a lower boundary on the size of prey suitable for larger scorpions (small prey are difficult for larger animals to manipulate).

The *P. mesaensis* study indicated age-specific differences in diet (Polis 1979, 1984). The direct correlation between predator and prey size (Fig. 7.4) indicates that each of the three age classes in the population consumes a different average-size prey. Consequently, some prey are only important in the diet of particular age classes. As an example, the tenebrionid beetle *Eusattus muricatus* formed 30 percent and 1 percent of the diet for adults and juveniles, respectively. The small tenebrionid *Batulius* sp. composed 7 percent of the juvenile diet but was excluded from the diet of both intermediate and adult animals. In fact, adult animals ate 14 unique species; both the intermediately aged and the youngest animals ate 20 unique species of prey (Polis 1979).

Although most scorpions are opportunistic predators, some species appear to concentrate on certain types of prey. Shachak (1980) and Krapf (1986a) reported that 20 to 60 percent of the prey remains from the burrows of *Scorpio maurus palmatus* were isopods. Similarly, the intertidal scorpion *Vaejovis littoralis* feeds primarily (over 80 percent of diet) on isopods abundant in the littoral zone (Due and Polis 1985). Australian *Isometroides vescus* specialize on burrowing spiders. Main (1956) collected 18 specimens of *Isometroides* from the burrows of ten species of spider; of these scorpions, it was concluded that 16 had eaten

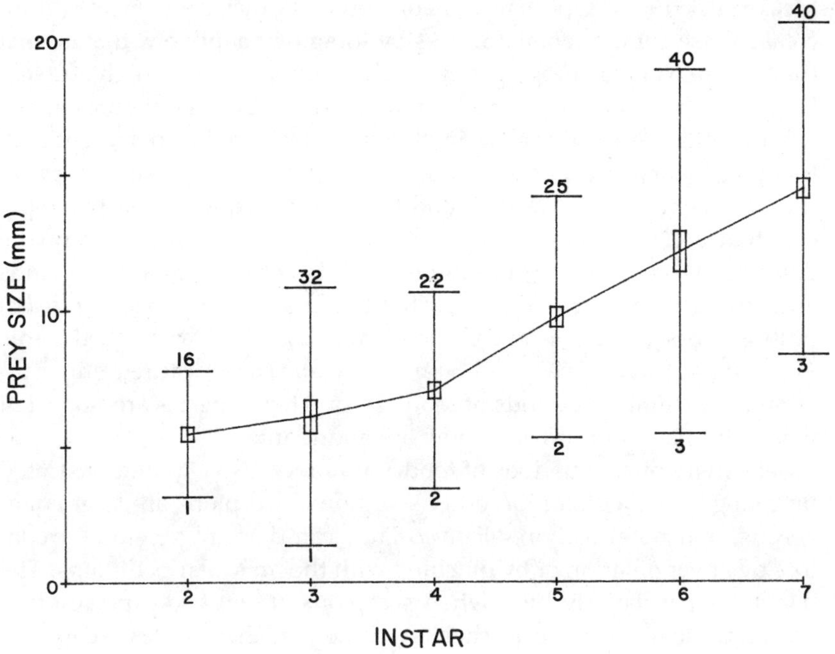

Fig. 7.4. The relationship between the body sizes of predator *(Paruroctonus mesaensis)* and prey, showing that the average prey size increases as predator size increases. For each instar, the following information about prey size is given: mean, standard error (boxes), standard deviation (vertical lines), and range (numbers).

the inhabiting spider. The other two scorpions were in the upper portion of the burrows of *Gaius* spiders, which pull down a lining from the top of the burrow to form a protective sock; evidently this is an effective defense against *Isometroides*. Similarly, Koch (1977) reported that *I. vescus* collected from the burrows of ten species of spider feeds solely on burrowing spiders. He, too, discovered that a spider, *Anidiops*, escapes scorpion predation by sealing itself into the lower portion of the burrow with a rock. No other scorpions have been observed to concentrate on a particular type of prey.

Optimality theory (Pyke 1984) has been used to predict both prey selection and diet changes by consumers under changing regimes of food availability. The ability of scorpions to forage optimally was analyzed by Polis (1988). Growth and reproduction of *P. mesaensis* are directly affected by food availability (Polis and Farley 1979b, Polis and

McCormick 1987). Optimality theory predicts that consumers can increase these components of fitness by foraging adaptively, that is, that they can maximize energy gain under conditions of both high and low food availability. During periods of food stress, several foraging characteristics change significantly in the direction predicted by optimal-foraging theory: scorpions feed less, eat significantly smaller prey on the average, are less selective, and feed more frequently on prey species that are noxious (insects with chemical defenses) or potentially dangerous (scorpions). For example, insects with repugnatorial glands and strong defensive odors (e.g., the tenebrionid stink beetle *Eleodes*) or those possessing poison and repellent body chemistry (e.g., the ants *Pogonomyrmex californicus* and *Solenopsis xyloni*) are captured only during the midsummer periods of food stress. These species are not eaten when alternate, non-noxious prey are abundant.

Selectivity during periods of food abundance also was demonstrated by changes in predation on other scorpions. Scorpions are large, dangerous, and potentially retaliatory. They can damage potential predators by envenomation or by pinching with the muscular pedipalps. The danger is particularly high when scorpions attempt to capture other scorpions that are close to their own size. In these cases, scorpions have been observed to become prey themselves while attempting to eat other scorpions. *Paruroctonus mesaensis* frequently eat both conspecific (9.1 percent and 5.0 percent of the total diet at two sites in southern Californian deserts) and heterospecific (6.6 percent, 9.2 percent) scorpions (Polis 1980a, 1980b; Polis and McCormick 1987). Both cannibalism and intraguild predation showed highly significant inverse correlations with indexes of prey abundance. Conspecifics were most frequent (13.3 percent) in the *P. mesaensis* diet during periods when prey were least available, but were not observed eaten when alternate (insect) prey were most abundant. Heterospecific scorpions formed 35 percent of all prey when prey were at their lowest availability, and only 3.2 percent at the highest.

Effects on Prey Populations

The density of scorpion populations may be particularly high in areas of great food abundance (see Chapter 6). Newlands (1978a) reported that there may be more than one *Cheloctonus jonesii* per square meter in such situations. *Vaejovis littoralis* may reach a density of more than 12 per square meter in the fresh *Sargassum* drift lining some cobblestone beaches in the Gulf of California (Due and Polis 1985); this beach wrack is teeming with isopods and other suitable prey. Several South American species live in bromeliads and forage near the water collected in the base of the epiphyte (González-Sponga 1978a). González-Sponga suggested that this is advantageous because the concentration of prey is higher in the epiphyte than elsewhere. However, the local abundance of *Scorpio maurus* and *P. utahensis* apparently is not correlated with prey availability (Bradley 1983, Shachak and Brand 1983).

Scorpions are potentially important consumers in some communities. They maintain sufficiently dense and stable populations to represent a significant link in the production and energy transfer of certain ecosystems. Indeed, in some North American desert communities, scorpions appear to be the dominant insectivorous predators (e.g., *P. mesaensis, Vejovoidus longiunguis*, some *Hadrurus*). Williams (1971c) suggested that they may be the top predator on some islands in the Gulf of California that have a depauperate vertebrate fauna. They may also be a key predator in more localized ecosystems, such as termite mounds (*Tityus fasciolatus*: Lourenço 1978) and the litter around desert perennials (*Superstitionia donensis*: Williams 1980).

Few specific data have been collected on the impact of scorpions on their prey communities. Birula (1917a) claimed that predation by *Mesobuthus gibbosus* seasonally decimates the population of a large ground beetle (*Procurstes*) in Crete. Shachak (1980) estimated the effect of predation by *Scorpio maurus palmatus* on their isopod prey (*Hemilepistus*) over a 5-year period in Israel; he calculated that scorpions annually ate an average of 10.9 percent (range, 1.7–19.3 percent) of the entire isopod population. Shorthouse (1971; Marples and Shorthouse 1982) determined the energy budget and production of a population of *Urodacus yaschenkoi*, finding that these scorpions eat an annual average of 7.90 kg/ha of prey. This represents 98,400 ants or 31,570 medium-sized spiders removed per hectare per year. Since the estimated biomass of *U. yaschenkoi* is not extraordinarily high (1.2–1.85 kg/ha) in comparison with some species, it is likely that some scorpions remove much more prey from their ecosystems. Assuming that *Hadrurus* and *P. mesaensis* have metabolisms and overall assimilation and utilization efficiencies similar to those of *U. yaschenkoi, Hadrurus* will process an estimated 106–160 kg of prey per hectare per year, and *P. mesaensis*, 8.6–38.5 kg/ha per year.

Undoubtedly, such predation exerts a tremendous impact on prey population dynamics. It is likely that some particularly vulnerable species of prey must either coevolve various escape tactics (temporal, behavioral, or morphological) or avoid altogether those habitats with high densities of scorpions (Brownell 1977a, b).

Long-term research on *P. mesaensis* has quantified the impact of predation by this scorpion on the behavior, distribution, and abundance of some of its prey species (Polis and McCormick 1986a, b, 1987). When over 6,000 *P. mesaensis* individuals (3 kg) were removed from 300 replicated quadrats (each 10×10 m = 100 m^2), the densities of prey scorpions and spiders increased dramatically, compared with densities in unmanipulated control quadrats. The number of *P. luteolus* increased 600 percent; *Vaejovis confusus* increased 150 percent. All spider species combined doubled in density within the removal quadrats. These data suggest that predation by *P. mesaensis* is the key factor that caused the rarity of these two species of scorpions and limited the spider population in this desert system. Surprisingly, the density of all prey insects combined did not increase in the areas where *P. mesaensis* were removed.

Polis and McCormick argued that, in general, the spatial and temporal distribution of smaller scorpion species and younger age classes of all species have coevolved to avoid predation from larger scorpions. In the *P. mesaensis* system, both the two smaller species and *P. mesaensis* age classes were active in microhabitats and time periods that were markedly different from those of adult *P. mesaensis*. Only more research will determine the effect of scorpion predation on prey population in other ecosystems and habitats.

Nevertheless, it is important to note that scorpions may exert little influence in some communities. First, they occur in low densities in many habitats. Second, even in habitats where they are relatively dense (e.g., *Vaejovis littoralis*, 12/m^2 in the intertidal), the small size of some species may make them insignificant predators. Further, although they may be dense, they are relatively inactive, and as a consequence few individuals are present on the surface to capture prey on any one night. Finally, scorpions move little and generally exhibit low metabolism (Chapter 8). Thus, although their biomass is greater than many other species', they may not eat as much prey per unit of biomass as do other predators.

Predators

There is much information available about the predators of scorpions. In a tabular form, Polis et al. (1981) summarized 150 taxa that prey on scorpions and presented data on the predators of *Paruroctonus mesaensis*; many of the conclusions about the predators of scorpions in this chapter are based on information compiled in that paper. Scorpions are prey of both vertebrates and invertebrates. A large variety of vertebrates was found to feed on scorpions, primarily from the analyses of stomach contents. Vertebrate predators of scorpions include birds (37 percent of all vertebrate predators), lizards (34 percent), mammals (18 percent), frogs and toads (6 percent), and snakes (5 percent). Invertebrate predators include other arachnids (black widow spiders, wolf spiders, and solifuges), centipedes, and some insects (primarily ants); but through cannibalism and intraguild predation, other scorpions are probably the most significant invertebrate predators.

Scorpion-scorpion predation, both cannibalism (Fig. 7.5) and intraguild predation (Fig. 7.6), occurs in many scorpion communities throughout North America (Polis 1979, 1980b; Polis et al. 1981; Polis and McCormick 1987). Other scorpions reported to be cannibalistic in the natural environment include the African *Isometrus maculatus* (Probst 1972) and the Australian *Urodacus manicatus* (Smith 1966). Most other reports of cannibalism and intraguild predation are from the laboratory.

Both maternal and mate cannibalism have been reported for scor-

Fig. 7.5. An adult *Paruroctonus mesaensis* feeding on an intermediately aged animal of the same species, an example of cannibalism.

Fig. 7.6. An adult *Paruroctonus mesaensis* feeding on an adult *Vaejovis confusus*. This is an example of scorpion intraguild predation in the Colorado Desert of southern California.

pions (Polis et al. 1981). Maternal cannibalism, the eating of newborns by the mother, was not actually observed in the natural environment and may be a laboratory artifact. In mate cannibalisms, one animal eats the other before or after mating. In *P. mesaensis*, it is the large females that eat the smaller males, and this type accounted for 18 percent of all cannibalisms in the population of this species (see Chapter 4). Many species show a sex ratio biased in favor of females (Chapter 6), and many of these sex ratios have been attributed to the cannibalism on males by females.

In almost every case of cannibalism and intraguild predation, the larger scorpion is the successful predator. Size determines the outcome of these encounters, because the larger scorpion has a longer tail, and although the tail may be held for a time in the smaller scorpion's chelae, the larger scorpion will eventually sting the smaller scorpion. As a consequence of the direct relationship between the age and size of *P. mesaensis*, adults formed 53 percent of all intraguild predators, and the youngest age classes formed 80 percent of all cannibalistic prey.

Interestingly, Shulov and Levy (1978) demonstrated that *Leiurus quinquestriatus* is resistant to its own venom. For death to occur, an individual would need the quantity of conspecific venom that is released in 18 stings. Cannibalism does occur, however; the larger animal turns the smaller one over and places the sting directly into an abdominal ganglion, producing immediate death.

Cannibalism contributes significantly to the diet of *P. mesaensis* (Polis 1979, 1980b): conspecifics were the fourth most frequent prey item in the diet, and 9 percent of all diet items. If prey are evaluated by biomass, then conspecifics may be regarded as the most important item, composing approximately 28 percent of the diet. Intraguild predation is an important source of mortality in some groups of scorpions (Polis 1979, Polis et al. 1981, Polis and McCormick 1987). Other species of scorpions formed from 8 to 22 percent of the diet of four species.

Predator Characteristics

Predators of scorpions are often behaviorally predictable. Most scorpions are active nocturnally, and consequently a large number of their predators are also nocturnal. Owls are important avian predators. Among mammals, rodents and other carnivores are nocturnal predators of scorpions. Some predators seek out scorpions in their diurnal retreats: for example, the lizard *Varanus gouldii* hunts for *Urodacus hoplurus* in its burrow (Koch 1970); baboons search under rocks during the day for scorpions (Eastwood 1978a). Many scorpion predators resist counterattack by being immune to the effects of scorpion venoms (e.g., the snake *Chionactis occipitalis* and the mongoose *Herpestes edwardsi*) or by breaking off the metasoma to avoid being stung (e.g., the grasshopper mouse *Onychomys*, the meerkat *Suricata suricata*, and the baboon *Papio*).

Much of our information about the predators of scorpions is anecdotal natural history, but some investigations have been quantitative. Scorpions are rare in the diet of some lizard populations: only 2 of 1,869 specimens of the lizard *Uta stansburiana* had eaten scorpions (Knowlton and Thomas 1936), and 1 of 6,471 specimens of *Typhlosaurus lineatus* (Huey et al. 1974). Conversely, some predators specialize on scorpions and hence are potentially important sources of mortality. The lizards *Pygopus nigriceps* and *Nucras tessellata* are specialists: 53.2 percent of the *N. tessellata* diet is scorpions. Some owls are scorpion specialists at times: 50 percent of the diet of the nestling elf owl *Micrathene whitneyi* is scorpions (Parrish 1966), and 70 percent of the pellets of the burrowing owl *Speotyto cunicularia* contained scorpion remains (Glover 1953). In some areas the grasshopper mice *Onychomys* are scorpion specialists: Horner et al. (1965) detected scorpion remains in 37 percent of the *Onychomys torridus* scats they examined.

Defensive Adaptations

Several defensive adaptations of scorpions decrease the likelihood of predation. Cryptic coloration and temporal avoidance of predators reduce the risk of prey detection. Venom and sound production (stridulation) also function in predator deterrence.

Animals that are similar to their surroundings in color or apparent texture are difficult for predators to locate. The coloration of *Heteronebo bermudezi* very closely matches its white sand habitat, and in general most Cuban scorpions are cryptically colored (Armas 1976, 1980). Showing even greater adaptability, many species are phenotypically variable in coloration. In Baja California, populations of *Paruroctonus grandis* and *Vaejovis hoffmanni* are darker on dark soils and lighter on light soils (Williams 1970a). Dark races of *Hadrurus concolorous* are located in volcanic areas, and rusty-reddish races in sedimentary and wind-deposited soils (Williams 1970c). Similar color differences occur among local populations of many other species (Stahnke and Calos 1977; Williams 1980; W. D. Sissom, pers. comm. 1982).

Many scorpions spend most of their time in burrows, crevices, or other hiding places inaccessible to many predators. Polis (1980a; see also Chapter 5) discussed scorpions that are time minimizers, species that spend a minimal amount of time in foraging and other activities outside of the burrow, thus decreasing the risk of predation. For example, on the average, *P. mesaensis* is active 20 percent of all nights available for foraging (Polis 1980a). Many other scorpions are reported to be rarely active above ground (Chapter 5). Age-specific patterns of seasonal and nightly surface activity are apparently important in reducing cannibalism in *P. mesaensis* and other species (Polis 1980b; Chapter 5). Cloudsley-Thompson (1959) suggested that ever-present diurnal predation in desert areas exerted an intense selective pressure that forced scorpions to exist nocturnally.

Venom and the sting are used both defensively and offensively. Scorpion venoms, their toxicity, and their physiological effects are discussed in Chapter 10. Heatwole (1967) proposed that venom is more significant in prey capture (see above) than in predator deterrence, although some scorpions do not sting their prey at all (Schultze 1927, Newlands 1969a). Most scorpions attempt to sting in response to threat stimuli. Thus, it appears that venom is used both to subdue prey and to injure predators.

In general, offensive and defensive behavior is related to the morphology of the scorpion. In response to threat, some buthid scorpions (e.g., *Centruroides*, with slender pedipalps and highly toxic venoms) assume a posture that includes the extension of the pedipalps and the curling of the metasoma so that the sting is positioned over the most anterior region of the body (see Fig. 7.7). Some buthids may successfully defend themselves by striking their opponent with a powerful blow; in some instances, the force alone sufficiently stuns the predator to allow the scorpion to retreat successfully (Newlands 1969a). During the strike, the sting may be so deeply embedded that the venom becomes effective rapidly. Perhaps because of the potency of the venom, the pedipalps are rarely used directly in buthid defense. Nevertheless, Alexander (1959b) stated that the pedipalps are important defensively because of the many sense organs located on them.

At the opposite end of the morphological gradient are the thick, powerful chelae characteristic of fossorial scorpions like *Opistophthalmus*, *Scorpio*, and *Anuroctonus* (see Chapter 6). The strength of the chelae is impressive. Alexander (1959b) reported that a female *Opistophthalmus* weighing 4.9 g was able to pick up a 100-g weight with one pedipalp. These burrowing scorpions depend primarily on the large chelae in both defensive and offensive behavior, since in these species the mesosoma is poorly developed and the venom is fairly weak. In a defensive situation the chelae protectively shield the anterior region of the scorpion, the tail is raised, and the sting may be used in quick pokes at the

Fig. 7.7. Defensive posture as illustrated by the iurid scorpion *Hadrurus hirsutus*.

enemy. In some fossorial scorpions the chelae are used to seal the entrance of the burrow and prevent intrusion from potential adversaries (e.g., *Liocheles waigiensis* from Queensland).

In addition to the direct envenomation of predators, venom may be sprayed over a short distance by some species of the genera *Parabuthus* (Newlands 1974a) and *Hadrurus* (Parrish 1966, Polis et al. 1981) in response to threatening stimuli. If the venom should contact sensitive tissue, such as the eye of a vertebrate predator, then this behavior might also serve to deter predation.

Some scorpions hiss (stridulate) in a distinct and audible response to a threat stimulus (Constantinou and Cloudsley-Thompson 1984). Stridulation results from the rubbing of two body parts (see Fig. 3.28*B–G*). Dumortier (1963) discussed four types of scorpion stridulatory mechanisms: pedipalp–walking leg, chelicera-cephalothorax, metasoma-sting, and pecten-sternite. Species of *Pandinus* and *Heterometrus* rub the coxae of the pedipalp and the first walking leg together to produce sound. In many species of *Opistophthalmus*, noise arises from the friction between the chelicera and cephalothorax (this is similar to the sound production of mygalomorph spiders). Several species of *Parabuthus* and *Androctonus* create sound by scraping the sting along the metasoma and/or the mesosoma; such scraping is usually concurrent with the dripping of venom. The fourth type, found in *Rhopalurus* from Brazil, occurs when the striated teeth of the pecten brush a granulated portion of the sternite (Lucas and Bücherl 1971). Rosin and Shulov (1961) distinguished between stridulation and other methods of sound production. *Scorpio maurus* emits a sound by striking the ground with the distal half of the metasoma, a nonstridulatory mechanism.

The evolutionary significance of sound production appears to be interspecific communication. Although not found in more than a few genera, where it does occur it is clearly useful in deterring predators. An example is provided by *Opistophthalmus latimanus*, which does not stridulate in any aspect of courtship, mating, or maternal behavior; confronted with a variety of small mammalian predators, three of ten *O. latimanus* individuals used stridulation for its obvious protective value (Alexander 1958). The stridulation mechanisms appear to have developed from attack, flight, and behavioral displacement responses to threatening stimuli (Alexander 1958). Rosin and Shulov (1961) also considered sound production (by *S. maurus*) to be a threatening response to predators.

Parasites

In addition to being involved in predator and prey relationships, scorpions are hosts for nematode and mite parasites. These parasites and their scorpion hosts are summarized in Table 7.3. Although spiders, a taxonomically related and ecologically similar group, frequently serve

TABLE 7.3
Parasites of scorpions

Parasite	Host	Reference
NEMATODA		
larvae	*Parabuthus granimanus* Pocock	Millot & Vachon 1949
Mermithidae larvae	*Urodacus hoplurus* Pocock, *U. similis* L. E. Koch	Koch 1977
ARTHROPODA: Acari		
mites	*Cazierius gundlachii* (Karsch)	Armas 1980
	Opisthacanthus cayaporum Vellard	Lourenço 1982
Cheyletidae	*Centruroides* spp.	Armas 1980
Gamasidae (?)	scorpions	Birula 1917a
Pterygosomidae		
Pimeliaphilus isometri	*Isometrus* sp.	Cunliffe 1949
	Uroplectes carinatus Hewitt	Eastwood 1978b
P. joshuae	*Hadrurus arizonensis* Ewing, *Vaejovis confusus* Stahnke	Newell & Ryckman 1966
	Centruroides exilicauda (Wood), *Paruroctonus mesaensis* Stahnke, *P. vachoni* Stahnke, *Superstitionia donensis* Stahnke, *Vaejovis spinigerus* (Wood)	Berkenkamp 1973
P. rapax	*Vaejovis punctatus* Karsch	Beer 1960
Trombidiidae		
larval mites	scorpions	Millot & Vachon 1949
	Vaejovis carolinianus (Beauvois)	Benton 1973
Drythraeidae		
mites	*Tityus mattogrossensis* Borelli	Lourenço 1979a
	Australian scorpions	Koch 1977
Leptus sp.	*Urodacus manicatus* (Thorell)	Southcott 1955
Trombiculidae		
Eutrombicula	*Ananteris balzani* Thorell, *Bothriurus araguayae* Vellard, *Tityus fasciolatus* Pessôa	Lourenço 1978, 1982
Acaridae		
Caloglyphus	*Tityus fasciolatus* Pessôa	Lourenço 1982

as hosts for insect eggs and larvae, there are no reports of scorpions as hosts for insect parasitoids.

Nematode larvae have been found in the cavities of both the mesosoma and the metasoma. Koch (1977) found living scorpions loaded with these parasites. The adult nematode of this species is free-living in water or soil, and only the larvae are parasitic. Benton (1973) found no internal parasites during dissections of more than 800 *Vaejovis carolinianus*.

The pectines and membranes in articulations of the chitin are particularly susceptible to parasitism by mites (Millot and Vachon 1949; Koch 1977; Lourenço 1978, 1982; Armas 1980). However, larval trombidiid mites are "fairly common" on the heavily sclerotized plates of *V. carolinianus* (Benton 1973). Berkenkamp (1973) recognized differences in the primary feeding sites of mites on North American Vaejovidae and Buthidae. The vaejovids were infected in the fourth and fifth metasomal segments; *Centruroides*, the investigated buthid, in the second through fourth coxae. The number of mites per host is variable. Berkenkamp (1973) reported a maximum of 23 *Pimeliaphilus joshuae* on an adult female *V. spinigerus* and determined that the infestation was a function of

Fig. 7.8. A scorpion *Tityus fasciolatus* with trombiculid mites. (From Lourenço 1982)

host size. In contrast, Lourenço (1982) observed 468 trombiculid mites on a female *Tityus fasciolatus*, although the usual range was 20 to 30 per host (Fig. 7.8); evidently these mites did not greatly affect the parasitized scorpion.

The percentage of the scorpion population parasitized also varies. Berkenkamp (1973) noted that it increased from June to September. Lourenço (1982) reported that 194 of 466 examined scorpions (42 percent) were infested, and Southcott (1955) found that 3 percent of a *Urodacus manicatus* population were hosts of the mite *Leptus*. But *Pimeliaphilus isometri* parasitized only one of 150 *Uroplectes carinatus* (Eastwood 1978b). Although Berkenkamp stated that three species of genus *Pimeliaphilus* (including *P. joshuae*) are obligate scorpion parasites, neither Lourenço (1978) nor Berkenkamp found any of the parasitic mites restricted to one host species. For example, *P. joshuae* parasitizes seven scorpion species; *Eutrombicula*, three species. There is one report (Lourenço 1982) of a mite (*Caloglyphus*) that occurs on both scorpion and nonscorpion hosts.

ACKNOWLEDGMENTS

We wish to thank David Sissom, Wilson Lourenço, Stanley Williams, and Herbert Levi for reviewing this chapter.

8

Environmental Physiology

NEIL F. HADLEY

The adaptations of scorpions to their environment have been extensively investigated. There are both biologically significant and practical reasons for this interest among environmental physiologists. Scorpions were among the first animals to occupy terrestrial habitats and thus exhibit adaptations that have evolved over millions of years. They are also a conspicuous faunal element in many ecosystems, particularly deserts, which offer some of the most rigorous conditions found on the earth's surface. Scorpions can be collected in fairly large numbers using ultraviolet-light detection, a technique that also permits the monitoring of physiological responses of coded individuals under free-roaming conditions. They can be maintained inexpensively for long periods in the laboratory. Finally, some species are large enough to permit injections or withdrawal of body fluids or microsurgery; these techniques are often very difficult or even impossible to perform on many other arthropod species.

The behavioral, morphological, and physiological adaptations of desert scorpions were reviewed by Hadley (1974). The coverage here builds on that review and emphasizes subsequent studies that have supported or extended our understanding of these adaptations and that have employed innovative techniques or modern analytical instrumentation to examine adaptive responses at the cellular and subcellular levels. Discussion will again center primarily on desert species that have been the subject of most of the recent investigations; however, coverage will include adaptations of scorpions to other stressful environments or adaptations exhibited by desert species that are likely utilized by species living elsewhere.

Behavioral Mechanisms

Body size is an important factor in determining the rate of heat and water flux between an organism and its environment. Radiation, convection, evaporation, and metabolism are all essentially proportional to the surface area of an animal. Since scorpions, like most other arthropods, are small animals, they possess a relatively large ratio of surface area to body mass. As a result, they can experience rapid heat gain and high rates of water loss. Their small size, however, is advantageous in that it permits scorpions to exploit burrows, rock crevices, and various types of cover not available to large animals. The microclimate within these sites is much less stressful than that on the surface, especially in desert regions during the day.

Many species of scorpions inhabit deep burrows, which they dig themselves or which have been excavated by lizards or small rodents (see Chapter 6). Burrow depths vary from 30 to 100 cm or more depending on the species and the size of the scorpion. Factors that apparently influence burrow-site selection include soil hardness and texture, the presence of shading, adequate soil moisture, protection against sand and debris being blown into the burrows, and protection from inundation from sheet flooding. Koch (1978) reported that burrows of *Urodacus* constructed beneath surface objects such as rocks were usually shallow (less than 10 cm), whereas burrows in "open ground" were deep and spiral in form. The deep spiral character was viewed as an adaptation for the maintenance of suitable levels of moisture and temperature.

Although burrows provide refuge from harsh surface conditions, few measurements have been made of the actual temperatures and humidities experienced by scorpions within their burrows. Because of the twisting or spiraling nature of most burrows, it is difficult to place thermocouples or thermistors into the deeper burrow chambers without disrupting the burrow structure. To overcome this problem, Hadley (1970b) attached thermocouples to the scorpions and allowed them to carry the sensors down into burrows. The relative humidity inside the burrow was determined by tethering individual scorpions with nylon thread to which thin strips of humidity-indicator paper were tied and, after sufficient time for equilibration, quickly removing the scorpion for

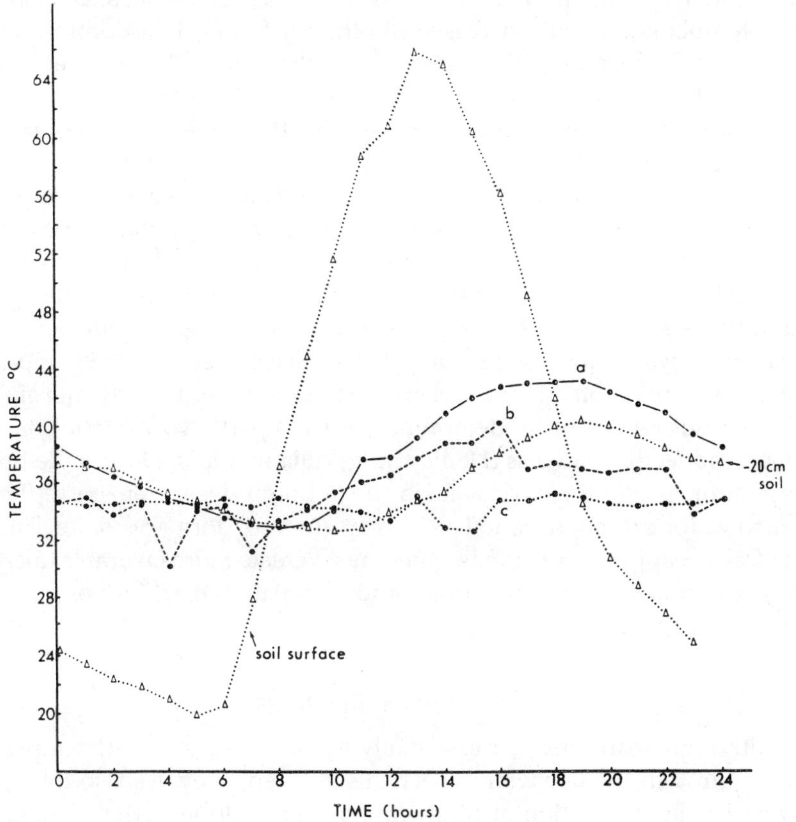

Fig. 8.1. Burrow temperatures experienced by three scorpions (*a,b,c*) at various depths over a 24-hour period (circles). Surface and −20 cm soil temperatures during the same period are included (triangles). All scorpions remained in the burrows throughout the 24 hours. (From Hadley 1970b)

the reading. Depths were confirmed by subsequent excavation of the burrow. The results of this study indicated that diurnal fluctuations in temperature affecting scorpions were largely dependent on the depth of penetration and scorpion movements within the burrow (Fig. 8.1). Scorpions occupying shallow terminal burrows experienced temperatures approaching 45° C, whereas an individual at 40 cm experienced an almost-constant 34° C (Hadley 1970b). Relative humidities at depths of 20 to 35 cm ranged from 55 to 70 percent, compared with humidities of less than 5 percent on the surface during the day and near saturation at night. Warburg and Ben-Horin (1978) also reported decreasing burrow temperatures with increased depth, but found that scorpion burrows located in arid regions in northern Israel are still warmer than comparable burrows constructed in more mesic habitats.

Most scorpions, especially those species that inhabit hot desert regions, complement the burrowing adaptation by being active on the surface only at night (see Chapters 5 and 6). A few species may appear at the entrance to their burrows during the day, where they have been observed to stilt to reduce the heat load (Alexander and Ewer 1958). In tropical forests, where the canopy and dense underbrush create light conditions similar to those of exposed habitats at night, some species may be active during the day. Strictly nocturnal species usually appear on the surface soon after sunset and remain there until early morning, the time of retreat being determined at least partly by climatic conditions and feeding success (Hadley and Williams 1968). During the hottest summer months, scorpions in the Sonoran desert remain in their burrows for extended periods and may enter a "dormant" state. They typically reappear soon after summer rains create more favorable microclimatic conditions on the surface and stimulate the activity of insects and other prey.

Temperature Relations

Although scorpions can essentially avoid stressful desert temperatures through the behavioral mechanisms cited, they are also tolerant of and able to function at high temperatures. Upper lethal temperatures of scorpions typically range between 45° and 47° C (Cloudsley-Thompson 1962, 1963; Hadley 1970b), with values exceeding 50° C possible on a short-term basis. These temperatures are generally several degrees above those of other desert arthropods; however, because of variations in the length of exposure, relative humidity, and previous thermal history of the species in question, accurate comparisons are difficult to make. It also appears that the thermal tolerance of desert scorpions is somewhat greater than that of their nondesert counterparts, but the extent to which this difference is due to genetic factors or acclimatization has not been fully established. Cloudsley-Thompson (1962) reported an enhanced temperature resistance in the Sudanese scorpion *Leiurus quinquestriatus* when preconditioned at high temperatures for 24 hours before testing. I have noted a similar increased tolerance in scorpions collected in the summer over those collected during the cooler months. Comparative studies using carefully controlled experimental designs are needed to strengthen our understanding of this aspect of the thermal biology of scorpions.

Although most adaptations of desert scorpions pertain to high ambient temperatures, nighttime and/or seasonal cold in many arid regions can create equally stressful conditions. Low temperatures are frequently encountered in deserts at high latitudes and in deserts containing mountains within their boundaries. Behavioral avoidance is once again the paramount coping mechanism, but apparently more complex adjustments also exist. One of the most spectacular of these is the ability to supercool, a process typically associated with high-latitude and polar species, whereby the animal's circulatory fluid (hemolymph) can be lowered well below the freezing point without solidification or crystallization and without any subsequent damage to the body tissue. Body temperatures of −5.8° C have been recorded for *Vaejovis* sp. (Cloudsley-Thompson and Crawford 1970), and of −5.0° to −9.5° C for *Diplocentrus peloncillensis** (Crawford and Riddle 1974, 1975). Another scorpion species found in the southwestern United States, *Paruroctonus utahensis*, increased its supercooling capacity from a mean of −8.6° C in August to −11.9° C in January (Riddle and Pugach 1976); the substantial seasonal depression of the supercooling point of this species' hemolymph is apparently associated with the cessation of feeding in the fall.

The physiological or biochemical basis for the ability of these scorpions to supercool is not clearly known. Increased levels of cryoprotectants such as glycerol and sorbitol in the hemolymph, which are typically observed in insects inhabiting extremely cold environments, were not noted in the above investigations. The adaptive significance of supercooling in scorpions is also open to question. Microenvironmental studies of the winter habitat of *P. utahensis* indicate that this species may be periodically subjected to subzero temperatures, and the ability to supercool would thus contribute to its survival. In contrast, the Sudanese scorpion *L. quinquestriatus*, which exhibits a supercooling point of −7.4° C, is never subjected to temperatures this low. Because

*Crawford and Riddle (1974, 1975) and Crawford and Wooten (1973) reported working with *D. spitzeri* Stahnke, but the population they studied, from the Peloncillo Mts. of New Mexico, was later described as *D. peloncillensis* Francke. See Francke (1975).

its supercooling capacity is greater than that necessary to survive climatic extremes, it may be a taxonomic rather than an adaptive feature (Cloudsley-Thompson 1973b).

A recent study by Whitmore et al. (1985) provides a new perspective on the phenomenon of cold hardiness in scorpions. They reported that the scorpion *Centruroides vittatus,* which is periodically exposed to freezing temperatures in its natural habitat, survives the cold by tolerating limited freezing of body tissues. This species contains potent ice-nucleating agents in the gut, which promote ice formation in this body compartment. The chemical nature of these ice-nucleating agents was not confirmed, but they are likely proteinaceous compounds. Unlike the species discussed above, *C. vittatus* exhibits only a superficial capacity to supercool. Freeze tolerance is well documented in insects; however, this is the first observation of a scorpion's using this strategy to withstand subzero cold.

Water Relations

All animals, regardless of their habitat, continually face the problem of maintaining sufficient body water and the proper balance of ions for essential biological activities. The problem is magnified for terrestrial species, especially desert animals, because of the increased drying power of the air coupled with a limited and unreliable source of water. The small size of scorpions, and thus their high ratio of surface area to body mass, further compounds the problem, since evaporation correlates directly with surface area. It is not surprising, therefore, to find that scorpions have foregone the use of water as a means of temperature regulation, but have instead evolved a suite of mechanisms for conserving limited body-water supplies while at the same time being able to tolerate fairly severe desiccation during unfavorable environmental conditions.

The rates of water loss for a variety of scorpions from five families are given in Table 8.1. Inconsistencies in experimental technique, age and hydration state of the test animal, and length of exposure, as well as the often small sample sizes, require that comparisons between species be made with considerable caution. Nevertheless, the values for scorpions as a group are comparable and, for many species, lower than those for other arthropods occupying similar habitats (Hadley 1970a). Moreover, there is a definite trend for lower transpiration rates in the

TABLE 8.1
Water loss of selected scorpions in dry air

Taxon	Habitat	Temperature (°C)	Water loss (mg·cm^{-2}·hr^{-1})	Reference
BUTHIDAE				
Androctonus australis	xeric	30	0.025	Cloudsley-Thompson 1956
Buthus occitanus	xeric	30	0.025	Cloudsley-Thompson 1956
Centruroides exilicauda	xeric	30	0.049	Toolson & Hadley 1979
Hottentotta hottentotta	xeric/mesic	25	0.131	Toye 1970
H. judaica	xeric	30	0.033	Warburg et al. 1980b
Leiurus quinquestriatus	xeric	30	0.025	Warburg et al. 1980b
CHACTIDAE				
Euscorpius germanus	mesic	30	1.00	Cloudsley-Thompson 1956
DIPLOCENTRIDAE				
Diplocentrus peloncillensis	mesic	30	0.270	Crawford & Wooten 1973
Nebo hierichonticus	mesic	30	0.064	Warburg et al. 1980b
SCORPIONIDAE				
Opistophthalmus capensis	mesic	30	0.125	Robertson et al. 1982
Pandinus imperator	mesic	25	0.233	Toye 1970
IURIDAE				
Hadrurus arizonensis	xeric	30	0.028	Hadley 1970a
			0.031	Toolson & Hadley 1977
VEJOVIDAE				
Uroctonus apacheanus	mesic	30	0.245	Toolson & Hadley 1977

more xeric species. For example, the scorpion with the highest water-loss rate, *Euscorpius germanus*, inhabits damp localities and is generally regarded as a "water-loving" species (Cloudsley-Thompson 1956). When tested under identical experimental conditions, the water loss of the montane scorpion *Uroctonus apacheanus* was approximately an order of magnitude greater than that of the xeric-adapted *Hadrurus arizonensis*. A similar trend in water loss but with smaller differences was observed in Warburg et al.'s (1980b) study of xeric versus mesic scorpions living in the northern part of Israel.

Cuticular Water Loss

Many factors contribute to the low rates of water loss for scorpions, but perhaps none is more important than their virtually impermeable integument. The scorpion cuticle is a noncellular, multilayered membrane overlying the epidermal cells. It consists of a thin, outermost epi-

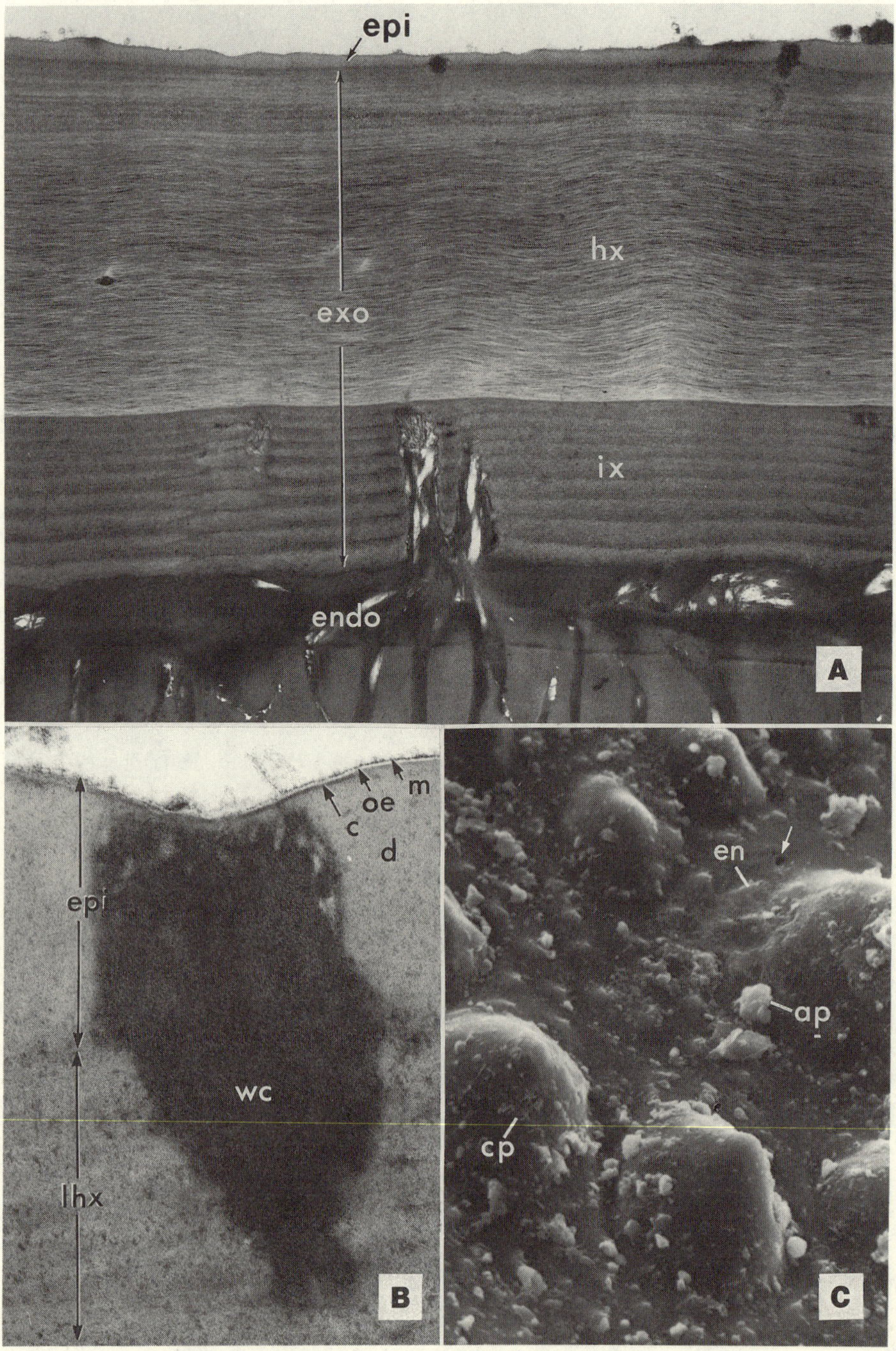
epi
hx
exo
ix
endo
A
epi
oe
m
c
d
wc
lhx
B
en
ap
cp
C

cuticle and a much thicker procuticle, which is typically subdivided into an outer sclerotized exocuticle and an inner endocuticle (Fig. 8.2). The structural integrity of the cuticle is provided by the chitin-protein complex present in the procuticle. Although all layers of the cuticle as well as possibly the epidermal cells (see Riddle 1981) contribute to waterproofing, for scorpions the principal barrier appears to be lipids (waxes) associated with the epicuticle. These lipids may be associated with specific sublayers of the epicuticle, either as free molecules or bound to protein, and/or simply deposited on the cuticle surface. Experimental evidence supporting the role of lipids in restricting water efflux is based on increased transpiration rates following lipid extraction with organic solvents or abrasion (see below), and on the reestablishment of permeabilities comparable to those observed for intact scorpions when extracted lipids are plated onto artificial membranes. The advent of modern analytical instrumentation such as the gas chromatograph and mass spectrometer and the probing capacity of the electron microscope have made it possible to analyze the chemical nature of these lipids and to provide information on their location within the epicuticle.

Several studies have attempted to correlate the chemical composition of scorpion epicuticular lipids with their cuticular permeability. The types of lipids present in desert species seem to be similar to those in nondesert species; however, the patterns of the lipid composition in desert forms appear to confer improved waterproofing. Cuticular lipids

---

Fig. 8.2. Scorpion cuticle. *A*, A transverse section of the cuticle of the sclerotized dorsal sclerite of *Hadrurus arizonensius* (×8,570): *epi*, epicuticle; *exo*, exocuticle; *endo*, a small portion of the endocuticle. The exocuticle can be further divided into an outer hyaline exocuticle (*hx*) and the inner exocuticle (*ix*). *B*, The cuticle at higher magnification (×78,750): *epi*, epicuticle; *lhx*, lamellate hyaline exocuticle; *wc*, a wax canal. The epicuticle is subdivided into four layers: *m*, an outer membrane; *oe*, an unstained outer epicuticle; *c*, a cuticulin layer; *d*, a dense homogenous layer. The wax canal penetrates the base of the cuticulin layer but is covered by the outer membrane and outer epicuticle. (From Hadley & Filshie 1979) *C*, Scanning electron micrograph of the external surface of an untreated sclerite cuticle (×630): *ap*, amorphous particles; *en*, filmlike encrustations; *cp*, small crystalline projections, most numerous on the sides of tubercles. The arrow points to the opening of a dermal gland duct. (From Hadley & Filshie 1979) accounted for 0.03 to 0.09 percent of the wet weight of *Paruroctonus mesaensis*, a vaejovid scorpion that inhabits sandy areas adjacent to dry riverbeds. Hydrocarbons were the most abundant lipid class; cholesterol, free fatty acids, and alcohols were also detected (Hadley and Jackson 1977). Analysis by gas-liquid chromatography has indicated that the hydrocarbon fraction is composed of saturated *n*-alkanes (straight-chain) and branched alkanes, ranging from 21 to over 39 carbon atoms in length. Large molecular size and saturation are two features that should account for enhanced cuticular impermeability if we assume that the lipid/permeability relationships established for plasma membranes and artificial bilayers also apply for lipids associated with the scorpion cuticle.

Subsequent studies have provided further evidence that the abundance and chemical composition of cuticular lipids enhance the effectiveness of the water barrier. The desert scorpion *Hadrurus arizonensis*, which has a very low transcuticular water loss (Table 8.1), has a hydrocarbon surface density of 5.8 μg per square centimeter. In contrast, the hydrocarbon surface density for the montane scorpion *Uroctonus apacheanus*, whose permeability is approximately ten times greater, is only 2.9 μg per square centimeter (Toolson and Hadley 1977). Furthermore, epicuticular lipids of *H. arizonensis* are characterized by higher proportions of long-chain branched hydrocarbons and long-chain saturated free fatty acids. The differences in cuticular permeability that occur on a seasonal basis can also be linked to cuticular lipid composition. Area-specific water loss rates for the scorpion *Centruroides exilicauda* are significantly lower in the summer than in the winter. A high percentage of long-chain epicuticular hydrocarbons in summer animals probably accounts for much of this increased permeability (Fig. 8.3; Toolson and Hadley 1979). These findings indicate that the cuticular permeability of scorpions can be altered to meet environmental requirements and that predictable changes in epicuticular abundance and composition are in part responsible.

The electron microscope has been used to establish the general location of lipids associated with the epicuticle of the scorpion (Filshie and Hadley 1979, Hadley and Filshie 1979). Scanning electron micrographs of the cuticle of *H. arizonensis* show the presence of amorphous particles, crystalline projections, and large dermal gland-duct openings on the surface of untreated cuticles (Fig. 8.2). Immersing the cuticle either in hexane or in a mixture of chloroform and methanol to remove lipids

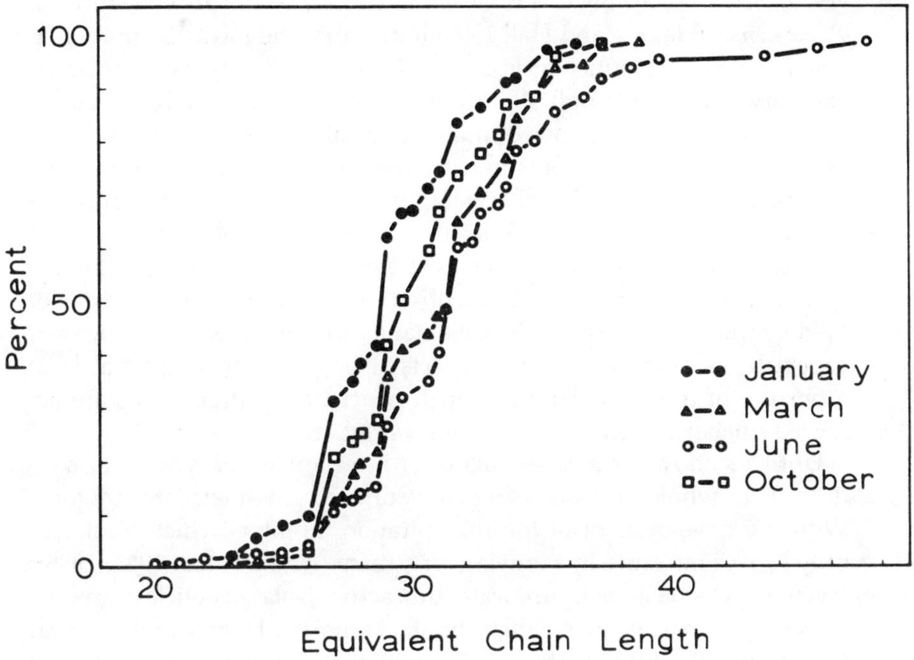

Fig. 8.3. Seasonal changes in the cumulative frequency distribution for epicuticular hydrocarbons of *Centruroides exilicauda*. The ordinate for a given equivalent chain length (ECL) represents the percentage of hydrocarbon molecules with an ECL less than or equal to that particular ECL. (From Toolson & Hadley 1979)

produced only minor changes in surface appearance; the amorphous particles remained, but the crystalline projections were removed, and the openings to the wax canals that penetrate the epicuticle became faintly visible. The improved resolution of surface detail probably results from the removal of a thin lipid film from the cuticle; however, neither solvent system produced any significant changes in the number, thickness, or general appearance of the epicuticular layers when viewed in vertical section.

The ultrastructural investigations clearly demonstrate the complex system of pore and wax canals that is believed to be the route by which lipids are transported from their site of synthesis to the outer cuticle. The synthesis of cuticular lipids in scorpions, unlike that in insects, has received only limited study. An earlier study by Ross and Monroe

(1970) of the scorpion *C. exilicauda* and more recent investigation of *P. mesaensis* (Hadley and Hall 1980) found that the incorporation of labeled acetate into cuticular lipids, and into hydrocarbons specifically, was very low (activity only slightly above background). Hall and Hadley (1982) subsequently showed that measurable amounts of cuticular and hepatopancreatic lipids and hydrocarbons were synthesized by *P. mesaensis* from ^{14}C-labeled acetate, propionate, mevalonate, and leucine when animals were acclimated to 35° C and incubated for at least 72 hours, but the *de novo* synthesis rates under the new experimental regime were still well below those observed for insects. This low rate of lipid synthesis in scorpions is consonant with their low metabolic rate, especially in comparison with insects. It is likely, then, that at least a portion of the cuticular and hepatopancreatic hydrocarbons are acquired unchanged from the scorpion's prey.

Until recently, all measurements of scorpion water loss were conducted on whole animals using conventional gravimetric techniques. With the development of the transpiration monitor, which electronically senses moisture in a moving air stream (Hadley et al. 1982, 1986), it is now possible to measure water flux across isolated cuticle segments or through small areas of cuticle in intact animals. Hadley and Quinlan (1987) used the transpiration monitor along with miniature ventilated capsules that could be attached to the cuticle surface to measure the permeability of arthrodial membrane (pleural cuticle) in the scorpion *H. arizonensis* and compared this with rates of water loss through sclerotized cuticle (sternite) of the same species. Despite differences in surface morphology and fine structure, the permeability of pleural cuticle was not significantly higher than that of the sternite. Hexane treatment of pleural cuticle increased its permeability ninefold, but had little effect on the permeability of the sternite. A strong base in combination with chloroform:methanol was required before marked increases in the permeability of sclerotized cuticle were noted. The authors concluded that lipids provide the principal barrier to transpiration in both cuticle types, but that the lipid barrier on the surface of arthrodial membrane is more labile. Without this lipid barrier, transpiration across the scorpion arthrodial membrane would lead to rapid dehydration, especially in gravid females during hot, dry summer months.

Respiratory Water Loss

The integument of scorpions, although relatively impermeable, accounts for most of the water lost at temperatures normally encountered in nature. Still, the respiratory membranes through which oxygen is absorbed from the gaseous phase are moist, thus representing a potential source of dehydration for any land arthropod (Edney 1977). In scorpions, the water loss associated with gas exchange, though most likely nominal at low temperatures, increases as temperatures rise, until at high or sublethal temperatures it may be the predominant source of transpired water. Unfortunately, there have been no direct measurements of respiratory transpiration in scorpions; the proposed relationship is based entirely on correlations obtained from independent plots of gravimetrically determined water loss and manometrically determined gas exchange (Hadley 1970a, Crawford and Wooten 1973).

The low rates of respiratory transpiration characteristic of scorpions can be attributed to an overall low level of metabolism and to modifications of the respiratory apparatus that minimize water loss without sacrificing metabolic efficiency. Rates of oxygen consumption below the levels predicted for poikilotherms on the basis of their body size (Anderson 1970) have been reported for most scorpions studied (Sreenivasa Reddy 1963, Dresco-Derouet 1964b, Hadley and Hill 1969, Hadley 1970a, Crawford and Wooten 1973, Riddle 1978, Yokota 1979, Robertson et al. 1982). Perhaps the most convincing evidence of the low metabolic demands in scorpions is Millot and Paulian's (1943) observation that individuals with only one of eight book lungs functional continued to live and feed for six months following the operation.

The site of gas exchange is also an important factor. Because the cuticular lipid layer, which so effectively restricts water loss across the integument, also restricts the exchange of oxygen and carbon dioxide, scorpions evolved a respiratory system in which the respiratory surfaces (book lungs) are located within the body cavity and open to the outside through narrow spiracular openings. This arrangement not only greatly reduces the effective area for evaporation, but also lengthens the diffusion pathway for water vapor, thus increasing resistance to the outward passage of water. Some scorpions also have muscles that can be used to regulate the aperture of the spiracles and the chamber area in which the respiratory "leaves" lie. It is not known if the muscular control of spiracular opening and closing is used by scorpions to further limit water loss when conditions dictate the need for conservation.

In addition to helping reduce respiratory water loss, low metabolic rates are an adaptive means of conserving energy. This is especially important when scorpions are forced to remain in their burrows for extended periods without access to food (preformed water) during stressful periods of the year or are active on the surface when temperatures exceed optimal levels. The significance of the latter is best documented for the desert grassland scorpion *Paruroctonus utahensis* (Riddle 1978, 1979). This species exhibited a compensatory reduction in oxygen consumption during a seven-day period following a temperature increase from 14° to 34° C. A lower metabolic rate than that predicted from Q_{10} relationships is adaptive in that it reduces the utilization of food reserves at elevated habitat temperatures. When test temperatures were returned to 14° C, the metabolic rates remained slightly below those found for scorpions previously acclimated to 14° C. This response was also considered to be adaptive in that at this low temperature scorpions either would not be able to feed or would do so ineffectively (Riddle 1979).

The water-loss and metabolic rates of scorpions mentioned thus far are all based on laboratory measurements. Such determinations can provide valid and useful comparative information about these physiological processes if the experimental design is sound; however, they suffer because the test animals are confined to conditions that are often artificial and ecologically meaningless. Thus, there is some question whether data obtained in this manner accurately represent the water loss or metabolism of scorpions in their natural environment. To answer this question, investigators have employed radiotracer techniques to monitor rate functions in unrestrained animals in the field. One of the most promising of these techniques, which uses doubly labeled water ($^3HH^{18}O$), provides data on body water content and water flux as well as oxygen consumption.

The first field application of the doubly labeled water technique in an invertebrate was performed on the scorpion *Hadrurus arizonensis* (King and Hadley 1979). The daily mean metabolic rates of two groups of free-roaming scorpions were 0.326 and 0.329 ml of oxygen per gram, approximately two to three times the previously reported laboratory-determined metabolic rates at similar temperatures. Water-turnover rates and water-loss rates of field scorpions were also much higher than those for scorpions maintained in the laboratory, with the highest values observed shortly after periods of rainfall, which presumably increased food availability and hence water input. The higher rates are not unexpected, since the doubly labeled water technique measures "gross" water loss and also integrates total metabolism over a period during which the animals are free to exhibit all levels of activity. In this particular study, however, carbon dioxide production determined by the $^{3}HH^{18}O$ method differed substantially from that measured by standard manometric procedures. A laboratory study by Buscarlet et al. (1978), using the doubly labeled water technique on locusts, reported a validation error of less than 10 percent, suggesting that the problem is one of technique refinement rather than technique principle. Although this and related techniques are indeed ideally suited for providing a better understanding of the complex interrelationships between scorpions and their environment, their future use is clouded by the high cost of ^{18}O and the limited facilities available for analyzing this isotope.

Excretory Water Loss

The ability to eliminate nitrogenous end products and excess electrolytes in a minimum volume of fluid is an important prerequisite to inhabiting hot dry environments. This function is the responsibility of the excretory system, which consists principally of the Malpighian tubules, the hindgut or ileum, and the rectal segment of the hindgut. The excretory process involves the secretion of urine into the lumen of the Malpighian tubules and its eventual discharge into the gut. The active secretion of potassium appears to play a central role in the formation of the tubular fluid. In insects, and perhaps all terrestrial arthropods, the rectum is the principal site at which the osmotic and ionic compositions are dramatically changed. Mechanisms for moving water against high osmotic gradients in the rectum and the importance of the rectal wall design were discussed in detail by Phillips (1977). The end result is usually the production of dry fecal material and a urine that is hypertonic to the hemolymph. The extent of water reabsorption, however, depends largely on external and internal moisture conditions. Under desiccating conditions or when partially dehydrated, arthropods typically extract a higher percentage of water from their feces.

Little information is available about the mechanism of water transport in the arthropod gut, apart from studies of the insect rectum. In scorpions the rectum is considerably reduced in size, occupying only the last half of the terminal tail segment. In contrast, the ileum extends the entire length of the tail. Using a flow-through technique developed by G. Ahearn, the ileum was removed from *Hadrurus arizonensis* and perfused with solutions of various ionic compositions and strengths to measure the movement of water across the ileal membrane (Ahearn and Hadley 1976, 1977). Net water transport from the gut to the hemolymph occurred in the absence of osmotic or hydrostatic pressure gradients. Moreover, concentrations of sodium ion in the lumen favored the transport of water from the gut into the hemolymph, whereas this transport was inhibited by a high luminal concentration of potassium ion. Since the concentration of luminal sodium ion is high relative to potassium ion in dehydrated scorpions (Fig. 8.4), this ratio would favor an increased net flow of water to the hemolymph, and thus reduce the further depletion of critical body water. Conversely, the luminal concentration of potassium is higher than that of sodium in hydrated scor-

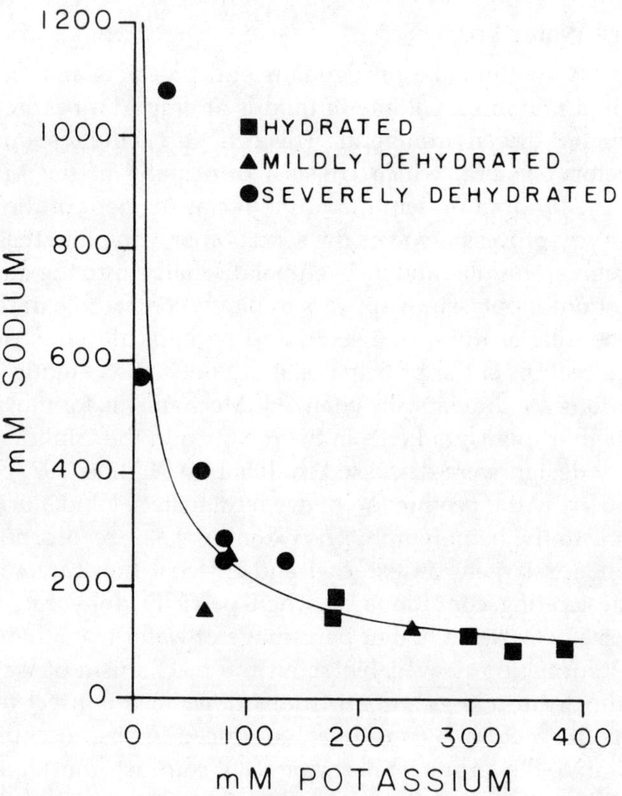

Fig. 8.4. The effect of the hydration state of *Hadrurus arizonensis* on ileal sodium and potassium concentrations. Each point represents the ion concentrations of an individual ileum in millimoles per liter. (From Ahearn & Hadley 1977)

pions. This ionic ratio would inhibit water transport into the hemolymph, causing the elimination of excess water at a time during which conservation is not required.

The principal nitrogenous excretory product of most scorpions is guanine, with uric acid generally second in importance (Horne 1969). These purines are ideally suited for animals in terrestrial environments where water is scarce. The essentially nontoxic and insoluble nature of these compounds allows these to be eliminated in crystalline form, which minimizes the concomitant loss of water. It has been suggested that guanine is actually more efficient than uric acid as an end product, in that it contains 25 percent more nitrogen per molecule and is less soluble. Recently, Yokota and Shoemaker (1981) reported that in *Paruroctonus mesaensis* xanthine accounted for 92 percent of the excreted nitrogen, with much smaller quantities of hypoxanthine also detected. This is the first animal shown to be primarily xanthotelic. Xanthine, like uric acid, contains one less nitrogen and is more soluble in water than either uric acid or guanine. Despite these considerations, xanthine can be excreted almost dry without exerting significant osmotic pressures.

Dehydration Tolerance and Osmoregulation

High temperatures and desiccating atmospheric conditions, combined with the inability to replenish lost body water, can produce severe temporary water deficits in scorpions despite their efficient mechanisms for restricting water loss. Like most terrestrial arthropods, scorpions are quite tolerant of losses of body water. Survival following losses of body water amounting to more than 30 percent of their original wet weight has been observed for *Hadrurus arizonensis* (Hadley 1974), whereas a few *Paruroctonus utahensis* individuals were desiccated to weights 40 to 50 percent of their initial levels (Riddle et al. 1976). Most insects are able to regulate their hemolymph osmotic pressure and body-fluid composition in spite of the reduction in hemolymph volume that occurs during dehydration. Most scorpions, in contrast, appear simply to tolerate the deviation from the norm. In *P. utahensis*, water loss during desiccation results in a steady elevation in hemolymph osmolality (Riddle et al. 1976). Warburg et al. (1980b) examined changes in hemolymph osmolarity in four scorpion species subjected to seven days of desiccating conditions (0 percent relative humidity and 37° C). Hemolymph osmolarity increased greatly in all four species but at a much greater rate in the mesic species. The maximum change occurred in the mesic *Scorpio maurus fuscus,* from the mean value for controls of 554 milliosmoles per gram to 1,122 milliosmoles per gram in individuals that had lost 17.37 percent of their initial body weight. Only the southern African scorpion *Parabuthus villosus* shows good osmotic and ionic regulation following prolonged desiccation (Robertson et al. 1982).

Water Gain

Of the several potential sources of water available to scorpions for replenishing the water lost through transpiration and excretion, three are most important: preformed water, atmospheric water vapor, and metabolic water. Preformed water can be further categorized as water from their prey, standing water in their habitat, and capillary water from moist substrates. Not all of these potential sources are typically available to a scorpion at one time, nor is the relative contribution of each avenue, if the source is used at all, of equal importance.

The most important source of water for scorpions is their food (see Chapter 7). Scorpions are strictly carnivorous, feeding primarily on insects and arachnids whose body mass consists of between 50 and 80 percent water. As long as sufficient prey are available, no other source of water is necessary. Most scorpions, especially if dehydrated, will drink water under laboratory conditions; however, bulk water is seldom available in nature and, if present, is probably not used. An exception is the Sonoran desert scorpion *Centruroides exilicauda*, which has been observed to drink from irrigation canals; laboratory experiments have shown that this species can consume water at a rate of 10 mg per minute (Hadley 1971). Nevertheless, its drinking behavior appears to be facultative, because C. *exilicauda* is also successful in extremely xeric habitats.

The ability of scorpions to absorb liquid water from moist substrates remains questionable. Hadley (1970a) found no significant increases in body weight when dehydrated *Hadrurus arizonensis* individuals were placed on wet sponge strips for 24 hours. Similar results were found by Crawford and Wooten (1973) for *Diplocentrus peloncillensis* adults; they reported, however, that second-instar individuals of this species are apparently able to take up enough water from moist soil in 24 hours at least to replenish water lost prior to the test. Severely desiccated

Paruroctonus utahensis adults of both sexes were able to regain approximately 80 percent of their predesiccation weight after exposure to moist or wet soil for 100 hours (Riddle et al. 1976). Decreased osmolarity of the hemolymph as compared with values for desiccated individuals indicates that at least a portion of this substrate water is incorporated into the body fluids. Such uptake probably occurs through the mouth rather than through the general integument, but evidence for this mechanism or any special ones has yet to be demonstrated.

The other potential sources of moisture are relatively unimportant. No scorpion species, regardless of its hydration state, has been shown to absorb water vapor from subsaturated atmospheres. Water formed from the oxidation of organic materials in the food enters the total water pool of the scorpion and thus represents an avenue of water gain. When food is abundant, metabolic water is probably of minimal importance, considering the low metabolic rates of scorpions and the large amount of preformed water present in the prey. Metabolic water may, however, assume a more significant role in scorpions unable to feed because of climatic conditions or the absence of prey. During these periods the scorpion must rely on energy and water released from the oxidation of fat reserves. It is not known if a scorpion can actively regulate the amount of metabolic water produced during dry conditions or reduced feeding. Nevertheless, because scorpions can excrete dry fecal material and can decrease the amount of excreta produced with prolonged times of starvation, any metabolic water that is produced will be effectively utilized.

The ability to obtain water or at least to minimize water loss is particularly critical to newborn scorpions, which remain on their mother's back until their first molt. These larvae do not actively feed or drink, and their cuticle is probably fairly ineffective in restricting water loss. To compensate, embryonic *C. exilicauda* exhibit a marked increase in water content in the final phase of development, so that at birth their body water content is over 80 percent (Toolson 1985). These water stores are used to balance transpiratory losses and to help ensure the successful completion of the first molt. There are also experimental data that suggest a trophic exchange can occur between the mother and her offspring prior to the first molt (Vannini et al. 1985a). Tritiated water injected in the mother appears in the first-instar larvae. Since the larvae never participate in their mother's meal and are never in a position whereby oral uptake of food or water from the mother is possible, it is speculated that either transpired water containing the radiolabel is absorbed from the vapor state by the larvae, or else newly synthesized cuticular wax that contains the incorporated tritium is transferred to the young as a result of contact. Thus, the mother may provide additional resistance to dehydration either by waterproofing the larvae or by replenishing their water loss.

ACKNOWLEDGMENTS

Portions of the research reported in this chapter were supported by National Science Foundation Grants PCM77–23808, PCM80–21566, and PCM84–01552.

9

Neurobiology

THOMAS M. ROOT

The scorpion nervous system is both an interesting and an important subject for neurobiological study, since the structure and function of this nervous system present a series of contrasts. As one of the first land arthropods, scorpions have preserved an ancient nervous-system organization, but one that is functionally capable of integrating highly sophisticated senses, and producing marvelously coordinated movements in locomotion, burrowing, and attacking prey. Compared with other invertebrates, arachnids generally have highly cephalized nervous systems representing the fusion of many segmental ganglia. Yet, in the scorpions, in addition to a large nervous-tissue mass in the cephalothoracic region (prosoma), one also finds a long ventral nerve cord with segmental ganglia. This contrasts with the highly cephalized nervous system of spiders, for example, and probably represents the primitive condition. Thus the scorpion nervous system is at once primitive, yet cephalized. How might this gross organization be reflected in the way the nervous system functions? Has the segmental arrangement of nerve cells been maintained in the brain, or have more integrated neural centers developed? These are questions that are only beginning to be studied.

The archaic scorpions also present the comparative neurobiologist with some interesting challenges. Scorpions may have descended from eurypterids, the dominant predators of the Silurian seas, and, as arthropods, they share lineage with an extremely diverse phylum. Yet arachnids have also probably undergone hundreds of millions of years of evolution separate from other arthropods, and therefore scorpions perhaps represent a close approximation to an early arthropod common ancestor.

A primitive arthropod would be expected to exhibit little behavioral sophistication. Yet the rapidity, complexity, and dexterity seen when a scorpion orients to, attacks, stings, and consumes its prey form but one example of its sophisticated behavioral repertoire. One therefore soon realizes that its relative structural simplicity and primitiveness do not necessarily imply physiological or behavioral crudeness. It is such contrasts as these, and their suitability to a variety of studies, that make scorpions unusual and valuable subjects of study.

The literature on scorpion neurobiology is very patchy and uneven. After initial interest in the scorpion nervous system in the 1800's and early 1900's subsided, only an occasional report appeared until about 1960. Since then, work has been continued along different and independent avenues by research groups in India, France, Germany, Israel, the United States, and a few other countries. Although this diversity makes for interesting reading, it unfortunately also makes reviewing the literature a complicated and risky process. Different research laboratories have, out of necessity, concentrated their work on only one or a few different scorpion species, thereby making interspecific comparisons difficult. Many papers exist only in preliminary format or as initial investigations that were later abandoned. Others represent the work of comparative biologists, using one or another scorpion system to study a specific problem in an area only tangential to the subject of this chapter. Finally, much of the current research is truly exciting and potentially valuable, but is only in its infancy. A review of scorpion neurobiology a decade hence would likely report a subject with a totally different character.

Sensory Systems

In 1909, Cecil Warburton wrote what is probably still the standard textbook statement on scorpion senses: "The only sense well developed seems to be that of touch" (Warburton 1909: 298). From our current perspective it appears that his summary is mostly true, but somewhat inaccurate. It is correct, of course, because the mechanical senses one thinks of as "touch" are indeed exceptionally well developed in the scorpion. These senses include its unusual ability to detect wind currents and ground vibrations, as well as the more typical sensitivity to contact stimuli, cuticular stresses, and a variety of proprioceptive influences. But Warburton's summary is at least misleading, because although the scorpion's mechanical senses are dominant, they are not the only well-developed ones. There is evidence for chemical and thermal sensitivity. The eyes, though probably not as good as the image-forming organs of some other arthropods, have a sensitivity in the dark-adapted state surpassing that of the human eye. Scorpions are also equipped with unusual sensory organs, such as photosensitive nerve cells in the tail, a distance- and direction-sensitive prey-detection system in the legs, and a pair of pectines, appendages unique to scorpions, which may have a variety of functions, including substrate-texture selection.

Photoreceptors

Scorpions have a pair of median eyes and up to five pairs of lateral eyes, depending upon the species. It is not immediately obvious whether all these eyes are functional or ornamental, for scorpions usually give little indication of a visual sense. A wave of the hand near the head region, casting a shadow over the eyes, seems to have little behavioral effect. Movement of objects in the vicinity of a stationary scorpion seems to go unnoticed, for the animal remains entirely motionless. What, then, is the role of the eyes? A hundred years have elapsed since the first published studies of the scorpion visual sense, and yet the question is still only partially answerable. Work within the last few years suggests that the eyes may serve less to provide visual images than to offer an extremely sensitive detector of faint light spots for navigation, and for timing of circadian rhythms.

ANATOMY OF THE MEDIAN AND LATERAL EYES. The median eyes are large, roughly three times the size of the lateral eyes, and they lie close together along the midline in the center of the prosoma. The lateral eyes are located in right and left groups along the anterolateral edge of the prosoma. Both the median and the lateral eyes are structurally simple, consisting of an overlying cuticular lens, a vitreous body specialized from hypodermis (absent in the lateral eyes), and a multilayered retina contributing axons to the optic nerve, which passes to the supraesophageal ganglion of the brain. It is believed that irregularities in lens structure limit the image-forming capabilities of the eyes (Carricaburu 1969).

Although the structure of the eyes was initially described about a century ago (Graber 1879, Lankester and Bourne 1883, Parker 1886, Police 1907, Scheuring 1913), improvements in microscopy since that time have considerably improved our understanding of their microscopic structure. Recent studies include those of Bedini (1967) on *Euscorpius carpathicus*, Machan (1966, 1967) on *Centruroides exilicauda*, Belmonte and Stensaas (1975) on *C. exilicauda* and *Hadrurus arizonensis*, and Fleissner et al. (Fleissner and Schliwa 1977; Schliwa and Fleissner 1977, 1979, 1980; Fleissner and Siegler 1978) on *Androctonus australis*.

The retina of the median eyes consists of multiple cellular layers, and is bounded distally by a preretinal membrane and proximally by a postretinal membrane. Although these are simple eyes, the retinal

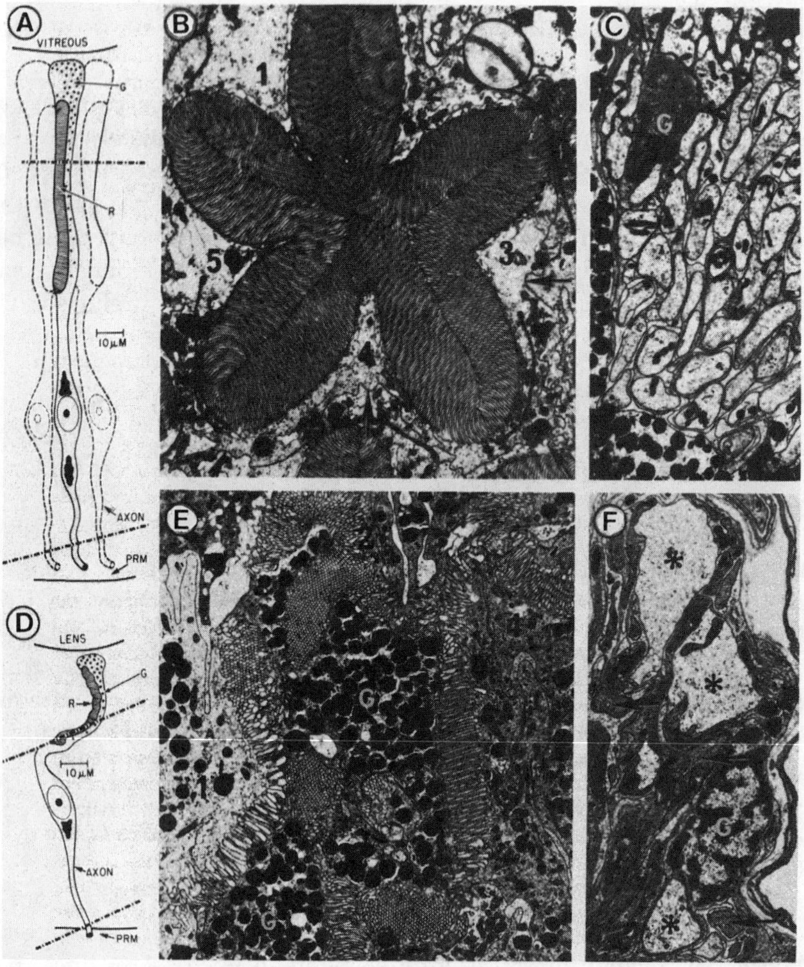

structure is reminiscent of a compound eye, and may represent remnants of ancestral compound eyes. Bedini (1967) noted four retinal layers. He described the first layer as containing pigment cells, but G. Fleissner (pers. comm., 1984) believes that this layer was misinterpreted in cross sections, and that the pigment is actually contained in a lower layer of retinula cells, presumably fixed during preparation in eyes adapted to daylight conditions. The second layer contains the rhabdomeres of retinula cells; the third contains the basal portions of the retinula cells, and the fourth contains nerve fibers from these receptor cells. Bedini provides evidence that the retinula cell contains a centriolar structure, common in arthropod eyes, and does not contain the ciliar structure characteristic of echinoderms and chordates.

The rhabdom of the median eyes of all scorpions thus far examined is typically composed of the adjacent rhabdomeres of five retinula cells, giving this fused rhabdom a characteristically pentagonal appearance in cross section (Fig. 9.1*A, B*). Five visual cells could therefore function

---

Fig. 9.1. Photoreceptor anatomy of the median and lateral eyes. *A*, Schematized scale drawing of photoreceptors of the median eye. The rhabdomeric (*R*) process is long, contains pigment granules (*G*), and is regularly apposed to the processes of other photoreceptors (retinula cells). The axon emerges from the base of the cell and unites with axons of other photoreceptors near the postretinal membrane (*PRM*) to form an optic nerve. Dashed lines indicate the planes of section for the two photographs to the right. *B*, Rhabdomeres arising from five photoreceptors (numbered) unite in the median eye to form a pentagonal rhabdom. Nonrhabdomeric surfaces of adjacent photoreceptors (arrows) not separated by glial cell processes are united by zonulae occludentes (*inset*). (×4,700) *C*, Axons of photoreceptors in the median eye are roughly uniform in size and are separated by processes of glial cells (*G*). (×4,700) *D*, Schematized scale drawing of a single photoreceptor (retinula cell) in the lateral eye. Rhabdomeric portions of these cells are short and irregular. The axon arises from the base of the cell and penetrates the postretinal membrane in small bundles, which later form an optic nerve. Dashed lines indicate planes of section for the two photographs to the right. *E*, Rhabdomeric portions of photoreceptors (numbered) in the lateral eye give rise to irregular aggregates of microvilli. Glial cells containing pigment granules (*G*) separate most photoreceptor processes. (×4,700) *F*, Photoreceptor axons in the lateral eye range from large, lightly staining profiles (asterisks) to small electron-dense processes (arrows). Axons in this small bundle near the postretinal membrane are ensheathed by processes of glial cells (*G*). (×4,700) (From Belmonte & Stensaas 1975) as an optical unit comparable with an ommatidium of insect eyes. In the scorpion eye, this unit is better termed a retinula. The observations of Scheuring (1913) suggested that each retinula cell gave rise to a single nerve fiber, but this could not be confirmed by Bedini. Belmonte and Stensaas (1975), however, reported that the retinula cells give rise to separate axons, which do not branch or synapse with one another before leaving the eye through the optic nerve. Recent evidence based upon reconstructions from serial thin sections and morphometric analysis have shown that each median-eye retinula of *A. australis* contains an arhabdomeric cell, and therefore strongly suggests that the retina is organized similarly to that of *Limulus* (Schliwa and Fleissner 1977, 1979; Fleissner and Siegler 1978). The arhabdomeric cell lies at the base of the retinula cells and sends a dendritic process distally (Fig. 9.2). The dendrite ends at the base of the rhabdom, with conspicuous finger-like evaginations that extend into each of the retinula cells. Fleissner and Schliwa (1977) also discovered that the retinula is supplied by numerous neurosecretory fibers originating in the central nervous system. The efferent fibers of between 10 and 12 neurosecretory cells on each side, suspected of mediating circadian pigment movements in the retina, make synaptoid connections with the visual cells of a retinula.

Although the lateral eye is structurally very similar to the median eye, there are some important differences. The lateral eye lacks a vitreous body, and, further, has rhabdoms composed of from two to ten retinula cells. The rhabdoms consequently have a highly variable appearance in cross section, and do not form regularly repeating units (Fig. 9.1*D, E*). "Closed" rhabdoms are formed with many retinula cells close together; in "open" rhabdoms, formed with fewer cells less closely applied to one another, there is space between the cells. Scheuring (1913) and Machan (1966) reported that in the median eye pigment is located both within the retinal cells and in separate pigment cells, but that in the lateral eye pigment is found only in the retinal cells. Bedini (1967), however, provided clear evidence for pigment cells in the retina of both median and lateral eyes in *Euscorpius*; Belmonte and Stensaas (1975) also found pigment cells in the eyes of *Centruroides* and *Hadrurus* (Fig. 9.1*E*), and Schliwa and Fleissner (1980) described pigment cells in *Androctonus* (Fig. 9.3*A*). In some cases, the discrepancy could be due to species differences, but Machan (1967) and Belmonte and Stensaas (1975) all used *C. exilicauda* in their studies. Certainly, more electron-microscopic studies on different species would help clarify the issue.

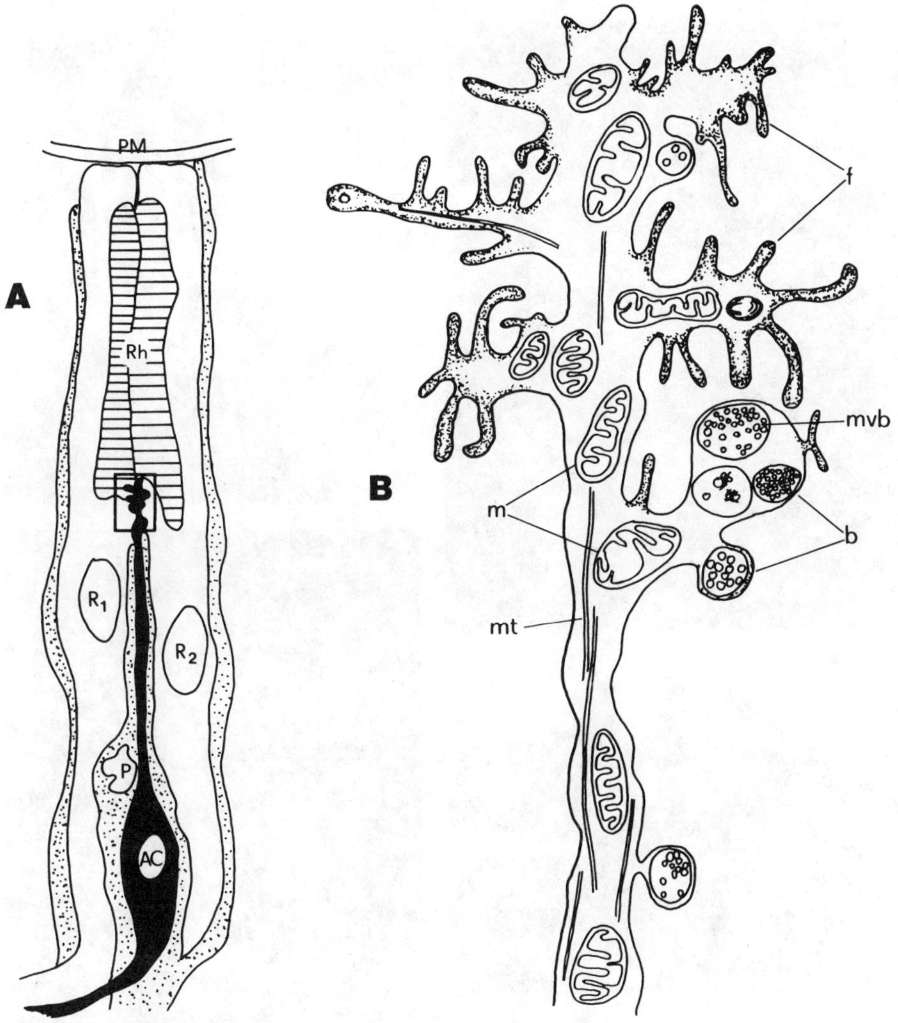

Fig. 9.2. Diagrammatic view of a retinula unit of the median eye of *Androctonus australis*. *A*, The retina lies beneath the lens and preretinal membrane (*PM*). The rhabdom (*Rh*) consists of the rhabdomeres of five adjacent retinula cells, two of which are shown (R_1, R_2). A pigment cell (*P*) and an arhabdomeric cell (*AC*) are also shown. The framed area indicates the region of contact between the arhabdomeric cell and the rhabdom, which is illustrated in view *B*. *B*, The contact region shows numerous fingerlike (*f*) and bulbous (*b*) extensions, mitochondria (*m*), multivesicular-like body organelles (*mvb*), and microtubules (*mt*). (From Schliwa & Fleissner 1979)

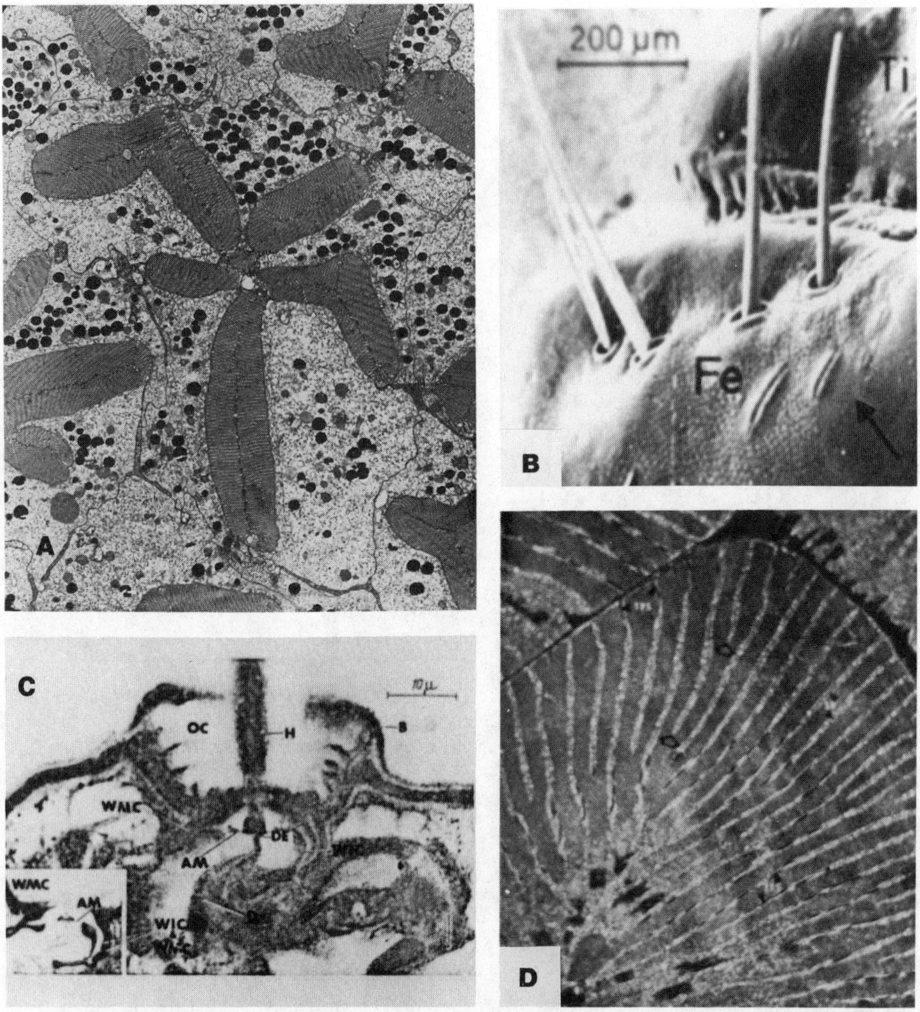

Fig. 9.3. Electron-micrographic views of photoreceptor, mechanoreceptor, and myofibril anatomy. *A*, Electron micrograph through the median eye of *Androctonus australis*. One complete retinula unit is in the center of the micrograph, surrounded by several other retinula units seen only partially. The rhabdom is star-shaped in cross section, reflecting the contribution of five rhabdomeres from separate retinula cells. Also visible are intervening pigment cells with electron-dense pigment granules. (×4,620) (M. Schliwa, unpubl.) *B*, Scanning

As in the median eye, Schliwa and Fleissner (1980) have also found neurosecretory fibers and arhabdomeric cells in the lateral eye of *A. australis* (Fig. 9.4). In fact, more arhabdomeric cells were found per rhabdom than in the median eye. This suggests that although the dioptric quality and retinal anatomy indicate poor acuity, spatial discrimination may be assisted by the presence of large numbers of arhabdomeric cells.

PHYSIOLOGY OF THE EYES. The physiology of the scorpion eye has been investigated on a number of different levels. Machan (1968b) and Fleissner (1968) examined the spectral response of scorpion eyes. They found that the lateral eye exhibited peak sensitivity near 371 nm (ultraviolet), and another, smaller, peak at about 490 to 520 nm (blue-green); but ultraviolet sensitivity was demonstrable only after extended dark adaptation.

Belmonte and Stensaas (1975) recorded from the optic (ocellar) nerves, and found that illumination of either the lateral or the median eyes produced a depolarizing wave upon which a series of spikes was superimposed. The spikes were tetrodotoxin-sensitive, indicating the involvement of sodium in spike generation. Since Belmonte and Stensaas observed that optic-nerve axons arose directly from photoreceptor cells, and since they did not find evidence for axon collaterals, synaptic contacts, or second-order cells, they suggested that the photoreceptors produced both the depolarizing receptor potential and the series of spikes. Photoreceptors of most other arthropods respond to illumination only with a depolarizing receptor potential.

---

electron micrograph of grouped slit sensilla (arrow) on the femur of the leg of *A. australis*. This group represents group 4 on the posterior surface of the right fourth leg seen in Fig. 9.6. *Fe*, femur; *Ti*, tibia. (×495) (Reprinted with permission from Barth & Wadepuhl 1975) *C*, Trichobothrial structure as revealed by a longitudinal section: *AM*, articulating membrane; *B*, bothrium; *D*, dendrite; *DE*, dendrite ending, *H*, hair shaft; *OC*, outer chamber; *WIC*, wall of inner chamber; *WMC*, wall of middle chamber. (×660) (Reprinted with permission from Hoffmann 1967) *D*, Low-power electron micrograph of a muscle fiber in cross section from *Paruroctonus mesaensis*. This figure illustrates the branching myofibrils (open arrows), intervening sarcoplasmic reticulum (*SR*), and transverse tubules (*TTS*). (×1,320) (From Root & Bowerman 1981)

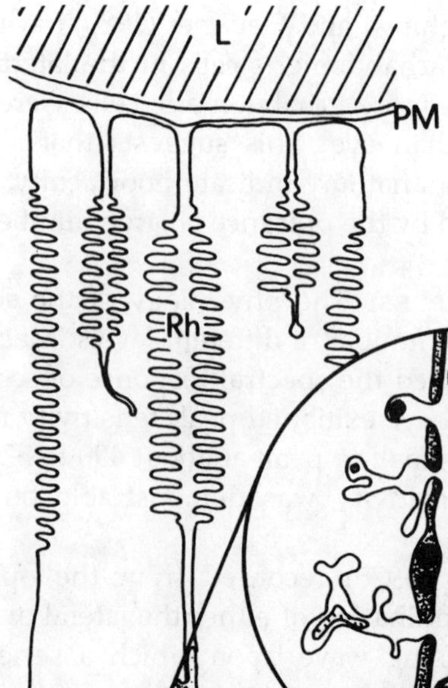

Fig. 9.4. Diagrammatic view of a retinula unit of the lateral eye of *Androctonus australis*. The rhabdom (*Rh*) consists of a variable number of retinula cells, two of which are shown (*R*). An arhabdomeric cell (*A*) sends a dendrite (*D*) between the retinula cells, where it makes contacts with their basal regions. The structure of these contacts is shown in the inset. The dendrite contains numerous mitochondria, vesicular organelles, and longitudinally arranged microtubules. *L*, lens; *PM*, preretinal membrane. (From Schliwa & Fleissner 1980)

Electroretinograms (ERGs) have been recorded by placing an electrode on the surface of the eye and recording the response to a test flash of light (Machan 1966, 1967; Carricaburu and Nait 1967; Carricaburu and Cherrak 1968; Fleissner 1968, 1972, 1974; Fouchard and Carricaburu 1970a; Ramakrishna 1977). Since this electrode is recording the mass response from the entire retina, the resulting ERG is a composite, and often complex, electrical "slow" wave that has its origins in the retinal cells and their interactions. In general, it appears that the ERGs of the scorpion eye are similar to those of other arthropod eyes.

Carricaburu and Cherrak (1968) recorded ERGs from the median and lateral eyes of *Androctonus australis*. They found that the ERG for both eyes had a complex wave form, with an initial quick negative "on" wave followed by a positive "off" wave of variable length, and, in the median eye, a later negative wave. They proposed that the ERG of the median eye was caused by the superimposition of three component waves: a diphasic "on/off" wave (D wave) producing the negative ("on") component and the positive ("off") component, a slow positive wave (P wave) lasting the duration of the stimulus, and a slow negative wave (N wave) rising and falling over the entire time course of the ERG. The ERG of the lateral eye is suspected of being composed of only a diphasic "on/off" wave and a positive wave, since the lateral-eye ERG did not exhibit the large slow negative wave seen for the median-eye ERG.

The complexity of the ERG and the recordings of a receptor potential with superimposed spike trains imply that the retina consists of multiple cell types, or operates via complex interactions between the photoreceptors, or both. Although it was previously believed that only the retinula cells contributed axons to the optic nerve, the recent discovery of neurosecretory and arhabdomeric cells helps explain the complexity of the ERG, since both cell types exhibit synapselike contacts with the retinula cells. Schliwa (1979) has also found arhabdomeric cells in the eye of the harvestman *Opilio ravennae*. He noted that arhabdomeric cells (or eccentric cells) may be more common in chelicerate eyes than was previously anticipated, since they are also found in the lateral and median eyes of *Limulus* (Smith et al. 1965, Jones et al. 1971), and since there is reason to suspect they also occur in other arthropods in which atypical retinal cells have been noted. It is apparent that electrophysiological studies of scorpion eyes should be reinterpreted in light of the new knowledge of arhabdomeric cells. Recently, Fleissner (1985) made intracellular recordings from receptor cells and arhabdomeric cells to determine their contribution to the ERG and the spike activity recorded in the optic nerve.

The form of the ERG for both the lateral and the median eyes changes in response to dark adaptation (Machan 1967, 1968a; Carricaburu and Cherrak 1968; Fouchard and Carricaburu 1970a, b; Fleissner 1974), indicating a change in visual sensitivity. Machan (1968a) observed that the lateral eyes adapt more rapidly than the median eyes. She found that, after overnight dark adaptation, the sensitivity of the median eye increased by approximately three log units, and she stated that sensitivity of the lateral eye increased much faster. An indication of this was provided by comparisons of ERGs from dark-adapted median and lateral eyes (Machan 1968a, figs. 2, 3). No data were presented to indicate the extent of the difference. Fouchard and Carricaburu (1970b) noted a similar phenomenon in *Buthus occitanus*. When the animal was in a light-adapted state, they saw very little difference in the forms of the ERGs for the lateral and the median eyes. As the animal adapted to the darkness, however, the wave form of the ERG in the lateral eye changed considerably, but the ERG of the median eye did not. Machan (1968a) implicated movements of retinal pigment ("shielding pigment") in these observed changes in ERG. After dark adaptation, histological sections of eyes revealed that the pigment had moved proximally (i.e., beneath the rhabdoms), exposing them entirely, and that this movement occurred faster in the lateral eyes.

Fleissner (1977c) was able to determine the sensitivity of the eyes of *A. australis*, and found that in the dark-adapted state the lateral eyes were about one log unit more sensitive than the median eyes. He calculated that in the dark-adapted state the lateral eye is probably among the most sensitive of arthropod eyes. He speculated that despite the eye's poor imaging capabilities, its sensitivity might allow for orientation to the moon and stars such as that demonstrated by Linsenmair (1968).

From the work of a number of investigators, it is now apparent that scorpions also use visual cues for timing the circadian rhythmicity of locomotor activity (Wuttke 1966; Cloudsley-Thompson 1973a, 1975; Fleissner 1974; Geetha Bali 1977). Fleissner (1977a, b, d) followed changes in visual sensitivity and found that both the lateral and the median eyes exhibited cyclic changes in sensitivity, although this was more pronounced in the median eyes (Fig. 9.5; the behavioral evidence

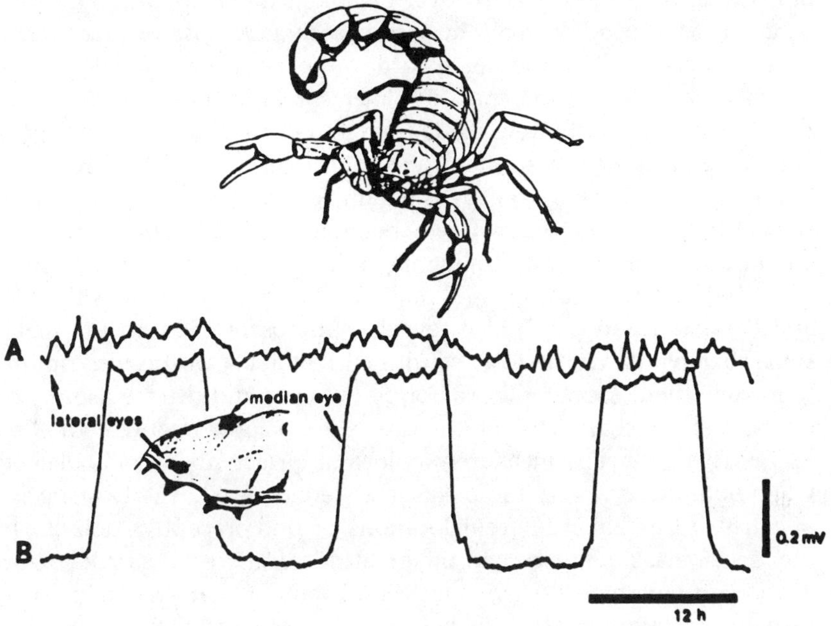

Fig. 9.5. Circadian rhythmicity of the visual system of the *Androctonus australis*. Changes in the sensitivity of the lateral eyes (*A*) and median eyes (*B*) are indicated for three circadian cycles. Eye sensitivity was determined by measuring the amplitude of the electroretinogram (ERG) to identical test flashes of light in dark/dark conditions. Vertical bar, ERG; horizontal bar, time scale. (From Fleissner 1977d)

for circadian rhythmicity is discussed in Chapter 5). Recent work on the physiological control of visual sensitivity is focused on the central nervous pathways mediating these cyclic changes (Fleissner and Fleissner 1978). For example, by backfilling the optic nerve with lucifer yellow, the somata of the neurosecretory efferent fibers were located in the brain (Fleissner and Heinrichs 1982), and when cuts of this tract were made at various levels, the rhythmicity could be interrupted (Fleissner 1983).

METASOMAL NEURAL PHOTORECEPTOR. A quite unusual, but not unique, photoreceptive system has been found in the scorpion metasoma (tail). First described by Zwicky (1968) for the scorpion *Urodacus*, it has since also been described for *Heterometrus* by Geetha Bali and Pampathi Rao (1973). Zwicky (1968, 1970b) observed that scorpions with their eyes painted over exhibited the same photonegative behavior as control animals in a choice chamber. Recordings made from various levels of the nerve cord in the metasoma revealed spike activity when the telson and fifth metasomal segment were illuminated. When attempts were made to exclude light scattering into more anterior ganglia (the telson and fifth metasomal segment do not contain ganglia, but are served by the ganglion in the fourth metasomal segment), the electrical activity was maintained. Cutting the nerve cord anterior to the illuminated segments, however, abolished all activity.

Studies of other scorpions (*H. indus* and *H. fulvipes*) by Geetha Bali and Pampathi Rao (1973) also revealed photosensitivity in the metasoma. Behavioral-choice tests confirmed the ability of these scorpions to mediate photonegative behavior by the eyes and the metasoma, or by the metasoma alone when the eyes were painted over (Geetha Bali 1977). This behavior, measured as locomotor activity, was abolished when both the eyes and the metasoma were covered. The two species of *Heterometrus* have different locations of photoreceptive activity in the metasoma. *H. fulvipes* had the greatest sensitivity in ganglia 5 and 6, and cutting the nerves from ganglion 7 did not alter spike activity in response to illumination of the tail. *H. gravimanus* had the greatest photoreceptive sensitivity in the segments posterior to the last ganglion, including the telson, but activity could also be recorded when ganglia 5, 6, and 7 were illuminated. In addition to this difference, Zwicky (1970a) found the metasomal photoreceptor of *Urodacus* to have a peak spectral sensitivity at 480 nm (blue), and a smaller peak at 400 nm (violet); Geetha Bali and Pampathi Rao (1973) report that for the different species of *Heterometrus* they examined the peak sensitivity was at 568 nm (green-yellow).

It therefore appears that there may be considerable variation in the location of the light sense in the tails of different scorpion species. Whether such variability is a general phenomenon common to all scorpions remains to be seen. A number of questions about the role of the metasomal photoreceptor also remain to be answered. For example, is it possible that this sense mediates the timing of circadian rhythms? Fleissner's experiments suggest that the eyes are involved, but he did not test the metasomal light sense (Fleissner 1977a). Further, Fleissner (1977b) points out that the exceptional sensitivity of scorpion lateral eyes requires cautious interpretation of behavioral experiments in which the lateral eyes are covered over to prevent light from reaching them (e.g.,

Zwicky 1970b, Geetha Bali 1977), since a small amount of light may in fact be penetrating the light barrier.

Chemoreceptors

Undoubtedly, scorpions are well supplied with chemoreceptors, which are common in other arthropods, including arachnids. Typically, arthropods have receptors located near the mouthparts and on appendages, and these receptors are capable of distinguishing the chemical composition of food sources and other chemical stimuli in the environment. Arthropods also have internal chemoreceptive elements to monitor the composition of the hemolymph and other body fluids. Unfortunately, very few chemoreceptors have been identified in scorpions. Instead, much of what we believe has been inferred from behavioral experiments.

When *Leiurus quinquestriatus* is exposed to a humidity gradient, it has a preference for moist areas (Abushama 1964). Sreenivasa Reddy (1959) suggested that the apparent permeability of the sensilla basonica of the pectines made them likely candidates for humidity detection. Although this is an attractively simple suggestion, Cloudsley-Thompson (1955) found that behavioral evidence for moisture detection by the pectines in *Buthus occitanus* and *Androctonus australis* was lacking, and he concluded that the pectines were not humidity receptors. (Further discussion of responses to humidity are presented in Chapters 5 and 8.)

Abushama (1964) suggested that the humidity receptors lie in the small hairs of the tarsal segment of the legs. When the tarsus was covered with nail polish, or when this segment was amputated, the scorpions no longer chose a humid region in the gradient; but, paradoxically, they did not exhibit a random-choice pattern either! Instead, they preferred a dry environment. If the tarsal hairs are implicated in humidity reception, their function would have to be interpreted in light of Cloudsley-Thompson's (1955) findings, since presumably the legs contacted the moistened filter paper in his experiments with dessicated scorpions. It is possible that humidity reception may be the function of the specialized tarsal organs described on each leg of *A. australis* by Foelix and Schabronath (1983). The structure of these organs suggests a chemoreceptive function, and the authors note that similar functions have been demonstrated for tarsal organs in spiders.

In contrast to humidity preference, for which the receptors involved have not been convincingly located, there is better evidence that taste receptors are located on the chelicerae. Food-rejection behavior has been observed when food was treated with distasteful compounds. Abushama (1964) observed that scorpions will reject food treated with concentrated sodium chloride solution. Alexander and Ewer (1957) observed that the scorpion *Opistophthalmus latimanus* will reject food treated with tinctura quassiae, made from the bitter ash: when tinctura quassiae was dropped onto food a scorpion was feeding upon, it immediately rejected the food and began cleaning the chelicerae. When conspicuous cheliceral platelets were removed, however, the scorpions did not reject the treated food.

Abushama (1964) suggested that cheliceral hairs were responsible for taste sensitivity, but was unable to demonstrate this function. Pampathi Rao (1964), describing a series of studies in his laboratory, successfully recorded electrical activity from these short cheliceral hairs (the Type A hairs of Venkateswara Rao 1963). Some hairs were sensitive to sodium chloride, sucrose, glycine, and glutamate, besides being sensitive to mechanical displacement.

Abushama (1964) also examined the response of scorpions to odorous substances. When given a choice between a compartment treated with chemical odors and another compartment left untreated, scorpions moved away from compartments treated with naphthalene, clove oil, or acetic acid, but showed equal preferences for untreated compartments and those treated with odors of other scorpions or odors of cockroaches. She observed that scorpions with their pedipalps painted over lost sensitivity to airborne chemicals, and so speculated that the smaller hairs distributed widely over the pedipalps were responsible. Currently, however, there is no physiological evidence to support or disprove this hypothesis. In addition, Foelix and Schabronath (1983) have described the structure of chemoreceptive hairs on the tarsi of the legs, so a chemical sense may be widely distributed.

Thermoreceptors

Although it is presently unclear what sensory organs might be involved, there is little doubt that scorpions are able to sense and react to changes in environmental temperature, perhaps through changes in internal body temperature (see Chapters 5 and 8).

Since thermoreceptors are commonly found on the antennae of insects, it would be of interest to locate the thermoreceptors of scorpions, which lack antennae. In some insects, however, thermoreceptors are located on the legs; and the ganglia of the crayfish and of the cockroach also exhibit a general thermal responsiveness. In the scorpion, Alexander and Ewer (1958) found that different body regions responded when a heated needle was held near them, with the greatest sensitivity located in the telson and pedipalps. It was suggested, however, that internal thermoreceptors could also lead to thermoresponsive behavior. Therefore, it does not seem necessary to speculate that organs such as the pectines are thermoreceptive, since a general thermal sensitivity may be present.

Mechanoreceptors

More than any other modality, the sensation of mechanical stimuli is the best-developed scorpion sense. Among the many different mechanoreceptors are slit sense organs on the cuticle, a wide variety of sensory hairs and joint receptors, muscle receptor organs, and several other types of proprioceptors. In addition, groups of several organs make up both the tarsal receptors involved in prey detection, and the pectines, which may be involved in the sensation of substrate texture.

SLIT SENSE ORGANS. Slit sense organs are uniquely arachnid cuticular mechanoreceptors. As the name suggests, they are slit-shaped receptors; they appear not so much as a deep slit in the cuticle, but as an elongate thinning of it, which gives the appearance of a slit. These tiny, barely visible receptors can be found singly, or in small, seemingly unorganized groups, or in ordered groups termed lyriform organs. Lyriform organs consist of a number of slit sense organs aligned roughly parallel in a pattern resembling a lyre. Although most abundant and most extensively studied in spiders, slit sense organs have also been found in other arachnid orders. Among arachnids, single slit sense organs, grouped slits, and lyriform organs are widely distributed on the cephalothorax (prosoma), abdomen (opisthosoma), chelicerae, pedipalps, and all other appendages.

The existence of lyriform organs has been known for almost a century (Gaubert 1890, 1892; Hansen 1893; McIndoo 1911). Their function as cuticular-stress sensors was not convincingly demonstrated, however, until the work of Pringle (1938, 1955), who noted their similarity to the campaniform sensilla of insects. Campaniform sensilla are also cuticular receptors, made up of a thin cuticular region with an attached dendrite, but they differ in that, instead of having an elongate slit shape, they are round or elliptical. The functional similarity of slit sense organs to campaniform sensilla was clarified when Pringle showed that both the campaniform sensilla and the slit sense organs respond to displacement or deformation of the cuticle.

Largely through the work of Barth (1972a, b, 1973) on spiders, the role of slit sensilla and lyriform organs as cuticular-stress detectors has been well established. The mechanical properties and physiological responses of slit sensilla and lyriform organs in spiders have likewise been thoroughly investigated (Barth 1972a, b, 1981; Barth and Picklemann 1975; Barth and Bohnenberger 1978; Bohnenberger 1978, 1981). It appears that they are involved in proprioceptive reflexes in the leg (Seyfarth 1978) and in kinesthetic orientation (Seyfarth and Barth 1972). Selective ablation of lyriform organs on the spider leg, however, appears to have no major effect upon smooth, normal walking, but they may be involved in more subtle proprioceptive feedback systems during walking, or in other behaviors (Seyfarth and Bohnenberger 1980). Walcott (1969) has also implicated lyriform organs in vibration reception, and Brownell and Farley (1979b) have shown that a slit sensillum on the tarsus of the leg is capable of distinguishing substrate vibrations (as described later in this section).

In scorpions, the slit sense organs that are located on the legs have been the most studied, although slit sensilla have also been observed on the prosoma, opisthosoma, and telson. Pringle (1955) examined the slit sense organs of an amblypygid and of the scorpion *Heterometrus swammerdami*, and was the first to demonstrate their mechanosensitivity. He recorded from small nerve trunks in the scorpion leg, and found that they responded to cuticular deformations produced with a small needle, but did not respond to application of chemicals.

Barth and Wadepuhl (1975) described the location of slit sensilla on the legs of the scorpion *Androctonus australis*. They observed that isolated slits were widely distributed on the lateral surface of all leg segments, and grouped slit sensilla were usually found on the lateral surface of the leg and near the joints (Figs. 9.3*B*, 9.6). This general pattern is comparable to that found in other arachnids; but Barth and Staghl (1976) noted that, in comparison with those of other arachnids, in scorpions the isolated slits are more common and less regular in their occurrence, and that lyriform organs are absent. The authors suggested that the scorpion arrangement may be more primitive. The slits on the scorpion leg are also generally oriented parallel to its axis, in the appropriate position to detect normal loading stresses, so it is assumed that

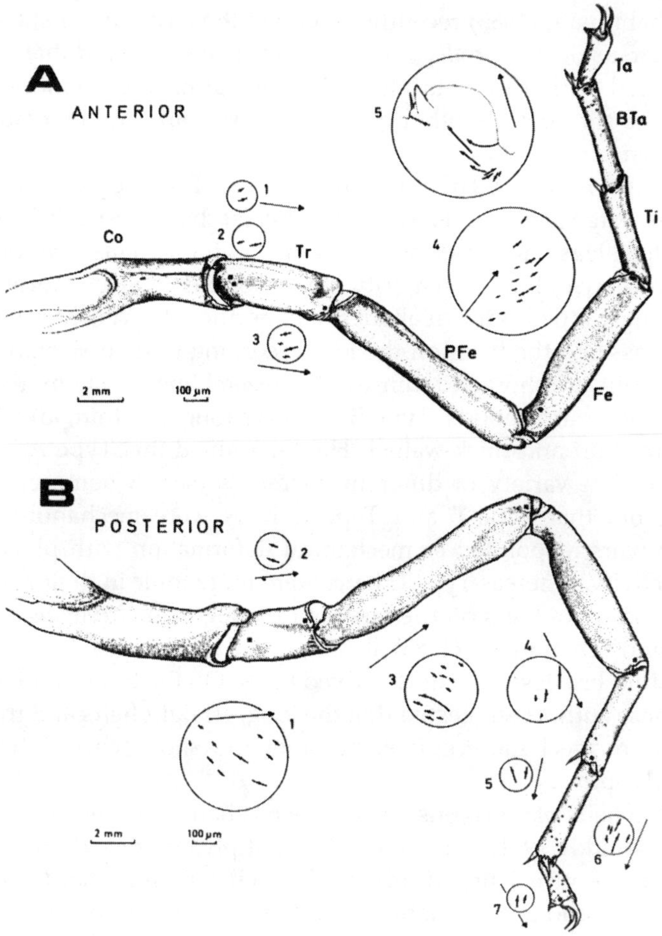

Fig. 9.6. Diagrammatic view of the fourth right walking leg of *Androctonus australis* in anterior (*A*) and posterior (*B*) views, showing the distribution of isolated slits longer than 30 μm (–) and of grouped slits (■). The insets show the details of groups, drawn to scale. The dendrite attachment sites are indicated by circular dilations of the slits. Each arrow runs parallel to the axis of its respective leg segment, pointing toward the distal end. (The leg segments are named according to Barth & Wadepuhl 1975; different names in Fig. 9.9 are noted.) *Co*, coxa; *Tr*, trochanter; *PFe*, prefemur (femur in Fig. 9.9); *Fe*, femur (patella in Fig. 9.9); *Ti*, tibia; *BTa*, basitarsus; *Ta*, tarsus. (From Barth & Wadepuhl 1975)

scorpion slit sensilla have a function similar to those of spiders. Foelix and Schabronath (1983) recently described the structure of slit sensilla on the tarsus of *A. australis* in more detail. They observed that each slit is supplied with two neurons, one innervating the membrane that covers the slit, and the other entering the slit but not innervating the outer membrane.

SENSORY HAIRS AND TRICHOBOTHRIA. The body of many scorpions is extensively covered with cuticular hairs, especially on the pedipalps, legs, and metasoma. Some of these hairs are probably chemoreceptive, but the majority are undoubtedly mechanoreceptive. Surprisingly little is known about their specific role. Venkateswara Rao (1963) classified the hairs on the legs according to their size and form: Type A hairs are short and thin-walled, resembling the chemoreceptive hairs of other arthropods; Type B hairs are long and thin; and Type C hairs are short and thick-walled. He determined that Type A hairs responded to a variety of different chemicals (see "Chemoreceptors," above), but that Type B and Type C hairs were mechanoreceptive. Type B hairs responded to mechanical deformation with phasic electrical activity, whereas Type C hairs were more tonic in their response. In their study of hairs on the tarsus of the leg in *Androctonus australis*, Foelix and Schabronath (1983) also observed three types: long, straight hairs; short bristles; and short, curved hairs. On the basis of their structure, these authors suggested that the long, straight hairs and the short bristles are mechanoreceptive, but that the short, curved hairs are chemoreceptive.

Certain hairs of scorpions are strictly mechanoreceptive (Foelix 1976, Sanjeeva Reddy 1971), but seem to be multiply innervated, unlike most mechanoreceptive hairs in insects. In their structural study of tarsal hairs, Foelix and Schabronath (1983) found seven neurons innervating each mechanoreceptive hair. Sanjeeva Reddy recorded the electrical responses of the four to five sensory cells that constitute the sensory element of certain hairs on the metasoma. Since he was able to show that all these sensory cells responded to mechanical stimuli, it was not necessary to presume different modalities for the different cells comprising this receptor system.

Movement of the hairs on the metasoma and telson by air puffs or mechanical displacement caused spike discharges (Babu and Sanjeeva Reddy 1967), which were correlated with the position, movement, and velocity of the stimulated hair (Sanjeeva Reddy 1969, 1971; Geetha Bali 1980). The function of these hairs in tail movements, however, is currently unknown. It is possible that hairs at arthropod leg joints may serve to determine joint position (Pringle 1963).

Trichobothria are cuticular hairs distinguished from other hairs by their very long, slender appearance and their location in conspicuous sockets (bothria). First described for spiders (Dahl 1883), trichobothria have since been found in most other arachnid orders and in insects. In scorpions they are restricted to the pedipalps, where their number and position are so exceptionally consistent that they have been used as a taxonomic tool.

The structure of scorpion trichobothria has been described for *Euscorpius carpathicus* by Hoffmann (1965, 1967) and for *Mesobuthus eupeus* by Ignat'yev et al. (1976). The trichobothria are 5 to 8 μm in diameter, and 0.5 to 1.5 mm long. Ignat'yev et al. distinguish two types of trichobothria, one approximately 0.5 mm long, and the other 1 to 1.5 mm long. The hair shaft extends from a cup-shaped chamber guarded by a thickened cuticular ring (Fig. 9.3C). Beneath the hair lie two additional chambers, and the base of the hair is attached to a membranous extension of the exocuticle. Receptor cells are attached to the base on one side of the hair. Hoffmann (1965) found only one receptor cell in *Euscorpius*, but Ignat'yev et al. found six or seven in *Mesobuthus*, as well as synaptic contacts between some unidentified efferent nerve fibers and the receptor cells.

It is now clear that the trichobothria of scorpions are mechanoreceptors that respond predominantly to weak air currents. This was initially established by Hoffmann (1967), and confirmed by Linsenmair (1968) and Ignat'yev et al. (1976). In response to air pulses, the hair moves in a specific plane independent of the direction of the air current. This plane is in the direction of the sensory-cell dendrites. Ignat'yev et al. suggest that this plane of movement is due, at least in part, to the articulating membrane of the trichobothrium. Figure 9.7 illustrates activity from a single trichobothrium recorded by Hoffmann (1967) in response to oscillations of the hair, and to maintained deflection by air currents. With oscillations in movement (Fig. 9.7*A*), a burst of spikes was recorded from the receptor. The impulse frequency and the number of impulses in each burst were correlated with the duration and velocity of movement. Upon prolonged deflection of the hair toward the dendrite (Fig. 9.7*B*), a burst of spikes was observed that gradually decreased in frequency and eventually stopped. In *M. eupeus* and *M. caucasicus* different-sized spikes were seen in electrical recordings from a trichobothrium with tungsten electrodes. This is consistent with the

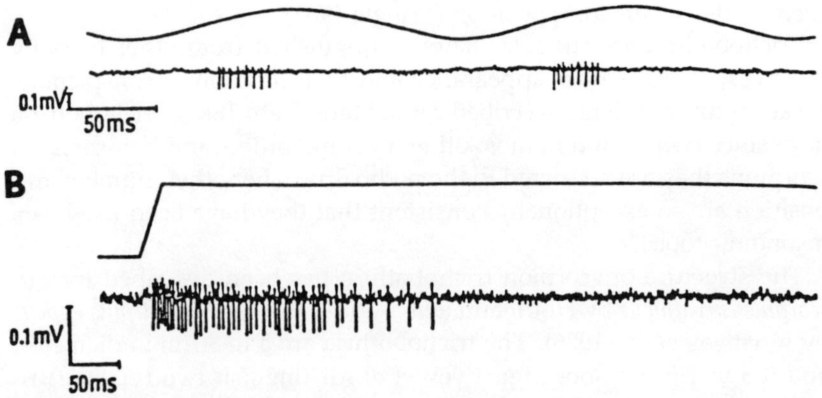

Fig. 9.7. Spike activity recorded from trichobothria of *Euscorpius carpathicus*. *A*, When recordings were made from a nerve of the trichobothrium (lower trace) while it was displaced at a rate of 5 Hz (upper trace), a burst of spikes was seen during the rising phase of the movement. *B*, When a constant deflection was applied to the trichobothrium, spike activity persisted for about 1/4 second before ending. (From Hoffmann 1967)

observation that six or seven receptor cells are located in each trichobothrium. Similar phasic and phasic-tonic responses were also observed (Ignat'yev et al. 1976). Since each trichobothrium has a directional sensitivity to air currents, and since a number of trichobothria with different orientations are located on the pedipalps, they may serve to detect the direction of prevailing air currents. Linsenmair (1968) has demonstrated that scorpions are capable of orienting to wind currents and suggests that the trichobothria mediate this behavior.

TARSAL SENSE ORGANS AND PREY-DETECTION BEHAVIOR. A very encouraging approach to studying the role of cuticular mechanoreceptors in behavior was recently made by Brownell and Farley in their investigations of prey detection by the nocturnal desert scorpion *Paruroctonus mesaensis*. In a series of behavioral experiments, they demonstrated that scorpions do not respond to nearby visual stimuli or airborne sound vibrations when detecting prey, but are exceptionally capable of responding to substrate vibrations up to half a meter away (Brownell and Farley 1979a, b, c). These experiments used simulated vibrations of a natural prey species, the burrowing cockroach *Arenivaga*, produced by movements of a wooden dowel in a sandy substrate. This stimulus reliably and repeatedly elicited prey-seeking orientation responses in the scorpions. For a wide range of stimulus locations, turning movements of the scorpion toward the stimulus were very accurate. Usually only one turn was sufficient to cause the scorpion to orient directly toward the stimulus source. Brownell (1977a, b) found that these scorpions detect two types of substrate vibration in this orienting behavior. An analysis of the physical properties of substrate vibrations in desert sand showed that low-frequency surface waves, termed Rayleigh waves, and higher-frequency compressional waves could be transmitted through a sandy substrate, and detected by the scorpions.

Brownell and Farley (1979a) discovered that two different receptors on the legs were responding to substrate vibrations. These were the tarsal hairs and a slit sensillum (Fig. 9.8*A*). The tarsal hairs are a group of cuticular hairs that contact the substrate when the leg is on the ground. The slit sensillum, termed the basitarsal compound slit sensillum (BCSS), lies near the basitarsus-tarsus joint, and is equivalent to the grouped slits described by Barth and Wadepuhl (1975) for *Androctonus australis* (see Fig. 9.6*A*: grouped sensilla no. 5 on the anterior leg surface). When they recorded from nerves in the tarsus and basitarsus while producing substrate vibrations with an electromagnetic vibration probe, they observed a burst of spikes. Within this burst one large, lower-threshold action potential and one small, higher-threshold action potential could be distinguished. Similar findings were obtained when a burrowing cockroach was used as a stimulus (Fig. 9.8*B*). Recordings from restrained animals with mechanical stimulation of the tarsus and basitarsus showed that the large action potentials probably originated from the tarsal hairs, and that the small action potentials probably originated from the BCSS. The tarsal hairs responded to vibrations up to 15 cm away, and the BCSS responded to vibrations up to 50 cm away, distances comparable to those observed in prey-detection tests (Brownell and Farley 1979b). The tarsal hairs responded better to compressional waves, whereas the BCSS responded better to Rayleigh waves. Therefore, each leg contains a long-distance detection system, tuned to Rayleigh waves (the BCSS), and a short-distance detection system, tuned to compressional waves (the tarsal hairs).

Observations of leg placement during orientation showed that all eight legs are spatially organized in a circular array, and that the BCSS in each leg probably participates in vibration localization, since ablation

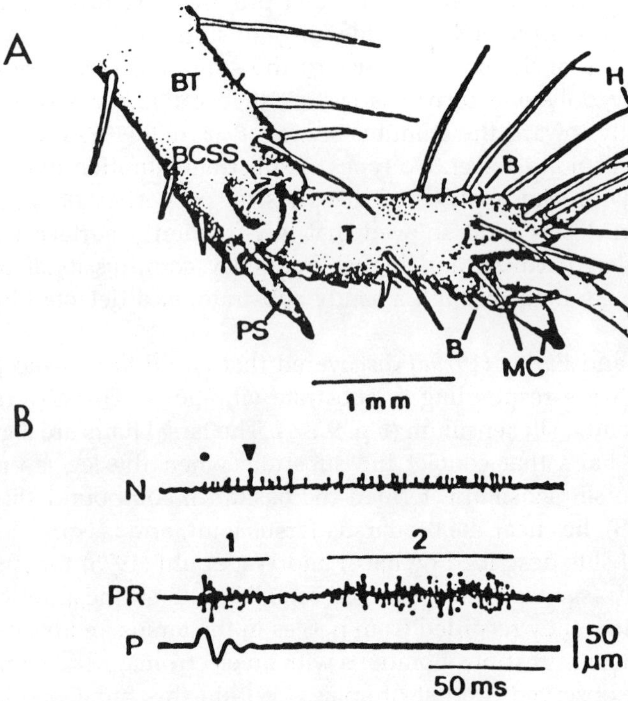

Fig. 9.8. Substrate vibration receptors in the scorpion leg. *A*, Diagram of the basitarsus (*BT*) and tarsus (*T*) segments at the end of the scorpion leg, illustrating the location of the tarsal hairs (*H*) and basitarsal compound slit sensillum (*BCSS*). Other indicated structures are: *B*, small and large bristle hairs; *LC*, *MC*, lateral and medial claws; *PS*, pedal spur. *B*, Response of the BCSS and tarsal hairs to movements of a burrowing cockroach. A nerve recording (*N*) occurred in response to waves generated by movement of an electromagnetically driven probe (*P*) on the surface of the sand and recorded by a piezoelectric receiver (*PR*, *1*). A burrowing cockroach in the sand nearby reacted to the disturbance with a startle response also recorded (*PR*, 2) by the receiver. Both events evoked large action potentials (▼) in the sensory neurons from hairs, and smaller action potentials (•) from the BCSS. The probe was 15 cm from the point of tarsal contact with the sand, and 5 cm from the burrowing cockroach. Vertical bar, probe displacement; horizontal bar, time scale. (From Brownell & Farley 1979a)

of the BCSS on some legs causes a predictable error in orientation (Brownell and Farley 1979c). The distance-detection system apparently has two components. Vibration stimuli presented within 15 cm reliably elicit accurate orientation and walking toward the stimulus. For stimuli 15 to 30 cm away, the movement response was more variable, even though the initial direction of the orientation was accurate. Since the electrophysiological recordings demonstrated activity in the tarsal hairs with stimuli up to 15 cm away but the BCSS were activated with stimuli at greater distances, there is a strong link between the two aspects of prey detection (orientation response and movement response) and the physiological responses of the two receptors in each leg.

JOINT RECEPTORS. The leg joints are serviced by small branches of the main leg nerve (MLN) and by the dorsal leg nerve (DLN), which is a branch of the MLN passing along the dorsal surface of the leg. Along its path, the DLN innervates sensory structures, including sense cells near the joints (Fig. 9.9).

Slit sense organs and cuticular hairs near the joints certainly contribute to the DLN, but other sensory elements have also been recognized. Pampathi Rao and Murthy (1966) and Pampathi Rao (1964) reported on some sense cells at the femur-patella joint of *Heterometrus fulvipes* that were innervated by the DLN. Although they describe these cells as scolopophorous, Laverack (1966) suggests that this term is inappropriate. He cites the work of V. P. Rao (1963), which indicates that scolopale-like structures were not evident in *H. fulvipes*. In his own study of joint receptors in *Hadrurus hirsutus*,* Laverack (1966) also indicated that a scolopale was absent, and believed that these elements should be classified as the Type 2 sensory elements of Pringle (1955). Type 2 structures are associated with connective tissue, and the DLN clearly gives off small branches to the hypodermis at the joints (Root, unpubl.). Further, Weltzin and Bowerman (1980) were able to trace the paths of sensory nerves to these joints using the cobalt-backfilling technique. The receptor cells at the femur-patella joint were viewed in detail, and the authors described a cluster of about ten cell bodies at the joint. The dendritic branches from these cells attached to the hypodermis at the joint.

Despite differences in opinion about the structure of these joint re-

*Because Laverack (1966) reported no precise locality for his specimens, it is quite probable that he was instead working with *H. arizonensis*: see Williams (1970a).

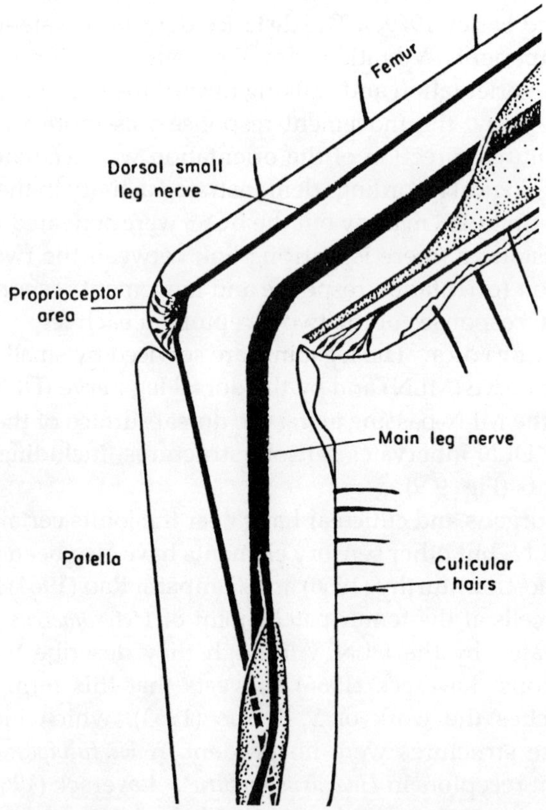

Fig. 9.9. Diagram of the femur and patella of the leg of *Hadrurus hirsutus*. The main leg nerve runs along the midline of the leg through the femur-patella joint, and gives off a branch to proprioceptors at the patella-tibia joint. The dorsal small leg nerve supplies proprioceptor regions at the femur-patella and patella-tibia joints. Shown beneath the main leg nerve in the femur is the ventral small leg nerve, which supplies proprioceptors at the femur-patella joint. (From Laverack 1966)

ceptors, there is basic agreement on their role in joint movements. Most studies have used imposed movements of the femur-patella joint in isolated legs while nerve recordings were being made from nerves upstream from the joints. Pampathi Rao (1964) and Pampathi Rao and Murthy (1966), studying the scorpion *Heterometrus fulvipes*, and Laverack (1966), studying the scorpion *Hadrurus hirsutus*, indicated that the joint receptors were directionally sensitive. Some units fired only during flexion of the joint; the firing of others was restricted to extension movements. In addition to studies on this one leg joint, Pringle (1955) recorded spike activity from four different joints in *Heterometrus swammerdami*, and Bowerman (1976) recorded spike activity from several joints in *Paruroctonus mesaensis*. They identified many directionally sensitive units from each joint studied. For example, Bowerman identified 10 to 15 flexion-sensitive and 17 to 25 extension-sensitive units at the femur-patella joint. Since he recorded from the MLN, it is likely that he sampled cuticular hairs and other unidentified receptors in addition to joint receptors. Recordings from smaller nerve bundles supplying the joint receptors revealed tonic, phasic, and phasic-tonic units when imposed movements were made at the femur-patella joint (Weltzin 1981, Weltzin and Bowerman 1980). It appears from all these studies that the joint receptors are capable of monitoring joint position, joint movement, and the rate of joint movement, as reflected in the spike activity of tonic and phasic units.

Bowerman and Larimer (1973) described the structure and physiology of a joint-receptor system at the patella-tibia joint of the pedipalp in the scorpion *Centruroides vittatus*, which functionally resembles that of leg-joint receptors. The patella-tibia joint is located proximal to the chela (or claw), and exhibits both dorsal/ventral movements and flexion/extension movements.

The joint-receptor system consists of two separate components, PT-1 and PT-2. Located near the dorsal surface of the patella is the PT-1 component, consisting of eight to ten sensory cells (bipolar neurons), most of which are sensitive primarily to dorsal bending of the tibia with respect to the patella. Located near the ventral surface, and closer to the patella-tibia joint, lies the PT-2 component. This system consists of about six to eight bipolar cells that also attach to the articular membrane. Most of these cells are sensitive primarily to ventral bending of the joint. Tonic, position-sensitive units and phasic, movement-sensitive units were characterized (Fig. 9.10). Except for one midrange tonic unit, all the position-sensitive units exhibited maximal sensitivity at one point in the range of joint movement, indicating that the range of joint movements is fractionated for different sensory cells. Despite maximal sensitivity at one particular position, units from both systems respond to either flexion or extension and either dorsal or ventral bending; thus each unit is bidirectionally sensitive. Bowerman and Larimer (1973) propose that both the PT-1 and PT-2 components interact to determine patella-tibia joint position for each plane of movement.

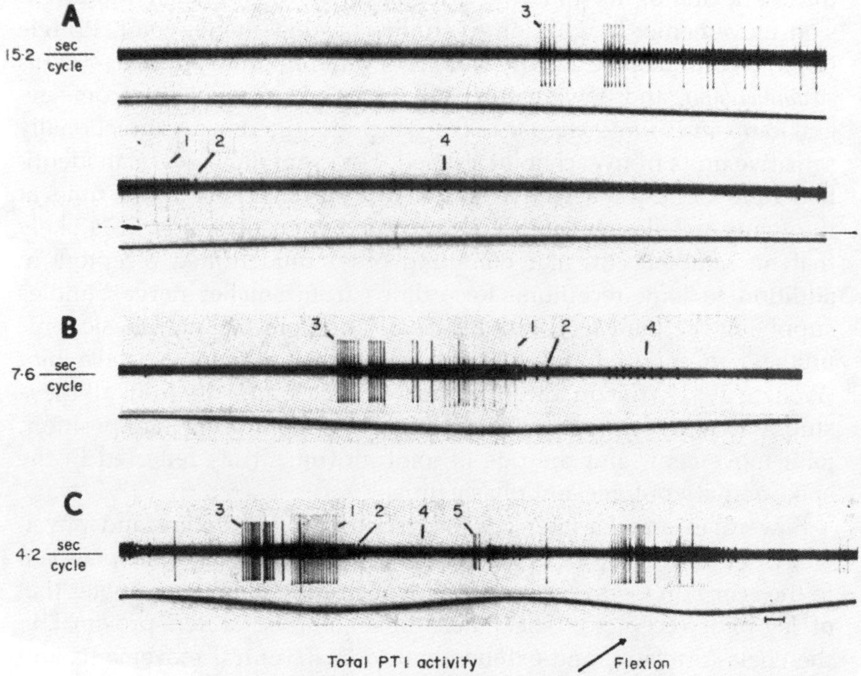

Fig. 9.10. Electrical recordings from patella-tibia joint receptors from the pedipalp of *Centruroides vittatus.* These records illustrate PT-1 activity evoked by flexion and extension of patella-tibia joint at three different rates of movement: *A,* 15.2 sec per cycle; *B,* 7.6 sec per cycle; *C,* 4.2 sec per cycle. Each cycle consisted of a complete extension and flexion of the joint from 80° to 180°. In each trace, the movement monitor is shown below the electrical record. Five different units could be distinguished on the basis of spike height (*1, 2, 3, 4, 5*). Units 1, 2, and 3 were all extension-sensitive, and units 4 and 5 were both flexion-sensitive. Bar: 200 msec. (From Bowerman & Larimer 1973)

METASOMAL-MUSCLE RECEPTOR ORGANS. Pampathi Rao and Murthy (1966) noted the presence of a receptor organ in the metasoma (tail) of *Heterometrus fulvipes,* which consisted of modified muscle fibers, and was sensitive to bending of the tail and telson (sting). Two separately innervated divisions were briefly described: Organ I responded primarily to extension of the next distal tail segment; Organ II responded predominantly to flexion of the next distal segment.

Bowerman (1972a) provided a detailed report of the structure and physiology of these paired muscle receptor organs (MROs) in metasomal segments I through IV in the scorpion *Centruroides gracilis*. He found that each MRO consists of a modified muscle component situated anteriorly in the segment, with a tendon connecting each muscle component to the next distal segment (Fig. 9.11).

The medial division of each receptor consists of modified muscle fibers with long sarcomeres (7–9 μm); the lateral division contains modi-

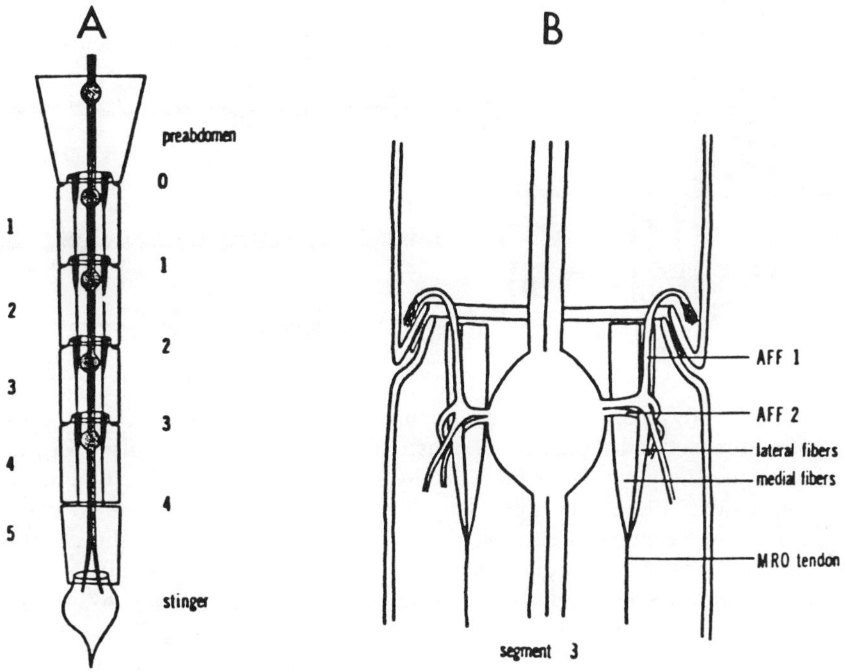

Fig. 9.11. Morphology of metasomal-muscle receptor organs (MROs). *A*, Diagram illustrating the location of the paired MROs in metasomal (postabdominal) segments I through IV of the scorpion *Centruroides gracilis*. The MROs lie alongside the ventral nerve cord, attached at their proximal ends to the preceding metasomal segment, and at their distal ends to the next distal segment via a long tendon. The numbering of the segments and joints is shown on the left and right, respectively. *B*, Diagram illustrating the morphology of MROs and the nerve supply to the two parts of each MRO. The Afferent 1 nerve (*AFF 1*) supplies the lateral fibers of the MRO, and the Afferent 2 nerve (*AFF 2*) supplies the medial fibers. (From Bowerman 1972a)

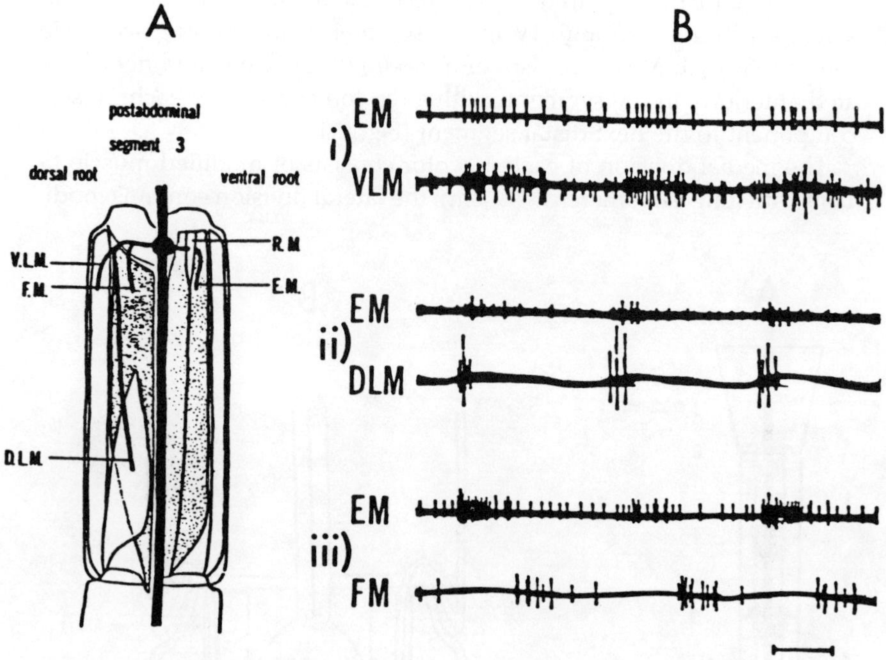

Fig. 9.12. Physiology of metasomal-muscle receptor organs (MROs). *A*, Diagram from metasomal segment III, illustrating the sites of motor-output recording for the study of reflexes mediated by MRO stimulation. The ventral nerve root, shown only on the right side of the diagram, contains the extensor-muscle and receptor-muscle motoneurons. The dorsal nerve root, shown only on the left side of the diagram, contains motoneurons to the flexor muscles, the dorsolateral muscles, and the ventrolateral muscles. *B*, Electrical recordings from the metasomal muscles during stretch and release of the ipsilateral MRO. Recordings from the extensor muscles (*EM*) are shown in the upper trace of each record. The lower trace represents recordings from the ventrolateral muscles (*VLM*) in *i*, dorsolateral muscles (*DLM*) in *ii*, and the flexor muscles (*FM*) in *iii*. The oscillations in the lower trace also provide a monitor of the MRO stretch and release, with an upward deflection indicating stretch. The magnitude of each reflex is variable, probably dependent upon the level of MRO input as well as upon the central state of excitability. Bar: 200 msec. (From Bowerman 1972b)

fied muscle fibers with short sarcomeres (3–4 μm). In addition to this difference, the two divisions also have different innervations and sensitivities to movements of adjacent metasomal segments.

The Afferent-1 nerve supply passes to the lateral (short-sarcomere) muscle division (Fig. 9.12*A*), and consists of approximately ten to twelve cells distinguishable in electrical recordings on the basis of spike height and response patterns. This Afferent-1 system responds to a variety of imposed movements, including flexion, extension, and rotation of the proximal (next anterior) joint, and to imposed stretching and release of the MRO tendon. The Afferent-2 nerve, which appears to innervate the medial (long-sarcomere) muscle division, is composed of approximately six sensory cells. It responds only to movements of the distal (next posterior) segment. In the Afferent-1 system, a single tonically active unit was identified; its activity was proportional to the degree of MRO stretch (i.e., flexion of the distal segment). Six other Afferent-1 tonic units were found to be sensitive to MRO relaxation (i.e., extension of the distal segment). In addition, a phasically sensitive stretch unit, three phasically sensitive relaxation units, and several units that were sensitive to movements of either the distal or proximal segments were described. In the Afferent-2 system, all units were phasic-tonic, and responded only to stretching of the MRO by flexion of the next distal segment. It is therefore apparent that each MRO responds to complex changes in movement of the adjacent segment, and that MROs probably are important proprioceptors for fine control of movement of the metasoma.

In his analysis of muscle reflexes evoked by stretching and releasing the MRO tendon, Bowerman (1972b) found that unilateral stretching of an MRO elicited bilateral reflex excitation of the major tail muscles in that segment (Fig. 9.12*B*). This effect was enhanced by bilateral activation of both MROs, and appeared to be mediated almost exclusively by the Afferent-2 sensory system (which, as described above, consists only of phasic-tonic stretch-sensitive units). Bowerman suggests that this system is utilized as a negative-feedback reflex system to maintain the position of the metasoma at each joint by resisting imposed flexion of the next distal segment.

The response patterns of the single tonically sensitive stretch unit in the Afferent-1 system, and of the phasic-tonic stretch-sensitive units of the Afferent-2 system, reminded Finlayson (1976) of a lepidopteran MRO. In addition, although there are some differences, the dual organization and sensitivity of the scorpion metasomal MRO and its function in maintaining position of the tail is similar to that of the crayfish abdominal MRO. Thus the design of this proprioceptive system is fundamentally similar in the three major arthropod groups.

Pectines

Despite many different attempts to discover the function of the pectines, these unique scorpion appendages have remained enigmatic for a century. The pectines are paired comblike appendages, originating from the ventral surface near the coxae of the last pair of legs. Each consists of a long basal element to which a row of teeth are attached. On these teeth are cylindrical projections, termed sensory pegs, which have been the subject of most of the recent work. Carthy (1966, 1968) described the fine structure of sensory pegs (Fig. 9.13): each consists of a thin cuticular cylinder projecting above the surface of the cuticle, and is supplied by numerous sensory cells. The hypodermal cell, lying beneath the cuticular surface of the peg, has a complex membrane concentric with the peg cuticle, but with numerous pleats, to which the sensory cells are attached by way of ciliary strands. Surrounding the sensory cells are many sheath cells, having at their apical ends microtubules that contact the cuticular cylinder or the pleated membrane. Carthy suggested that these sheath cells may produce components of the membrane near the surface. Bending of the cuticle projecting above the surface was theorized to cause deformation of the pleated membrane and therefore stretching of the ciliary projections.

Although much of the earlier speculation about the function of the pectines has been refined, it still seems possible at this point that they serve more than one purpose. The literature on pectines is incredibly frustrating and confusing: it is not my purpose here to provide a critical review of all this work, but rather to illustrate the variety of approaches to study of the pectines, and the variety of opinions regarding their functions. This subject has been reviewed by Millot and Vachon (1949), Cloudsley-Thompson (1955), and Carthy (1968); interested readers are referred to those reports.

Cloudsley-Thompson (1955) has written an especially readable and entertaining account of the early work on pectines, and provides a list of the functions attributed to them by various investigators. These include tactile sensitivity (Pocock 1893b), equilibration and audition (Gaskell 1902), chemoreception and stimulation during mating (Schroeder 1908), respiration (Ubisch 1922), braces during live birth (Schultze 1927; but males also have pectines), olfaction (Lawrence 1953), and copulation

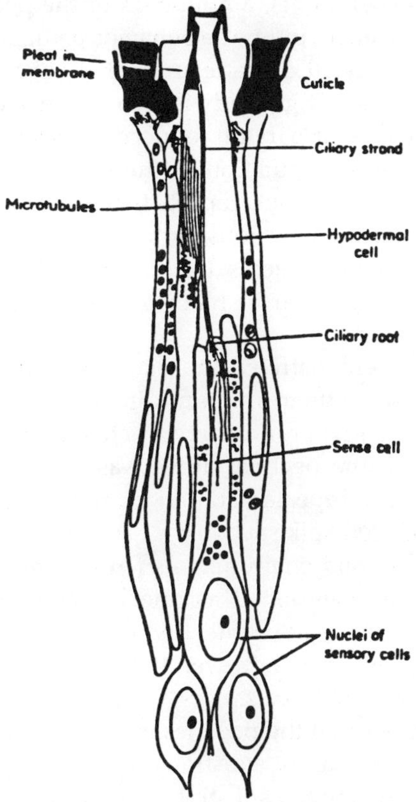

Fig. 9.13. Diagram of the structure of a sensory peg from the pectines of *Leiurus quinquestriatus*. The exposed portion of the peg is shown at the top of the diagram as a thin-walled fold in the cuticle of the pecten. Attached to the peg is a pleated membrane beneath the cuticle. One pleat is diagrammed, and the attachment of a ciliary strand from a sensory cell is shown. In addition to the hypodermal cell and the sensory cells, a number of sheath cells are drawn, with connections through microtubules to the cuticular surface and the membrane pleat. (From Carthy 1968)

(Lankester 1883). In some species, the form of the pectines may be different in females and males. Undoubtedly, the location of the pectines near the genital aperture and book lungs, their contact with the substrate during walking, and the response of scorpions to ground vibrations have led to some of this speculation on pectinal function. Unfortunately, very little of the early work was supported by physiological data. Much of this work was confined to behavioral observation; and in some subsequent experiments, amputation of the pectines appears to have had no adverse effect upon the behavior being examined.

Cloudsley-Thompson (1955) made preliminary observations suggesting that the pectines detect ground vibrations, since covering the pectines with cellulose glue or paraffin reduced the scorpion's response to ground vibrations, and amputation of the pectines abolished the response. Carthy (1968), however, reports that scorpions with their pectines amputated remained responsive to ground vibrations. As was mentioned earlier, it now seems likely that a basitarsal compound slit sensillum (BCSS) and tarsal hairs on the leg mediate vibration sensitivity.

Hoffmann (1964) demonstrated the mechanosensitivity of the pectinal pegs by stimulating them and recording their electrical responses. No response was seen when a variety of chemicals or water was applied, but activity in the pectinal nerve was dramatic when the peg projections were bent. Repeated stimuli up to frequencies of approximately 200 Hz produced spike activity.

Scorpions prefer humid environments (Abushama 1964); hence humidity reception was mentioned as one possible pectinal function, since the pectines often make contact with the substrate during locomotion. As noted previously, however, the evidence for this sensitivity is contradictory (see "Chemoreceptors," above).

Carthy (1968) argues that the pectines of *Leiurus quinquestriatus* function in the selection of sandy substrates with a particular grain size. This species is known to prefer substrates with fine textures (Abushama 1964). The mechanosensitivity and fine structure of the pegs would seem to support Carthy's contention. He proposes that the pegs on each tooth comprise a sensory field, and that, depending upon the size of the area that is stimulated, a scorpion should be able to determine whether a single large grain of sand or a number of smaller ones is stimulating a sensory field. A recent preliminary report confirms the observations of Abushama and Carthy, demonstrating a sand preference in scorpions, and an altered preference when the pectines were covered with wax (Harrington and Root 1983). In addition, Ivanov and Balashov (1979) and Foelix and Müller-Vorholt (1983) recently examined the structure of the pectines in more detail. They found that structurally the pectinal pegs may be capable of both a chemoreceptive and a mechanoreceptive function.

There is certainly more theoretical speculation than strong, convincing data for most of the suggested pectinal functions. Convincing information would best be provided by a combination of well-controlled physiological and behavioral studies.

Motor Systems

We are currently enjoying a marked advance in our understanding of invertebrate motor systems, because invertebrate preparations have recently become more common and favored tools for studying the neural control of behavior. The motor system (muscles, motoneurons, and related components controlling movement) is often the logical site to begin studying the physiology of a behavioral pattern. As a result, the organization and actions of muscles, neuromuscular function, and the structure of motor programs for invertebrate behaviors are becoming better understood, particularly such simple behaviors as locomotion, feeding, escape, and defense (see Kennedy and Davis 1977 for a review).

Much attention to behavior and motor-system function has been devoted to arthropod preparations, but primarily to insects and crustaceans rather than to arachnids. As a result, we have not seen a comparable increase in our understanding of arachnid motor systems. Several scorpion behaviors, however, are being examined, and perhaps because of this work a better understanding of the motor system will follow. For example, orientation toward prey was described earlier (see "*Tarsal sense organs and prey-detection behavior*," above); from this work it appears that two sets of leg receptors detect substrate vibrations caused by nearby prey. The actual stinging movements during prey capture may be controlled by giant neurons in the scorpion central nervous system (as described in "*Giant neurons and the control of stinging behavior*," below). Unfortunately, very little is known about the motor systems in these two behaviors.

The system controlling leg movements during walking is probably the best-understood scorpion motor system at this time, since the control of locomotion has recently been studied quite intensely. This work, however, is also in the initial stages. In the present paper, following a general discussion of scorpion muscles and neuromuscular systems, the control of scorpion locomotion will be discussed to illustrate some features of motor-system organization and function.

Muscular System

In general, the organization of the scorpion muscular system is quite insectlike. Both the appendicular and trunk muscles are grouped into antagonists that are attached directly to exoskeletal elements, apodemes, or, in the prosoma, an endosternal plate. In most cases muscle segmentation roughly corresponds to the segmental organization of the body and appendages. The fusion of the cephalic and thoracic regions of the prosoma, however, has caused the musculature in that region to become complex. The early descriptions of muscles provided by Lankester (1884, 1885b), Beck (1885), and Lankester et al. (1885) are accurate. A good general account of the scorpion muscular system is provided by Snodgrass (1952), and the reader is referred there for further information.

Three specific muscular systems should be considered further, since they have been the subject of physiological work. The muscles of the pedipalp chela (claw), the muscles of the metasoma (tail), and the muscles of the walking legs have been studied in greater depth than other scorpion muscles.

The movable finger, or tarsus, of the pedipalp chela is closed by two muscles, one that has its origin in the tibia, and another that has its origin in the patella and sends a long tendon through the tibia to the tarsus (Snodgrass 1952). Both muscles insert upon an apodeme of the tarsus. There are no opener muscles for the tibia, and it is believed that the chela opens as a result of closer-muscle relaxation accompanied by elasticity of the cuticle at the inner angles of the hinge joint between the tibia and tarsus (Mathew 1965, Alexander 1967).

The musculature of the metasomal segments was described by Bowerman (1972b), and the muscles of the telson (sting) were described by Mazurkiewicz and Bertke (1972). Metasomal segments I through IV each contain pairs of flexor muscles, extensor muscles, ventrolateral muscles, and dorsolateral muscles. These muscles control flexion, extension, and rotational movements of each joint. Reflex excitation of these muscles by paired muscle receptor organs (MROs) located in each segment was investigated by Bowerman (1972a, b; see above, "*Metasomal-muscle receptor organs*"). Mazurkiewicz and Bertke (1972) report that venom from the ducts of paired venom glands in the telson is released by a muscle that separates the two glands, causing the contents of the ducts to be expelled from the aculeus (sting). They also observed that the venom gland was enclosed by an epithelium that had close dendritic contacts, suggesting that myoepithelial cells may assist contraction of the gland; but this possibility was not investigated further.

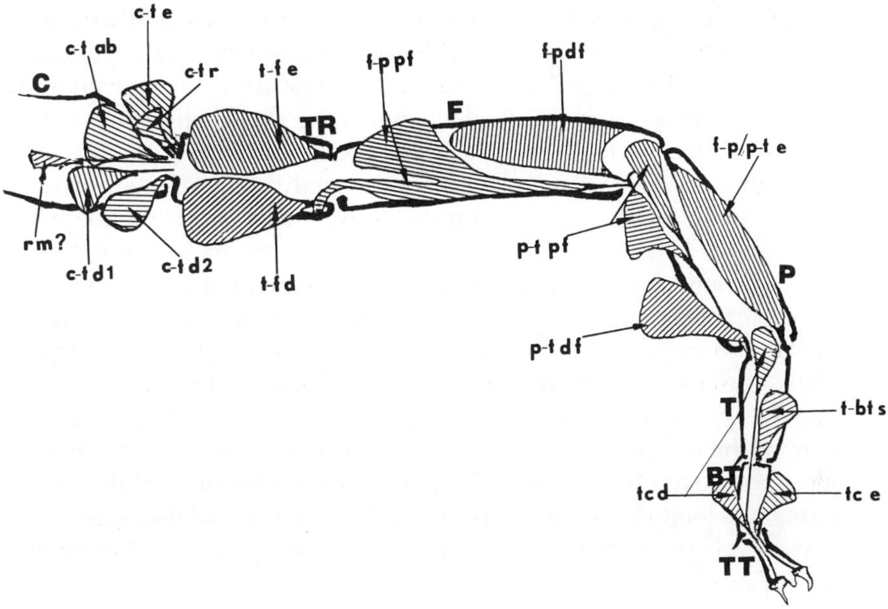

Fig. 9.14. Schematic diagram illustrating the leg muscles of *Paruroctonus mesaensis*. The muscles are shown only in their approximate shape and location. Leg segments: *C*, coxa; *TR*, trochanter; *F*, femur; *P*, patella; *T*, tibia; *BT*, basitarsus; *TT*, tarsus (telotarsus). Muscles: *c-t ab*, coxa-trochanter abductor; *c-t d1*, coxa-trochanter depressor 1; *c-t d2*, coxa-trochanter depressor 2; *c-t e*, coxa-trochanter elevator; *c-t r*, coxa-trochanter rotator; *f-p df*, femur-patella distal flexor; *f-p pf*, femur-patella proximal flexor; *f-p/p-t e*, femur-patella/patella-tibia extensor; *p-t df*, patella-tibia distal flexor; *p-t pf*, patella-tibia proximal flexor; *rm?*, receptor muscle?; *t-bt s*, tibia-basitarsus straightener; *tc d*, tarsal claw depressor; *tc e*, tarsal claw elevator; *t-f d*, trochanter-femur depressor; *t-f e*, trochanter-femur elevator.

The muscles of all eight legs are similar, reflecting the homologous segmentation of all the legs. Because the terminology for arachnid leg segments is quite confusing, Couzijn (1976) reviewed the morphology and homologies of scorpion leg segments (see also Chapter 2). He proposed that they be named as follows, proceeding distally: coxa, trochanter, femur, patella, tibia, tarsus (composed of a basitarsus and telotarsus), and apotele (tarsal claw). This scheme is used here, with exceptions or alternative names noted where applicable (see Fig. 9.14).

Some spiders lack extensor muscles at certain leg joints, and a hydraulic mechanism for leg extension has been proposed (Parry and Brown 1959a, b). There is little evidence that scorpion legs are similar. The leg is well equipped with muscles (Manton 1958, Bowerman and Root 1978). Bowerman and Root examined the musculature and external morphology of the legs in *Hadrurus arizonensis* (Fig. 9.14). The coxa is the first segment of each leg, but it is rigidly attached to the body, providing no movement, so no muscles are found operating the joint between the body and coxal segments. The coxa-trochanter joint, however, is monocondylic, permitting a wide range of movements, and six separate muscles control it. All remaining joints are planar-hinge joints, and are operated by a simple pair or a small group of antagonistic muscles, save for the tibia-basitarsus joint, which has only one muscle. This joint is thought to be homologous with the tibia-tarsus joint of the pedipalp chela (see above), which also has only one movement controlled by muscles. The antagonistic movement of the tibia-basitarsus joint of the leg is also possibly caused by joint-cuticle elasticity, perhaps in conjunction with hydrostatic pressure (Alexander 1967, Couzijn 1976).

Muscle Structure and Neuromuscular Physiology

Scorpion neuromuscular systems are poorly understood, as is true for chelicerates in general. Only one review of chelicerate neuromuscular systems has been published (Fourtner and Sherman 1973). In that review, the authors pointed out a possible difference betwen the neuromuscular systems of chelicerates and those of the better-understood crustaceans and insects (mandibulates). It appears from the limited amount of information available on chelicerates that their muscles and neuromuscular systems may, in general, be more primitive and more simply organized than the mandibulates'.

Crustacean and insect muscles are known to be extremely diverse. A wide range of muscle-fiber types has been described, and there is usually a correlation among their tension properties, innervation, and structural features, such as sarcomere length. For example, phasic ("fast") and tonic ("slow") fibers have been clearly defined in a variety of animals. Phasic fibers are typically characterized by short sarcomeres (2–4 μm), rapid "twitch" contractions, and large excitatory postsynaptic potentials (EPSPs). In contrast, tonic fibers typically have longer sarcomeres (6–14 μm), slow, "nontwitch" contractions, and small-amplitude, facilitating EPSPs. Even more revealing is the fact that some mandibulate muscles consist of a range of fiber types with diverse structural and mechanical properties, either intermixed within the muscle or segregated into populations of similar types (Dorai Raj and Cohen 1964, Jahromi and Atwood 1969, Hoyle and Burrows 1973, Hoyle 1978). Therefore, compared to those of other invertebrates, insect and crustacean muscles can be very specialized.

Diversity of muscle-fiber structure has not been demonstrated in most chelicerates, and the few studies of scorpion muscle structure seem to support the idea that individual chelicerate muscles are indeed composed of structurally similar fibers (Fourtner and Sherman 1973). Auber (1963a, b), Fountain (1970), Mazurkiewicz and Bertke (1972), Bowerman (1972a), and Gilai and Parnas (1972) reported that single scorpion muscles were composed of fibers with little variability in sarcomere length. For the muscles examined, sarcomere length ranged from 2 to 10 μm. In another study, Root and Bowerman (1981) examined seven different leg muscles, and reported that each muscle was composed of fibers with similar structural properties. The sarcomere lengths for all seven muscles ranged from only 2 to 5 μm. In general, the fine structure of individual muscle fibers coincides with the fine structure of other arthropod muscle fibers. The muscles of the pedipalps and legs have tubular muscle fibers with an extensive sarcoplasmic reticulum (*SR*) and transverse tubule system (*TTS*: Figs. 9.3*D*, 9.15). Dyads are found at the sarcomere near the A band (Fig. 9.16*A*). Contractile filaments are arranged in hexagonal patterns, with from 9 to 13 thin filaments surrounding a single thick filament. The synapses between motoneurons and muscle fibers are simple contacts on the surface of the fiber (Fig. 9.16*B*). All fibers of a muscle were quite similar in these respects. Although more work on different species is necessary, these findings suggest that scorpion muscles, like other chelicerate muscles, are more nearly uniform than mandibulate muscles.

Recent evidence, however, suggests that the near-uniformity of sarcomere lengths in these muscles may be misleading. In a preliminary study of the trochanter-femur elevator and trochanter-femur depressor muscles of *Paruroctonus mesaensis*, Root and Gatwood (1981) observed clear differences between muscle fibers. When sections of these muscles were stained for three enzymes (α-glycerophosphate dehydrogenase, succinic dehydrogenase, and myosin ATPase), lightly staining, darkly staining, and intermediately staining fibers were observed. Since these

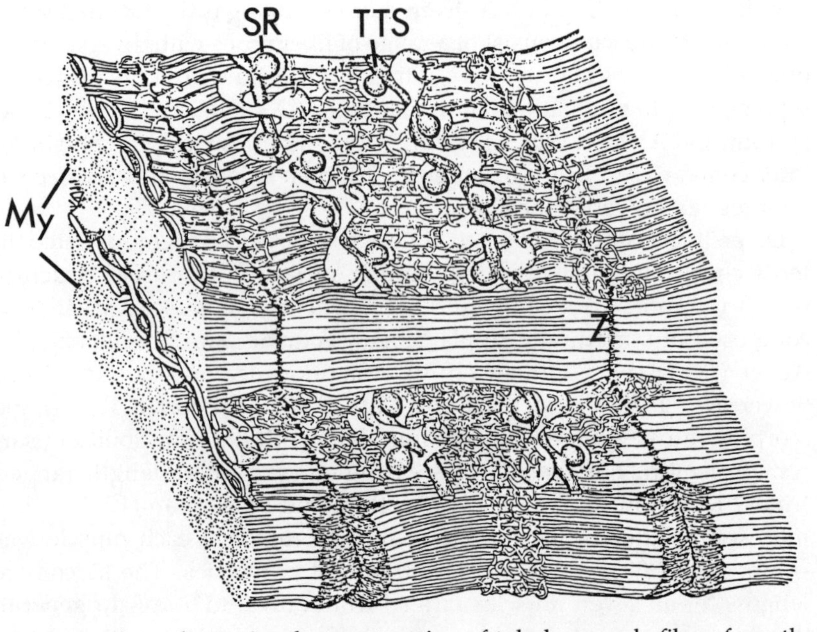

Fig. 9.15. Three-dimensional reconstruction of tubular muscle fibers from the claw-closer muscle of the pedipalp in *Leiurus quinquestriatus*. Portions of two myofibrils (*My*) are shown, one reaching the cell membrane, with a broader Z line (Z), the second cut to show the relation between the transverse tubular system (*TTS*) and the sacroplasmic reticulum (*SR*). (From Gilai & Parnas 1972)

enzyme differences may reflect different abilities to produce tension and to provide energy during contraction, muscle-fiber function may be more diverse than is indicated by fine structure.

Some preliminary work has also been done on the biochemistry of scorpion muscles. Vijayalakshimi and Kurup (1976) described various free amino acids and proteins in scorpion muscle, and there is evidence for both glycolytic and Krebs-metabolic pathways (Sinha 1966, Kaur and Kanungo 1967, Devarajulu Naidu and Venkatachari 1974, Satyanarayanam et al. 1975, Jayaram et al. 1978).

Despite the structural uniformity of scorpion muscles, different motoneurons have been found. Electrophysiological studies indicate that different types of postsynaptic potentials (PSPs) can be elicited at different thresholds with electrical stimulation of motor nerves. Fountain

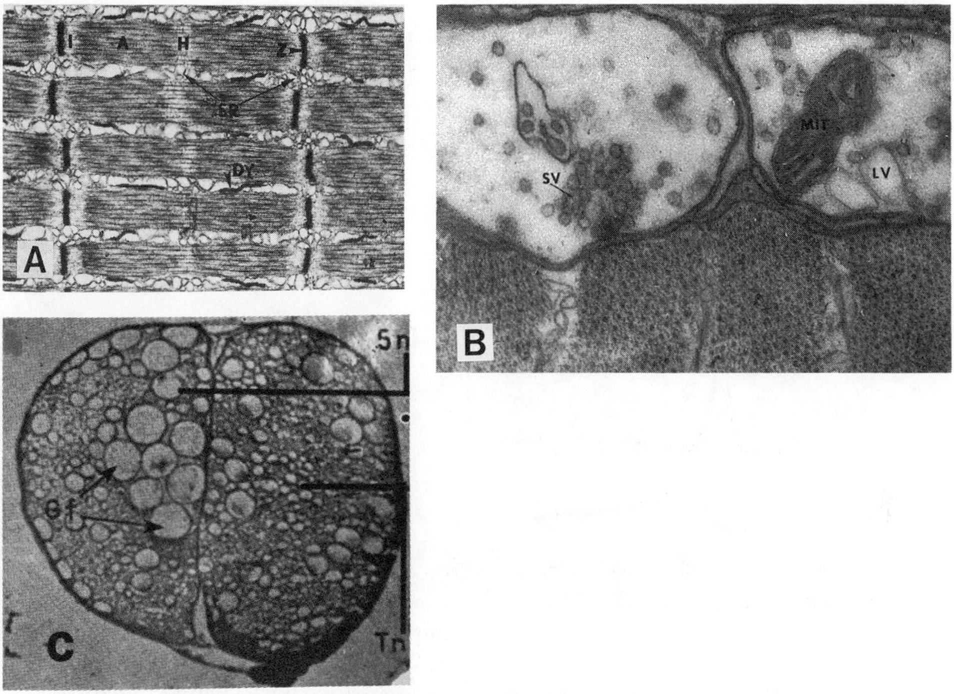

Fig. 9.16. Micrographic views of muscle-fiber, neuromuscular-junction, and nerve-fiber structure. *A*, Electron micrograph of a muscle fiber in longitudinal section from *Paruroctonus mesaensis*. Individual sarcomeres are shown, with the Z line, A, I, and H bands indicated. Sarcoplasmic reticulum (*SR*) intervenes between the myofibrils, and dyads (*DY*) can be seen near the mid-A-band region. (×11,600) (Reprinted with permission from Root and Bowerman 1981) *B*, Electron micrograph from a leg muscle in *P. mesaensis*, showing two contiguous axons on the surface of a muscle with nearby myofilaments cut in cross section. These neuromuscular junctions show the simple structure typical of arthropod motoneuron contacts. In one axon, synaptic vesicles (*SV*) can be seen grouped near the presynaptic membrane. In the other axon, a mitochondrion (*MIT*) and two large vesicles (*LV*) are also visible. (×43,500) *C*, Transverse section of the fifth segmental nerve (*5n*) on the left and telsonic nerve (*Tn*) on the right, from the seventh abdominal ganglion, showing the location of some giant fibers (*Gf*) in nerve 5n. (×400) (From Yellamma et al. 1980)

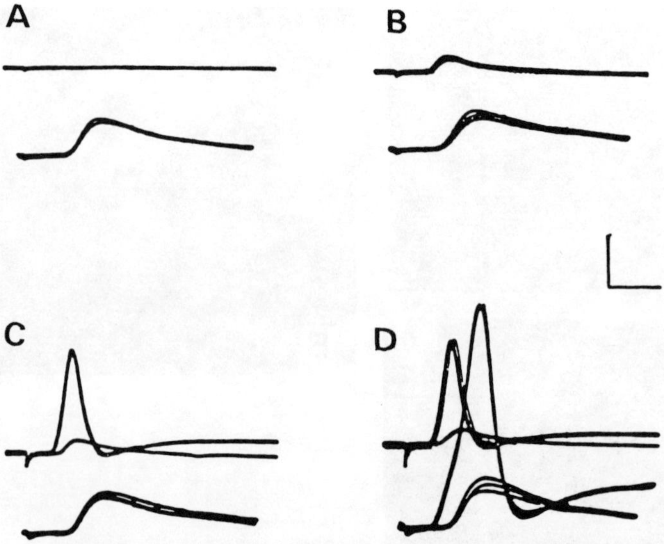

Fig. 9.17. Intracellular electrical recordings from fibers of the pedipalp-chela long closer muscle (LCM) of *Leiurus quinquestriatus* with electrical stimulation of motor nerves. Each frame shows superimposed traces from a fiber in bundle 1 (LCM1) in the lower trace, and bundle 2 (LCM2) in the upper trace. *A*, Stimulation of the Type A axon induces a local excitatory postsynaptic potential (EPSP) only in LCM1. *B*, A small increase in the stimulus intensity elicits a similar response in LCM2. *C*, A further increase in stimulus intensity activates a Type B axon, which elicits a spikelike response (SLR) in LCM2. *D*, A further increase in stimulus intensity activates another Type B axon, causing an SLR in LCM1. Horizontal bar, 10 msec. Vertical bar: upper trace, 20 mV; lower trace, 8 mV. (From Gilai & Parnas 1970)

(1970) gave evidence for three distinct excitatory motoneurons supplying the tibia extensor muscle in the leg of the scorpion *Centruroides gracilis*. Gilai and Parnas (1972) studied the two pedipalp-chela closer muscles of the scorpion *Leiurus quinquestriatus*, and found that each muscle received four to six excitatory motoneurons. Each muscle fiber was innervated by two of these motoneurons. One of them caused a small EPSP, and the other induced a large spikelike potential (Fig. 9.17). Root and Bowerman (1979) described two different motoneurons innervating fibers of certain leg muscles in *P. mesaensis* (Fig. 9.18), similar to those observed by Gilai and Parnas. One of these motoneurons elicited small EPSPs, which upon repeated stimulation became a graded-

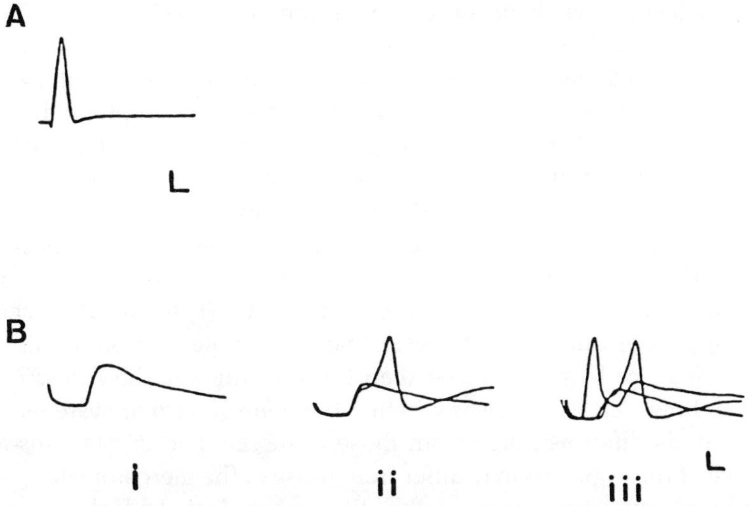

Fig. 9.18. Intracellular electrical recordings from muscle fibers of the leg femur–patella proximal flexor muscle in *Paruroctonus mesaensis*. Two different motoneurons could be distinguished on the basis of threshold, latency, and synaptic potentials. *A*, Some muscle fibers were innervated only by a single fast motoneuron, which produced a spike potential of constant amplitude and wave form with a very short latency from stimulation. *B*, Other muscle fibers were innervated by both a single slow motoneuron and a single fast motoneuron. At low thresholds, the slow motoneuron elicited a small depolarizing potential (*i*). With repeated stimulation at the same intensity, this potential facilitated into a graded spike (*ii*). At higher intensities, a fast motoneuron was also stimulated, which elicited a spike in addition to the graded spike potential of the slow motoneuron (*iii*). Horizontal bars, 5 msec; vertical bars, 10mV. (From Root 1979)

spike potential (similar graded spikes have been observed in *Limulus* muscles: see Fourtner and Pax 1972). The other motoneuron caused large nonfacilitating spikes. Gilly and Scheuer (1984) studied the relationship between the generation of muscle spikes and muscle contraction in the pedipalp long-closer muscle of three different scorpions (*Uroctonus mordax, Vaejovis glimmei, C. exilicauda*). The authors found that the muscle spikes were basically similar to those of vertebrates, but, as in many other invertebrates, the scorpion muscle required extracellular calcium for normal contraction. The authors suggested that extracellular calcium may be necessary to trigger the release of intracellular calcium in the sarcoplasmic reticulum, the step that initiates the molecular events of contraction in the muscle cell.

Another striking difference between the mandibulate and chelicerate neuromuscular systems is the occurrence of peripheral inhibitory motoneurons. Although peripheral inhibition is a common phenomenon in crustaceans and insects, it is surprisingly rare among the arachnids, having been observed in only three studies of spiders (Brenner 1972, Ruhland 1976, Maier et al. 1987); but voltage-clamping, transmitter-substance, and ion-substitution experiments have not yet been performed. Finally, chelicerate muscles tend to be innervated by more motoneurons than the muscles of mandibulates (Fourtner and Sherman 1973). These differences suggest that chelicerate neuromuscular systems may be less specialized than those of the mandibulates. At this time, however, it is unclear whether the neuromuscular systems of the arachnids differ as much from those of insects and crustaceans as all three of these groups may differ from those of the merostomates (horseshoe crabs), since many of the differences described above pertain more to the better-studied merostomates than to the chelicerates as a group. The uncertainty is doubtless due to a lack of information about arachnids, and in particular about non-araneid neuromuscular systems.

Therefore, although muscle fibers of scorpions may be structurally less diverse than those of insects and crustaceans, there may be a comparable level of biochemical and neuromuscular diversity. Unfortunately, since very few muscles have been studied adequately in scorpions, this possibility remains untested. Much of the work on scorpion motor systems is preliminary, so it is really impossible to judge the similarities or differences between scorpion muscles and those of other arthropods.

Neural Control of Locomotion

In the study of the neural control of behavior, invertebrate preparations have recently become particularly valuable, for they often offer the investigator the advantages of economy, simplicity, and durability. Locomotion studies have been especially fruitful, since the cyclic repetition of leg movements affords the investigator an opportunity to collect many cycles of similar behavior over short periods of time. Accordingly, important advances in our understanding of how nervous systems function to produce and modify behavior have been made by examining locomotion in such invertebrates as the locust, stick insect, cockroach, crayfish, and crab. Although work on arachnids has been limited, through recent work we now have information on locomotion in the scorpion (see Root 1985 for a review).

DESCRIPTION OF LOCOMOTION. Beginning with behavioral studies of locomotion in normal scorpions, Bowerman (1975a) was able to describe the stepping pattern of the legs during forward walking, backward walking, stopping, starting, and turning. Most of the following information is based on data obtained during normal forward walking. High-speed cinematography showed that during walking, the scorpion moves its eight legs in a pattern that Bowerman termed "alternating tetrapods." In this pattern, legs I and III on one side and legs II and IV on the opposite side step together while the remaining four legs remain on the substrate; then the cycle is reversed. Thus legs L-I, L-III, R-II, and R-IV usually comprise one tetrapod, and legs R-I, R-III, L-II, and L-IV comprise the alternate tetrapod. When he examined the latency for stepping between legs, Bowerman observed slight delays between the legs of a stepping tetrapod. As a result, the four legs of an alternating tetrapod do not all move at the same instant, but are delayed slightly to provide a smooth, stable movement (Fig. 9.19*A*). This pattern contrasts with the strict alternation of leg sets seen in some insects, which produces quite jerky movements (Graham 1972, Burns 1973). Studies of stepping patterns in other arachnids have shown basic similarity to the stepping patterns of scorpions (Wilson 1967, Land 1972, Moffett and Doell 1980; see Bowerman 1977 for a review).

Each leg of the scorpion goes through a similar cycle of movement: a swing phase, during which the leg is brought up and swings forward, and a stance phase, during which the leg is placed down on the substrate and pushes rearward. Despite this basic similarity in movement, Root and Bowerman (1978) found that each leg accomplished its cycle differently (Fig. 9.19*B*). As it is brought forward, leg I steps quite high, almost parallel to the ground, before being placed down. Leg II also steps high, but legs III and IV remain close to the ground as they are brought forward. Raising legs I and II may assist their tactile function when objects are in the way. The stepping ranges for the legs also describe separate paths alongside the body when viewed from above (Fig. 9.19*C*). These leg placements provide a very stable arrangement during forward walking in an animal that must balance a long pair of pedipalps anteriorly and a long tail posteriorly. Root and Bowerman also measured changes in joint angles for the major leg joints during

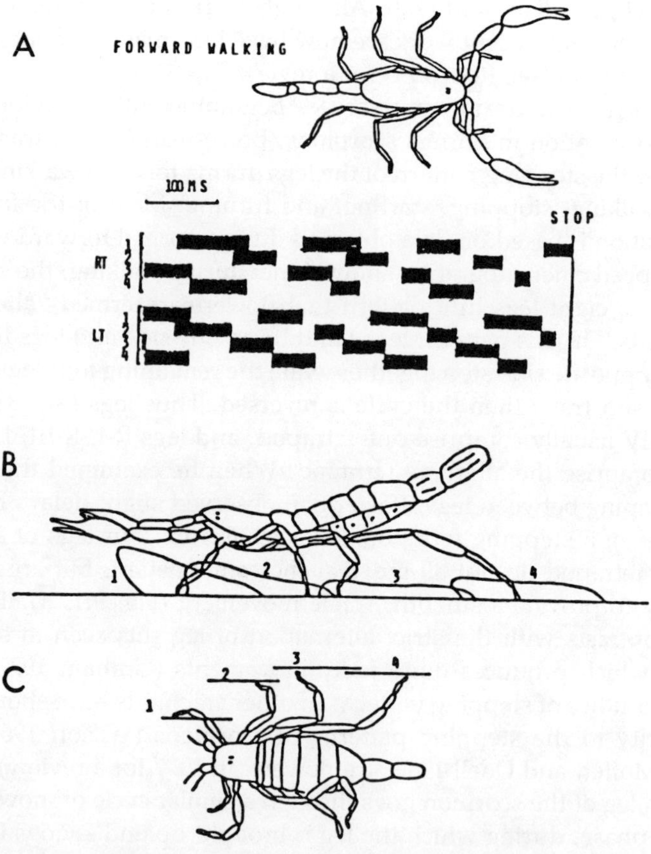

Fig. 9.19. Behavioral analysis of locomotion in the scorpions *Hadrurus arizonensis* and *Paruroctonus mesaensis*. *A*, A sequence of forward walking, with the durations of leg contact with the substrate indicated by white spaces, and the intervening dark blocks indicating the duration each leg is off the substrate. For most of the time, the animal is supported by four legs, either legs Right I and III and Left II and IV, or Right II and IV and Left I and III (alternating tetrapods). (R. Bowerman, unpubl.) *B*, Lateral view, illustrating the movements of four legs. Leg I, and to a lesser extent leg II, is raised upward as it is brought forward. Legs III and IV remain close to the surface of the substrate. (Reprinted with permission from Root & Bowerman 1978) *C*, Overhead view, illustrating the pathways of movement for four legs. Each leg maintains a roughly linear path parallel to the body and not interfering with adjacent legs. (From Root & Bowerman 1978)

walking, and found that some joint movements are quite different for each of the four legs on a side, reflecting the different way each leg steps. For example, the principal joints of the first leg extend during the swing phase, whereas those of the second leg initially flex, then extend. The third and fourth legs both flex during their swing phases, though to different extents. During the stance phase, the first leg is flexed rearward, but the second leg is first flexed and then extended; and the third and fourth legs are extended.

In addition to describing locomotion behavior, these behavioral studies have allowed some conceptualization of the neural mechanisms underlying the control of locomotion. Each leg goes through a unique pattern of movement, suggesting that separate neural oscillators control each leg. The strong coordination between adjacent pairs of legs, however, implies that the oscillators for adjacent legs are tightly coupled. The motor program must also be able to direct homologous muscles in different legs in quite distinct actions, since joint-angle changes differ in the legs, and the neural circuitry must be able to account for slight changes in velocity, direction, or substrate texture.

This system is obviously complex: though leg movements are rather specific in normal forward walking on a smooth surface, coordination between legs changes readily during other behaviors. For example, in the stepping sequence illustrated in Figure 9.19*A*, as the scorpion stopped walking, leg R-I stopped in the middle of its swing phase, and legs R-III and L-II had shortened swing phases prior to the stop. Bowerman (1981b) observed other reorganizations of stepping movements prior to stopping, including termination of all leg movements in mid-range, and the failure of legs to step when expected. Other variations in leg stepping patterns were observed in turning movements, readjustments to a standing posture after locomotion, and backward walking. These variations suggest that the walking movements of each leg are controlled by a central oscillator, but that the coordination between legs can change upon demand from the central nervous system (CNS) for alternative behaviors. As an example, in recent studies of burrowing behavior, we found that legs I, II, and III have rhythmic movements similar to those in walking, but the order in which these legs are used depends upon the phase of burrow construction, and leg IV always remains on the ground (T. Root, pers. obs.). In addition to the central program for leg movements, it is equally clear that peripheral feedback mechanisms are also operative. When one or two legs were surgically removed in order that they could no longer contribute sensory feedback to the CNS, the scorpion was immediately able to alter the stepping relationships to produce a new stable walking pattern (Bowerman 1975b).

ELECTROMYOGRAPHIC ANALYSIS. In an effort to describe the motor program for freely walking scorpions, Bowerman (1981a) and Bowerman and Butler (1981) recorded simultaneously from the trochanter-femur depressor muscle, from the trochanter-femur elevator muscle, and from a sensory nerve containing fibers from touch receptors of the apotele (tarsal claw) on the end of the leg. Since the elevator and depressor muscles cause the major lifting and lowering movements of the leg, and since the tarsal-claw touch receptors are active only when the leg is on the ground, the resultant electrical record represents a description of the swing-phase muscle (elevator muscle) and the stance-phase muscle (depressor muscle), and the behavioral record of stepping (touch receptors). Figure 9.20*A* illustrates the general features of a recording from a freely walking animal for four cycles of movement. Depressor activity alternates with elevator activity, and, as would be expected, the tarsal-claw touch receptors are active when the depressor muscle is active (i.e., when the leg is on the ground). In over 98 percent of the sequences examined (more than 200 total), walking was always initiated by activity of the depressor muscle prior to activity in the elevator muscle. Therefore, depressor activity in all four legs of an alternating tetrapod must be initiated simultaneously, and the duration of this initial burst determines when the switch to the swing phase will begin. Over a range of walking speeds, depressor- and elevator-burst durations changed in different ways (Fig. 9.20*B*). At slow walking speeds (step-cycle durations greater than 600 msec), the elevator-burst duration remains roughly constant, but the depressor-burst duration increases considerably. At step-cycle durations less than 600 msec, both the depressor- and elevator-burst duration decrease, with the slope of the depressor burst being the greater of the two. This finding agrees with an earlier behavioral observation (Bowerman 1975a) that at high rates of stepping the duration of the stance phase decreased more than that of the swing phase.

Another difference between depressor-muscle activity and elevator-muscle activity was observed at the termination of each burst, that is, at the switch between stance and swing (Bowerman 1981a). A plot of the latency between depressor-burst termination and elevator-burst onset (switch from stance to swing) against cycle time for the depressor (Fig.

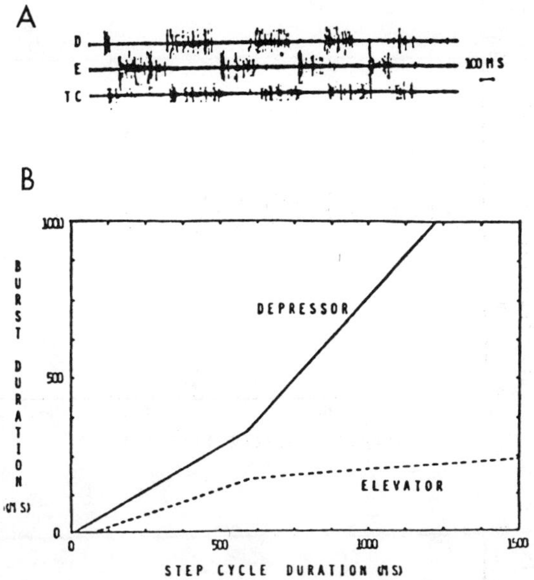

Fig. 9.20. Electromyograms (EMGs) from *Paruroctonus mesaensis* during walking. *A*, Simultaneous electrical recordings from the trochanter-femur depressor muscle (*D*, upper trace), trochanter-femur elevator muscle (*E*, middle trace) and tarsal-claw touch receptors (*TC*, bottom trace). The two muscles controlling depression and elevation of the leg exhibit alternating bursts of spikes. The tarsal-claw touch receptors monitor the time the leg is in contact with the substrate. *B*, Regression lines illustrating the relationship between depressor-EMG burst duration and step duration and between elevator-EMG burst duration and step duration. At slower rates of stepping (i.e., step durations greater than about 600 msec), the elevator burst remains roughly constant; the depressor burst changes linearly with changes in step duration. (From Bowerman 1981a)

9.21*A*) indicates that the latency is roughly constant for step-cycle times greater than 600 msec, but that it decreases for cycle times shorter than 600 msec. There is never an occasion when both the depressor and the elevator are activated together. In contrast, a plot of the latency between elevator-burst termination and depressor-burst onset (swtich from swing to stance) against cycle time for the elevator (Fig. 9.21*B*) revealed that the time for this transition did not change consistently with changes in stepping rates, and that sometimes both muscles were coactivated. Thus, the control of the motor system is asymmetrical. The switch is controlled loosely between swing and stance, but tightly between stance and swing.

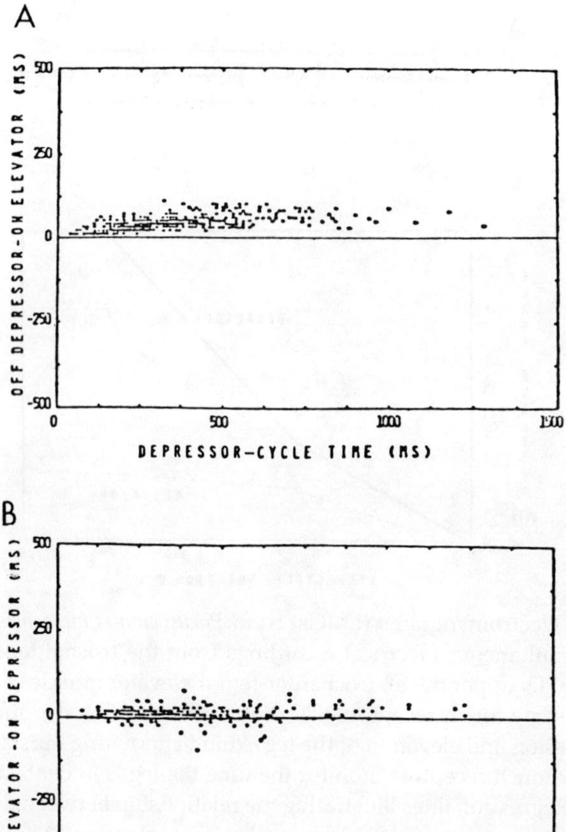

Fig. 9.21. Analysis of electromyograms (EMGs) recorded during walking in *Paruroctonus mesaensis*. *A*, Composite scattergram illustrating the relationship for latency between depressor-burst termination and elevator-burst onset, plotted against depressor-cycle duration. For step durations greater than 600 msec, the latency is generally constant. There are no periods of elevator–depressor coactivation, which would be indicated by negative values. *B*, Composite scattergram illustrating the relationship for latency between elevator-burst termination and depressor-burst onset, plotted against elevator-cycle duration. The negative points indicate occurrences of elevator–depressor coactivation. (From Bowerman 1981a)

MOTONEURON AND PROPRIOCEPTIVE ACTIVITY. In the first published account of intracellular recordings from the scorpion CNS, Bowerman and Burrows (1980) were able to categorize a number of different motoneurons from leg ganglia in the subesophageal ganglion. Many of these cells were filled with cobalt chloride to study their cytoarchitecture, and were also characterized electrophysiologically. A scorpion was pinned ventral side up, and the subesophageal ganglion exposed. Recordings were made from the cell bodies of motoneurons while the legs were free to move. Motoneurons supplying eight different muscles were categorized. Some motoneurons produced rapid twitches of the muscles they innervated; others produced slower movements, and still others produced intermediate-level movements. Apart from confirming the location of leg motoneurons, this study also revealed that motor programs can be recorded from the scorpion CNS. Therefore it is now possible to examine directly the central mechanisms responsible for locomotion. Since the leg ganglia in the scorpion lie in the large, fused subesophageal ganglion, the walking-leg control centers may be unlike those of most insects and crustaceans, in which separate thoracic ganglia control the legs or wings.

It has also been possible to study the generation of proprioceptive feedback from leg receptors during locomotion (Weltzin and Bowerman 1980; Weltzin 1981; R. Weltzin and T. Root, pers. obs.). Electrical activity in the dorsal leg nerve, which is primarily a sensory nerve, was recorded with fine-wire electrodes simultaneously with high-speed cinematography of joint movements in the legs. In this way, leg-joint angles were correlated with the specific activity of sensory cells identified by spike height in the nerve record. A recording of two of these units is illustrated in Figure 9.22. In this record, unit 2 fired during extension of the femur-patella joint, and generally remained active throughout extension. Unit 3 also fired during extension, but generally only after extension was a third to a half complete. Unit 3 also responded to femur-patella flexion, and therefore exhibited bidirectional sensitivity. These findings agree with an earlier account of leg-receptor function in dissected preparations (Bowerman 1976), which indicated that receptors could provide the CNS with information about the position of leg joints and their direction and rate of movement. The later studies of Weltzin and Bowerman (1980) and Weltzin (1981), however, indicated the danger of extrapolating electrophysiological data from restrained and dissected preparations to freely behaving animals, since

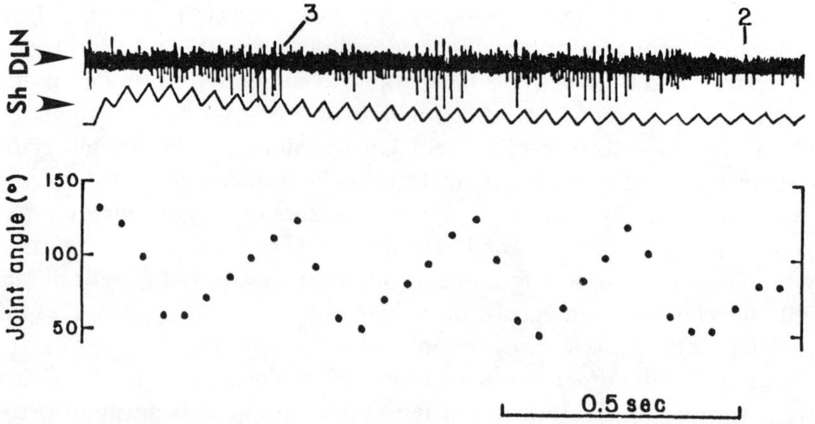

Fig. 9.22. Electrical recordings of proprioceptive activity during walking in *Paruroctonus mesaensis*. Three simultaneous traces are indicated: the top trace is a neural recording from the dorsal leg nerve (*DLN*); the middle trace is a record of shutter movements of a movie camera (*Sh*); the bottom trace is a plot of the femur-patella joint angle of leg IV from the film records (each dot represents the joint angle when the camera shutter was opened). These three simultaneous measurements allowed comparisons of proprioceptor spikes in the DLN (specific spikes of units 2 and 3 are indicated) with movements of the femur-patella joint. Both units, 2 and 3, had spike bursts during extension of the joint, though at different phases in the extension (see text). (From Weltzin 1981)

some of the sensory units recorded during imposed leg movements in a dissected preparation responded differently from those same units recorded during unrestrained walking.

Central Nervous System

The gross anatomy of the scorpion CNS has been described in basic form for well over a century. Early works on the morphology and histology of the scorpion nervous system were published by Dufour (1858), Saint-Remy (1886a, b, 1887), Police (1901, 1902, 1904), McClendon (1904), Haller (1912), and Buxton (1917), with subsequent reviews by Hanstrom (1923, 1928), Gottlieb (1926), Werner (1934), and Henry (1949). Bullock and Horridge (1965) noted some of the discrepancies in questions of nervous-system development and segmentation in these earlier reports. Babu (1961a, 1965) has published the most detailed study of scorpion nervous-system histology, utilizing reconstructions from serial sections to establish the location of major cell groups and nerve pathways in the CNS of *Heterometrus*.

The scorpion CNS is highly cephalized, but less so than in most other arachnids. It consists of a large cephalothoracic mass surrounding the gut in the prosoma, from which a long ventral nerve cord with ganglia extends almost the entire length of the animal (Fig. 9.23). In contrast, in *Limulus* and most spiders the brain region is dominant, with

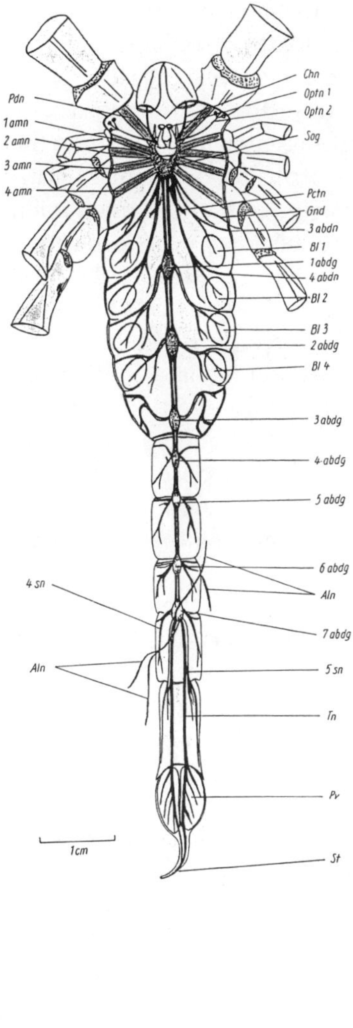

Fig. 9.23. The central nervous system of *Heterometrus* sp. At the anterior end of the scorpion lies the cephalothoracic mass, which is further illustrated in Figure 9.24. From the cephalothoracic mass arise the major nerves to the chelicerae (*Chn*, cheliceral nerve), eyes (*Optn 1*, median optic nerve; *Optn 2*, lateral optic nerve), pedipalps (*Pdn*, pedipalpal nerve), legs (*1 amn, 2 amn, 3 amn, 4 amn*), pectines (*Pctn*, pectinal nerve), genital region (*Gnd*, genital nerve), and nerves to anterior mesosomal segments including their corresponding book lungs (*3 abdn, 4 abdn*, abdominal [mesosomal] nerves 3 and 4; *Bl 1–4*, book lungs 1–4). The ventral nerve cord runs from the subesophageal ganglion (*Sog*) to the fourth metasomal segment, where it divides into two nerve branches. The nerve cord contains seven abdominal ganglia (*1 abdg – 7 abdg*, abdominal ganglia 1–7; three in the mesosoma and four in the metasoma). Each ganglion has paired segmental nerves (e.g., *4 sn*, fourth metasomal segmental nerve). The last ganglion (*7abdg*, seventh abdominal or fourth metasomal ganglion) gives rise to long nerves to the gut (*Aln*, alimentary canal nerves), the last metasomal segment (*5 sn*, fifth segmental nerve), telson (*Tn*, telsonic nerve), poison vesicle (*Pv*) and sting or aculeus (*St*). (From Babu 1965)

concomitant reduction of the ventral cord to one abdominal ganglion, or even none at all.

The scorpion CNS is ensheathed by a fibrous outer layer of connective tissue and a cellular inner perineurium (Babu and Venkatachari 1966). The connective-tissue sheath is composed of collagen fibrils embedded in a mucopolysaccharide matrix (Baccetti and Lazzeroni 1969). The perineurium acts as a blood-brain barrier, with numerous tight junctions between cells in the perineurium selectively restricting the passage of molecules into the CNS (Lane and Harrison 1980).

Anatomy of the Cephalothoracic Mass

The cephalothoracic mass is easily distinguishable in the prosoma as a large nervous-tissue mass lying ventral to the esophagus, and a smaller nervous-tissue mass lying dorsal to the esophagus. It appears that the dorsal mass develops separately from the ventral mass, and later fuses to it (Laurie 1890); but there is disagreement concerning the number of neuromeres contributing to the cephalothoracic mass. This confusion is no doubt due to unusual changes in body segmentation during development. For example, the number of precheliceral neuromeres early in development has been disputed. Lankester (1904) and McClendon (1904) believed that there was one; Brauer (1894, 1895) and Saint-Remy (1887), two; and Patten (1890), three. Further, the initial segmentation of this nervous tissue becomes obscured later in development, when the segments in the prosoma fuse. In addition, the seventh postoral segment disappears, so that the first abdominal (mesosomal) segment in the adult is, in fact, the eighth postoral segment (Snodgrass 1952). Consequently, there have been different reports concerning the segmentation of the nervous system, since the anterior neuromeres form the ventral mass. Depending upon the author and the species, from 9 to 11 anterior ganglia contribute to the ventral mass (Brauer 1894, 1895; McClendon 1904; Buxton 1917; Kästner 1940; Henry 1949).

The dorsal portion of the cephalothoracic mass is more noticeably equivalent to the "brain" of other arthropods, but comparisons of its histological structure with those of the proto-, deutero-, and tritocerebrum of other arthropods are not precise. Moreover, since arachnids lack antennae, a deuterocerebrum is not defined in the adult. As a result, many different names have been given to divisions of the cephalothoracic mass (Table 9.1). The most logical divisions, based upon histological information, seem to be as follows: the supraesophageal

TABLE 9.1
Terminology for the divisions of the cephalothoracic mass

Dorsal division	Ventral division	Reference
"Brain" = protocerebrum (supraesophageal ganglion) and tritocerebrum (cheliceral-stomatogastric region)	Subesophageal ganglion	Bullock & Horridge 1965
Supraesophageal "brain"	Subesophageal ganglion	Kästner 1940
Supraesophageal ganglion	Subesophageal ganglion	Babu 1965
Dorsal cerebral ganglion	Ventral subesophageal ganglion	Henry 1949
Cephalic or dorsopharyngeal ganglion	Thoracic or ventropharyngeal ganglion	Birula 1917a
"Brain" = proto-deutero-cerebrum and trito-cerebrum	Cephalothoracic mass	Holmgren 1916, Hanstrom 1923

NOTE: The dorsal division lies above the gut; the ventral, below. The divisions used in the present paper are those of Bullock & Horridge 1965.

ganglion is equivalent to the protocerebrum; the circumesophageal connectives are derived from the tritocerebrum; and the subesophageal ganglion, representing the fusion of several ganglia, contains neural centers controlling segments of the cephalothoracic region (Fig. 9.24*A*, *C*).

The major nerves, tracts, and neural centers of the cephalothoracic mass have been described by Babu (1965). The protocerebrum contains four bilateral optic centers. From two of these, paired median optic (ocellar) nerves pass to the median eyes dorsally (*Optn 1*, Fig. 9.24*A*, C); from the other two proceed paired lateral optic (ocellar) nerves, which subdivide and then pass to the lateral eyes (*Optn* 2, Fig. 9.24*A*, C). A single neuropile region lies at the base of each median optic nerve, but at least two neuropile regions lie at the base of each lateral optic nerve (Fig. 9.24C). Both median and lateral optic centers send tracts to the central body (*Cb*, Fig. 9.24C), which also receives inputs from other sensory and motor centers, and therefore probably functions as a sensorimotor integration center. Also within the protocerebrum are globuli-cell regions (*Gl*, Fig. 9.24C), which contribute fibers to the glomeruli and to the mushroom bodies (*Cpt*, Fig. 9.24C). The mushroom bodies are thought to correspond to the corpora pedunculata of insects. The globuli-cell regions are probably higher-order association and integration centers. Within the supraesophageal ganglion a few giant cells have also been described (Saint-Remy 1887, Gottlieb 1926,

Babu 1965), the largest of these being approximately 47 to 50 μm in diameter. These are the Type C cells of Babu (1965) and the Gl cells of Habibulla (1970). Habibulla (1971a) found that these cells are probably neurosecretory.

The largest internal centers of the tritocerebrum are the cheliceral ganglia, located in the circumesophageal region at the bases of the paired cheliceral nerves, which control the movements of the mouthparts. The single median rostral ganglion (*Rst*, Fig. 9.24*B*, *C*) has been

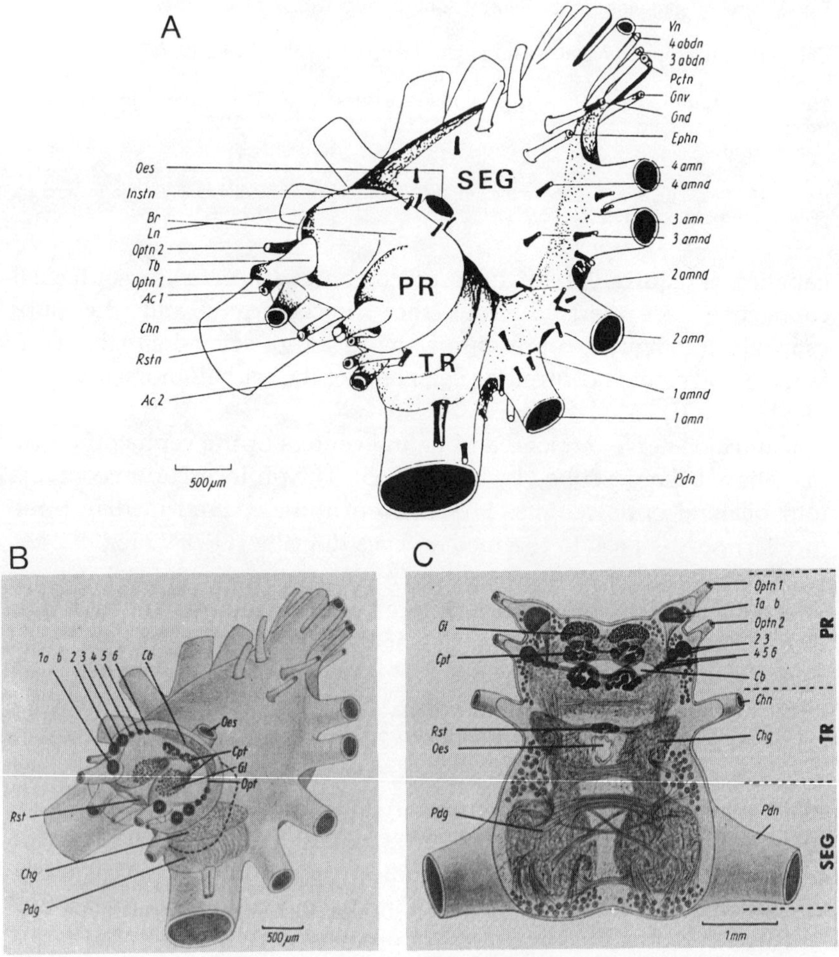

described as the sympathetic ganglion, analogous to the stomatogastric ganglion of other arthropods. It is believed to control the functions of the anterior regions of the gut. Depending upon the species, one or more pairs of stomatogastric nerves (rostral nerves) serve the anterior gut.

The subesophageal ganglion gives rise to many paired peripheral nerves, most notably large pedipalp nerves, four pairs of leg (pedal) nerves, abdominal (mesosomal) nerves, and nerves leading to the ventral nerve cord (see Fig. 9.24*A*). Other, smaller nerves from this region pass to the genital area, pectines, book lungs, and nearby muscles. Major centers in the subesophageal ganglion include cell-body collections giving rise to these nerves. Thus we find pedipalp ganglia, leg ganglia, and groups of cells at the bases of the other nerves. For the pedipalp and leg centers, separate large and small cell groups have been described. Recent studies using cobalt-chloride backfilling of leg nerves (Weltzin and Bowerman 1980) and horseradish-peroxidase labeling of motor cells from the leg (Root 1980) have confirmed that motoneurons are located in the leg ganglia. Groups of giant cells (up to 110 μm in diameter) are located in the cheliceral, pedipalp, and leg ganglia of *Heterometrus* (Babu 1965).

Fig. 9.24. The cephalothoracic mass of *Heterometrus fulvipes*. *A*, Anterodorsal oblique aspect with the three principal regions indicated (*PR*, protocerebrum; *TR*, tritocerebrum; *SEG*, subesophageal ganglion). The major nerves from these regions are indicated. *B*, Same perspective as view *A*, but with the principal neuropile masses and globuli masses of the protocerebrum and tritocerebrum indicated. In the subesophageal ganglion, only the pedipalpal ganglion (*Pdg*) is indicated. *C*, Transverse section through the anterior end of the cephalothoracic mass. The protocerebrum, tritocerebrum, and subesophageal ganglion are indicated at the right. Major cell masses, neuropile regions, and tracts are indicated. *1a*, *b*, *2–6*, median and lateral optic nerve masses; *3*, *4 abdn*, mesosomatic segmental nerves 3 and 4; *Ac 1*, *2*, accessory nerves 1 and 2; *1–4 amn*, ambulatory motor nerves (leg nerves) 1–4; *1–4 amnd*, four dorsal ambulatory nerves; *Br*, brain (supraesophageal ganglion); *Cb*, central body; *Chg*, cheliceral ganglion; *Chn*, cheliceral nerve; *Cpt*, corpora pedunculata; *Ephn*, ephemeral nerve; *Gl*, globuli cells; *Gnd*, dorsal gonadal nerve; *Gnv*, ventral gonadal nerve; *Instn*, intestinal nerve; *Ln*, lateral nerve; *Oes*, esophagus; *Opt*, optic commissure; *Optn 1*, median optic nerve; *Optn 2*, lateral optic nerve; *Pctn*, pectinal nerve; *Pdg*, pedipalpal ganglion; *Pdn*, pedipalpal nerve; *Rst*, rostral ganglion; *Rstn*, rostral nerve; *Tb*, tubercular body; *Vn*, ventral nerve cord. (From Babu 1965)

Bowerman and Burrows (1980) filled individual motoneurons in the leg-nerve centers of the subesophageal ganglion with cobalt chloride to study their cytoarchitecture. They stained 74 cells, which innervated eight different leg muscles. The distribution of these cells is shown in Figure 9.25. The cells were located in the ventral cortex of the ganglion, and were divided into two groups: a group anterior to the tract formed by fibers of the leg nerve, and a group posterior to this tract. The anterior and posterior groups of adjacent ipsilateral leg centers overlapped. The cytoarchitecture of two motoneurons is shown in Figure 9.26. The cell-body diameters ranged from 10 to 50 μm. The cell body gives rise to a single process (2–9 μm), which enters the central neuropile of the ganglion, where many parallel branches are given off. The major segment from which these branches arise is usually less than 12 μm in diameter, and tapers to approximately 2 to 5 μm prior to entering the leg nerve. It was not possible to distinguish individually identifiable cells on the basis of morphology.

Two other centers in the subesophageal ganglion should be mentioned. The central ganglia, as their name suggests, lie along the midline, near the ventral surface. They receive connections from sensory tracts arising in the pedipalps and legs, and connect with many of the major incoming and outgoing tracts in the subesophageal ganglion. The ventral association center runs along the central ganglia, and exhibits fine-fiber arborizations and complex syncytial-like contacts, especially evident near the pedipalps and legs. Although virtually nothing is known about their function, the connections from the central ganglia and ventral association center are suggestive of higher-order association and integration.

The cephalothoracic mass also contains numerous neurosecretory centers. Habibulla (1970, 1971a) mapped out the major neurosecretory regions in the CNS of the scorpion *Heterometrus swammerdami*. He identified nine groups of neurosecretory cells of the basis of their staining properties. Two groups of cells are distributed as paired sets in the brain (Fig. 9.27*A*). Seven other paired sets are metamerically arranged in the subesophageal ganglion (Fig. 9.27*B*). Since no such metameric arrangement has been described for insects, Habibulla suggests that it is primitive in the scorpion.

Anatomy of the Ventral Nerve Cord

The subesophageal ganglion is believed to represent the concentration of a number of ganglia, since the mesosomal and metasomal re-

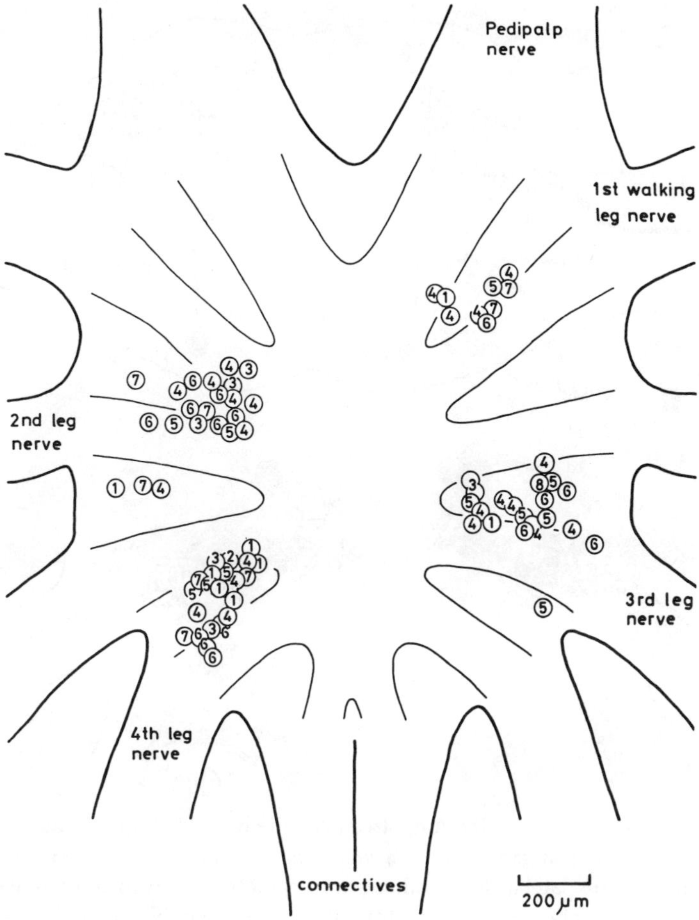

Fig. 9.25. Ventral view through the entire subesophageal ganglion of *Paruroctonus mesaensis* showing the positions of the motoneuron cell bodies to the walking legs that have been physiologically and morphologically characterized. The drawing is a composite of several whole mounts. To simplify the drawing, the motoneurons innervating the first and third leg are drawn only on the right side of the diagram; those innervating the second and fourth leg are drawn on the left side. The map shows the positions of all the neurons stained in 18 scorpions. The motoneurons are numbered according to the muscle they innervate: *1,* coxa-trochanter depressor; *2,* coxa-trochanter elevator; *3,* trochanter-femur depressor; *4,* femur-patella flexor; *5,* patella-tibia flexor; *6,* double extensor of patella and tibia; *7,* tibia-tarsus I straightener; *8,* tarsal-claw depressor. (From Bowerman & Burrows 1980)

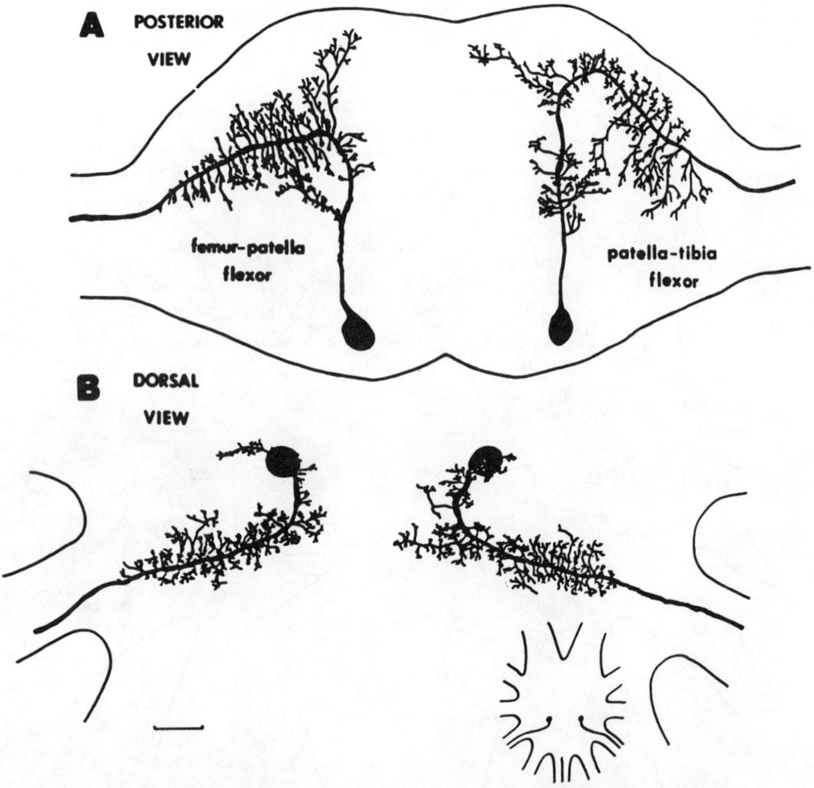

Fig. 9.26. Camera lucida drawings from cobalt-chloride-stained motoneurons in the subesophageal ganglion of *Paruroctonus mesaensis*. A motoneuron innervating the femur-patella flexor muscle and a motoneuron to the patella-tibia flexor muscle are shown. These motoneurons innervate the left and right third legs, respectively. *A*, Posterior view; *B*, dorsal view. In both views, the cell body and primary neurites and axons are shown. Calibration: 100 μm. (From Bowerman & Burrows 1980)

gions of the scorpion ventral nerve cord contain fewer ganglia than there are segments. Although there are variations depending upon the species, three ganglia are typically found in the mesosoma (preabdomen), and four in the metasoma (postabdomen, or tail). Each ganglion is a fusion of two symmetrical hemiganglia, but the connectives (paired nerves between ganglia) usually remain separated. As a result of the concentration of ganglia into the subesophageal ganglion, all but the

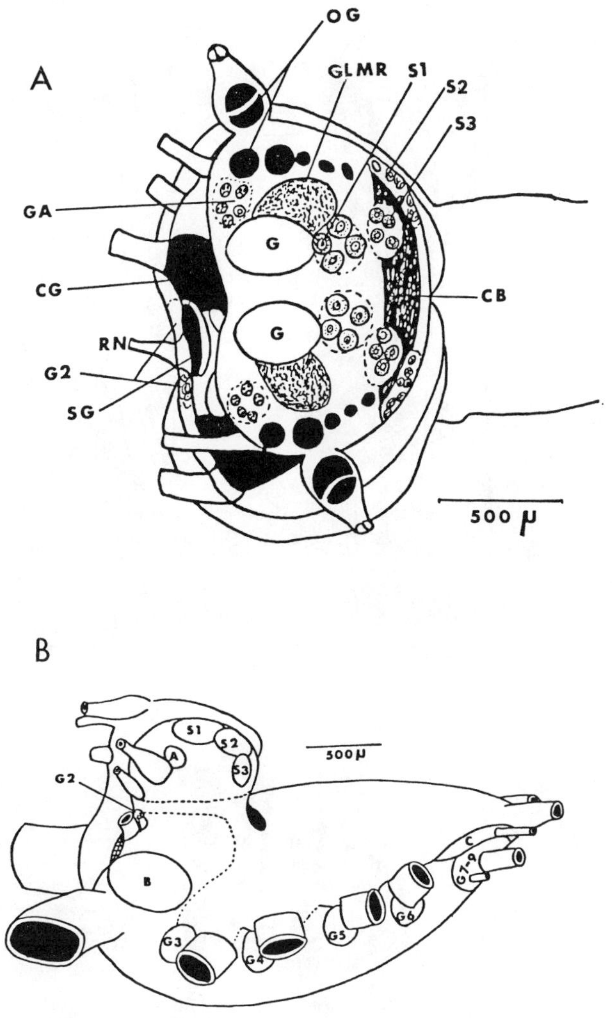

Fig. 9.27. Neurosecretory cells in the cephalothoracic mass of *Heterometrus swammerdami*. *A*, Dorsal view of cephalothoracic mass, showing the neurosecretory cells of the protocerebrum. Cell groups include those named for specific CNS regions (*OG*, optic ganglion; *GLMR*, glomeruli; *CB*, central body; *CG*, cheliceral ganglion; *SG*, sympathetic ganglion), and other nonspecific cell groups (*S1*, *S2*, *S3*, *G*, *GA*, *G2*). Also identified is the rostral nerve (*RN*). *B*, Lateral view of cephalothoracic mass, showing the location of major neurosecretory cell masses. Shown here, and also visible in view *A*, are groups G2, S1, S2, and S3 in the protocerebrum. Not shown in *A*, but visible in this view, is group A in the protocerebrum, and groups B, C, and G3–G9, which are located near the principal nerves of the subesophageal ganglion. (From Habibulla 1970)

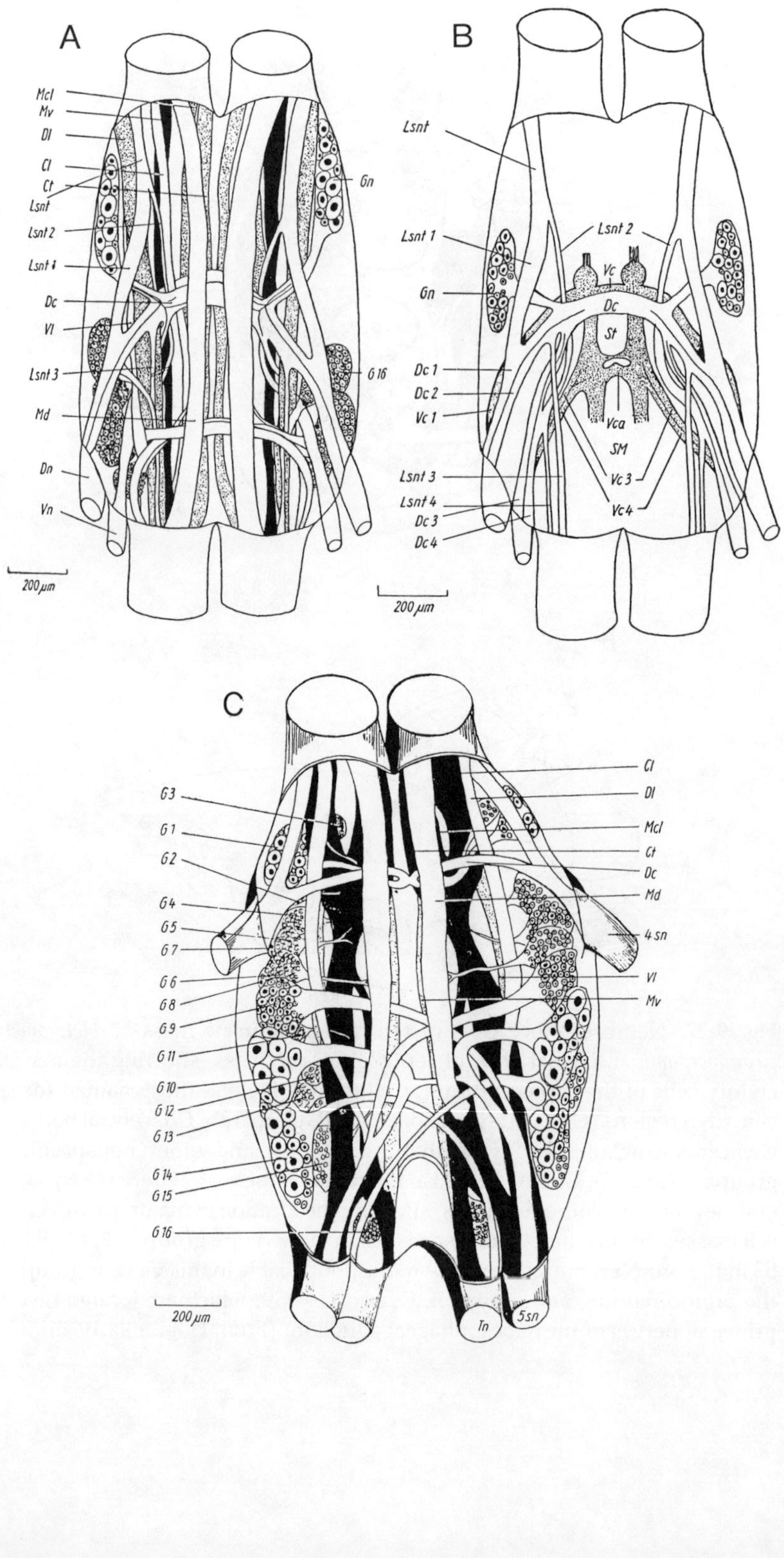

A
Mcl
Mv
Dl
Cl
Ct
Lsnt
Lsnt 2
Lsnt 1
Dc
Vl
Lsnt 3
Md
Dn
Vn
Gn
G16
200 μm
B
Lsnt
Lsnt 1
Gn
Lsnt 2
Vc
Dc
St
Dc 1
Dc 2
Vc 1
Vca
SM
Lsnt 3
Lsnt 4
Dc 3
Dc 4
Vc 3
Vc 4
200 μm
C
G3
G1
G2
G4
G5
G7
G6
G8
G9
G11
G10
G12
G13
G14
G15
G16
200 μm
Cl
Dl
Mcl
Ct
Dc
Md
4sn
Vl
Mv
Tn
5sn

last three mesosomal segments are supplied by nerves arising there (Fig. 9.23). The first three metasomal segments are supplied by individual segmental ganglia, but the last metasomal ganglion supplies the fourth and fifth metasomal segments and the telson (sting). The first six ganglia in the ventral nerve cord give rise to paired dorsal and paired ventral peripheral nerves. The dorsal nerves from each ganglion supply the muscles, receptors, and body-wall structures on the dorsal side, and the ventral nerves supply the corresponding structures on the ventral side. In addition to these paired segmental nerves, the seventh ganglion also gives off another pair of nerves (*5sn*, Fig. 9.23) to the fifth metasomal segment, a pair of telsonic nerves (*Tn*) to the rectum and the telson, and a pair of nerves (*Aln*) to the gut.

Histologically, the first six ganglia are similarly organized. Seven major fiber tracts run through the central neuropile of each ganglion, whereas the cell bodies are located in the surrounding cortex, or "rind." Babu (1965) described the major concentrations of neurons and the major fiber tracts within each ganglion. (The details of his analysis are not included here, but essential features of his work will be presented.) The major cell concentrations and tracts of the first abdominal (mesosomal) ganglion are shown in Figure 9.28, views *A* and *B*. The organization of the last (seventh) ganglion, which is thought to be a composite of the last two ganglia, and is more complex, is shown in Figure 9.28*C*. Neuron-cell bodies lie ventrally within the ganglion, connected to a centrally located neuropile and to each other via commissural and

---

Fig. 9.28. Structure of the ventral nerve-cord ganglia of *Heterometrus* sp. There are seven nerve-cord (abdominal) ganglia: three mesosomal ganglia and four metasomal ganglia. *A*, *B*, The major cell masses and fiber tracts of the first free ganglion (the first mesosomal ganglion) (*A*) and the fiber-tract connections of the lateral segmental nerves (*B*): *Cl*, centrolateral tract; *Ct*, central tract; *Dc 1, 2, 3, 4*, dorsal commissures 1, 2, 3, and 4; *Dl*, dorsolateral tract; *Dn*, dorsal nerve; *Gn*, group of cells; *Lsnt 1, 2, 3, 4*, laterosegmental nerve tracts 1, 2, 3, and 4; *Mcl*, midcentral tract; *Md*, middorsal tract; *Mv*, midventral tract; *SM*, ventral sensory associative mass; *St*, association-center tract; *Vc*, ventral commissure; *Vca*, ventral commissure a; *Vc 1, 3, 4*, ventral commissures 1, 3, and 4; *Vl*, ventrolateral tract; *Vn*, ventral nerve. *C*, The last ganglion (fourth metasomal or seventh abdominal ganglion), showing the major cell masses (*G1* through *G16*) and the major tracts and their connections to the fourth segmental nerves (*4sn*), fifth segmental nerves (*5sn*), and telsonic nerves (*Tn*). Other structures are indicated as in views *A* and *B*. (From Babu 1965) longitudinal tracts (Fig. 9.28*A, B*). Most sensory-cell bodies are located peripherally near the receptors, and send their fibers into the ganglion through the paired nerves of each segment.

Bilaterally symmetrical groups of nerve cells can be recognized, reflecting the double organization of the nerve cord and ganglia. Two major centers among the many groups of cells are one (*Gn*, Fig. 9.28) consisting of large cells that contribute to the dorsal commissure (*Dc*) and segmental nerves (*Dn*), and another (*G16*) made up of smaller cells that contribute to the ventral commissure (*Vc*) and segmental nerves (*Vn*). The studies of Venkateswara Rao (1963) and Babu and Venkatachari (1966) have described the microanatomy of these cells. Two broad classes of cells were distinguished: large Type D cells, which are presumed to be motoneurons, and small Type B cells, which are probably interneurons. From 450 to 650 cells were estimated in each of the first six ganglia, and about 790 cells in the last (seventh) ganglion. In the mesosomal and metasomal ganglia of *Heterometrus* sp., Babu (1961b) found numerous giant cells. Approximately 30 to 40 of these cells, with diameters up to 78 μm, were observed in each of the first six ganglia, and 50 to 60 were found in the seventh, the largest being 110 μm in diameter. The vast majority of cells in all ganglia were Type B cells.

Multiple pairs of longitudinal tracts originating in the subesophageal ganglion pass from the connectives through the central region of each abdominal ganglion. The central neuropile contains a large association center receiving and sending fibers from and to the sensory- and motor-cell groups both ipsilaterally and contralaterally, from and to the segmental dorsal and ventral nerves of that region, and from and to adjacent ganglia. The double connectives between ganglia contain axons with diameters ranging from 1 to 30 μm, but 50 to 60 percent of the fibers are between 3 and 6 μm in diameter. Overall, this organization is very reminiscent of the ganglia of the insect nerve cord.

Babu (1961b) and Babu and Venkatachari (1966) reported on numerous "giant fibers" in the ventral nerve cord of *H. fulvipes* and *H. swammerdami*. Giant fibers have been reported in many different invertebrates: they generally seem to mediate quick behaviors, such as escape, by virtue of their large diameters and their consequently fast conduction velocities. Giant fibers typically have much greater axon diameters than other nerve fibers. In these scorpions, within the ganglia of the ventral nerve cord giant fibers were observed in all the major longitudinal tracts. The fibers range from approximately 15 to 40 μm in diameter; the largest of these innervate the fifth metasomal segment and the telson. The cell bodies of these fibers are located either in the abdominal ganglia or in the subesophageal ganglion. Other giant fibers, approximately 31 μm in diameter, are located in tracts of the subesophageal ganglion, and primarily enter the leg and pedipalp nerves. In contrast to the unicellular giant fibers of the ventral nerve cord, these are multicellular or syncytial fibers. (Similar syncytial fibers have been observed in annelids, cephalopod mollusks, and in the lateral giant fibers of crustaceans.) Since both sets of giant fibers pass through the central ganglia of the subesophageal ganglion, interconnections between fibers to the pedipalps, legs, and telson seem possible. Babu (1961b) proposed that this system of giant fibers integrates quick movements of these appendages during defensive or stinging behaviors.

Cobalt-chloride backfills of the fifth segmental nerve revealed the cytoarchitecture of some of the giant fibers innervating the fifth metasomal segment (Yellamma et al. 1980). Two groups of giant neuron-cell bodies were observed in the cortex of each half of the last ganglion (Fig. 9.29). Each cell consisted of a large soma (50–62 μm in diameter), and a narrow connecting process (about 13 μm), leading to a profuse branching network in the neuropile and then giving off a large process (about 40 μm in diameter) from the ipsilateral nerve root, presumably innervating the musculature of the last segment. Cross sections of the fifth segmental nerve showed some of these giant-fiber axons (Fig. 9.16C). A number of smaller sensory-neuron processes, also filled in these preparations, had extensive branches in the neuropile, closely associated with those seen from the giant cells.

Physiology of the Cephalothoracic Mass

Studies of neural function in the cephalothoracic mass have been concerned almost exclusively with biochemical effects upon either behavior or ongoing electrical activity. Two exceptions are noteworthy.

Pampathi Rao (1964) and Selvarajan (1976) attempted to localize functions in the cephalothoracic mass by electrical stimulation of various regions. Most regions of the supraesophageal ganglion did not respond to stimulation. However, stimulation of the cheliceral ganglion in the tritocerebrum elicited feeding-like movements of the chelicerae, and stimulation of the rostral ganglion caused contraction of the pharynx. In the subesophageal ganglion, movements of the pedipalps, walking legs, and pectines were observed when the authors stimulated the centers for these areas, located near the origins of their respective nerves. Complex movements of the mesosoma, metasoma, and telson were

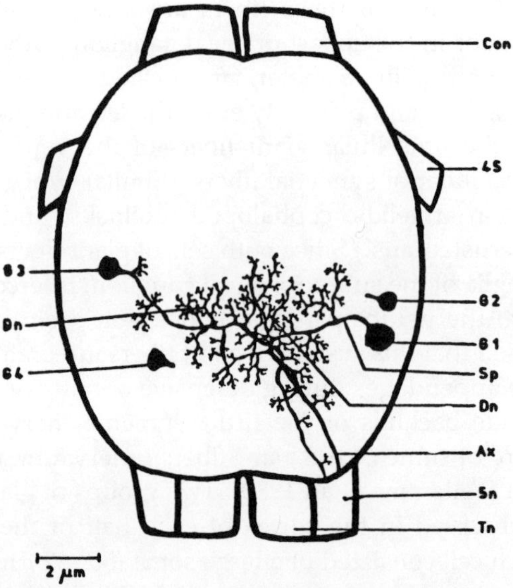

Fig. 9.29. Motor neurons and a giant neuron in the central nervous system of *Heterometrus fulvipes:* camera lucida drawing of motor neurons in the last ganglion (fourth metasomal or seventh abdominal ganglion), shown in the dorsal view. The cells were backfilled with cobalt chloride through the fifth segmental nerve. Two ipsilateral cells (*G1, G2*) and two contralateral cells (*G3, G4*) are shown, with the major neurites of G1 and G3 indicated. The axon from the G1 cell constitutes a giant fiber. *Ax,* axon; *Con,* connective; *Dn,* dorsal neurite; *Sp,* spiking zone; *Tn,* telsonic nerve; *4S,* fourth segmental nerve; *5n,* fifth segmental nerve. (From Yellamma et al. 1980)

seen with stimulation of regions posterior to the pectinal masses. But in most regions of the cephalothoracic mass, fiber-association regions (glomeruli, optic ganglia, central body, central ganglion, and ventral association center) did not respond to stimulation.

At the opposite end of the spectrum, intracellular recordings have been made from cells of the subesophageal ganglion. Bowerman and Burrows (1980) were able to insert glass microelectrodes through the sheaths of the subesophageal ganglion and to penetrate motoneuron cell bodies of the leg nerves. Single cells were characterized as motoneurons on the basis of electrical activity that was correlated with simultaneously recorded activity in muscles or on the basis of antidromically recorded spikes elicited by stimulation of specific muscles (Fig. 9.30). The researchers found that spikes recorded in the cell bodies of these motoneurons had amplitudes of 2 to 4 mV, suggesting that the spike-initiating zone lies distant from the cell body. Synaptic activity consisting of both depolarizing and hyperpolarizing potentials was also observed, and the pattern of synaptic activity could be influenced by imposed movements of the leg, eliciting reflex excitation or inhibition. Furthermore, simultaneously recorded activity in two separate cells occasionally showed that some synaptic activity was common to both. It was concluded that physiologically these motoneurons were quite similar to other arthropod neurons.

The biochemistry of the cephalothoracic mass has been studied in a variety of different approaches. To date, there have been published reports of the cephalothoracic mass containing acetylcholine and acetylcholine esterase (Corteggiani and Serfaty 1939; Venkatachari and De-

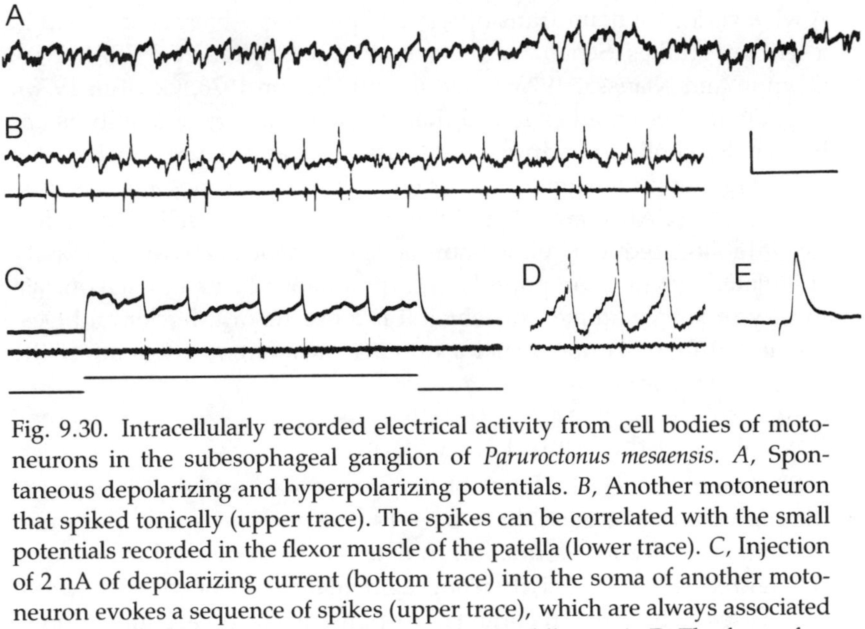

Fig. 9.30. Intracellularly recorded electrical activity from cell bodies of motoneurons in the subesophageal ganglion of *Paruroctonus mesaensis*. *A*, Spontaneous depolarizing and hyperpolarizing potentials. *B*, Another motoneuron that spiked tonically (upper trace). The spikes can be correlated with the small potentials recorded in the flexor muscle of the patella (lower trace). *C*, Injection of 2 nA of depolarizing current (bottom trace) into the soma of another motoneuron evokes a sequence of spikes (upper trace), which are always associated with potentials recorded in the flexor muscles (middle trace). *D*, The form of an orthodromic spike recorded intracellularly (upper trace) shown more clearly at higher gain. The spikes are associated with potentials recorded in the flexor muscle (lower trace). *E*, An antidromic spike from a motoneuron cell body elicited by stimulation of motoneuron terminals in the muscle. Vertical bar (voltage): *A, D, E*, 3.5 mV; *B*, 1.2 mV; *C*, 8 mV. Horizontal bar (time): *A, B, C*, 200 msec; *D*, 100 msec; *E*, 16 msec. (From Bowerman & Burrows 1980)

varajulu Naidu 1969; Vasantha et al. 1975, 1977) and various free amino acids (Pasantes et al. 1965, Habibulla 1971b).

Given the prevalence of neurosecretory cells in the CNS, it is likely that the CNS exerts humoral influences upon visceral activity (Pampathi Rao and Habibulla 1973), but virtually nothing is known about these possible effects. Raghavaiah et al. (1977, 1978) have extracted a factor from the cephalothoracic mass of *Heterometrus fulvipes* that influences carbohydrate metabolism. Upon injection, it causes hyperglycemia, but also paradoxically raises the level of carbohydrate in the hepatopancreas. The source and mechanism of action of this factor are unknown.

Goyffon et al. (1974, 1975) recorded spontaneous activity in the cephalothoracic mass, and have studied the biochemistry of the brain by injecting various pharmacological agents into the blood system and noting subsequent changes in spontaneous activity of brain cells. A wide variety of neurotransmitters and precursors have been tested in a series of studies (Schantz and Goyffon 1970; Goyffon et al. 1974, 1980; Goyffon and Niaussat 1975; Goyffon and Vachon 1976; Goyffon 1978). In general, these studies found that spontaneous brain activity is enhanced by cholinergic drugs and glutamate, and depressed by catacholinergic and tryptaminergic drugs. Glycine, taurine, and γ-amino butyric acid (GABA) are all inhibitory; glycine was a particularly effective inhibitor, reducing either normal spontaneous activity or activity stimulated by other compounds. *l*-Tryptophan is excitatory, in contrast to 5-hydroxytryptamine, for which it is a precursor. Mercier and Dessaigne (1970, 1971) noted that a number of different drugs affect the behavior of scorpions.

Physiology of the Ventral Nerve Cord

Electrophysiological studies of the ventral nerve cord, like those of the cephalothoracic mass, have been for the most part descriptive and preliminary. All the information to date has been derived from extracellular recordings of activity from segmental nerves or connectives. In most cases activity was recorded in response to electrical stimulation of nerves, or to mechanical or photic stimulation of nearby structures.

Venkatachari and Babu (1970) recorded from nerves of the second abdominal (mesosomal) ganglion, and examined the firing patterns of 40 motoneurons. In a similar study, Babu and Venkatachari (1972) examined the firing patterns of 80 interneurons. They recorded activity from the left connective between the first and second abdominal (mesosomal) ganglia in response to electrical stimulation of nerves and connectives from both ganglia, or mechanical stimulation of nearby cuticular structures. In both these studies, response categories for these motoneurons and interneurons were described. Both spontaneously active and silent fibers were noted, and many of the cells often responded to a broad range of stimuli. Vereshchagin et al. (1971) described the activity of some neurons in ganglia of *Mesobuthus caucasicus*, and found a wide variety of responses. All these authors suggest that these neurons are fundamentally similar in their response characteristics to those of other arthropods.

The ventral nerve cord exhibits spontaneous activity (Pampathi Rao 1963, 1964; Venkatachari 1971). Pampathi Rao recorded along the length of the ventral nerve cord, and noted that the number of spontaneously active cells increased toward the cephalothoracic mass; thus, much of this spontaneous activity is probably due to descending fibers. Pampathi Rao also noted that the intensity of spontaneous activity followed a daily cycle, correlated with the circadian rhythmicity of locomotion, of enzymatic activity of heart muscle (see Jayaram et al. 1978), and of neurosecretory activity in the brain. This cycle is influenced by various extracts of the cephalothoracic mass (Pampathi Rao and Gropalakrishnareddy 1967).

Another property of the ventral nerve cord seems peculiar but has also been demonstrated in a few other invertebrates: Geetha Bali and Pampathi Rao (1973) and Zwicky (1968) have described photosensitivity for some metasomal ganglia and the telsonic nerves (see "*Metasomal neural photoreceptor*," above). Geetha Bali has further examined the influence of this neural photoreception upon electrical activity in the ventral nerve cord. In addition to the discovery that many different neurons in the nerve cord exhibit photosensitivity (Geetha Bali 1974, 1975, 1976a, b, 1979), it was observed that some of these same cells also responded to mechanical stimulation of the first metasomal segment (Geetha Bali 1976c). Responses to more than one stimulus have also been observed in other invertebrates, and presumably the pattern of neuron spikes permits the CNS to discriminate between them, or else this information is compared with information from other receptors to determine what is being sensed. The significance of this system in the scorpion is unknown at this time.

GIANT NEURONS AND THE CONTROL OF STINGING BEHAVIOR. In an attempt to understand the function of the giant fibers in the fourth metasomal ganglion (seventh abdominal ganglion; *7abdg*, Fig.

9.23), Yellamma et al. (1980) recorded from the segmental nerve of that ganglion (*5sn*, Fig. 9.23) while electrically stimulating the ipsilateral connective. They were able to distinguish giant-fiber spikes in the divisions of this nerve that innervate the musculature of the last tail segment (fifth metasomal segment) and the telson. They also recorded from the telsonic nerve (*Tn*, Fig. 9.23), which innervates the telson, and observed giant-fiber spikes. (The close anatomical relationship between neuropile processes from the giant fibers and sensory fibers from the fourth and fifth metasomal segments and the telson was noted previously: see "Anatomy of the Ventral Nerve Cord.") Consequently Yellamma et al. suggest that these giant fibers act as final motoneurons in a monosynaptic reflex pathway from receptors in the telson and the most distal metasomal segments, allowing quick activation of the musculature in these same segments during offensive and defensive tail movements.

As described previously, Babu (1965) noted seven paired longitudinal tracts running through each ganglion in *Heterometrus*. Within each, a few large ("giant") fibers run multisegmentally through the ventral nerve cord, presumably mediating fast stinging relfexes. Venkatachari (1975) recorded large, compound action potentials at either end of the cord with conduction velocities of 2.5 to 5.0 meters per second, presumably representing the giant fibers described by Babu, which run through the cord. While stimulating the cord at one end and moving the recording electrode progressively toward the other, Venkatachari noticed that the number of peaks in the compound action potential decreased or increased, depending upon the position of the recording electrode. He therefore concluded that this conduction pathway is actually an asymmetrical system of fibers, not all of which run the entire length of the cord. He categorized the action potentials into two groups: anteroposterior (AP) and posteroanterior (PA), according to the direction in which component fibers (action potentials) decreased. He suggested that some of the fibers in each system make synaptic contacts with sensory and motor fibers in each ganglion, and that therefore this ventral nerve-cord conduction pathway consists of multiple parallel fibers providing rapid conduction from either the cephalothoracic mass or the fourth metasomal ganglion to the other metasomal segments.

Sanjeeva Reddy and Pampathi Rao (1970a, b) reported that tactile input from single mechanoreceptive hairs on the telson was capable of activating components of the PA system in *Heterometrus fulvipes*. More extensive tactile stimulation was necessary to cause a behavioral response. Therefore it seems plausible that the longitudinal conduction system might serve to activate quick tail movements in response to tactile stimulation.

It is unclear, however, whether the giant fibers in each segment mediate quick stinging movements during prey capture and defensive behavior. The sequence of movements during stinging behavior is quite complex (Bub and Bowerman 1979, Capser 1985), but the defensive strike is simpler and easier to elicit, so it has been studied in more depth. In their study of defensive strike behavior in six species, Palka and Babu (1967) found that lateral lesions in the cephalothoracic mass abolish strike movements to touch stimuli on the same side, but not on the opposite side. Cutting one of the connectives leading from the cephalothoracic mass to the ventral nerve cord had no observable effect upon the strike response to touch on either side. Cutting both connectives abolished the quick strike, although tail movement in the direction of the touch was still weakly elicited. These results suggest involvement of the cephalothoracic mass in generation or maintenance of the strike. The authors also noted, however, that the defensive strike's resistance to fatigue, its gradation, and its independence from pedipalp movements at low stimulation intensities suggest a neural system more complex than single giant fibers running the length of the ventral nerve cord. Therefore the giant fibers in each segment are possibly involved in local reflexes, and in segmental activation of muscles during the defensive strike. Whether they are active during normal stinging movements needs to be clarified.

The picture that emerges from this discussion of the scorpion CNS is one of a highly bilateral nervous system in which all the special senses and appendages are controlled by the cephalothoracic mass, and the ventral nerve cord remains as a series of control centers for the mesosoma, metasoma, and telson. The one major behavioral function mediated by the ventral nerve cord seems to be the control of the metasomal segments and the telson, although these may be ultimately controlled in some species by giant fibers passing rearward from the subesophageal ganglion.

Conclusions

Overall, the scorpion nervous system is poorly understood in comparison with the nervous systems of other arthropods, but there is tremendous potential for exciting discoveries in the near future. For example, the CNS in most arachnids is highly cephalized, but the scorpion CNS has a primitive, ganglionated nerve cord in addition to a large cephalothoracic mass: therefore the scorpion CNS is of interest from a comparative perspective. Furthermore, although it is a relatively simple animal, the scorpion exhibits a variety of sophisticated and intriguing behaviors, such as burrowing, the detection and stinging of prey, and elaborate courtship rituals. Because of its large size, its ease of laboratory maintenance, and its demonstrated usefulness as a neurophysiological preparation, the scorpion appears to be a valuable system for studying the neural basis of behavior in arachnids. Significant advances have recently been made in this area from studies of visual function, locomotion, prey detection, and stinging movements.

Unfortunately, although the gross anatomy and histology of the CNS have been fairly well studied, virtually nothing is known about the function of individual nerve cells, not to mention simple neural networks. One requirement for successful interpretation of CNS function has been demonstrated: it is possible to make intracellular recordings from single neurons in the cephalothoracic mass, and thus neural connections can be mapped. But it is not yet clear whether individually identifiable neurons can be found from one animal to the next; if they can be, that would greatly facilitate the study of neural circuits.

Certain scorpion sensory systems have attracted considerable attention, and so are better understood than others. The anatomy and physiology of the eyes have been examined from different perspectives with general success. Mechanoreceptors, particularly joint receptors in the pedipalps and the legs, muscle receptor organs in the metasoma, trichobothria, and slit sense organs have also been described both anatomically and physiologically, although their functions in behavior remain to be elucidated. Other sensory systems are in need of further clarification. For example, the metasomal photoreceptor is an intriguing adaptation, possibly mediating the entrainment of circadian rhythms, but this and other suggested functions need to be evaluated in view of recent work on the eyes. The pectines are in need of further well-conceived and definitive experiments to define their role. Finally, other sensory systems, such as thermoreception and chemoreception, are known only from behavioral work.

Almost nothing is known about the physiology of most scorpion motor systems. The muscles do not appear unusual, although they may be less varied in structure than those of insects and crustaceans.

Except for a few muscles, the physiology of neuromuscular systems remains undescribed. With the recent interest in such behaviors as locomotion, prey detection, and stinging, it is likely that much more work will be done on motor systems.

There is currently a sincere effort by a number of laboratories to describe different features of the scorpion nervous system, such as visual processing, the control of locomotion and burrowing, the function of the pectines, and neuron cytoarchitecture. These experiments will undoubtedly lead to better information about many of the systems described in this chapter.

ACKNOWLEDGMENTS

I gratefully acknowledge the assistance of Lorraine Root in the preparation of the manuscript, and of Drs. H. Peter Wimmer and Duncan J. McDonald for their reading and criticism. I also wish to thank all the many authors who graciously allowed me to reprint their published or unpublished work, and those who offered comments or suggestions.

10

Venoms and Toxins

J. MARC SIMARD AND

DEAN D. WATT

Although envenomations by poisonous animals, including scorpions, have been scourges of humankind since antiquity, only recently have we learned about the chemical nature and physiological effects of the toxins present in the venoms. Historically, interest in scorpion venoms stemmed largely from their importance as health hazards to humans. The oldest medical and literary writings make reference to scorpions: "My father hath chastised you with whips, but I will chastise you with scorpions" (1 Kings 12.11). The neurotoxins in scorpion venoms are proteins and represent some of the most powerful poisons known. A comparison of the median lethal doses (LD_{50}) for a number of potent toxins from a variety of sources is given in Table 10.1. Expressed on a molar basis, the lethality of the scorpion toxins is exceeded only by certain bacterial toxins and is comparable to that of the snake neurotoxins. Although cyanide is generally considered potent, on a molar basis it is 10^5 times less toxic than some scorpion toxins (Table 10.1).

Recently, a new basis of interest in scorpion toxins has developed with the introduction of techniques for isolating and characterizing the individual neurotoxins in the venoms. Studies on both the chemical nature of the toxins and their interactions with receptor sites on target cells thus have gained prominence. Only recently have we begun to explore the multiple interactions between the scorpion toxins and their molecular targets, the ion channels of excitable cells. Not only has our understanding of the mechanism of action of the scorpion toxins grown but, in addition, we have learned more about the molecular na-

TABLE 10.1
Comparative molar toxicities of selected toxins in mice

Toxin	Source[a]	Dose[b]
SCORPION TOXINS		
Toxin I	*Androctonus australis*	2.5×10^{-9}
Toxin II	*A. australis*	1.2×10^{-9}
BACTERIAL TOXINS		
Toxin type A	*Clostridium botulinum*	1.7×10^{-16}
Toxin type B	*C. botulinum*	0.6×10^{-16}
Tetanus toxin	*C. tetani*	2.0×10^{-16}
SNAKE TOXINS		
Taipoxin	*Oxyuremus scutellatus*	0.048×10^{-9}
Sea-snake toxin	*Enhydrina schistosa*	0.8×10^{-9}
β-Bungarotoxin	*Bungarus multicinctus*	0.88×10^{-9}
Crotoxin	*Crotalus durissus terrificus*	1.7×10^{-9}
Notexin	*Notechus scutatus*	1.8×10^{-9}
Siamensistoxin (a cobratoxin)	*Naja naja siamensis*	9.6×10^{-9}
MISCELLANEOUS TOXINS		
Batrachotoxin (Colombian arrow poison)	*Phyllobates bicolor*	3.7×10^{-9}
Tetrodotoxin	*Spheroides rubripes* (Japanese fugu fish)	25.0×10^{-9}
Saxitoxin	*Gonyaulax catanella* (in poisonous shellfish)	28.0×10^{-9}
d-Tubocurarine (Indian arrow poison)	*Chondrodendrum tomentosum*	290.0×10^{-9}
Potassium cyanide	Chemical synthesis	1.5×10^{-4}

SOURCE: Karlsson 1973.
[a] Other sources of toxin may be known.
[b] Dose expressed in moles per kilogram: LD_{50} except for bacterial toxins, which are expressed as minimum lethal dose; doses are intravenous except for batrachotoxin, which is subcutaneous.

ture and the physiological functioning of the ion channels themselves. Additional fertile areas of study relate to the phylogeny of these toxic proteins and the roles they play in the biology of the scorpion. In this chapter some of these exciting developments in the molecular biology of scorpion venoms are reviewed. In addition, an overview is provided of the medical significance of scorpions and their venoms.

Chemical Characterization

Venoms from scorpions are complex mixtures, in part because they are apocrine secretions (Keegan and Lockwood 1971; see also Chapter 2). Characteristically, the venom from a single species contains multiple low-molecular-weight basic proteins (the neurotoxins), mucus

(5–10 percent), salts, and various organic compounds (e.g., oligopeptides, nucleotides, amino acids: Miranda et al. 1970; Watt et al. 1978; Grishin 1981; Possani et al. 1981 a, b; Grégoire and Rochat 1983; Martin and Rochat 1984b; McIntosh and Watt 1972).

Composition

An example of the complex composition typical of scorpion venoms is demonstrated in Figure 10.1 by the elution profile obtained on chromatography of whole venom from *Centruroides exilicauda* (= *sculpturatus*) on the ion-exchange medium carboxymethylcellulose (CMC: Babin et al. 1974, Watt et al. 1978). Twelve zones of elution are obtained. Approximately 50 percent of the total weight of the venom is eluted in zones 1 and 2, which together contain nucleotides (absorption at 260 nm: dashed line, Fig. 10.1), at least two acidic proteins, and various low-molecular-weight salts and organic substances (Pool 1979). Zones 3 through 12 each contain multiple proteins. The neurotoxins thus far isolated have been obtained from zones 4 through 12. Isolation of individual toxins requires rechromatography of a specified zone on other chromatographic media. For example, rechromatography of zone 4 on DEAE-Sephadex yields three different elution peaks: one contains Toxin var3; another, Toxin var4; and the third peak, an unidentified protein. The potency of the isolated toxins increases from left to right across the profile, with the maximum toxicity (in chicks) shown by Toxins III and IV from CMC zone 11 (Watt et al. 1978). Notably, the apparent basicity of the proteins in each zone also increases from left to right across the elution profile. Toxins var3 and var4 from CMC zone 4 are the least basic of the toxic fractions, as evidenced by their early elution from CMC and their slower migration rates during disc-gel electrophoresis at pH 4.5. A summary of the protein composition of the various CMC zones is given in Table 10.2.

The amino-acid compositions are known for about 70 scorpion toxins (Babin et al. 1974, 1975; Zlotkin et al. 1978; Grishin 1981; Lazarovici and Zlotkin 1982; Martin and Rochat 1984b). These toxins contain only the 20 *l*-amino acids common to proteins. Quantitatively, however, the amino-acid compositions are somewhat unusual when compared with other proteins. The scorpion toxins have a high content of cysteine—eight residues with four disulfide bridges per molecule, which is an unusual amount of cysteine for single-chain globular proteins of these low molecular weights. Considerable amounts of glycine, tyrosine, dicarboxylic amino acids and their amides, and lysine are also present.

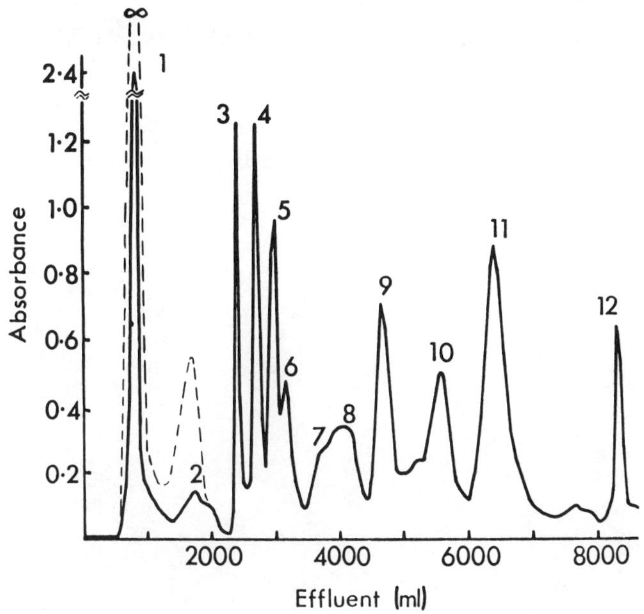

Fig. 10.1. Chromatography on carboxymethylcellulose (CMC) of venom from *Centruroides exilicauda*. CMC (CM–11; Reeve Angel, Clifton, N.J.) with a capacity of 0.6 meq/g and column size 4.0 × 55 cm is used. Elution is monitored by determining the absorbance at 280 nm to give the elution profile in the figure (solid line). Absorbance at 260 nm (dashed line) is also determined on the first 250 fractions, beyond which the 260/280 absorbance ratios show essentially no nucleic acids. The high absorbance at 260 nm in zones 1 and 2 is believed to result from nucleic acids and nucleotides. See Babin et al. (1974) for elution protocol. (After Babin et al. 1974)

Notably absent in most scorpion toxins are methionine and, to a lesser extent, histidine.

A puzzling question that has not been satisfactorily addressed concerns the presence of multiple toxins in a given venom. Scorpions use venoms for immobilization of prey and protection against predators (see Chapter 7). These two purported functions during evolution might have exerted selection pressures that favored development of complex venoms. Results from studies with isolated toxins have demonstrated a significant degree of host specificity of certain toxins. For example, the venom of one species may contain one toxin that is preferentially toxic to insects, another to crustaceans, and still another to mammals (Zlotkin

TABLE 10.2

Summary of elution zones from carboxymethylcellulose (CMC) chromatography of venom from Centruroides exilicauda

CMC zone number[a]	Approximate percent of total weight[b]	Proteins present	
		Unidentified	Identified[c]
1	15	none	2
2	25	?	none
3	5	3	none
4	5	1	Toxin var3[d]
4	3		Toxin var4[d]
5	3	4	Toxin var2[d]
6	3	1	B137-1
7, 8	5	3	Toxin var5[d]
			Toxin var6[d]
9	3	3	Toxin var1[d]
10	4	3	Toxin V[d]
11	6	1	Toxin I[d]
			Toxin III
			Toxin IV
12	4	5–6	Toxin VI
			Toxin VII
TOTAL	81	24–25?	15

[a] See Fig. 10.1 for details of chromatographic procedure.
[b] Mucus comprises about 3 percent of the dry weight; the remainder of the weight (16 percent) is present in the valleys between the zones.
[c] The designation "var" in a toxin number signifies an isolate that has a toxicity in chicks, $LD_{50} > 5.0$ mg/kg. The chromatogram number (e.g., B137-1) of an isolate is retained until the isolate is fully characterized.
[d] Amino-acid sequences known.

et al. 1971, 1976; Watt et al. 1978; Grishin 1981). In some instances, venoms with multiple toxins may have been favored because of synergy among the toxic components, as has been demonstrated with the Insect Toxins from *Scorpio maurus palmatus* (Lazarovici et al. 1982). Alternatively, venoms might be repositories of evolutionally changing proteins for which few or no selection pressures have existed. In addition, it is noteworthy that the neurotoxin composition of pooled venom may vary when a single species inhabits different geographic locations (Miranda et al. 1970; Grishin 1981; W. Catterall, pers. comm. 1984; Watt, unpubl. data). Whether each individual scorpion produces the large number of neurotoxic components is not known. Thus, the possibility that subspecies occur and contribute to the complexity of pooled venom cannot be excluded. The problem of polymorphism in these neurotoxic proteins is amenable to experimentation but awaits further clarification.

Many scorpion venoms, unlike most spider and snake venoms, either are lacking in enzymes or demonstrate very low levels of enzyme activity (McIntosh 1969, Balozet 1971, Zlotkin et al. 1978). An exception is venom from *Heterometrus scaber,* which contains several enzymes, including acid phosphatase, ribonuclease, 5′-nucleotidase, hyaluronidase, acetylcholinesterase, and phospholipase A (Bhaskaran Nair and Kurup 1975). Other components reported to be present in some scorpion venoms include serotonin (5-hydroxytryptamine) and factors that are inhibitors of enzymes. Serotonin does not appear to alter the toxicity of the venom, but it may contribute to the pain of a sting (Adam and Weiss 1958). The isolated toxins themselves, however, also induce local pain, as evidenced by the responses of test animals to injections of pure toxins (D. D. Watt, unpubl. data). The red scorpion of India, *Mesobuthus tamulus* (formerly *Buthotus tamulus*), contains a protein that inhibits proteases (e.g., trypsin: Chhatwal and Habermann 1981). An inhibitor of the angiotensin-converting enzyme, angiotensinase, has been identified in venom from *C. exilicauda* (Longenecker et al. 1980), as has a succinate-dehydrogenase inhibitor from *H. fulvipes* (Reddy et al. 1980). Little is known about the nature of the enzyme inhibitors in scorpion venom, since they have not been isolated and are not well characterized. The possible contribution to the toxicity of scorpion venoms by enzymes and enzyme inhibitors has not been investigated.

Morphology

The toxins are single-chain basic proteins containing 60 to 70 amino acids. Exceptions are the short-chain (35–40 residues) toxins from *Mesobuthus eupeus* (= *Buthus eupeus;* Grishin 1981) and Noxiustoxin from *Centruroides noxius* (Possani et al. 1982). The peptide chain is cross-linked by four intrachain disulfide bridges (Kopeyan et al. 1974). Multiple disulfide bridges in proteins of this molecular size maintain compact folding and confer exceptional molecular stability. Consequently, these toxins are exceedingly resistant to many changes in the environment that are normally hostile to most proteins. For example, they are not inactivated by extreme changes in pH, by denaturing agents of various kinds, by many common proteolytic enzymes, or by a wide range of temperatures. On the other hand, the toxins are inactivated by reagents that reduce disulfide bridges (e.g., mercaptoethanol and dithiothreitol) and react with amino and carboxyl groups (Watt 1964, Watt and McIntosh 1972, Rochat et al. 1979). The molecular stability of the toxins

Table 10.3

Amino-acid sequences of selected scorpion toxins

```
A:                 10        20        30        40        50        60        70
                    |         |         |         |         |         |         |
CsE v3    --KEGYLVKKSDGCKYGCLKLGENEGCDTECKAKNQGGSYGYCYAF-----ACWCEGLPESTPTYPLPNKSC
CsE I     --KDGYLVEK-TGCKKTCYKLGENDFCNRECKWKHIGGSYGYCYGF-----GCYCEGLPDSTQTWPLPNK-CT
CsE V     -KKDGYPVD-SGNCKYECLK---DDYCNDLCLER--KADKGYCYWGKV---SCYCYGLPDNSPTKT-SGK-CNPA
CsE v5    --KDGYPVD-SKGCKLSCVA---NNYCDNQCKMK--KASGHCYAM-----SCYCEGLPENAKVSDSATNIC
CnII 14   --KDGYLVD-AKGCKKNCYKLGKNDYCNRECRMKHRGGSYGYCYGF-----GCYCEGLSDSTPTWPLTNKTC
Css II    --KEGYLVSKSTGCKYECLKLGDNDYCLRECKQQYGKSSGGYCYAF-----ACWCTHLYEQAVVWPLPNKTCN
Css III   --KEGYLVSKSTGCKYECLKLGDNDYCLRECKQQYGKSSGGYCYAF-----ACWCEALPDHTQVW-VPNK-CT
Ts VII    --KEGYLMD-HEGCKLSCFIR-PSGYCGRECGIK--KGSSGYCAWP-----ACYCYGLPNWVKVWDRATNKC

B:
AaH I     -KRDGYIVY-PNNCVYHCVPP-----CDGLCKKN--GGSSGSCSFLVPSGLACWCKDLPDNVPIKDTSRK-CT
AaH III   -VRDGYIVN-SKNCVYHCVPP-----CDGLCKKN--GAKSGSCGFLIPSGLACWCVALPDNVPIKDPSYK-CHS
AaH II    -VKDGYIVD-DVNCTYFCGR---NAYCNEECTKL--KGESGYCQWASPYGNACYCYKLPDHVRTKG-PGR-CH
Amm V     -LKDGYIID-DLNCTFFCGR---NAYCDDECKKK--GGESGYCQWASPYGNACWCYKLPDRVSIKE-KGR-CN
Be M9     -ARDAYIAK-PHNCVYECYNP-KGSYCNDLCTEN--GAESGYCQILGKYGNACWCIQLPDNVPIRI-PGK-CH
Be M10    -VRDGYIAD-DKDCAYFCGR---NAYCDEECKK---GAESGKCWYAGQYGNACWCYKLPDWVPIKQKVSGKCN
Be I2     -A-DGYVKG-KSGCKISCFLD--NDLCNADCKYYG-GKLNSWCIPDKSG--YCWC--PNKGWNSIKSETNTC
Bom III   -GRDGYIAQ-PENCVYHCFPG--SSGCDTLCKEK--GATSGHCGFLPGSGVACWCDNLPNKVPIVVGGEK-CH
Bot I     -GRDAYIAQ-PENCVYECAQ---NSYCNDLCTKN--GATSGYCQWLGKYGNACWCKDLPDNVPIRI-PGK-CHF
Bot III   -VKDGYIVD-DRNCTYFCGR---NAYCNEECTKL--KGESGYCQWASPYGNACYCYKVPDHVRTKG-PGR-CN
Lqq IV    GVRDAYIAD-DKNCVYTCGS---NSYCNTECTKN--GAESGYCQWLGKYGNACWCIKLPDKVPIRI-PGK-CR
Lqq V     -LKDGYIVD-DKNCTFFCGR---NAYCNDECKKK--GGESGYCQWASPYGNACWCYKLPDRVSIKE-KGR-CN
Os III    GVRDGYIAQ-PHNCVYHCFPG--SGGCDTLCKENG-ATQGSSCFILGRG-TACWCKDLPDRVGVIVDGEK-CH
                           |----J----|                |----B----|
```

Table 10.3

(continued)

	Sequence
C:	10 20 30 40 50 60 70
AaH IT	KKNGYAVDSSGKAPECLLSNYCNNQCTKVHYADKGYCCLLSCYCFGLNDDKKVLEISDTRKSYCDTTIIN
D:	
Amm P_2	–CGPCFTTDPYTESKCATCCGGRGKCVGPQCLCNRI
Be I_1	MCMPCFTTRPDMAQQCRACCKGRGKCFGPQCLCGYD
Be I_5A	MCMPCFTTDPNMAKKCRDCCGGNGKCFGPQCLCNR
E:	
Cn NOX	TIINVKCTSPKQCSKPCKELYGSSAGAKCMNGKCKCYXN

SOURCES: CsE v3 (corrected sequence), Fontecilla-Camps et al. 1982; CsE I, Babin et al. 1975; CsE V (corrected sequence), Simard & Watt, this paper (courtesy J. E. Mole, unpubl.); CsE v5, J. E. Mole & D. D. Watt, unpubl.: CnII 14, Possani et al. 1985; CssII, Martin et al. 1987; Css III, Habersetzer-Rochat & Sampieri 1976; Ts VII, Bechis et al. 1984; AaH I, Martin & Rochat 1984a; AaH III, Kopeyan et al. 1979; AaH II, Rochat et al. 1979; Amm V, Amm P_2, Rosso & Rochat 1985; BeM_9, Volkova et al. 1984b; BeM_{10}, BeI_2, BeI_1, Grishin 1981; Bom III, Vargas et al. 1987; Bot I, Grégoire & Rochat 1983; Bot III, el-Ayeb et al. 1983; Lqq IV, Kopeyan et al. 1985; Lqq V, Kopeyan et al. 1978; Os III, Volkova et al. 1984a; AaH IT, Darbon et al. 1982; BeI_5A, Arseniev et al. 1984; Cn NOX, Possani et al. 1982.

NOTE: **A**, Long-chain toxins, New World; **B**, long-chain toxins, Old World; **C**, long-chain "Insect Toxin"; **D**, short-chain toxins, Old World; **E**, short-chain toxins, New World. In **A** and **B**, amino-acid sequences are arranged with the cysteine residues (*C*) in register with those of Toxin variant 3 (*CsE v3*), leaving gaps when necessary. Numbering at the head of the table is for every tenth residue of Toxin variant 3, including gaps. Recurring homologous sequences are enclosed in boxes. The jutting (*J*) and blunt (*B*) loops are identified by dashed lines. Those sequences in **C**, **D**, and **E** show limited homology with toxins in **A** and **B**. *CsE, Centruroides sculpturatus* Ewing (now *C. exilicauda* [Wood]), Toxins variant 3, I, V, and variant 5; *Cn, C. noxius*, Toxin II 14 and short-chain toxin NOX (noxiustoxin); *Css, C. s. suffusus*, Toxins II and III; *Ts, Tityus serrulatus*, Toxin VII; *AaH, Androctonus australis* Hector, Toxins I, III, II, and "Insect Toxin" (*IT*); *Amm, A. m. mauretanicus*, Toxin V and short-chain toxin P_2; *Be, Buthus eupeus* (now *Mesobuthus eupeus*), Toxins M_9, M_{10}, and I_2, and short-chain toxins I_1 and I_5A; *Bom, B. occitanus mardochei*, Toxin III; *Bot, B. occitanus tunetanus*, Toxins I and III; *Lqq, Leiurus q. quinquestriatus*, Toxins IV and V; *Os, Orthochirus scrobiculosus*, Toxin III. Included in the Old World toxins are those from mid-Asian species, Be and Os. *A*, alanine; *C*, cysteine; *D*, aspartate; *E*, glutamate; *F*, phenylalanine; *G*, glycine; *H*, histidine; *I*, isoleucine; *K*, lysine; *L*, leucine; *M*, methionine; *N*, asparagine; *P*, proline; *Q*, glutamine; *R*, arginine; *S*, serine; *T*, threonine; *V*, valine; *W*, tryptophan; *X*, Cn NOX penultimate residue undetermined but either aspartate or asparagine; *Y*, tyrosine.

would be advantageous to the scorpion, which may use its venom only infrequently and under conditions that would inactivate less stable proteins. Compact folding also facilitates rapid movement of the toxins to target sites.

Listed in Table 10.3A and 10.3B are amino-acid sequences of selected toxins from venoms of several species of scorpions from the New World (Table 10.3A) and the Old World (Table 10.3B). The sequences of four short-chain toxins and one long-chain toxin with somewhat different primary structures are given at the bottom of the table. Listings in Table 10.3A and 10.3B are arranged with reference to the sequence of Toxin var3 from *Centruroides exilicauda*, the first scorpion toxin for which the three-dimensional structure was determined at high resolution by x-ray crystallography (Fontecilla-Camps et al. 1980). Sequences are aligned, leaving gaps as necessary, so that the cysteine residues and other recurring amino acids are in register. Four structural features may be distinguished:

1. The scorpion toxins constitute a family of homologous proteins with conservation of certain amino acids (boxed areas in Table 10.3). Examples are residues 3 through 8, 41 through 44, 52 through 55, and 58 through 60. In these segments, if there are amino-acid substitutions from one toxin to another, the substituted amino acid is usually very similar in chemical structure (conservative as opposed to radical substitution). Conserved amino-acid sequences imply that these regions of the molecule have withstood evolutionary selection pressures and consequently may be associated with the biological function of the protein, either by constituting an "active site" or by being involved in the proper folding of the molecule.

2. Evident from Table 10.3A and 10.3B is the spatial conservation of certain segments: the loops between cysteines 14 through 18 and 27 through 31 always contain three amino acids. Likewise, a single residue, tryptophan or tyrosine, connects cysteines 53 through 55 in all sequences.

3. In certain cases, interspecies sequence homology may be greater than intraspecies homology. For example, toxins from two different species, Toxins Lqq V and Amm V, are highly homologous, differing in three amino-acid residues. In contrast, two toxins from the same species, Bot I and III, show 33 amino-acid substitutions.

4. With the exceptions of CsE V and CsE v5, toxins from New World species (Table 10.3A) of scorpions generally resemble one another

more than they resemble toxins from Old World species (Table 10.3B). Distinct differences between Old World and New World scorpion toxins are observed in the lengths of the loops designated J and B. The J loop is long in the New World and short in the Old World scorpion toxins. The reverse is observed for the B loop. Toxin V and Toxin var5 from *C. exilicauda* are exceptions to this structural dichotomy. Both toxins have short J loops, but Toxin V has a longer B loop than Toxin var5. The structural features in the J and B loops of Toxins V and var5 suggest that these two toxins may be phylogenetically more closely related to the Old World scorpion toxins than are other New World toxins that have been sequenced thus far.

Scorpions are a unique group for studies of molecular phylogeny. They are ancient, and their existence spans the time period for the separation of the continents. Venom from an individual species contains many small homologous proteins. Thus, at the molecular level, evolutionary changes are recorded that can be studied both within and between species. The strategy of comparing primary structures of proteins to establish phylogenetic relationships has been exploited successfully with other homologous proteins (e.g., cytochrome C, myoglobin, the postsynaptic snake toxins, and many enzymes: see Creighton 1984). Similar studies are under way with the scorpion toxins and promise to be equally revealing (Erickson 1978, Ivanov 1981, Dufton and Rochat 1984, Possani et al. 1985).

In comparison with most of the scorpion toxins listed in Table 10.3, the so-named Insect Toxins (Table 10.3C) and the short-chain toxins (Table 10.3D, E) have considerably different primary structures. These differences, however, are not necessarily related to variations in either biological action or host specificity. The adjacent cysteine pairs (–CC–, boxed sequences) are atypical for the scorpion toxins, although they commonly occur in the snake toxins. The three toxins listed in Table 10.3D have eight cysteine residues and four disulfide bonds, whereas Toxin Cn NOX (Table 10.3E) has six cysteines and presumably three disulfide bonds. Hence these small peptides are very tightly folded molecules (Bystrov et al. 1983).

Conformation

The presence of homologous regions in the sequences of a family of proteins suggests that their biological activity is in some way dependent upon these regions. The ultimate determinant of biological activity, however, is the specific folding of a given protein into a functional three-dimensional structure (conformation). The conformation of one scorpion toxin, Toxin var3 from *Centruroides exilicauda,* has been determined at high resolution (Fontecilla-Camps et al. 1980, 1982; Almassy et al. 1983). Space-filling models of Toxin var3 (Fig. 10.2*A*, *B*) highlight regions of conserved amino acids and hydrophobicity. In Figure 10.2*A* the conserved sequences 1 through 4 and 51 through 53 (positions 3–6 and 53–55, Table 10.3A) are enclosed within the outlined area and form a continuous surface patch with tyrosine 58 (position 65, Table 10.3A), located at the lower right (arrow). Many of the hydrophobic residues in Toxin var3 form a continuous patch (shaded area, Fig. 10.2*A*) on one surface of the molecule. If the model is rotated 180°, a largely hydrophilic surface with narrow bands of hydrophobic residues is observed (Fig. 10.2*B*). The concentration of conserved and hydrophobic amino acids on the surface of Toxin var3 may be associated with binding and biological activity. Whether individual amino acids located within this patch serve a unique function in the biological action of the toxin has not yet been determined. The conformation reported for Toxin var3 is believed to be typical for the scorpion toxins in general. It is evident from the model that the toxin is a very compact, tightly folded structure, which undoubtedly is maintained by the four disulfide bridges. For details about the conformation of Toxin var3, the reader is referred to Fontecilla-Camps et al. (1982), Almassy et al. (1983), and Zell et al. (1985). It is noteworthy that results from nuclear-magnetic-resonance studies with Insect Toxin I_5A, a 36-amino-acid toxin from *Mesobuthus eupeus,* show elements of secondary and tertiary structure closely resembling those of Toxin var3 (Bystrov et al. 1983).

Recently, the three-dimensional structure of a second scorpion toxin

---

Fig. 10.2. Space-filling model of Toxin var3 from *Centruroides exilicauda*. All hydrophobic amino acid residues are darkened. *A*, Hydrophobic residues form a continuous patch across this side of the molecule. Tyrosine 58 lies on edge in the lower right-hand segment (arrow); seven of the eight aromatic amino-acid residues lie in this patch. Conserved residues are outlined by a dashed line. *B*, The model rotated 180° about the vertical axis. Strips of buried hydrophobic residues project through this side of the molecule. The light residues demonstrate the predominantly hydrophilic character of this side of the molecule. (From Fontecilla-Camps el al. 1982)

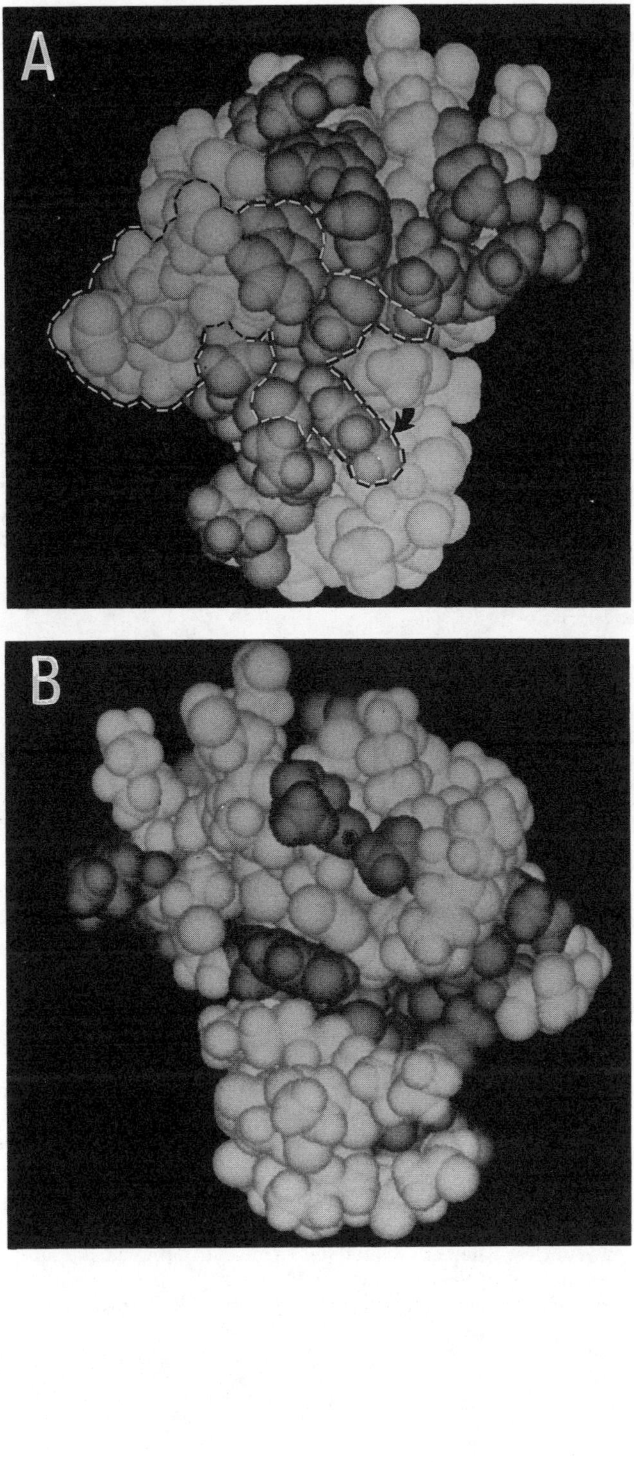
A
B

has been determined, Toxin II from *Androctonus australis** (Fontecilla-Camps et al. 1988). The core structure of this toxin is similar to that of Toxin var3 from *C. exilicauda*, in that this core also contains the α-helix and antiparallel β-sheet, and three of the four disulfide bridges. There is a marked difference, however, in the orientation of the loops protruding from the core. (See Fontecilla-Camps et al. 1988 for further details.)

Biological Action

Scorpion venoms are poisonous to a broad spectrum of animals, including humans, rodents (rats, mice, guinea pigs), rabbits, chicks, fish, arthropods (fly larvae, crickets, moths, mealworms, isopods), and many others. Notably resistant to the venom are certain spiders (the tarantula *Aphonopelma smithi*), some species of locust, and scorpions themselves (see also Chapter 7). The resistance of the scorpion to its venom results from toxin neutralization by the hemolymph. If venom is injected into a nerve ganglion, however, the scorpion rapidly dies (Shulov 1955).

Toxicity

Table 10.4 is a compilation of acute toxicity data in the mouse for various scorpion venoms. Although toxicity data of this sort may have little relevance to the biology of a scorpion in its natural habitat, these data do have relevance in classifying the importance of various species of scorpion as hazards to humans. As shown in the table, several species of scorpion from North Africa and the Middle East have venoms with high toxicities to mice. In the Americas, venoms from species of *Centruroides* and *Tityus* are the most poisonous.

Electrophysiology

Numerous studies have been carried out on the effects of scorpion envenomations *in vivo* and on the effects both of venoms and of isolated toxins on *in vitro* tissue preparations. In principle, the results of such investigations are best understood in the context of the action of

*Many venomologists studying *A. australis* have referred to the species in their literature as "*Androctonus australis* Hector" (*sic*). Since *A. australis* was in fact first described by Linnaeus, it is unclear whether such references mean to designate the subspecies *A. australis hector*, or are erroneously naming one Hector as the first to describe the species *A. australis*. Resolving that question is beyond the scope of this paper.

TABLE 10.4

Toxicities in mice of venom from various scorpion species

Species	Dose[a]
SCORPIONS OF MEDICAL IMPORTANCE	
Androctonus amoreuxi	0.75
A. australis	0.32
A. crassicauda	0.40
A. mauretanicus	0.31
Buthus occitanus tunetanus	0.90
Centruroides exilicauda	1.12
C. limpidus tecomanas	0.69
C. santa maria[b]	0.39
Leiurus quinquestriatus	0.25
Parabuthus transvaalensis	4.25
Tityus serrulatus	0.43
SCORPIONS NOT OF MEDICAL IMPORTANCE	
Hadogenes sp.	2,000–2,667
Hadrurus arizonensis	168
Pandinus exitialis	40

SOURCES: scorpions of medical importance, Zlotkin et al. 1978; *Hadogenes*, Bücherl 1971; *Hadrurus*, *Pandinus*, Tu 1977.

[a] LD_{50} expressed in mg per kg of mouse; venom given subcutaneously, except from *Hadrurus* (administered intraperitoneally).

[b] Zlotkin's *C. santa maria*, apparently deriving from Balozet, must be regarded as a *nomen nudum*.

the toxins on their main targets, excitable cells. Our current understanding of the fundamental biological action of the scorpion toxins has come largely from electrophysiological studies on isolated nerve and muscle cells using the voltage-clamp technique. Voltage clamping allows the quantitative study of the membrane ionic currents that contribute to the excitability of the cell.

The principal molecular targets of the scorpion toxins are the voltage-dependent Na^+ and K^+ channels that are found in many excitable tissues and are responsible for the action potential in nerves. Normally, Na^+ channels undergo transitions between (multiple) nonconducting and conducting states. In the classical view formulated by Hodgkin and Huxley (1952), the turning on and turning off of the Na^+ current is governed by voltage-dependent rate processes referred to as activation and inactivation. Under voltage-clamp conditions, a step depolarization of the membrane is associated with a transient Na^+-inward current. The process that governs the turning on of this current is referred to as activation and is responsible for the initiation of an action potential. The process that turns off the Na^+-inward current during maintained depolarization and is in part responsible for termination of the action potential is referred to as inactivation. An outward K^+ current also contributes to terminating the action potential. (For an introduction to these basic concepts of electrophysiology, see Junge 1976.) It is notable that there are different toxins present in scorpion venoms that affect all three mechanisms regulating the action potential (the activation and inactivation of Na^+ currents as well as the K^+ currents) and that individually and cooperatively these effects result in an increase in excitability.

The majority of studies concerning the effects of scorpion toxins on excitable cells have focused on Na^+ currents. Fewer studies have examined the effects on K^+ currents (see below), and there are essentially no data on other types of channels. Most of the studies have been carried out using traditional electrophysiological preparations, for example, the squid axon or the frog or toad node of Ranvier (Meves et al. 1986). In a few instances, however, frog muscle fibers, cockroach or crayfish axons, tunicate eggs, and mouse neuroblastoma cells have been used. The results of these studies allow a grouping of the observed effects of the toxins into three general classes. The first class consists exclusively of effects on the process of Na^+ inactivation. The second class involves principally an effect on the process of Na^+ activation, with a minor effect on inactivation and a nonspecific reduction of peak inward currents. Finally, in the third class, only a reduction of the inward Na^+ or outward K^+ currents is observed. Although many scorpion toxins fall neatly into the first or second class when tested in certain preparations, toxins from either group may fall into the third class when tested in other preparations (see below for examples). Thus, it may be preferable to specify classes of effect, rather than classes of toxin with a specific effect.

In the first class, the observed effect of certain toxins consists exclusively of an alteration in the inactivation of Na^+ channels. Inactivation is slowed and usually is made incomplete. Thus a significant inward current may be present at the end of a depolarizing pulse (Fig. 10.3). The result is a prolongation of the action potential to many times its normal duration. This effect was first observed by Koppenhöfer and Schmidt (1968a, b) using whole venom from the Old World scorpion *Leiurus quinquestriatus quinquestriatus* and has come to be known as the classical effect of the scorpion toxins. It has since been observed with

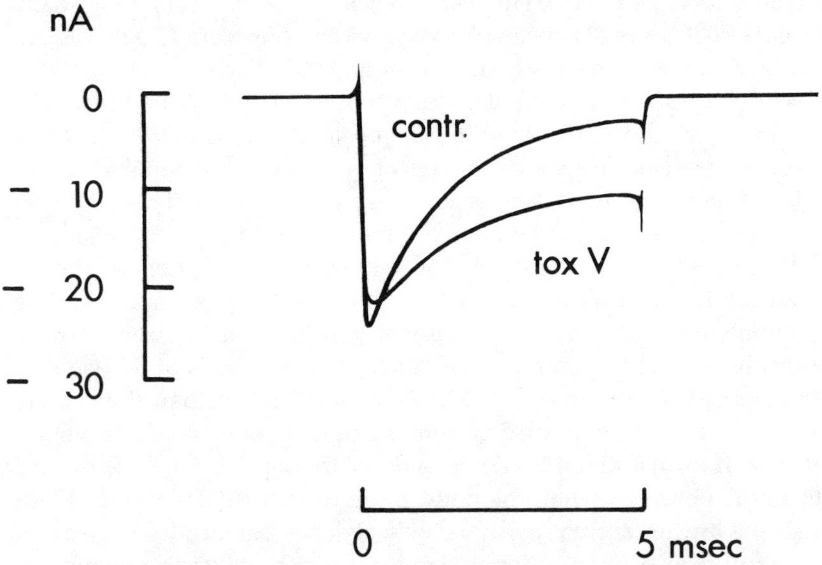

Fig. 10.3. Effect of 3.3 μg/ml Toxin V from *Centruroides exilicauda* on the inward current of the voltage-clamped frog node of Ranvier. Note the slowing and incompleteness of inactivation following treatment with toxin. Holding potential, −92 mV; test potential, −22 mv. (After Hu et al. 1983)

venoms and several isolated toxins from other Old World species of scorpion, including *Androctonus australis, Mesobuthus tamulus,* and *M. eupeus* (Narahashi et al. 1972, Schmitt and Schmidt 1972, Mozhayeva et al. 1980, Pelhate and Zlotkin 1982, Wang and Strichartz 1982). Slowing of inactivation has been observed also with some isolated toxins from the New World species *Centruroides exilicauda* (Meves et al. 1982, Wang and Strichartz 1983). In addition, the effect and possibly the binding of toxins in this class appear to be voltage-dependent, since these parameters are decreased by a depolarization of the membrane brought about either by increasing $[K]_o$ or by a strong depolarizing conditioning pulse (Catterall et al. 1976; Catterall 1979; Mozhayeva et al. 1980; Meves et al. 1983, 1984a).

The second class of toxin effect on Na^+ channels has been observed mostly with venoms or toxins from New World species of scorpion. This effect has been studied most extensively using venom and isolated toxins from *C. exilicauda* (Cahalan 1975; Meves et al. 1982, 1984b; Wang and Strichartz 1982, 1983; Hu et al. 1983; Simard et al. 1986). Essentially similar effects have also been observed with toxins from *C. suffusus suffusus* and *Tityus serrulatus* (Couraud et al. 1982, Barhanin et al. 1983). Recently, a toxin with a similar effect has also been isolated from the venom of the Old World scorpion *L. quinquestriatus* (Rack et al. 1987). The characteristic feature of this effect is a transient, depolarization-induced shift in the voltage dependence of activation. This effect results in an increased tendency of the membrane to fire spontaneously and repetitively (Cahalan 1975). The shift in the voltage dependence of activation is such that inward Na^+ currents are observed at membrane potentials more negative than those at which inward currents would normally occur. For example, in an untreated frog node of Ranvier hyperpolarized to −92 mV (i.e., 22 mV more negative than the normal resting potential), no inward current is normally observed. After treatment with toxin and application of a depolarizing pulse, a large inward current is observed when the node is returned to −92 mV (Fig. 10.4). Since the inward current decays slowly over several hundred msec and is best observed after a depolarizing pulse, the shift in voltage dependence of activation is referred to as transient and depolarization-induced. The decay of the inward current (Fig. 10.4) is thought to reflect the slow return of Na^+ channels to the state of voltage dependence present before the depolarizing pulse.

In the example given above (at −92 mV), only one membrane potential was examined for this effect of toxin. Examining the toxin-treated node at other potentials yields a more complete picture of the effect, as illustrated in Figure 10.5. In the figure, the peak inward currents at different membrane potentials without (open circles) and with (solid circles) a conditioning depolarizing pulse are plotted as a function of membrane potential. It is evident that the conditioning pulse causes a shift to more negative potentials of the descending branch of the current-voltage curve. That is, as mentioned above, inward currents are observed at potentials more negative than one would expect. The magnitude of the transient depolarization-induced negative shift in voltage dependence of activation varies up to −60 mV, depending in part on the particular toxin studied, on its concentration, and on the strength of the depolarizing conditioning pulse (Cahalan 1975, Meves et al. 1982).

This effect on activation brings out clearly an important difference between the second class of toxins, affecting principally activation, and the first class, affecting inactivation. In the first class, depolarization

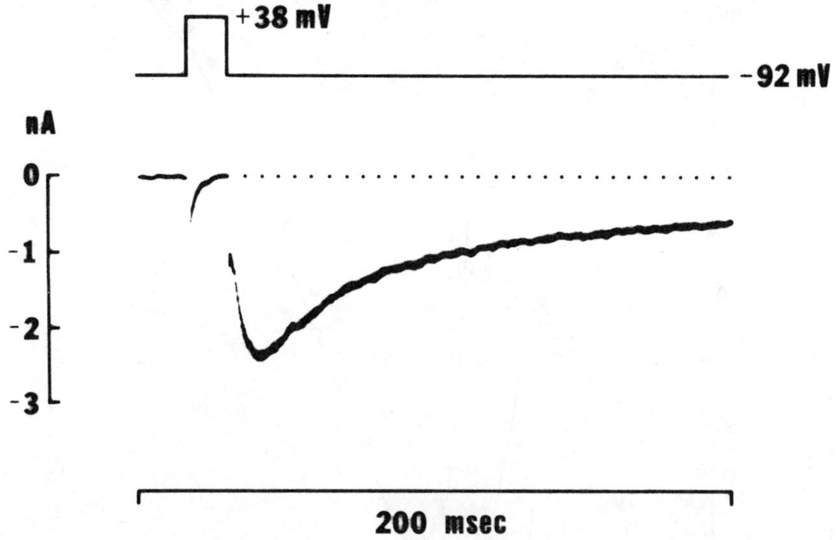

Fig. 10.4. Effect of 0.83 μg/ml Toxin IV from *Centruroides exilicauda* on the inward current of the voltage-clamped frog node of Ranvier. Following a 15 msec depolarizing pulse to 38 mV, the node was returned to the holding potential (−92 mV). Upon repolarization, a large inward current developed, which decayed only slowly. The inward current at −92 mV reflects the transient shift in voltage dependence of activation brought on by the depolarizing pulse in the Toxin IV–treated node. The rate at which the inward current turns on corresponds to the recovery of inactivation, produced by the depolarizing pulse. The decay of the inward current reflects the slow return of the activation process to the state of voltage dependence present before the depolarizing pulse. (After Meves et al. 1982)

removes the effect of the toxins; whereas in the second class depolarization manifests the action of the toxins. Toxins in the second class also make inactivation incomplete, although they cause little or no slowing of inactivation. Finally, these toxins also generally cause a significant reduction in magnitude of the peak inward currents during depolarizing steps. The last effect is considered to be nonspecific, however, and mimics the action of tetrodotoxin, a more potent Na^+-channel blocker.

The third class of effect involves simply the reduction of ionic currents, with no changes in the kinetics or voltage dependence of activa-

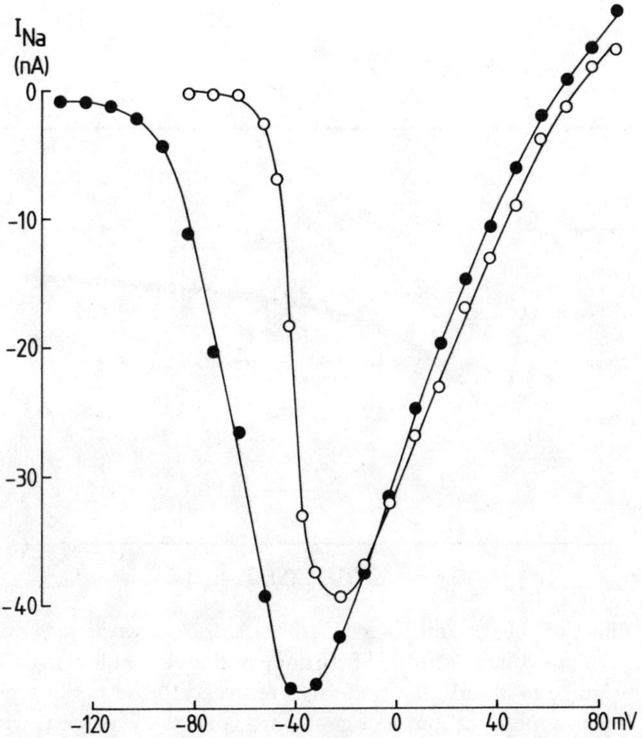

Fig. 10.5. Effect of a toxic isolate (0.2 μg/ml Toxin VI) from zone 12 of *Centruroides exilicauda* on the inward current of the voltage-clamped frog node of Ranvier. Note the 30 mV shift of the descending branch of the current-voltage curve to more hyperpolarized potentials when the test potential is preceded by a depolarizing conditioning pulse, 15 msec to E = 38 mV, followed by a 20 msec pause before each test pulse. Holding potential, −92 mV; TTX-insensitive currents subtracted. ○, toxin with no conditioning pulse; ●, toxin with conditioning pulse. (From Simard et al. 1986)

tion or inactivation. Either Na^+ or K^+ currents may be reduced. Toxin I from *A. australis,* normally considered as belonging to the first class (affecting inactivation of Na^+ currents) exhibits only a simple blocking effect when tested in the giant axon of *Sepia* (Romey et al. 1975). Similarly, Toxin II from *C. s. suffusus* and Toxin γ from *T. serrulatus,* which shift the voltage dependence of activation in both frog-node and neuroblastoma cells, respectively (Couraud et al. 1982, Barhanin et al. 1983), cause only a simple block of Na^+ currents in skeletal muscle (Jaimovich et al. 1982, Barhanin et al. 1984). An interesting cooperativity between two isolated fractions from *Scorpio maurus palmatus* has been demonstrated. By themselves, Insect Toxins I and II from this venom have little effect on the cockroach axon; yet combined and at lower concentrations they reversibly decrease both Na^+ and K^+ currents (Lazarovici et al. 1982).

Recently, increasing attention has been focused on the action of scorpion toxins on K^+ currents. Venoms or isolated toxins from several species block delayed rectifier K^+ channels in nerve preparations (Koppenhöfer and Schmidt 1968a; Meves et al. 1982; Carbone et al. 1982, 1983, 1987) or Ca^{2+}-activated K^+ channels in mammalian muscle or carcinoma cells (Miller et al. 1985, Leneveu and Simonneau 1986). A blocking effect on K^+ channels is particularly interesting because this would, in general, increase excitability, and thus facilitate the actions of toxins with the first two classes of effect.

Pathophysiology

The functional correlates of the fundamental effects of the scorpion toxins noted above are many. Not surprisingly, a plethora of physiological and neurochemical studies has been published, detailing the various effects of venoms and toxins from many species of scorpions on organs and tissues both *in vivo* and *in vitro.* A comprehensive review of these investigations is beyond the scope of this chapter. However, at the risk of oversimplification, certain salient features may be outlined. At the cellular level, as was mentioned previously, prolongation of the action potential and repetitive firing may be expected (e.g., Adam et al. 1966, Simard and Watt 1983). In addition, depolarization of cell membranes occurs (e.g., Narahashi et al. 1972, Lin Shiau et al. 1975). As a consequence, intracellular accumulation of Na^+ or Ca^{2+} ions may be observed (e.g., Catterall 1975, 1976; Couraud et al. 1976, 1980). As would be expected, the release of various neurotransmitters, including acetylcholine, norepinephrine, and γ-amino butyric acid (GABA), therefore may be documented (e.g., Gomez et al. 1973; Moss et al. 1974a, b; Freire-Maia et al. 1976; Couraud et al. 1982). In neuromuscular preparations, the facilitated release of neurotransmitters augments spontaneous miniature endplate potentials as well as endplate potentials after single shocks (e.g., Parnas et al. 1970, Rathmayer et al. 1977). Thus, baseline tension and the twitch responses of various muscles may be exaggerated (e.g., Adam and Weiss 1959, Warnick et al. 1976, Watt et al. 1982). Autonomically innervated organs are also affected, including heart, gut, and uterus, with the dominant effect observed generally being attributable to the dominant autonomic influence (e.g., Osman et al. 1972, Tintpulver et al. 1976, Grupp et al. 1980).

The cumulative effect in the whole animal subjected to envenomation is complex. Although, as implied above, many of the *in vivo* effects may be traced to alterations in the resting and action potentials brought about by the toxins, current knowledge is incomplete, and other possibilities cannot be excluded. After envenomation, organs and tissues especially affected are the cardiovascular system, lungs, visceral smooth muscle, uterus, various glands, and skeletal muscle. Nerve terminals of the autonomic system and adrenal medulla are depolarized, causing the massive release of neurotransmitters, including catecholamines. Envenomation thus results in a mixture of sympathetic, parasympathetic, and paralytic symptoms.

The onset and progression of symptoms from experimental envenomation of the mouse and chick are listed in Table 10.5. Responses in dogs to scorpion venom follow a similar pattern to those listed for the mouse and include parasympathetic symptoms such as lacrimation, salivation, sphincter relaxation, gastric distension, exaggerated peristalsis, bradycardia, and hypotension; and sympathetic symptoms such as mydriasis, perspiration, piloerection, hyperglycemia, tachycardia, and hypertension. Pathological studies of artificially envenomated dogs reveal extensive hemorrhages in the heart, kidneys, lungs, and intestines. There is general congestion of all organs. Histopathological findings include focal myocarditis and thrombotic occlusion of capillaries (Reddy et al. 1972). Many of the pathophysiological findings from scorpion envenomations are believed to be due to the massive release of catecholamines from the adrenal glands and sympathetic nerve endings. It would appear, therefore, that the complex pathophysiological responses to scorpion venoms do not result exclusively from a direct effect of the toxins on affected organs and tissues.

Because of speedy absorption and distribution of the toxins, death from scorpion envenomation can be dramatically rapid. The toxins bind quasi-reversibly, however, and recovery may be rapid and complete. Mechanisms of *in vivo* detoxification are poorly understood. Ismail et al. (1974) injected ^{125}I-labeled venom of *Leiurus quinquestriatus quinquestriatus* into female rats. High levels of radioactivity were found in the urine 15 minutes after administration. Scorpion toxins are not

TABLE 10.5

Responses of mice and chicks to venom of Centruroides exilicauda

MICE[a]	
	Immediate severe local pain evidenced by vocalization, favoring the limb, and licking the site of injection; rapid swelling of the involved limb within a few minutes after envenomation
	Restlessness, anxiety, hypersensitivity to noise and manual stimulation develop within 5–10 minutes after envenomation; salivation blocked by atropine
	Convulsions, especially in response to touch and loud noises, develop within 10–15 minutes after envenomation; apparent stupor with maintained responsiveness to touch and loud noises
	Rapid and labored breathing, muscle fasciculations and weakness, undulations of the tail, and various combinations of flaccid and spastic paralysis are late responses and may occur within 10–30 minutes after envenomation
	Of those dying, death occurs by respiratory arrest between 15 minutes and 24 hours (the maximum time of observation) after envenomation
CHICKS[b]	
	Symptoms as in mice: restlessness expressed by backing away and running forward; immediate pain at the site of injection expressed by lowering the head, backing away, and vocalization
	Salivation intense, the crop becoming engorged with viscous fluid; blocked by atropine
	Muscle weakness (drooping of the head, unsteadiness, lying on side); irreversible spasticity begins within 10 minutes after envenomation

SOURCE: Watt et al. 1974.

[a] Subcutaneous administration of LD_{50} dosage of venom to the medial aspect of the left rear leg; $LD_{50} = 1.40$ mg/kg.

[b] Subcutaneous administration to the dorsum of the neck; $LD_{50} = 0.15$ mg/kg.

readily hydrolyzed by the commonly occurring proteolytic enzymes. Whether the toxins are excreted as intact molecules or as degradation products remains to be determined.

Envenomations in Humans

Despite the historical and medical importance of scorpion envenomation in humans, a health problem that dates back to antiquity, accurate information on most aspects of this topic is fairly scarce and, it must be said, rather little progress has been made in treatment.

Distribution of Dangerous Species

Scorpions dangerous to humans are found in desert or semiarid regions throughout the world. Their habitats include areas where pro-

TABLE 10.6

Geographic distributions of selected medically important scorpions

Species	Geographic distribution
NORTH AMERICAN	
Centruroides exilicauda	SW U.S., N Mexico
C. l. limpidus	Mexico
C. noxius	Mexico
C. s. suffusus	Mexico
Tityus trinitatis	West Indies
SOUTH AMERICAN	
T. bahiensis	Argentina, Brazil
T. cambridgei	Guyana
T. serrulatus	Brazil
T. trinitatis	Venezuela
AFRICAN AND MIDDLE EASTERN	
Androctonus australis	Algeria, Egypt, Morocco
A. crassicauda	Israel, Iraq, Turkey
Buthus occitanus	Morocco, Algeria, Jordan
Hottentotta minax	Sudan
Leiurus quinquestriatus	Egypt, Israel, Turkey
Parabuthus sp.	s Africa
INDIAN	
Mesobuthus tamulus	s India

SOURCE: Keegan 1980 (where a more extensive listing can be found).

tection and fresh water are available, that is, areas frequently inhabited by humans. In the Americas, dangerous species of scorpion are found in the southwestern United States, Mexico, the east-central area of South America, and islands of the Caribbean. Elsewhere, areas of inhabitance span the equator to include northern and southern Africa, the Middle East, and the northern Mediterranean, and extend across Iran and into Soviet central Asia and India (see Chapter 6). Of the more than 1,400 species of scorpion throughout the world, few are of medical importance. Species of medically important scorpions belong to the family Buthidae: the genera include *Androctonus, Buthacus, Buthus, Centruroides, Leiurus, Mesobuthus, Parabuthus,* and *Tityus.* A listing of selected medically important scorpions and their geographic distribution is given in Table 10.6. A more extensive tabulation, with references, is given by Keegan (1980). Additional discussions may be found in Balozet (1971), Bücherl (1971, 1978), Efrati (1978), Shulov and Levy (1978), and Hershkovich et al. (1985).

Epidemiology

Accurate statistics on the incidence of scorpion envenomations worldwide are not available. Isolated reports have appeared providing morbidity and mortality data for victims entering health centers (Table 10.7). Such data, however, do not reflect the incidence of unreported envenomations and thus skew morbidity and mortality statistics. Exemplifying the high proportion of stings that may not find entry into reports from medical centers is the large series collected by Likes et al. (1984). In 438 cases of scorpion stings with diagnosed neurotoxic symptoms, 92 percent were managed conservatively at home. These data correlate with the observation that most stings in adults are not associated with systemic symptoms and thus are unlikely to be reported (Efrati 1978, Goyffon and Kovoor 1978). Adults are more often victims of scorpion stings than are children (Campos et al. 1980, Goyffon et al. 1982, Curry et al. 1983–84, Likes et al. 1984). However, since mortality and morbidity are greater in children (Bücherl 1971, Campos et al. 1980, Rimsza et al. 1980), series reported from treatment centers may inadvertently misrepresent the age incidence of envenomations.

There appears to be a seasonal variation in the incidence of envenomations, with a larger number of cases occurring during summer and early autumn (Goyffon and Kovoor 1978, Campos et al. 1980, Likes et al. 1984).

TABLE 10.7

Selected reported incidences of envenomation by medically important scorpions

			Incidence (yr)	
Principal species	Locality	Years	Stings	Deaths
1. *Androctonus australis*	Algeria	1942–58	1,260	24
2. *A. australis*	Tunisia	1967–77	2,672	12
3. *Mesobuthus tamulus*	India	1971–72	151	12
4. *Centruroides* sp.	Mexico	1940–49	—	1,800
5. *Centruroides* sp.	Mexico	1981	300,000 [a]	1,000[a]
6. *C. exilicauda*	U.S.	1981–82	970	0
7. *Tityus serrulatus*	Brazil	1938–45	843	16

SOURCES: 1, Balozet 1971; 2, Goyffon et al. 1982; 3, Santhanakrishnan & Raju 1974; 4, Mazzotti & Bravo-Becherelle 1963; 5, WHO Report 1981; 6, communicated by Arizona Poison Control Center, 1983; 7, Diniz 1971.

[a] Although these figures were reported in a recent WHO publication, there is reason to believe that the data are exaggerated and probably represent figures from 30 to 40 years ago. An estimated incidence of 100,000 stings per year (Possani et al. 1980) may be more realistic. Recent mortality statistics estimate about 800 deaths per year (L. D. Possani, pers. comm. 1988).

In areas with favorable habitats for dangerous scorpions, programs for protection of small children, eradication of the scorpion population, and cleaning up debris help diminish the incidence of envenomations. Although contact insecticides (DDT, BHC) at sufficiently high concentrations can be effective against scorpions, the environmental effects of high levels of these pesticides are generally undesirable. Specific recommendations on housing construction to minimize scorpion entry are given by Mazzotti and Bravo-Becherelle (1963) and Goyffon and Kovoor (1978). It is reasonable to expect that there will be a decline in the incidence of serious envenomations in epidemic areas as programs of public awareness are implemented.

Symptomatology

Symptoms from envenomations by scorpions are surprisingly uniform, regardless of the envenomating species (Yarom 1970, Goyffon and Kovoor 1978, Newlands 1978b, Campos et al. 1980, Keegan 1980, Rimsza et al. 1980). Sites of envenomation are most frequently the distal extremities, although any body part may be involved. Ocular envenomation by "spitting" has also been reported (Newlands 1978b). Local symptoms, lasting up to several hours, generally consist of burning pain, occasionally accompanied by swelling, erythema, or ecchymosis. Adolescents and adults invariably complain of pain, whereas children do so less frequently (Santhanakrishnan and Raju 1974, Rimsza et al. 1980).

Systemic symptoms may be absent, may develop within minutes, or may be delayed several hours. The development of systemic symptoms is related to several factors, including the body mass of the victim (with mortality being five to ten times greater in infants than in adults), the location of the sting on the victim's body (the degree of vascularity may be a factor), the amount of venom injected, and the physical health of the subject. Systemic symptoms in humans from scorpion envenomation may be complex, but, either directly or indirectly, are due largely to the action of the toxins on peripheral excitable tissues, as outlined in "Biological Action," above.

Sweating, pallor, tachypnea, tachycardia, and hypertension are frequent signs of early systemic involvement. Findings of autonomic and branchial motor involvement may follow, with salivation and production of frothy sputum; respiratory wheezing, stridor, and distress; blurred vision, roving eye movements, and oculogyric crises; and slurred speech, dysphagia, and pharyngeal spasms. Advanced symptoms include muscle twitching, motor hyperactivity, incoordination, increased tone, myoclonic twitches, and even opisthotonos. Muscle stretch reflexes may be either hyperactive or hypoactive. Extreme restlessness, which superficially may resemble seizure, is a frequent presenting sign in children (Santhanakrishnan and Raju 1974, Rimsza et al. 1980). Chest pain, which may be due to cardiovascular effects, as well as vomiting, gastric distension, diarrhea, abdominal cramps, and temperature dysregulation have also been reported. Late findings include an altered sensorium, cyanosis, hypotension, bradycardia, paresis, incontinence, seizures, and respiratory and cardiac failure. Some of the late findings may not result directly from envenomation, but may be secondary to general systemic failure precipitated by envenomation. Death following scorpion envenomation has been attributed to cardiovascular failure complicated by pulmonary edema (Yarom 1970) and to respiratory arrest (Balozet 1971). The specific cause of death may vary, given the overall complexity of the syndrome.

Balozet (1971) noted that after apparent clinical improvement, sudden reappearance of respiratory distress may occur. It has thus been suggested that patients with severe systemic involvement should be observed closely during the initial stages of recovery.

Characteristic findings in patients after scorpion envenomation are elevated levels of serum and urine catecholamines and their metabolites. Leukocytosis, hyperkalemia, hyponatremia, and hyperglycemia have also been noted. At autopsy, victims may be found to have diffuse hemorrhages and passive congestion in various organs, as well as focal myocarditis with acute infiltrates and fatty necrosis.

It is interesting to note that envenomations by spiders of the genus *Latrodectus* present clinical features comparable to those observed from scorpion envenomations. The symptomatology of latrodectism suggests that the principal sites of action of the venom are the somatic and autonomic nervous systems, with release of neurotransmitters and initial excitatory effects (Maretic 1983).

Clinical Management

Controversy remains regarding appropriate clinical management of scorpion envenomations. Unfortunately, no controlled studies have been performed on the treatment protocols that have been advocated. We have no direct experience with clinical management of envenomated patients and thus cannot advocate any particular treatment modality.

Symptoms of scorpion envenomation are generally self-limited, with resolution frequently occurring within several hours. Only in rare cases do symptoms extend beyond 72 hours. In most adults, symptoms are sufficiently mild to warrant treatment at home or on an outpatient basis, without hospitalization. Local hypothermia (an ice pack) and analgesics (e.g., aspirin or codeine) by mouth may be all that are required. Severe cases, usually in children demonstrating agitation and other systemic symptoms, may require hospitalization for close monitoring and supportive care. Nonhypnotic doses of phenobarbital may be useful in countering severe agitation (Rimsza et al. 1980). Sedation was the mainstay of therapy employed by Santhanakrishnan and Raju (1974) in their series of 301 Indian children hospitalized with systemic symptoms. Seven deaths were reported in this series. These authors used the "lytic cocktail" familiar to pediatricians, consisting of Demerol, Phenergan, and Thorazine (meperidine, promethazine HCl, and chlorpromazine HCl). Others have cautioned against the use of narcotic analgesics, especially meperidine (Greenberg and Ingalls 1963, Stahnke and Dengler 1964, Gotlieb 1966). A rational pharmacological basis for this warning is unclear, unless the point is to avoid doses large enough to cause respiratory depression. Use of other drugs, notably atropine and β-blockers, may not be unreasonable, depending on the severity of the target symptoms (Bawaskar 1984, Bawaskar and Bawaskar 1986, Freire-Maia and Campos 1987, Gueron and Ovsyshcher 1987). No controlled data have been presented regarding the beneficial effects of autonomic blocking agents in the treatment of scorpion envenomations. Anecdotal reports of the relief of symptoms in humans after administration of $CaCl_2$ (Abdullah 1957) have not been generally substantiated. Reports of treatment efficacy without adequate controls are complicated, of course, by the self-limited nature of scorpion envenomations.

Use of antiserum (antivenom) has been advocated (e.g., Balozet 1971, Delori et al. 1976, Newlands 1978b, Campos et al. 1980). Both technical and therapeutic factors have limited its utility. The toxins in scorpion venoms are not good antigens and may require a long period of immunization in order to produce an effective antiserum. Further, venoms

TABLE 10.8

Institutions producing antivenoms

Producer	Species
Antivenom Production Laboratory Arizona State University Tempe, Arizona 85281 U.S.A.	*Centruroides exilicauda*
Instituto Nacional de Higiene Av. M. Escobedo 20 Mexico D.F., Mexico	*C. noxius*
Laboratorio Zapata Mexico D.F., Mexico	*C. noxius, C. suffusus*
Laboratorios "Myn," S.A. Av. Coyoacan 1707 Mexico D.F. 12, Mexico	*C. limpidus* or *C. noxius, C. suffusus*
Instituto Butantan P. O. Box 65 05504 São Paulo, Brazil	*Tityus bahiensis, T. serrulatus*
Lister Institute of Preventative Medicine Elstree, Hertfordshire WD6 3AT England	*Androctonus australis, Buthus occitanus, Leiurus quinquestriatus*
Reyfik Saydan Central Institute of Hygiene Ankara, Turkey	*A. crassicauda, L. quinquestriatus*
Ministry of Health Department of Laboratories P. O. Box 6115 Jerusalem 91060, Israel	*L. quinquestriatus*
Institut Pasteur Place Charles-Nicolle Casablanca, Morocco	*A. mauretanicus*
Institut Pasteur d'Algérie Rue du Docteur Laveran Algiers, Algeria	*A. australis, B. occitanus*
Institut Pasteur 13 Place Pasteur Tunis, Tunisia	*A. australis, B. occitanus*
State Serum and Vaccine Institute Agouza Cairo, Egypt	*L. quinquestriatus*
Institut d'État des Sérums et Vaccins Razi Hessarek P. O. Box 656 Tehran, Iran	*A. crassicauda, Hottentotta saulcyi, Hemiscorpius lepturus, Mesobuthus eupeus, Odontobuthus doriae, Scorpio maurus*
Central Research Institute Kasauli, India	*M. tamulus*
South African Institute for Medical Research P. O. Box 1038 Johannesburg 2000, South Africa	*Parabuthus* sp.

SOURCES: Keegan 1980, WHO Report 1981, Chippaux & Goyffon 1983.

from different species of scorpions do not usually induce antisera that cross-react. The limiting therapeutic factor in treatment with antiserum concerns principally the risk of anaphylactic reaction to which the patient is exposed (Mazzotti and Bravo-Becherelle 1963). Preliminary assessment of experience in about 30 patients treated with goat antiserum to venom from *Centruroides exilicauda* revealed minor rashes in only two patients (D. Kunkel, pers. comm. 1983; Curry et al. 1983–84).

Although there are claims of decreased rates of mortality from scorpion envenomations with the use of antiserum (Galván Cervantes 1966, Campos et al. 1980), indications for use of antisera remain to be clarified by appropriate clinical trials. D. Kunkel (pers. comm. 1983) has found that in stings by *C. exilicauda* goat serum was of little use in mitigating either local or radiating pain and was of equivocal benefit in treating hypersecretion or symptoms of branchial motor involvement (see above), but was highly effective in treating advanced symptoms of motor hyperactivity frequently noted in children (Curry et al. 1983–84). Intramuscular administration is reportedly ineffective and may result in treatment failure.

A list of institutions currently preparing antiserum is given in Table 10.8 (see also Keegan 1980, WHO 1981, Chippaux and Goyffon 1983).

Summary

Scorpion venoms are an interesting lot. They are complex mixtures that contain a variety of low-molecular-weight, tightly folded globular proteins, the neurotoxins, as well as a large number of other constituents, largely cell contents, which accompany the apocrine secretion of the venom. The toxicities and compositions of the venoms vary widely, depending primarily on the species of origin.

The toxins comprise a large and distinct family of homologous proteins. Over 70 different toxins have been isolated from the venoms of different species of scorpion. The amino-acid sequences of about one-third of these isolates have been determined. Although there is considerable variability in the sequences, certain sequence segments are conserved. Frequently, amino-acid substitutions are conservative. The tertiary structure of Toxin var3 from *Centruroides exilicauda* demonstrates that the protein is a tightly folded globular molecule, stabilized by four disulfide bridges and containing a segment of α-helix and three short strands of antiparallel β-structure. The tertiary structure of Toxin var3 is believed to be prototypical for this family of homologous neurotoxic proteins.

The understanding of the physiological effects of the toxins is now sufficient for the symptomatology of envenomations in animals to be explicable at a molecular level. The toxins interact directly with the molecular machinery of excitable membranes, principally the voltage-dependent Na^+ channel. Voltage-clamp analysis reveals three classes of interaction between the toxins and the Na^+ channel. These interactions result in altered excitability manifested as depolarization, repetitive firing, and prolongation of the action potential. Uncontrolled release of neurotransmitters, most importantly from nerves to the adrenal medulla, results in what can be termed an adrenergic storm. Except for those organs sheltered by a diffusion barrier, such as that between the blood and the brain, virtually all major organ systems may be affected either directly by the toxins or indirectly by the released neurotransmitters. Thus is produced the complex pathophysiological syndrome observed in humans upon severe scorpion envenomation.

Although the importance of scorpions as environmental hazards is declining, scorpion stings continue to account for hundreds of deaths worldwide each year. Morbidity and mortality are higher in children. Effective management of envenomations hinges not only on appropriate and timely medical intervention when warranted, but also on prevention. Programs for reducing the populations of dangerous scorpions in urban and other inhabited areas have proved effective in lowering the incidence of envenomations.

This review has served to highlight certain aspects of our current understanding of scorpion venomology and toxinology. Numerous questions in these areas remain to be elucidated. This family of homologous toxic proteins will continue to provide scientists with unique tools for investigations in such diverse areas as studies on the Na^+ channel of excitable membranes, phylogeny of proteins, and structure–function relations of proteins.

ACKNOWLEDGMENTS

This work was supported in part by NIH Grant NS 12365 and by funds from the Department of Neurological Surgery, University of Florida, Gainesville. We gratefully acknowledge consultation provided by Donald Kunkel, M.D., Director of the Department of Medical Toxicology at St. Luke's Hospital, Phoenix, Arizona, and Co–Medical Director of the Arizona Poison Control Center; and by William Banner, M.D., Department of Pediatrics, University of Arizona, Tucson. We also thank Charles E. Bugg, Ph.D., University of Alabama, Birmingham; R. J. Feldman, Ph.D., Division of Computer Research and Technology at the NIH; and John E. Mole, Ph.D., University of Massachusetts, for their contributions. Invaluable secretarial and technical assistance was provided by Patricia Mancuso and Clare Manhart.

11

Field and Laboratory Methods

W. DAVID SISSOM,
GARY A. POLIS, AND
DEAN D. WATT

Scorpions exhibit a number of characteristics that make them excellent organisms for many types of research. Their suitability for various types of research has been noted repeatedly throughout the book. In this chapter, we outline methods used in different types of field and laboratory research.

Field Methods

Portable ultraviolet lights are the method of choice for scorpion detection, collection, and field research. Fluorescent lanterns, normally sold with white bulbs for camping, are easily equipped with UV bulbs for scorpion work (Stahnke 1972a). Good field lanterns are sold at camping-supply stores.

Scorpions fluoresce most brightly when exposed to near-ultraviolet light, radiating in the range 320–400 nm (3,200–4,000 Å). Scorpions will also fluoresce (but less brightly) in wavelengths as low as 250 nm. This "black light" is invisible unless it is reradiated at higher wavelengths in the visible range by fluorescent objects. Both Sylvania and General Electric manufacture a UV bulb that is useful in the field (model F8T5). This bulb emits a maximum intensity at 356 nm.

Although the wavelengths in the 320–400-nm range reportedly are not harmful to the human eye, shorter wavelengths are quite harmful. Fortunately, such high-energy short wavelengths cannot penetrate glass or most transparent plastics. Therefore, prescription or safety glasses are strongly advised for extended use of UV light in the field. Glasses also decrease the luminous glaze and occasional dizziness that occur when UV light is viewed. Luminous glaze is probably produced when the fluids within the eye fluoresce. One further note: the glass and plastic protective lenses of the portable lanterns act to decrease the intensity of emitted UV light. Therefore it is best to remove such lenses to achieve maximum UV emission.

Scorpions are best collected on dark nights when the moon is new or less than half full. On such nights, scorpions can be detected up to a distance of 15 m, and it is not uncommon to see 500 to 1,000 scorpions in good habitats. There are two reasons why scorpion collection is much poorer on nights brightly illuminated with moonlight. First, UV lanterns will detect scorpions only for a distance of 2 to 4 m because of the intense moonlight. Second, it appears that scorpion surface density is generally lower on bright nights (see Chapter 6). Thus collecting trips should be planned to coordinate with the period around the new moon.

Before the advent of UV-light detection, scorpions were collected either during daytime searches under bark, rock, and other surface debris, or with the aid of can traps. Can traps are unbaited, may or may not be covered, and may or may not contain preservative. The preservative is a 1:1 mixture of 70 percent ethyl or isopropyl alcohol and ethylene glycol (antifreeze). Ethylene glycol reduces the rate of evaporation of the fluid in the trap. Both of these methods capture a consistent but small number of specimens (e.g., Gertsch and Allred 1965). Our experience suggests a capture rate of less than 1 percent of the number observed with black light. It is further unfortunate that biases exist with can-trap and diurnal searches. Can traps usually catch a disproportionately greater percentage of vagile scorpions (i.e., mature males moving during the mating season) and species of errant scorpions (see Chapters 4, 6). Rock rolling and other diurnal searches likewise produce biased data because nonburrowing ("bark") scorpions are found much more frequently than scorpions that live in burrows.

Black-light detection is superior for other reasons. Black light usually does not disturb scorpions. Behaviors such as homing, orientation, feeding, and mating can be observed and quantified (e.g., Brownell 1977b; Polis 1979; Polis and Farley 1979a; Polis et al. 1985, 1986). Data relating to predation (e.g., prey identity and size, handling time) are easily tabulated. Courtship and mating behaviors occur normally if the observer does not bring the UV light too close. Spermatophores also fluoresce brightly and can be collected after mating. Further, with the use of unique marks (see below), the behavior and ecology of individuals can be studied. Thus home ranges, territorial dimensions, site fidelity, feeding rates, and surface activity can be determined.

Before closing this section on UV light, it is important to make two more recommendations. First, the use of UV light is potentially dangerous in habitats where poisonous snakes are abundant. This is particularly true for areas that are heavily vegetated. The use of snake chaps lined with metal cloth in combination with removing or (preferably) spraying snakes with fluorescent paint decreases some of the risk. Snake chaps are available in a variety of styles from companies that supply foresters and other outdoor workers. Second, when using UV light to study scorpions, it is better to record field data with mini- or microcassette tape recorders than to use a white light and record data directly in a field notebook. White light both disturbs scorpions and causes the observer's eyes to become less acclimated to dark vision, hence increasing risk during UV light operation. Further, tapes are rapidly transcribed.

Scorpions are easily marked in the field with enamel paint. Both fluorescent and regular paint are used. Some marks on adults were observed to last at least three years in the field. Unfortunately, immature animals lose their marks at ecdysis. Individuals are assigned a unique mark by using combinations of colors and dots placed on different locations on the mesosoma and metasoma. Care must be taken not to cover eyes or trichobothria nor to glue the legs together.

Marked scorpions can be used to estimate population density or size. Three useful methods of population estimation are the Jolly multiple mark-and-recapture procedure, a modified Moran or Zippen procedure of exhaustive sampling, and a direct count of the cumulative number of individuals marked during a sampling period. These methods as applied to scorpions are discussed by Polis and Farley (1980).

Additional field techniques have been used. For example, the use of UV light has also facilitated physiological research on scorpions in the field (see Chapter 8). Marked animals are experimentally treated and released. After a suitable period of living under natural conditions, animals are recaptured and analyzed. Using this technique, radiotracers (doubly labeled water) were injected into scorpions to monitor water loss in the field (King and Hadley 1979). Radiotracers likewise can be used for *in situ* research on energy budgets and osmoregulation. Environmental physiologists also use scorpions to determine microclimatic conditions within the burrow. Hadley (1970a) determined both burrow temperature and humidity by attaching thermocouples and humidity-indicator paper to scorpions and allowing these animals to use their burrows.

Field cages are also a useful technique for physiological and ecological research on scorpions. Such cages subject scorpions to field conditions and concurrently enable repeated recapture of individuals. For example, R. Bradley (see design in Bradley 1982) used large plastic buckets buried in the soil for research on growth, digestive physiology, and the patterns and determinants of surface activity. These buckets are large enough to allow scorpions to construct normal burrows. Thus captive scorpions experience the same ambient physical conditions as free-living conspecifics. When these buckets are covered with window screen, the amount of food available to individuals can be precisely controlled.

Scorpions are collected in the field with forceps or by hand. Painting the tips of the forceps with fluorescent paint enables collection without the need of white light. A more durable coat of paint on the forceps is obtained if paint is mixed with epoxy glue. Alternatively, it is possible to pick up scorpions in the field by quickly and carefully pinching the telson between the thumb and index finger. It is important that the sting be oriented forward between the two digits and that the telson be pinched on the sides. Further, care must be taken by the collector not to release the sting if the scorpion pinches with its pedipalps. This technique is not advised for species that are potentially lethal or that possess enlarged, crablike pedipalps. Such muscular pedipalps are quite capable of drawing blood.

Collected scorpions can be placed together in buckets or bottles or separately in individual bottles or in plastic sandwich bags. Both Ziploc bags and bags with wire ties are suitable. Collectors should use caution, since some scorpions can sting through plastic bags. Scorpions collected in groups must be separated soon after capture in order to avoid cannibalism.

Preservation of Collected Material

The old method for preserving scorpions was simply to drop the live animals into a jar or bottle of alcohol, wherein they were generally stored after sorting. This yielded undesirable results: the specimens contorted their bodies in death (usually with the legs folded underneath, the metasoma strongly arched, and the pedipalps and chelicerae drawn tightly toward the body), colors darkened and changed hues (especially when exposed to light for extended periods), and the scorpions were usually not well preserved internally. The limitations imposed by such specimens to scientific study are quite obvious. Finally, scorpions should never be pinned and dried, lest softer external structures (e.g., pectines) become distorted.

Williams (1968) and Newlands (1969b) devised a method for preserving scorpions that overcomes these problems. The steps involved are summarized below:

1. Specimens should be killed by heat shock—immersing the live scorpion in hot (90–99° C) water until the metasoma straightens (usually < 5 sec).
2. Fix the specimens in the following solution:

Formalin, commercial strength	12 parts
Isopropyl or *n*-propyl alcohol, 99 percent	30 parts
Glacial acetic acid	2 parts
Distilled water	56 parts

Fixation is accomplished by placing scorpions on their dorsal sides in flat trays, covered completely with the fixative. Large specimens should be injected with fixative or have their pleural membranes slit to aid fixative penetration. If specimens are for taxonomic purposes, the chelicerae should be pulled anteriorly until visible and the fingers of one chelicera opened; likewise, the fingers of one pedipalp chela should be opened. Williams (1968) recommends fixation for 12 to 48 hours, depending on the size of the specimens (large ones take longer). However, Ennik (1972) has obtained favorable fixation after 5 to 10 hours for most scorpions. Specimens should not be left in the fixative too long, because they become hard and brittle.

3. Rinse the scorpions in 50 percent isopropyl alcohol for 1 hour and store in 70 percent isopropyl alcohol. Specimens should be stored in the dark.

Ethyl alcohol (95 percent) may be substituted for isopropyl alcohol in the fixative; if so, the specimens should be stored in 80 percent ethyl alcohol when fixation and rinsing is completed.

Rearing and Maintenance in Captivity

It is obvious from the low success rates of laboratory life-history studies (see Chapter 4) that scorpions are difficult subjects for rearing in captivity. However, it is possible to maintain live scorpions for several years under laboratory conditions, especially if they are collected in older instars.

There are no stringent guidelines for housing scorpions. As long as food and water are provided, any type of container is adequate; the only necessity is that the container be escape-proof. It is usually desirable, however, to place scorpions individually in square-bottomed plastic containers with hinged lids to allow easy access for feeding and watering. The container should not be too large—the larger the container, the more difficult it becomes for the scorpion to catch its prey. For small to medium-sized (20–60 mm) scorpions, a container with a 7 cm × 7 cm bottom is sufficient; for larger scorpions, correspondingly larger containers are best. Containers of a variety of sizes are available from biological supply companies or other sources.

Before introducing the scorpion to the container, a substrate can be added. Depending upon the habits of the scorpion, the substrate selected may be sand (for psammophiles), soil (for most species), or rocks (for lithophiles). Under more elaborate conditions (i.e., with larger containers and deep layers of soil), scorpions have been induced to burrow in the laboratory (e.g., Williams 1966, Harington 1978).

Water should definitely be provided at regular intervals, usually once every week or two. Water can be introduced to the container in a small bottle cap or on a moistened bit of sponge. The bottle cap is preferable, because the lid of the container can be opened and the cap filled with a squirt bottle without disturbing the scorpion. If a sponge is used, care should be taken not to overload it; the leakage can drown small scorpions.

Scorpions will eat a wide variety of live insect prey under laboratory conditions, especially crickets, roaches, flies, moths, and mealworms. Although it is not necessary to feed them frequently (scorpions are known to survive for long periods without eating), it is common practice to offer them prey two to three times a week (e.g., Francke 1976b, 1979b). It is often difficult to find suitable prey for a small scorpion—if the prey is too large, the scorpion will flee from it. We have successfully fed very young scorpions on apterous *Drosophila* (available from biological supply companies). In general, Francke (1979b) suggests that greatest success in prey capture occurs when the prey offered is approximately the same length as the pedipalp chela of the scorpion. It is also noteworthy in this case that very young scorpions will often accept dead prey and have even been given bits of red meat (Baerg 1961). A last item to consider with respect to feeding is that uneaten remains or dead prey should be removed from the container as soon as possible. These remains mold rapidly, and very young scorpions have been reported to become passively entrapped in the fungal hyphae and die (Francke 1979b, Sissom and Francke 1984).

Mensuration

Between 20 and 30 standard measurements are usually given in scorpion taxonomic studies. Many of these are shown in Figure 11.1. Scorpion mensuration was standardized by Stahnke (1970), and, for the most part, his recommendations have been followed. Some mensural characters need further explanation:

TOTAL BODY LENGTH. There are many ways to measure body length. Some authors simply measure the linear distance from the anterior margin to the tip of the telson, usually with a ruler or calipers. This is oriented toward field work. Other authors, especially systematists, choose to measure each cuticular component of total length separately (carapace, mesosomal segments 1–7, metasomal segments I–V, and telson) and add them up. The latter method avoids errors in measurements of total length (see below).

TOTAL LENGTH OF PEDIPALP. Here only the individual lengths of the femur, patella, and chela are totaled (see Fig. 11.1*A*, *B*, *D*).

TOTAL LENGTH OF MESOSOMA. The recommended method (Stahnke 1970) is to sum the individual measurements of the tergites along the median line. This measurement includes the total sclerotized portion of each segment (pre-tergite + post-tergite). This method is preferred to an overall mesosomal measurement with a ruler, because it does not vary with the state of nourishment a specimen is in. Well-fed specimens have stretched intersegmental membranes; starved ones have overlapping tergites. It is obvious what these conditions will do to an overall measurement of mesosoma length or total body length.

TOTAL LENGTH OF METASOMA. This is the total of the five separate metasomal segment lengths. This measurement never includes the telson.

PEDIPALP-CHELA MEASUREMENTS. As recommended by Williams (1980), pedipalp-chela width (Fig. 11.1*C*) is taken with the animal's fixed and movable pedipalp fingers held in a vertical plane; pedipalp-chela depth (Fig. 11.1*D*) is taken with the pedipalp fingers held in a horizontal plane.

Of the numerous meristic characters used in scorpion taxonomy, only two need explanation here. The first is the pectinal tooth count. Usually the counts of both pectines are given for a single specimen, with the left side given first (e.g., 24–25). Ranges, means, or modes are given for the taxa when many specimens are available.

The second character is the spine formula for the tarsus. The tarsus is

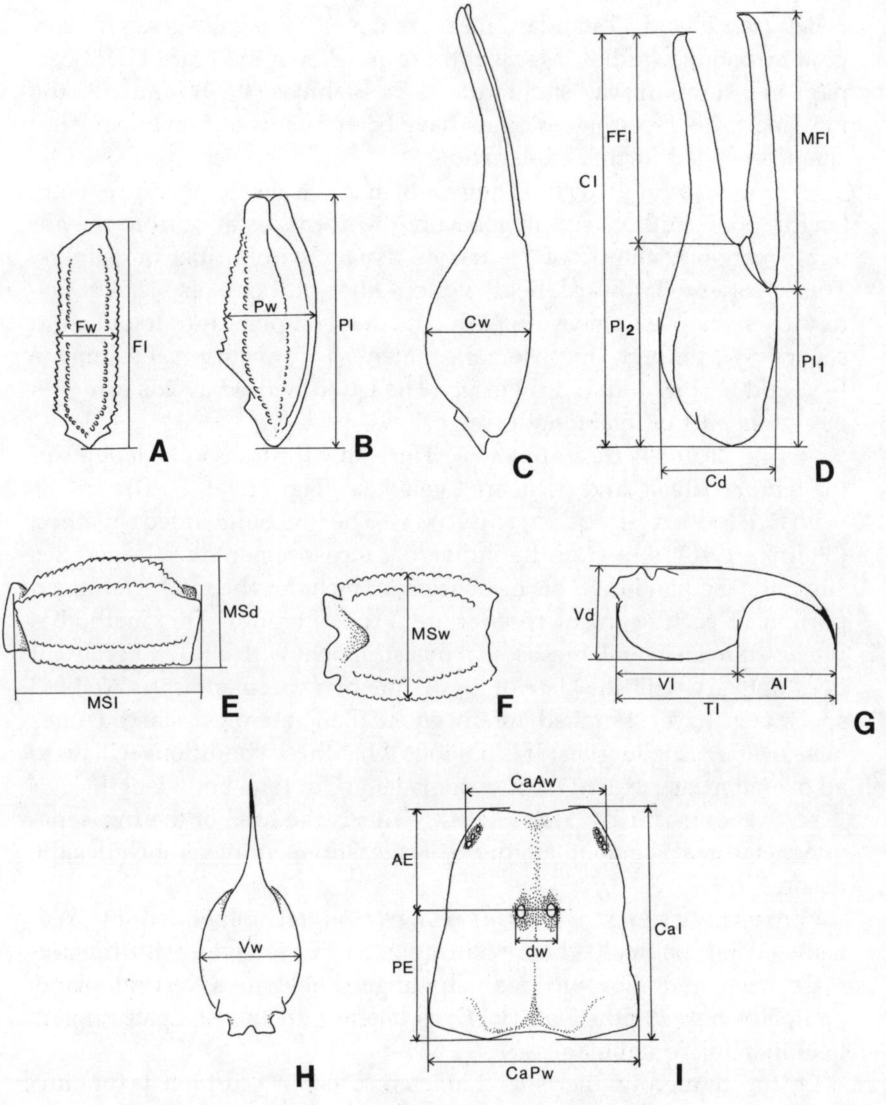

divided into two tarsomeres, I and II. Tarsomere II in many taxa (e.g., Diplocentridae, Bothriuridae, Scorpionidae, and some "chactoids") bears short, paired ventral or ventrolateral spines. The number of spines on each leg is represented in a spine formula, as follows:

$$\overset{\text{I}}{\frac{RP}{RR}\ \frac{LP}{LR}} : \overset{\text{II}}{\frac{RP}{RR}\ \frac{LP}{LR}} : \overset{\text{III}}{\frac{RP}{RR}\ \frac{LP}{LR}} : \overset{\text{IV}}{\frac{RP}{RR}\ \frac{LP}{LR}}$$

where the Roman numerals refer to the legs, from anterior to posterior: RP indicates the spines in the right prolateral row; LP, the left prolateral row; RR, the right retrolateral row; and LR, the left retrolateral row. The tarsomere-II spine formula is an important specific character.

Dissection

The scorpion body plan has changed little from its first appearance in the fossil record. Consequently, scorpions exhibit many primitive characteristics. There has not been so much reduction or specialization of segments as is evidenced in more advanced arachnids and arthropods in general (Williston's rule). Therefore, scorpion anatomy well illustrates the basic arachnid organization. The following dissection draws heavily from Dales (1970) and from Chapters 2 and 9 of this volume.

The adult body is composed of 18 visible segments: 6 on the prosoma, 7 on the mesosoma, and 5 on the metasoma (see Fig. 2.1). The mesosoma and metasoma combined are often referred to as the opisthosoma. The mesosoma contains most of the internal organs. Mesosomal segments consist of individual plates located both dorsally (tergites) and ventrally (sternites). These plates are separated by a flexible

---

Fig. 11.1. Scorpion mensuration. *A*, Pedipalp femur: *Fl*, femur length; *Fw*, femur width. *B*, Pedipalp patella: *Pl*, patella length; *Pw*, patella width. *C*, *D*, Pedipalp chela: *Cw*, chela width; *Cd*, chela depth; *Cl*, chela length; Pl_1, chela-palm length (or "underhand length"); Pl_2, alternative measurement of palm length; *FFl*, fixed-finger length; *MFl*, movable-finger length. *E*, *F*, Metasomal segment: *MSd*, metasomal-segment depth; *MSl*, metasomal-segment length; *MSw*, metasomal-segment width. *G*, *H*, Telson: *Al*, aculeus length; *Tl*, telson length; *Vd*, vesicle depth; *Vl*, vesicle length; *Vw*, vesicle width. *I*, Carapace: *AE*, distance from anterior margin to median eyes; *CaAw*, carapace anterior width; *CaPW*, carapace posterior width; *Cal*, carapace length; *dw*, width of ocular diad; *PE*, distance from median eyes to posterior margin.

pleural membrane laterally and intersegmental membranes between successive plates. The narrow metasoma is composed of complete rings of segmented exoskeleton. The telson, which follows the last metasomal segment, contains the venom glands and aculeus (sting).

The prosoma is characterized by a dorsal carapace, which normally bears a pair of median eyes and a variable number (0–5) of lateral pairs. The ventral surface of the prosoma is composed of a small sternum and the coxae of the appendages (Fig. 2.2). The coxae of legs I and II are modified anteriorly to form the coxapophyses, which combined with the pedipalp coxae form the ventral and lateral sections of the preoral cavity. The chelicerae form the dorsal section of this cavity, and the opening to the digestive tract is situated at its posterior end. The chelate chelicerae rip and macerate the food (see Chapter 7). Semimacerated and semidigested food passes into the preoral cavity. It is sucked into the opening of the foregut by the sucking action of the muscular pharyngeal pump. Only liquid food is actually digested; a dense brush of setae in the preoral cavity filters out particulate matter. Such solids (e.g., insect chitin) are discarded in the form of a pellet after feeding is completed.

The chelicerae are three-segmented, provided with a basal coxa, tibia (with its fixed finger), and tarsus (movable finger). The pedipalps are divided, from proximal to distal, into a coxa, trochanter, femur, patella, tibia (chela manus with its fixed finger), and tarsus (movable finger). The four pairs of legs possess eight segments each: coxa, trochanter, femur, patella, tibia, basitarsus, tarsus (with tarsomeres I and II), and apotele (see Fig. 9.4). The apotele possesses lateral claws (ungues) and a short median claw (dactyl). Setae of different lengths and thicknesses occur on all limbs. The morphology of the setae and tarsal claws of the legs exhibit adaptations to the scorpion's microhabitat (see "Ecomorphology" in Chapter 6). In addition to setae, a number of fine trichobothria are scattered on the pedipalps (see Chapter 2).

The seventh (pregenital, or first mesosomal) segment is absent in adult scorpions. The eighth (genital) segment bears a pair of ventral genital opercula, which cover the genital opening. Mature males of many species are distinguished from females and immature males by the presence of small genital papillae (crochets) underneath the medial edges of each operculum. The next ventral segment consists of the basal piece, from which the pectines arise (see Fig. 2.5). The pectines bear a number of teeth (dents). Both the size and number of teeth and the length of the pectines are sexually dimorphic in many species, being more strongly developed in the male. Pectines probably function in enabling the male to find a suitable place to deposit the spermatophore during courtship (see Chapter 4). Newer evidence demonstrates that they function as chemoreceptors as well (Foelix and Müller-Vorholt 1983).

The remaining segments of the mesosoma on the ventral aspect are provided with ventral sternites. The first four possess paired spiracles near the posterolateral corners. These spiracles may be slitlike to round, depending on the taxon. The spiracles lead to the book lungs, which are composed of vascularized sheets (lamellae) across which gases are exchanged (see Fig. 2.19).

The tubular metasomal segments possess setae, granules, and raised longitudinal ridges, called carinae (keels). Metasomal segments articulate with one another in the vertical plane, allowing the scorpion to bring the sting up and forward over the prosoma during defense or prey capture. The anus, which is surrounded by the anal papillae, is located ventrally between the fifth metasomal segment and the telson (see Fig. 2.6). The telson itself bears internal venom glands, which are surrounded by muscle. Contraction of these muscles forcibly ejects venom through the hypodermic needle–like aculeus. In some species, venom may actually be squirted many centimeters through the air.

Internal structures are best studied during a dissection starting on the dorsal surface. The scorpion should be securely pinned. Beginning at the last mesosomal segment, cut forward through the lateral pleural membrane on each side of the body. Next, carefully cut transversely through the tergites of the first and last mesosomal segments. Remove the entire dorsal aspect of the mesosoma (be careful not to damage the dorsal heart). Remove the carapace in a similar manner.

Much of the body cavity (prosoma and mesosoma) is filled with six pairs of saclike diverticula (the hepatopancreas). These function both in the production of digestive enzymes and for food storage. Remove the membranes from the top of the diverticula and the pericardial sac (over the heart). Remove any preservative that may have been injected into the specimen. The tubular heart has seven pairs of ostia on the dorsal aspect. Hemolymph flows into the pericardium, through the ostia, and into the heart. It is then pumped from the heart through an anterior and a posterior aorta. The anterior aorta passes through the muscular diaphragm, which separates the prosomal cavity from that of the mesosoma. The aorta then divides in the prosoma into two aortic arches, which pass over the esophagus. Each of these arches has two main branches, the inner and outer prosomal arteries. The smaller branches of the inner prosomal artery supply the brain, eyes, and chelicerae, and those of the outer prosomal artery supply the rostrum and the legs.

Return to the mesosoma. Remove the diverticula from one side of the body (be careful not to damage the gonadal tubules). The heart is suspended by lateral and ventral ligaments. The ventral ligaments attach both to the book lungs and to the top of the ventral blood sinus. These ligaments transmit heart movement to the ventral sinus, thus pumping hemolymph in and out of the book lungs. Hemolymph returns to the pericardial sac via pulmonary veins originating in the sinus.

Return now to the posterior portion of the aorta where it leaves the heart, and follow it posteriorly. The posterior aorta passes through each metasomal segment to the telson. To view the posterior aorta, carefully remove the dorsal surface of the metasomal segments by inserting fine scissors in the ends of the segments and cutting through the sides.

Next, remove the diverticula and various bundles of dorsoventral and longitudinal muscles in the prosoma to expose the arterial vessels leading into the legs and pedipalps. The supraesophageal ganglion and vessels are surrounded by a membranous endosternite, which connects with both the epistomal apodemes and the diaphragm. This membrane should be carefully removed to expose the pharynx, pharyngeal musculature, and supraesophageal ganglion. This ganglion is composed of a protocerebrum and a tritocerebrum (see Figs. 2.18, 9.21). Nerves from the protocerebrum innervate the eyes. Nerves to the chelicerae originate in the tritocerebrum. The cheliceral nerves are difficult to find; they are often torn away when the pharynx is exposed. A pair of circumesophageal connectives links the supra- and subesophageal ganglia. Before leaving the prosoma, note the coxal glands, which open into the coxae of the third pair of legs. Coxal glands concentrate excretory products from the prosomal cavity.

Once again, return to the mesosoma. Cut away and remove the arteries to the limbs, the anterior aorta, the heart, pericardium, remaining diverticular tissue, and muscle bundles. Take care not to damage the gonads. The digestive system may now be examined. In the prosoma, the mouth leads into the pharynx, with its associated pumping musculature. The pharynx is followed by the esophagus. The dilated stomach is located between the esophagus and the diaphragm. The intestine begins on the other side of the diaphragm and continues through the fourth metasomal segment. It receives five pairs of ducts from the diverticula (hepatopancreas) in the mesosoma. The stomach, intestine, and hepatopancreas are the major components of the mesenteron.

The proctodeum consists of the hindgut and anus. The hindgut is contained in the fifth metasomal segment; the anus is its terminus, situated between this segment and the telson.

In the mesosoma, the major excretory organs are the Malpighian tubules. There are two pairs arising in the seventh mesosomal segment. Excretory wastes are removed from the coelom and deposited into the digestive system. The hepatopancreas also serves as an excretory organ.

The male reproductive system consists of the paired membranous paraxial organs (containing the hemispermatophores) and the testicular tubules (see Fig. 2.28). The paraxial organs are connected ventromedially to the genital atrium, and give rise to the testicular tubules posteriorly. There are usually four longitudinal tubules connected to one another by transverse tubules, giving a reticular appearance. The anterior branches of the testes become the vasa deferentia, which are somewhat dilated where they connect to the medial sides of the paraxial organs. Seminal vesicles (and accessory glands, when present) empty into the vasa deferentia. Variations to this pattern exist (see Fig. 3.11*A, C, D*). Sperm from the testes is transferred to the capsule of the hemispermatophore by the seminal vesicle. During courtship the two halves of the hemispermatophore are extruded and fused together to form the spermatophore, which is glued to a substrate.

The female reproductive system is similar to that of the male (Fig. 3.13). The ovariuterus, with longitudinal and transverse ovarian tubules, also takes on a reticular appearance. In apoikogenic scorpions (see Chapters 2, 3), the small, spheroid ovarian follicles are connected directly to the ovarian tubules. Upon fertilization, the zygote enters the lumen of the ovariuterus, where development occurs. In katoikogenic scorpions (see Chapters 2, 3), the follicles are contained in numerous blind diverticula branching from the ovarian tubules. Fertilization and development occur in these diverticula. The anteriormost branches of the ovariuterus, which connect to the genital atrium, are sometimes swollen into seminal receptacles.

The last structures to be identified are the components of the ventral nerve cord. Remove the gonads, esophagus, and mesenteron. The ventral nerve cord is covered dorsally by a ventral artery, which arises as a branch of the outer prosomal artery. The segmental ganglia and associated nerve cord are exposed by removing this ventral artery. There are three free ganglia in the mesosoma (mesosomal segments 5–7) and four in the metasoma (metasomal segments I–IV). The anterior ganglia of the mesosoma are incorporated into the subesophageal ganglion, and the ganglion for the last metasomal segment is fused with that of the fourth.

Preparation of the Hemispermatophore for Systematic Study

To study the hemispermatophore, it is necessary to remove it from the scorpion's body. This dissection is done under a binocular dissecting microscope. First, a slit is made in the right pleural membrane. This slit is held open, and a pair of fine forceps is inserted to grasp the right paraxial organ near its connection with the genital atrium. The paraxial organ is teased loose from the genital atrium and the testicular tubules and lifted from the body (Vachon 1952, Lamoral 1979).

Once the hemispermatophore is removed, it is transferred to a watch glass containing some preservative (usually 80 percent ethyl alcohol: the watch glass should contain the same liquid that the specimen is preserved in). To view the fine structure of the hemispermatophore, the membranous paraxial organ is carefully dissected away with fine forceps and/or dissecting needles. This is the most widely utilized technique among scorpion systematists (Vachon 1952, Koch 1977, Lamoral 1979).

Often even the most careful dissection results in damage to the hemispermatophore. In addition, of course, the paraxial organ is lost to future study. Recently, at the suggestion of Mr. James C. Cokendolpher, one of us (WDS) has begun using clove oil as a clearing agent to facilitate study of the hemispermatophore. Some minor dissection may still be necessary to remove tissue clinging to the paraxial organ. The results have been very good to date, and the disadvantages mentioned above are circumvented. Clove oil is widely used as a clearing agent to study genitalia in other arthropods.

The hemispermatophore must be completely dehydrated before immersion in clove oil to prevent the formation of a cloudy precipitate. Assuming the scorpion is preserved in 80 percent ethyl alcohol, dehydration is accomplished by placing the hemispermatophore in successive rinses of 95 percent and 100 percent ethyl alcohol before transfer to clove oil. If the scorpion is preserved in some other medium, adjustments can be easily made. The hemispermatophore is adequately cleared after about 10 minutes, and all structures are easily viewed.

Whichever method is used, the hemispermatophore, upon completion of study, should be stored in a small shell vial permanently kept with the specimen from which it came (i.e., both the vial and the specimen in a larger vial). If the clove-oil method is used, the hemispermatophore should be rehydrated before storage.

Preparations Using Living Tissue

Living tissue is used in many types of physiological research. The following techniques are adapted from Ahern and Hadley (1976). Scorpions are anesthetized lightly with chloroform. The appropriate tissue is removed for *in vitro* studies or left *in situ* for *in vivo* research. The tissue is perfused with scorpion Ringer's solution. The ionic composition and osmotic pressure of this saline solution is based on data from flame photometry, chloride titration, and osmometry of hemolymph samples of several species of scorpion. This scorpion saline had an osmotic pressure of 550 mOsmol/l, a pH of 7.4, and consisted of the following ion concentrations in mM: Na^+, 281.5; K^+, 8.0; Ca^{2+}, 10.0; Mg^{2+}, 20.0; Cl^-, 260.0; HCO_3^-, 1.5; SO_4^{2-}, 44.0. When incubation media of altered ionic compositions were used, choline chloride was substituted for NaCl and KCl, and $KHCO_3$ for $NaHCO_3$; KCl concentrations were increased at the expense of those of NaCl (Na_2SO_4 replacing the altered NaCl), and raffinose was added to maintain approximately equal osmotic pressures on both sides of the epithelium.

Venom Collection and Purification

Scorpion venoms are used by biochemists, neurophysiologists, and toxicologists in a variety of research (see Chapter 10). In this section, methods for extraction and purification of scorpion venoms are presented.

Three different methods are used to obtain venom from the telson: trituration of the amputated telson macerated with water or saline solution; physical stimulation, either by allowing the scorpion to sting a suitable object (e.g., a sheet of parafilm) or by manual manipulation of the cephalothorax (prosoma) while immobilizing the telson with forceps; or electrical stimulation of the telson. Macerating the telson is a fast and easy method for collecting venom, but suffers the disadvantage of contaminating the venom with material from non-venom-producing cells. These contaminants may inactivate as well as dilute the toxins. The toxicity of such a preparation is low and decreases with storage time. This method is used quite extensively, however, in preparing injectable material for antivenom production. The second method, physical stimulation, has the advantage of yielding venom that is more potent; it may be 10 times more toxic than venom obtained by electrical stimulation. The disadvantage is that the method is slow and cumbersome. The last method, electrical stimulation, is used quite commonly. It is the method of choice. The disadvantage with this method is that the number of milkings is limited because the muscles of the gland eventually stop responding to electrical stimulation. Electrical stimulation allows an average of four milkings per scorpion. Small amounts of nonvenomous tissue fluid may also contaminate the venom. Following collection, the extracted venom is freeze-dried and stored frozen. Freezing permits indefinite storage without loss of activity. (One sample of venom stored by D. Watt was 12 years old and still as potent as had been originally.) When dried venom is reconstituted with water or saline, much of the mucus does not dissolve and may be removed by centrifugation.

The yield of venom per milking is variable and depends on the species, the number and frequency of milkings, and the scorpion's physiological condition. An average of 0.1 mg of dried venom per milking is obtained from *Centruroides exilicauda*. The mean yield from milkings of *Tityus serrulatus* is 0.62 mg per scorpion (Bücherl 1971). Much higher yields are obtained from species of large scorpions, although the toxicity of venoms from these species is usually low.

Specific components of the venom are isolated and purified by chromatography carried out on carboxymethylcellulose (CMC: see legend to Fig. 10.2 for details, and Babin et al. 1974 for elution protocol). For example, the elution profile of *C. exilicauda* (Fig. 10.2, solid line) shows 12 major peaks and shoulders. Each zone contains two or more proteins, with zones 4 to 12 containing the toxic, basic ones. Isolation of individual proteins requires rechromatography of a specified zone using other chromatographic media. For example, zone 4, upon rechromatography on DEAE-Sephadex, yields three peaks, each containing a unique protein.

The chromatographic media that have proved most useful for rechromatography of the CMC zones are DEAE-Sephadex A-25, CM-Sephadex C-25, and Amberlite CG-50, all ion-exchange media. A similar chart has been published for isolated toxins from venoms of North African scorpions (Miranda et al. 1970).

The criteria we use in establishing molecular homogeneity are migration as a single band in disc-gel electrophoresis (Babin et al. 1974), uniformity in the contour of the final chromatographic elution profile, amino-acid analysis or polypeptide sequencing, and migration as a single, uniform peak in reversed-phase, high-performance liquid chromatography (HPLC). Although HPLC with gradients of either acetonitrile (5–60 percent) or isopropyl alcohol (0–40 percent) with 0.1 percent trifluoracetic acid resolves impurities from semipurified toxins, these solvent systems are not adequate for resolution of whole venom (Watt, unpubl. data). Amino-acid analysis is important not only as a criterion of purity but also in identifying certain toxins with specific amino-acid deletions (see Chapter 10).

ACKNOWLEDGMENTS

We would like to thank James Cokendolpher, Sharon McCormick, and Marc Simard for information and suggestions. As always, Ms. Sherrie Hughes deserves a special thanks for her care and patience in typing this chapter.

12

Scorpions in Mythology, Folklore, and History

J. L. CLOUDSLEY-THOMPSON

Some 4,000 years ago, the astronomers of Babylonia (Chaldaea) observed that the paths of the sun and moon, and of the five planets then known, moved across a belt in the heavens about 16° in width. (To this belt or zone of sky, the ancient Greeks later gave the name "zodiac.") The stars contained in it were subsequently grouped into 12 constellations, each of which was graced with its own symbol and name. One of these was Skorpios or Scorpio, a constellation lying near to the center of the Milky Way and, from northern latitudes, to be seen low in the southern summer sky. It consists of a long-curving sweep of stars, thought to resemble the outline of a scorpion, with its sting marked by two brighter stars rather close together. Scorpio contains the brilliant red star Antares. It originally extended farther westward into the constellation now known as Libra (The Scales), whose stars represented the scorpion's claws. In Roman times, however, this grouping was changed, and Libra came to be regarded as a separate constellation.

Scorpions have influenced the imagination of the peoples of the Orient and of the Mediterranean since earliest times. According to legend, when Mithras, the Persian god of light, sacrificed the sacred bull whose blood was to fertilize the universe, the evil spirit Ahriman sent a scorpion to destroy the source of life by attacking the testicles of the animal (Cumont 1896–99; cf. Vermaseren 1978). Scorpions, as agents of the Devil, are prominent in monuments to Mithras, whose worship was very popular in Roman armies, persisting nearly anywhere they had been stationed until as late as the fifth century A.D., especially in North

Africa and Germany. In ancient Egypt, scorpions were frequently represented in tombs and on monuments. They are mentioned in the Ebers papyrus ("How to Rid the House of Scorpions") and in several passages of the Book of the Dead. The Talmud and the Bible also refer to scorpions as repugnant and formidable animals. According to Greek mythology, the great hunter Orion, son of Zeus, was killed by a scorpion that the goddess Artemis produced when he defied her. Consequently, when the two constellations were placed in the sky, then were separated as far as possible to prevent them from ever coming together again. Even so, Orion still flees before Scorpio: as Scorpio rises, Orion sets. Scorpio also terrified the Horses of the Sun while they were being driven by the inexperienced young Phaëthon, son of Helios. Losing control of the horses, he almost set the earth on fire, and was killed by Zeus with a thunderbolt.

Scorpio is the autumnal sign of the zodiac, the sign of the eighth month, whose god among the Babylonians was Marduk. In the beginning, according to the Babylonian Creation Tablets from the library of Ashurbanipal (ca. 650 B.C.; now in the British Museum), there were neither land, nor gods, nor men. There were only two elements, called Apsu and Tiamat. Apsu was male, the spirit of fresh water and the void in which the world existed; Tiamat was female, the spirit of salt water and of primeval chaos. She was a monstrous being, with a serpentine body, scaly legs, and horns on her head. She was possibly the first of the dragons, and undoubtedly the embodiment of evil. Her union with Apsu resulted in the creation of numerous oddly assorted progeny, including the first gods of the primordial universe. Disturbed by the horrifying appearance of these new gods, Tiamat consulted with Apsu through an intermediary named Mommu. Apsu himself complained to Tiamat about the unruly behavior and lack of respect shown by their loathsome offspring, and threatened to destroy them. Thereupon he was destroyed by Ea, one of the younger gods, who had guessed his intentions, and Marduk was created in his place. Tiamat then sought help from Ummu Khuber (or Melili), who became the mother of 6,000 devils, and from 11 other mighty helpers; one of these was the Scorpion Man, the guardian of Mount Mashu, the place of sunrise and sunset (cf. the Gilgamesh epic, Nineveh tablet IX). The combined forces of evil were placed under the command of Kingu: the scorpion has, therefore, been associated with wickedness and sin from earliest times, and with the sign of the Zodiac of which Marduk was the god.

The Maoris of New Zealand call this constellation the "Fishhook";

with it, their mythology says, Maui fished their islands from the underworld. Among the Berbers, in their imagery, the constellation is likened sometimes to a scorpion, sometimes to a palm tree. They identify Antares as a youth climbing the palm, who stopped halfway up to eye some pretty girls. Scorpio's helical rising occurs a few days before Christmas in Honduras and Nicaragua; hence the Miskito Indians of these countries have named Antares *Kristmas.* The Maya name for Scorpio means "Sign of the Death God."

Even before they were transferred to the zodiac, scorpions represented a religious symbol and were engraved on boundary stones, magical tablets, and seals. A Kassite boundary stone of Melishihu found at Susa and dating from the twelfth century B.C. has two signs for the equinoxes—the vernal one being an arc with degrees, and the autumnal sign with a scorpion and the word *n'ibiru,* "crossing."

Like the Chaldaeans, the ancient Egyptians also had their folk myths that explained the origin of the world. They believed that Osiris was murdered by his brother, Set (whom the Greeks later equated with Typhon), with the assistance of 72 wicked accomplices. On the famous

Fig. 12.1. The Scorpion Man of Mesopotamia, as figured on a harp found in the Royal Necropolis of Ur. The Scorpion Man was long important in Mesopotamian belief: at least 1,500 years after this portrayal in Sumer, a pair of Scorpion Men was sculpted to flank the gateway of the Temple Palace of Kapara, king of the Assyrian city of Guzana. (Ca. 2600 B.C.; courtesy of the University Museum, University of Pennsylvania.)

Fig. 12.2. The Egyptian scorpion-goddess Selkit (or Serqet), "friend of the dead," seen at right protecting the cabinet that contained the pharaoh's viscera in the tomb of Tutankhamen. The stylized scorpion atop her head is a sure mark of her identity. (Ca. 1350 B.C.; courtesy of the Egyptian National Museum.)

Metternich stele, a fine example of the Egyptian magical amulet of Horus on the Crocodiles, the story is told of how the faithful Isis, wife of Osiris (her tears caused the annual flooding of the Nile), took refuge in the papyrus swamps of the Delta to protect her child, Horus, from his father's killer. In her flight, she was accompanied by seven scorpions (because the crime had been committed in the month of Athyr [Hathor], when the sun is in the sign of the scorpion). One evening she came to the house of a woman who, alarmed at the sight of the scorpions, slammed the door in her face—whereupon one of the scorpions crept under it and stung the woman's child. When Isis heard the mother's sad lamentations her heart was touched; she laid her hands on the child and uttered powerful spells so that the poison was drawn out and the child lived.

At some time while Isis was away, the child Horus was also stung by a scorpion—in the heel, where, like Achilles, he was particularly vulnerable to his enemies. Isis came home to find him stretched lifeless on the ground. Apparently her magic was of no avail on this occasion, for she prayed to Ra, the sun god, for help, and he sent Thoth to teach her an exceptionally potent spell. No sooner had he uttered the words of power than the poison flowed from the body of Horus, and straightway he was restored to life. Thoth then ascended into the sky and took his place once more in the boat of Ra.

Writing of the scorpions of Coptos, Aelian said that, despite the fear of *Hedj'dj*, "The Destroyer," women in the temple precinct of Isis were able to walk barefoot, or even to sleep on the ground, without harm from their stings (*De Nat. Anim.* 10.23). Scorpions were believed to originate from the decaying corpses of crocodiles (ibid. 2.33), and in medieval Europe some people, retailing a tradition at least as old as Nicander's *Theriaka* (lines 791–96; cf. Pliny, *Hist. Nat.* 9.99), still believed that dead crabs turned into scorpions!

Paradoxically, Set came to the defense of Ra when Apep attempted to overthrow him. In chapter 39 of the Book of the Dead, he is said to threaten Apep with these words: "Back, villain! Plunge into the depths of the Abyss! If you speak, your face will be overturned by the gods. Your heart will be seized by the lynx; your reins will be bound by the scorpion."

According to Diodorus Siculus (1.87), writing in the first century B.C., falcons were honored in Egypt because they feed on scorpions. On later Gnostic steles, scorpions were connected with Sarapis-Hermes; they were elsewhere associated with Hermes' Roman counterpart, Mercury (Deonna 1959). A scorpion symbolized the kingdom of Syria Commagene on its coinage in the first century A.D., and the scorpion of Adiabene in Syria was so feared that it could be killed on the Sabbath, even if it did not attack (*Sabbat* 121b). The army of Cato of Utica, who served in opposition to Julius Caesar during the Roman Civil War, was said to have been menaced not only by dragons, basilisks, and other terrible monsters as it marched through the dry desert of North Africa,

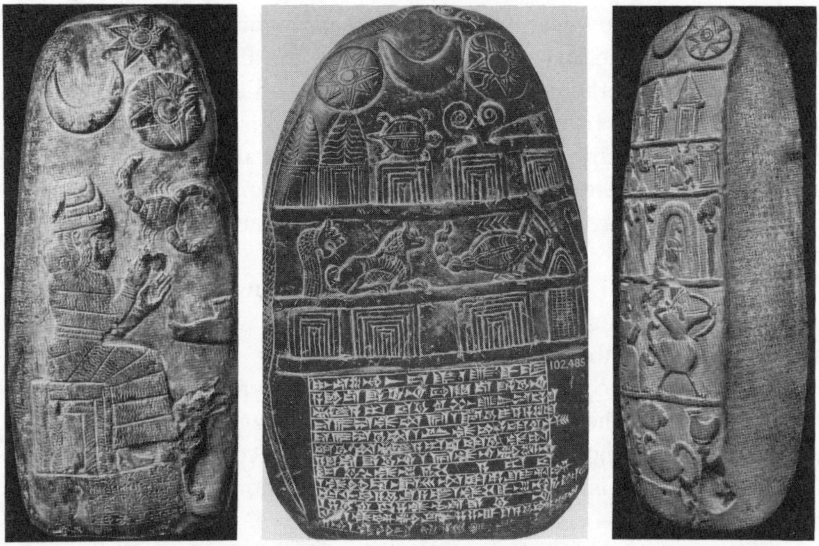

Fig. 12.3. Babylonian boundary stones of the Kassite period and shortly thereafter: the scorpion symbolizes the goddess Ishhara, one of many deities protecting the land grants or purchases marked by such stones. The scorpion and other symbols seen here remained in use on Babylonian boundary stones until at least the end of the eighth century B.C. Note also a Babylonian Scorpion Man in the next-lowest register of the stone at right. (Ca. 1300–1100 B.C.; courtesy of the Trustees of the British Museum, and of the Musée du Louvre.)

but also by solifugids and scorpions, which killed the soldiers with their deadly poison (Lucan, *Pharsalia* 9.608–949, esp. 735–847).

Both spiders and scorpions are mentioned in early Chinese writings. They are associated with the toad, the centipede, and the snake to form the group of the *Wu Tu,* the five poisonous animals. Scorpio was *Tsing Lung,* the beneficent Azure Dragon of a former Chinese zodiac, but, in the time of Confucius, that whole constellation took the name of the fire star, *Ta Who,* "Great Fire" (Antares), and was worshipped as a protection against fire. The great Chinese encyclopedia *Ku Chin T'u Shu Chi Ch'ang,* of 10,000 volumes, produced in Peking in 1726, devoted 14 pages to scorpions.

In biblical times, as today, scorpions were abundant in the wilderness south of Judah (Deut. 8.15). Rehoboam threatened to chastise his subjects not with whips, but with scorpions (1 Kings 12.11, 2 Chron. 10.14). By these, however, he may have meant whips armed with sharp points to make the lash more severe. Like the Babylonians, the Hebrews recognized the constellation Scorpio, and the scorpion became the emblem of the tribe of Dan. To the Akkadians, it was *Girtab*, "The Stinger." In the Euphrates Valley, where it probably originated, Scorpio's symbol represented a monster, half scorpion and half man. The human half belonged to the upper regions; the animal portion, to the underworld.

To the astrologer, Scorpio is an ill-omened constellation that brings in its wake cold, darkness, and storm, causing wars and exerting an evil influence on the affairs of men. Antares is said to be as baleful as Mars, which is considered to be its twin. A comet appearing in the constellation is believed to be a harbinger of plagues. In contrast, alchemists always rejoiced in the helical rising of Scorpio, for only when the sun was in this sign could the transmutation take place of base metal into gold. Scorpio and its natural "house," the eighth, have always been associated with the mystic life force in its critical phases of transformation—birth, reproduction, and death—especially sex and death, since both of these imply rebirth.

Fig. 12.4. Athenian black-figured amphora with the scorpion as the shield device of a legendary warrior: decorated shields are very often shown on Greek pottery of this style and era, but the depiction of a scorpion is somewhat unusual. (Ca. 520 B.C.; courtesy of the Trustees of the British Museum.)

Fig. 12.5. Amulets of Horus on the Crocodiles, used in Egypt from as early as the Nineteenth Dynasty (1320–1200 B.C.) and down through the Roman period: clenched in his fists, the child Horus holds scorpions, serpents, an oryx, and a lion. Such steles are typically covered with magical inscriptions meant to protect their owners from scorpions and other harmful creatures, and to cure the wounds they inflict; some, like the example at right, have basins built in, which presumably collected water poured over the stele in order to imbue it with magical powers for drinking or anointment. (Ca. 350–300 B.C.; courtesy of the Oriental Institute of the University of Chicago, and of the Egyptian National Museum.)

In countries where scorpions are found, pictures of them are often carried about as protection from their stings. It is not surprising, therefore, to find scorpions engraved on Hebrew Shiviti amulets. These large, handsome pieces of thick-gauge silver come from Persia. All are characterized by a superscription from the first half of Psalm 16.8 and the text "I have set the Lord always before me." The Tetragrammaton, which is the second word of the Hebrew verse, occupies the center of the superscription, engraved in large, bold *Ashuri* characters.

During his 13 years in Athens, Aristotle wrote more than 20 works of a biological nature, and in them he mentioned scorpions on a number

Fig. 12.6. The warrior-god Sadrafa, with his attributes the scorpion and the serpent, on a stele from the temple of Bel, Palmyra, datable precisely to May of A.D. 55: this deity, a forerunner of the Iranian divinity Mithras among Semitic peoples under Persian rule, was worshipped in Syria from at least as early as the sixth century B.C.; by the first century A.D. his cult was also kept in Sardinia and North Africa. (Courtesy of the Trustees of the British Museum.)

of occasions. He wrote that the scorpion carries its sting exposed and is the only *entomon,* "insect," with a long tail (*Hist. Anim.* 4.7; Pliny, *Hist. Nat.* 11.100). He evidently knew that scorpions are viviparous, for he said that the land scorpion produces a number of egglike grubs and broods over them (*Hist. Anim.* 5.26; Pliny, *Hist. Nat.* 11.86). On Pharos and elsewhere, the bite of the scorpion was not dangerous, he claimed, but in Scythia (roughly, the southwestern Ukraine), where scorpions were venomous as well as plentiful and of large size, the sting was fatal to men and to beasts, even the pig (*Hist. Anim.* 8.29; Pliny, *Hist. Nat.* 11.90); and in Caria, south of Anatolia, scorpion stings were lethal to natives of the country, but did no great harm to strangers (Aristotle, frg. 605 ed. Rose = Pliny, *Hist. Nat.* 8.229; cf. Aelian, *De Nat. Anim.* 5.14).

Scorpio maurus is the well-known *scorpio* of antiquity, of which Pliny the Elder wrote: "They are a horrible plague, poisonous like snakes, except that they inflict a worse torture by despatching their victim with a lingering death lasting three days, their wound being always fatal to girls and almost absolutely so to women, but to men only in the morning. . . . Their tail is always engaged in striking and does not stop practising at any moment, lest it should ever miss an opportunity" (*Hist. Nat.* 11.28, transl. Rackham).

Pliny repeated all that Aristotle and Nicander had written before him—thereby setting a trend that did not disappear for 17 centuries. He described scorpions as a *pestis inportuna,* "a horrible plague," the curse of Africa, and asserted that they could fly on a south wind, which supported their "arms" when they spread them out like wings (ibid. 11.88; Keller [1913, with further refs.] cites scorpionflies [*Panorpa communis*] and water scorpions [*Nepa cinerea*] as being possible candidates for the ancients' "flying scorpion"). As Theodore Savory (1961) has pointed out, Pliny's most interesting remark about scorpions was that some have a pair of stings (ibid. 11.87, cf. 163; cf. Aelian, *De Nat. Anim.* 6.20, 16.42; schol. *ad* Nicander, *Ther.* 781[b]). Bifurcation of the tail is an abnormality known to occur in a small percentage of scorpions, and "one wonders whether Apollodorus, to whom the statement is attributed, may perhaps have been recording the first observation of this phenomenon."

The dread of mad dogs, scorpions, and other venomous animals was extreme during the Middle Ages; every medical book and herbal abounded in preservatives from, and antidotes for, such perils to the traveler. Scorpion grass was so called from its supposed virtue as a

Fig. 12.7. Roman votive relief of the bull-slaying Mithras: the scorpion and all other elements of the central group here are standard conventions of Mithraic iconography. At the lower left, a small inset scene portrays Mithras and the bull before a grotto. (Ca. A.D. 300; courtesy of the Museo di Roma.)

cure—indicated unmistakably to the superstitious by the fact that its head of flowers and buds can be rolled into some more or less satisfactory resemblance to a scorpion's tail! In former times, it was popularly believed that an oil might be extracted from scorpions that was capable of healing a wound caused by the sting (Oyle of Scorpions distilled against Poysons). This idea was even entertained by Sir Kenelm Digby, the seventeenth-century diplomat and writer who secretly married "that celebrated beautie and courtezane" Venetia Stanley. On his beloved wife's early death, Digby withdrew to Gresham College, where, for two years, he lived like a hermit, diverting himself with the study of chemistry and natural philosophy. Another influential naturalist of the period, Dr. Thomas Moffatt (whose daughter was probably *the* Miss Muffet of nursery-rhyme fame) was also sympathetic to this unlikely hypothesis!

In an account of his travels, begun in the year 1626, Sir Thomas Herbert gave a description of scorpions in Cashan, a city of Persia: "But which rages there is no less violence in scorpio; not that in the Zodiak, but real scorpions which in numbers engender here. A little serpent of a finger long . . . the onely creature that stings with his tail, some flyes excepted. Of great terror is the sting; and so inflaming as with their envenomed arrow some die, few avoid madness, at least for a whole day; the sting proving most dangerous when the season is hottest, which is when the Dog-star rages" (*Travels*, p. 222). Ben Jonson was likewise indebted to Pliny for his knowledge of scorpions: "I have heard that aconite, being timely taken, hath a healing might against the scorpion's stroke" (*Sejanus* 3.3), he wrote, and, "Art can beget bees, hornets, beetles, waspes out of the carcasses and dung of creatures; yea, scorpions of an herbe, being ritely placed" (*Alchemist* 2.3). The reference to scorpions is paralleled by a statement of John Lyly, the sixteenth-century dramatist and novelist: "By Basill the Scorpion is engendered; and by means of the same hearb destroyed." Both items of information have been drawn from Pliny, of course, who wrote that "basill taken in wine with a little vinegar put thereto cureth the sting of land scorpions" (*Hist. Nat.* 20.132).

It was frequently claimed by medieval writers that the scorpion provided a cure for its own poison. Hence, in Elizabethan times, Lyly asserted that the scorpion "though he sting yet he stints the pain," and Thomas Lodge in *Rosalynde*, published in 1590, wrote, "They which

Fig. 12.8. Syro-Palestinian magical pendant of a type common between the third and sixth centuries A.D.: on the obverse, a mounted warrior, probably a Christian saint, transfixes an evil spirit with his lance; on the reverse, the scorpion and four other creatures (lion, ibis or stork, serpent, dog or cheetah) wound the Evil Eye, which is also pierced at the sides by nails and from above by a pitchfork. (From Bonner 1950, by permission of the University of Michigan Press.)

Fig. 12.9. Manuscript illuminations from a tenth-century Byzantine copy of Nicander's *Theriaka* in paraphrase, illustrating the "white," "red," and "dusky" scorpions there described. Besides these, Nicander lists four other types of scorpion; the entire assemblage represents an early attempt at systematic classification based on color, deriving from Apollodorus of Alexandria in the third century B.C. (Courtesy of the Pierpont Morgan Library, cod. M.652 fol. 379r.)

are stung by the scorpion cannot be recovered but by the scorpion." Twenty years later, Beaumont and Fletcher's *Philaster* had the statement "Now your tongues like scorpions, both heal and poison," and in the same authors' *Custom of the Country* we read, "And though I once despaired of women, now I find they relish much of scorpions; for both have stings, and both can hurt and cure too." Samuel Butler wrote in 1678, "'Tis true, a scorpion's oil is said to cure the wounds the venim made" (*Hudibras* 3.2.1029). Another medieval idea was that a scorpion is a kind of serpent with a face like a woman's, that puts on a pleasant countenance. Geckos have long been regarded as poisonous in Africa and the Middle East, and Pliny recommended pulverized scorpions as an antidote to their venom (*Hist. Nat.* 29.4).

A curious transposition of the legend of Cleopatra was created at the ancient Caesarea (modern Cherchell) in Algeria. There was once a palace there about which Qazouni, a Berber of the third century, told the following tale. This palace had been built by a king for his son, who the astrologers predicted would one day be killed by a scorpion. The palace was therefore constructed of stone to keep the scorpions out. (They were said to be unable to reproduce in such surroundings or to find their way in across the polished surface of the marble.) One day, however, a basket of ripe grapes, containing a hidden scorpion, was delivered. When the young prince helped himself to the grapes, he was stung and died. That this tale is connected with the death of the famous Cleopatra VII of Egypt through the bite of an asp is indicated by the fact that King Juba II, whose capital was Caesarea, had married her daughter, Cleopatra Selene.

In the twelfth century, Moses Maimonides, the foremost rabbinic scholar of his time, discounted the efficacy of bezoar and certain stones in the treatment of scorpion stings. In the fourteenth century, the Arab historian Ibn Khaldun recounted that seven centuries before his day Sidi Okba had decided Kairouan should be built where there was a marsh infested with ferocious and venomous pests. With the aid of the 18 companions of the Prophet who accompanied him, he invoked the protection of Allah, the All-Powerful. His wishes were granted, and, for 40 years, no serpents or scorpions were seen in *Ifrikya.*

During the fourteenth, fifteenth, and sixteenth centuries, the scorpion was often used as an emblem for the Jewish people, to symbolize perfidy (Bulard 1935). A painting by Fra Angelico, for example, in the Museum of San Marco in Florence, shows Christ carrying the Cross, surrounded by three heralds wearing tunics ornamented with scorpions.

Current misconceptions about scorpions are often as fanciful as the legends of past ages. One, which dates from the time of Paracelsus (Keller 1913), is that, if surrounded by a ring of fire, a scorpion will commit suicide by stinging itself to death. This must be nonsense, because no animal, other than man, could possibly have the imagination to realize that by self-destruction it might avoid unnecessary pain. Moreover, a suicidal instinct would inevitably become eliminated by natural selection during the passage of time, since it could convey no biological advantage and might, on the contrary, be disadvantageous. In any case, scorpions are immune to their own venom. What actually happens is that the unfortunate scorpion, lashing frantically with its sting, may inadvertently strike its own body, and this has been misinterpreted as suicide. It is also untrue that scorpions do not drink, or that black scorpions are more venomous than yellow ones. All really dangerous species belong to the family Buthidae, most of whose members are yellow or brown in color.

Occasionally, in the Sahara, one meets someone who will pick up scorpions with his hand, and the creatures will make no attempt to

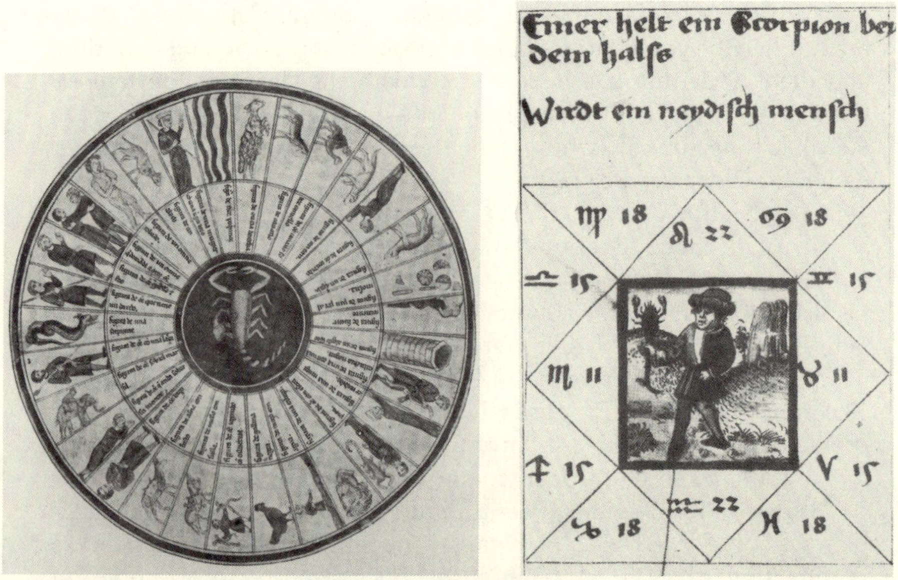

Fig. 12.10. Scorpion illustrations in astrological reference manuals: at the left, the sign of Scorpio with its paranatellonta (stars "rising beside" the sign), as tabulated in a Spanish codex of the fourteenth century; at the right, the "quadrature" of the eleventh degree of Scorpio, as shown in a fifteenth-century German manuscript. Through an ancient amalgamation of Mesopotamian, Greek, and Egyptian astrological traditions, each sign of the zodiac was subdivided into 30 parts, corresponding to the degrees of the ecliptic, and paranatellonta representing lesser astrological figures were identified for all 360. With each star upon its rising assigned by quadrature to the first of twelve celestial "houses," and its counterparts across the zodiac distributed among the other eleven, medieval astrologers used an elaborate system of interpretation to advise on matters these houses pertained to—livelihood, marriage, enemies, and so on—according to the star under which anyone was born. Note the scorpion handler also toward the bottom right of the left-hand illustration. (Courtesy of the Biblioteca Apostolica Vaticana, cod. Vat. reg. lat. 1283A fol. 7v; and of the Universitätsbibliothek, Heidelberg, CPG 832 fol. 64v.)

sting. Either the man will be wearing a piece of the dried root of a plant (often hibiscus) as a charm, or he will previously have drunk a decoction of it. Consequently, he handles the scorpions without nervousness, and thereby does not irritate them, because he has complete confidence in his magic. In some parts of the Sudan scorpions are placed in sesame oil, where they die and disintegrate. The oil is then kept as a remedy to be applied to the site of a sting. In other parts of the country, the afflicted area is rubbed with the charred toenail of a baboon; the ape's foot is dried and worn as a charm in readiness for use in an emergency.

In his book *A Cure for Serpents* (1955), Alberto Denti di Pirajno describes a lady magician of Tripolitania who could charm scorpions and

Fig. 12.11. Heralds wearing scorpion coats of arms, on a paneled chest door painted by Fra Angelico and his pupils. The scorpion is a conventional figure in Crucifixion scenes of the Renaissance, appearing in works by Jacopo Bellini, Donatello, Paolo Uccello, Albrecht Altdorfer, Bosch, and many other painters and sculptors. (Ca. 1450; courtesy of the Museo di San Marco, Florence.)

handled them as though they were harmless crickets. On one occasion, she placed one in her mouth. Its tail, quivering and lashing out in every direction, protruded between her lips, yet though the poisonous sting struck her chin and nostrils, it produced no harmful effect. West African pilgrims are much respected in the Sudan on account of their magical powers, in the demonstration of which they sometimes perform with snakes and scorpions. It is probable that in such cases the sharp point of the sting has been broken off so that the animal is incapable of causing any harm.

During the 2,000 years following his death, the statements and opinions of Aristotle were accepted uncritically, and it was not until the craft of printing spread across Europe that the long centuries of apathy and ignorance came imperceptibly to an end. The first truly objective and scientific investigation of scorpions was made by Francesco Redi (one of the early followers of the method of Descartes) in 1671. Redi studied the scorpions of Italy (which, he declared, were not venomous), the scorpions of Egypt, and those of Tunisia, whose sting, it was claimed, was often lethal. The illustrations in his book show, without

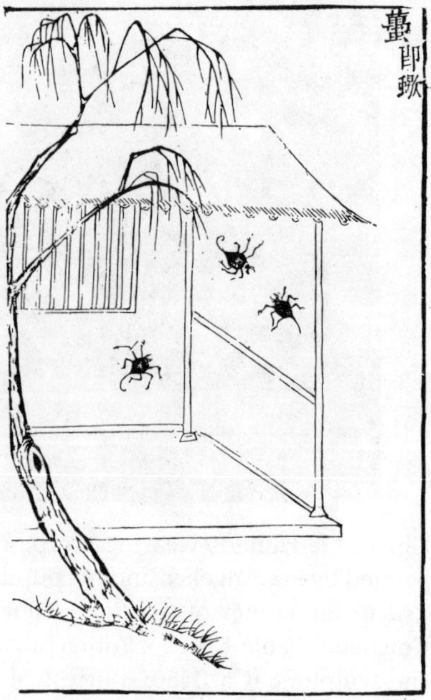

Fig. 12.12. Illustration for the entry *ch'ai,* "scorpion," in the *San Ts'ai T'u Hui* of Wang Ch'i, an illustrated lexicon of classical Chinese, published in 1607.

doubt, that the Tunisian species he studied was the dangerous *Androctonus australis,* whose sting can, indeed, kill a man.

The next author of significance to write on scorpions was Pierre-Louis de Maupertuis, whose "Expériences sur les Scorpiones" appeared in 1733. This was based on his own observations and contradicted many earlier superstitions, including that of the scorpion's suicide, which, even today, has not entirely been suppressed, as has already been mentioned.

Perhaps the most significant event in the history of scorpion biology was the publication in 1758 of the tenth edition of Linnaeus's *Systema Naturae.* This work, which standardized taxonomic nomenclature, marked the beginning of modern scorpion biology. Within the Insecta, Linnaeus listed the genus *Scorpio* with five species: *afer, americanus, australis, europaeus,* and *maurus.* It was not until 1810, however, that Latreille formally considered scorpions to represent a distinct order within the Arachnida, as they are regarded today.

A number of publications between Linnaeus's time and the mid-1800's served to increase dramatically the number of scorpion species known to science. Some of the more important workers of this period were C. DeGeer, W. F. Leach, F. G. Hemprich, C. G. Ehrenberg, J. F. W. Herbst, P. Gervais, and C. L. Koch. Koch, in particular, is best known for his twelve-volume work *Die Arachniden* (1836–45), in which many new scorpions were described.

The second half of the nineteenth century was even more productive in scorpion taxonomy, and workers began to turn their attention toward higher classification within the order. W. Peters (1861) produced the first comprehensive classification of scorpions that used many of the characters still today considered taxonomically important. Three other arachnologists emerged soon afterwards, each making significant contributions to classification: E. Simon, F. Karsch, and T. Thorell. The seventh volume of Simon's *Arachnides de France* (1879), Karsch's "Skorpionologische Beiträge" (1879a, b), and Thorell's "Études Scorpiologiques" (1876a) stand out as the major works of the period. Thorell (1876b) also published a paper dealing specifically with higher classification.

In 1888, R. I. Pocock began publishing extensively on the scorpions and other arachnids contained in the collection of the British Museum of Natural History. Many of the specimens he studied came from organized expeditions of the Museum to study animal life in the New

World, Africa, Australia, and tropical Asia. In the 20 or so years that Pocock studied scorpions, he described a considerable number of genera and several hundred species of scorpion. He also published some important monographs, including the arachnid volume in *The Fauna of British India, Including Ceylon and Burma,* edited by W. T. Blanford (1900), the arachnid volume of *Biologia Centrali-Americana* (1902b), and a work on the British fossil scorpions (1911).

About the same time, K. Kraepelin, in the Hamburg Museum, and A. A. B. Birula, in Russia, began making significant contributions to systematics. Kraepelin (1891, 1894) published his *Revision der Skorpione* in two parts, one dealing with the buthids and the second with the remaining families. In 1899, he published *Scorpiones und Pedipalpi* in F. Schulze's comprehensive treatise of the animal kingdom, *Das Tierreich.* This work was essentially a revised version of Kraepelin's two earlier studies; all three are significant in that they were the first treatments of scorpions on a truly worldwide basis, covering over 600 species. Birula (1917a, b) is best known for his fine work on the scorpion fauna of Russia and adjacent lands, but he also published on the species in Africa. Others making significant contributions prior to 1930 were E. Lönnberg, M. Laurie, N. Banks, A. Borelli, W. Purcell, J. Hewitt, and E. R. Lankester. Studies on scorpion morphology during this period by E. Pavlovsky (1924b, c, 1925) have also proved to be quite valuable to scorpion biology and systematics.

The next decade brought three major contributions to scorpion biology: F. Werner's (1934) *Scorpiones, Pedipalpi,* in Dr. H. G. Bronn's *Klassen und Ordnungen des Tierreichs;* Alfred Kästner's (1940) similar work in W. Kukenthal's *Handbuch der Zoologie;* and J. Millot and M. Vachon's (1949) chapter in volume 6 of P.-P. Grassé's monumental *Traité de Zoologie.* All three were outstanding summaries of the existing knowledge of scorpion biology; Werner's treatment considered systematics in more detail.

With the 1930's, a large number of regional studies began appearing, establishing a trend that continues today. Three major works of this type were C. C. Hoffmann's (1931, 1932) "Los Scorpiones de México," C. Mello-Leitão's (1945) *Escorpiões Sul-Americanos,* and M. Vachon's (1952) *Études sur les Scorpions,* which dealt with the scorpions of North Africa. Hoffmann's monographs were significant in that they were the first detailed treatment of the scorpions of North America.

Professor Max Vachon, of the Muséum National d'Histoire Naturelle, in Paris, stands out among scorpion systematists for his singular accomplishments. In over 130 publications, he has studied systematics and general biology of Old World scorpions. He has also been instrumental in evaluating taxonomic characters, and has as well contributed at least in part to new methods of study (e.g., hemolymph electrophoresis and karyology, with M. Goyffon, J. Lamy, R. Stockmann, and others). His major work on trichobothrial nomenclature (Vachon 1973) is still the standard used in scorpion systematics.

Although Vachon has been the dominant figure in scorpion biology, others cannot be slighted. Many recent workers have been productive in regional studies, revisions, character evaluation, and α-taxonomy. The contributions of P. R. San Martín, J. W. Abalos, E. A. Maury, M. A. González-Sponga, W. Bücherl, and T. Cekalovic have greatly augmented the work of Mello-Leitão in South America. Dr. Maury, in particular, has contributed greatly to our understanding of bothriurid systematics. W. R. Lourenço, also working in South America, has undertaken revision of the various species groups of the difficult genus *Tityus*. Dr. Lourenço currently has over 100 publications to his credit: in addition to his work on *Tityus*, he has published major revisions of *Ananteris*, *Rhopalurus*, and *Opisthacanthus*.

In North America, most of the work is very recent. Scorpions were largely neglected until H. L. Stahnke began working in the southwestern United States. S. C. Williams, who has collected and examined thousands of scorpions from Baja California, has done perhaps some of the most comprehensive systematic work on a regional basis with scorpions to date. He has also published a considerable amount on the scorpions of the southwestern United States. The work of O. F. Francke in North America and the Caribbean has been quite outstanding. Dr. Francke is considered the authority on the family Diplocentridae, and has published extensively on most of the other families and on general scorpion biology. R. W. Mitchell is well known for the description of the eyeless and depigmented cave scorpions of the genus *Typhlochactas* collected in Mexico by himself, J. Reddell, W. Elliott, and others. New species of cave scorpions are still being discovered as a result of their continuing efforts.

Several others have played an important role in the study of North American scorpions. W. Gertsch (better known for his spider work) and M. Soleglad, both singly and in collaboration, have contributed significantly to the knowledge of the Vaejovidae. W. D. Sissom, working initially with O. F. Francke, has recently begun taxonomic investigations of the North American fauna. R. M. Haradon, in a series of papers, has recently revised the genus *Paruroctonus*. Finally, L. F. de Armas has been working for a number of years on the scorpion fauna of the Caribbean region.

In the Old World, many have added to the accomplishments of Vachon. B. Lamoral (1979) published a monograph on the scorpions of Namibia, examining a large number of specimens he collected, as well as museum material, during the course of the study. R. F. Lawrence, J. Hewitt, and G. Newlands have also contributed to the knowledge of African scorpions. The revision of the genus *Heterometrus* by H. W. C. Couzijn (1981), the monograph on Australian scorpions by L. E. Koch (1977), the monograph on the scorpions of India by B. K. Tikader and

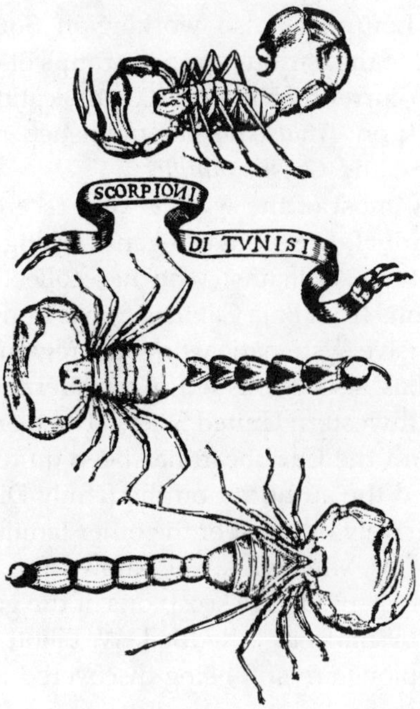

Fig. 12.13. Francesco Redi's drawing of *Androctonus australis*, from the Italian translation (Florence, 1688) of his *Experimenta circa Generationem Insectorum* (1671).

D. B. Bastawade (1983), and the study of the scorpions of Israel and the Sinai by G. Levy and P. Amitai (1980) are other major works that have appeared recently on Old World scorpions.

Scorpion fossils, though rare, have received considerable attention through the years. The first fossil scorpion was described by A. Corda in 1835. Descriptions of new fossil species continued, and it soon became necessary to classify them. Pocock's (1911) monograph, which has already been mentioned, was one of the first comprehensive attempts to do so. Contributions by L. J. Wills and A. Petrunkevitch greatly increased our knowledge of fossil taxa and relationships. Petrunkevitch (1953) published a major revision of European fossils, and followed this in 1955 with the arachnid section of the *Treatise on Invertebrate Paleontology*, edited by R. C. Moore.

Two other paleontologists have made noteworthy contributions: L. Størmer and E. N. Kjellesvig-Waering. Størmer (1963) described one of the largest known scorpions, *Gigantoscorpio willsi*, and discussed comparative morphology of the scorpions and eurypterids. Kjellesvig-Waering (1966a, 1969) clarified several misinterpretations concerning fossil taxa. More important, until his death he was at work on a huge manuscript (Kjellesvig-Waering 1986) that has revolutionized our ideas of fossil classification.

Progress in most other areas of scorpion biology lagged far behind systematics. A. Maccary (1810) wrote a description of the habits of *Buthus occitanus*. Sporadic observations on life habits of various species were published by Pocock, Pavlovsky, Simon, and others. Only three major syntheses of general scorpion biology appeared before 1920: J. H. Fabre's (1907; see also Fabre 1923) *Souvenirs Entomologiques*, which devoted almost an entire volume to the biology of *B. occitanus* in southern France (specifically discussing courtship behavior), C. Warburton's (1909) chapter "Arachnida Embolobranchiata" in volume 4 of *The Cambridge Natural History*, and A. A. B. Birula's (1917b) *Arthrogastric Arachnids of Caucasia*.

Several other works of consequence appeared in the next couple of decades. Already mentioned are the comprehensive reviews of scorpion biology by Werner (1934) and Kästner (1940). L. Berland's *Les Arachnides* (1932; published as volume 16 of the Encyclopédie Entomologique, série A), J. L. Cloudsley-Thompson's (1959) *Spiders, Scorpions, Centipedes and Mites*, and T. H. Savory's (1935, 1977) *Arachnida* contained considerable general information on scorpions.

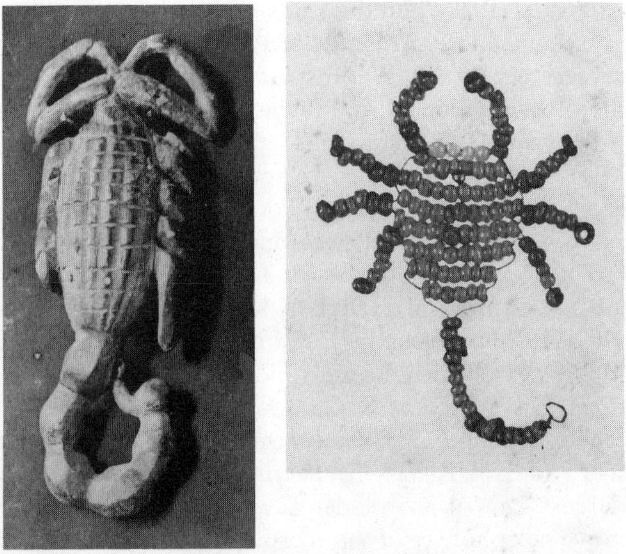

Fig. 12.14. Two Egyptian objects illustrating the scorpion's persistent appeal to the imagination in lands where it is commonly seen: at left, a stone figurine from the predynastic temple site at Hierakonpolis, where evidence of a ruler known as the Scorpion King was also found (ca. 4800 B.C.; courtesy of the Ashmolean Museum); at right, a glass-bead protective amulet from modern Cairo, said to be of a sort worn in the hair by women (ca. 1925; from Keimer 1929, by permission of *Kêmi*).

Scorpion reproduction and life history have been topics of considerable interest for many years. That scorpions transferred sperm via a spermatophore, however, was not discovered until the 1950's. In 1955 H. Angermann discovered the phenomenon in *Euscorpius italicus*; A. J. Alexander discovered it independently of Angermann's work in 1956 in *Opistophthalmus latimanus*. A. P. Mathew (1957), A. Shulov (1958), and Shulov and P. Amitai (1958) confirmed the results of Angermann and Alexander.

Embryology is a neglected area of scorpion biology, although several important contributions have been made. In the 1890's, M. Laurie (1890, 1891) and A. Brauer (1894, 1895) provided detailed accounts of the embryology of *Euscorpius* and *Heterometrus*. O. Pflugfelder (1930) studied embryology in *Liocheles australasiae*, and A. P. Mathew (1956) in *Heterometrus scaber*. Mathew (1948, 1960, 1968) also made contributions to the study of embryonic nutrition and early embryology in other species. A. Abd-el-Wahab (1952, 1954) studied embryology in *Leiurus quinquestriatus* in Egypt.

Birth behavior and life history have occupied the interests of scorpion biologists for the last 20 years. Notably, the works of R. Rosin and A. Shulov, S. C. Williams, H. Angermann, O. F. Francke, M. Haradon, J. T. Hjelle, G. Larrouy, and M. Vachon should be mentioned in regard to birth behavior. Post-birth development has been studied in the field by G. A. Polis, G. T. Smith, and D. A. Shorthouse, and in the laboratory by many biologists, including Vachon, Francke, Angermann, W. R. Lourenço, P. R. San Martín and L. Gambardella, M. A. Matthiesen, M. Auber-Thomay, P. Probst, R. Stockmann, and W. D. Sissom.

Great strides have been taken in the last 15 years in ecology, physiology, and venomology, largely as a result of improved techniques. The detailed observations of G. A. Polis, S. J. McCormick, E. B. Eastwood, P. R. San Martín, R. A. Bradley, and D. A. Shorthouse on natural populations of scorpions have led to an understanding of the importance of scorpions in ecological communities. The studies of environmental physiology by N. F. Hadley, C. Crawford, W. Riddle, J. L. Cloudsley-Thompson, L. Dresco-Derouet, and E. Toolson have shown how scorpions function in their environment and how they cope with conditions of the deserts that they so often inhabit. G. Fleissner, M. Schliwa, R. Bowerman, P. Brownell, and T. M. Root have made extraordinary efforts to understand the complex sensory systems in scorpions. Finally, J. L. Cloudsley-Thompson, C. Constantinou, M. Warburg, and G. Fleissner have made significant contributions to the study of circadian rhythms and their underlying mechanisms.

[illegible] of our [illegible] and [illegible] embryology [illegible] [illegible]

Pyle, [illegible] and [illegible] [illegible] Williams, M. [illegible], C. [illegible] [illegible] and M. [illegible] found [illegible] [illegible] to [illegible]. For [illegible] has been [illegible] in the field by [illegible], C. [illegible] and D. [illegible] [illegible] [illegible], W. R. [illegible] and [illegible]. M. [illegible] [illegible] and [illegible].

Studies of [illegible] this list is very incomplete [illegible] physiology [illegible] and [illegible]. The detailed descriptions of G. A. [illegible] [illegible]. A. [illegible] of [illegible] have [illegible] a [illegible] of [illegible] communities. The studies [illegible] physiology by [illegible] W. [illegible] C. Lindsey [illegible] and [illegible] have [illegible] how [illegible] and how [illegible] with conditions of the deserts [illegible] [illegible] have made extraordinary [illegible] the complex sensory systems in [illegible] [illegible] have made significant contributions to the [illegible] and [illegible] mechanism.

Appendix

Appendix: List of Synonyms

Some of the contributors to this volume have preferred generic or specific names used in older or original sources rather than the names currently accepted as valid. The synonymy of such names appears below.

JUNIOR SYNONYM	NAME CURRENTLY VALID
Buthotus Vachon	*Hottentotta* Birula
Buthus caucasicus (Fischer)	*Mesobuthus caucasicus* (Fischer)
Buthus confucius Simon	*Mesobuthus martensi* (Karsch)
Buthus eupeus (C. L. Koch)	*Mesobuthus eupeus* (C. L. Koch)
Buthus martensi Karsch	*Mesobuthus martensi* (Karsch)
Buthus minax C. L. Koch	*Hottentotta minax* (C. L. Koch)
Buthus tamulus (Fabricius)	*Mesobuthus tamulus* (Fabricius)
Calchas Birula	*Paraiurus* Francke
Centruroides sculpturatus Ewing	*Centruroides exilicauda* (Wood)
Diplocentrus scaber Pocock	*Cazierius scaber* (Pocock)
Heterometrus gravimanus (Pocock)	*Heterometrus indus* (DeGeer)
Heterometrus longimanus petersii (Thorell)	*Heterometrus petersii petersii* (Thorell)
Palamnaeus Thorell	*Heterometrus* Hemprich & Ehrenberg
Palamnaeus bengalensis (C. L. Koch)	*Heterometrus bengalensis* (C. L. Koch)

JUNIOR SYNONYM	NAME CURRENTLY VALID
Palamnaeus gravimanus (Pocock)	*Heterometrus indus* (DeGeer)
Palamnaeus swammerdami (Simon)	*Heterometrus swammerdami* Simon
Paruroctonus aquilonalis (Stahnke)	*Paruroctonus utahensis* (Williams)
Stenochirus Pocock	*Pocockius* Francke
Urodacus abruptus Pocock	*Urodacus manicatus* (Thorell)
Vaejovis mesaensis (Stahnke)	*Paruroctonus mesaensis* Stahnke

Reference Matter

Glossary

This glossary lists terms that appear regularly in this volume, defining them usually with regard to their application in scorpions even if they have various meanings in general biology or in reference to other arthropods. Of the anatomical terms used in this volume, only the more specialized are included below; for fully descriptive anatomical definitions, the reader is referred to Chapter 2.

abdominal endosternite. See *endosternite*.

abdominal plate. A ventral mesosomal plate, found only in fossil scorpions, that bore posterior gill slits leading into internal gill chambers. There were five abdominal plates on the ventral surface of these scorpions. Compare *sternite*.

accessory trichobothrium. Any trichobothrium regarded as not belonging to the fundamental trichobothrial number and pattern, and thus as representing an evolutionary gain of the trichobothria.

actograph. A machine that records the timing and intensity of an activity pattern of an organism.

alecithal ovum. An ovum that does not possess yolk. Compare *telolecithal ovum*.

allometry. Differential growth in cuticular structures, whether with reference to structural-growth rates differing between male and female (*sexual allometry*) or with reference to differing growth rates of a single structure at different periods in the life cycle (*ontogenetic allometry*: e.g., accentuated growth at the maturation molt).

allopatric distribution. The distribution of two or more populations or taxa with nonoverlapping geographic ranges. See also *disjunct distribution*; compare *sympatric distribution*.

amnion. The second of the two extraembryonic membranes appearing during embryonic development in apoikogenic scorpions, derived from the epiblast of the germ band. Compare *serosa*.

apoikogenic development. The type of embryonic development in which the fertilized ova develop in the ovariuterus. Compare *katoikogenic development*.

apomorphy. A derived character state. See also *synapomorphy*; compare *plesiomorphy*.

apophysis. A pronglike protuberance arising from the inner surface of the pedipalp-chela manus in some male bothriurids.

apotele. The distal segment of the leg, bearing the ungues and the dactyl.

arhabdomeric cells. In the eyes of scorpions, cells lying outside the rhabdom and carrying impulses from the eye to the brain.

axon. The long process often extending from the cell body (soma) of a neuron and carrying electrical impulses outward. Compare *dendrite*.

basidens. A small sclerite at the base of each pectinal tooth, on its dorsal surface.

book lungs. The paired respiratory organs of terrestrial (orthostern) scorpions, located in mesosomal segments 3 through 6.

capsule. The part of the spermatophore (or hemispermatophore) that bears the sperm packet. It is situated basally to the flagellum or distal lamina; its structure and ornamentation are variable and of considerable taxonomic value.

carapace. The sclerotized plate that covers the cephalothorax dorsally, upon which the eyes are located.

cell body. With reference to nerve cells (neurons), the portion containing the nucleus and the surroundng cytoplasm. Compare *axon, dendrite*.

cephalothoracic mass. The brain of a scorpion: a large neural structure in the cephalothorax, beneath the eyes. See also *deuterocerebrum, protocerebrum, tritocerebrum*.

cephalothorax. The anteriormost body region, covered dorsally by the carapace and bearing the chelicerae, pedipalps, and legs; also *prosoma*. See also *tagma*; compare *mesosoma, metasoma*.

character. Any feature useful for segregating a distinguishable organism into its proper category. Compare *character state*.

character state. Any of two or more alternative and homologous expressions of a character.

chela. Collectively, the tibia (manus and fixed finger) and tarsus (movable finger) of each pedipalp.

chelicera. Either of the two first postoral appendages, extending anteriorly from beneath the anterior margin of the carapace, which function in feeding and grooming.

Chelicerata. A major subphylum of arthropods, including the class Merostomata (horseshoe crabs and eurypterids), the class Arachnida (including spiders, scorpions, mites and ticks, harvestmen, and their allies), and the class Pycnogonida (sea spiders).

chemoreceptor. A receptor cell that is sensitive to chemical stimuli.

circadian rhythm. An endogenous oscillation with a natural period close but not necessarily equal to that of the solar day.

circumesophageal region. The neural region of the cephalothoracic mass extending laterally around the gut tube (esophagus). Compare *subesophageal ganglion, supraesophageal ganglion*.

cladistics. The method of systematic biology that attempts to produce a classification based on genealogical relationships, as inferred directly from cladograms, rather than on overall similarity or degree of morphological divergence.

cladogram. A diagram depicting the presumed historical sequence of genealogical branching within a monophyletic group, based on a nested pattern of synapomorphies.

connective. Any of the nerve fibers between ganglia in the nerve cord, which are usually grouped together into identifiable nerves.

coxapophysis. Any of the hollow anterior basal apophyses of the coxae of the first (*coxapophyses I*) and second (*coxapophyses II*) pairs of legs.

coxosternal region. The anteroventral surface of a scorpion, including the coxae of the legs, the pedipalps, and the sternum.

dactyl. The median claw of the apotele of each leg. Compare *unguis*.

dendrite. The fine process often extending from the cell body (soma) of a neuron and typically carrying incoming electrical activity received from other neurons. Compare *axon*.

determinate growth. Growth characterized by an increase in body size to a specific upper limit, after which growth ceases. Compare *indeterminate growth*.

deuterocerebrum. The middle region of the three major brain portions in arthropods, typically receiving inputs from the antennae; hence, its presence in arachnids is debatable. Compare *protocerebrum, tritocerebrum*.

disjunct distribution. Noncontiguous allopatric distribution, usually of widely separated populations or taxa. See also *generalized biogeographic track*.

dispersal event. Any movement of one or more individuals of a species to a place outside the normal range of that species, frequently resulting in an expansion of the range or the establishment of a new population at some distant location across a barrier. The new population may evolve in isolation into a new taxon. Compare *vicariance event*.

distal lamina. See *lamina*.

diverticulum. In katoikogenic scorpions, a blind outpocketing of the ovariuterus in which the oocytes are contained and the embryos develop.

doublure. The inwardly deflected posterolateral margin of the abdominal plate in fossil scorpions. The gill slits were found either between the doublure and the abdominal plate proper or along the posterior margin of the doublure.

ecdysis. The periodic shedding of the exoskeleton to accommodate growth: molting. See also *exuvium, instar*.

electromyogram (*EMG*). An electrical recording from a muscle, made by placing an electrode on the surface of or into the muscle to receive impulses from many nearby muscle fibers.

electroretinogram (*ERG*). An electrical recording from the retina of an eye, made by placing an electrode near the visual surface to record the activity from many retinal cells.

embryonal capsule. The first of the three extraembryonic membranes appearing during embryonic development in katoikogenic scorpions, deriving from follicle cells. Compare *embryonal envelope, trophamnion*.

embryonal envelope. The third of the three extraembryonic membranes appearing during embryonic development in katoikogenic scorpions, deriving from follicle cells. Compare *embryonal capsule, trophamnion*.

endogenous rhythm. Any periodic system that is part of the temporal organization of an organism; also *endogenous oscillation*.

endosternite. In scorpions, an internal skeleton lying horizontally above the subesophageal ganglion and joining the diaphragm posteriorly. An *abdominal endosternite* is also present, ventral to the ventral connectives of the subesophageal ganglion.

equilibrium species. Species living in a relatively stable environment, whose populations do not fluctuate greatly; also *K-selected species*. Compare *opportunistic species*.

eudesmatic. With reference to joints, having one or more tendon attachments at the base of a distal leg segment.

exploitation competition. The concurrent use by two or more organisms of a resource the supply of which is insufficient to meet all demands. See also *interference competition.*

exteroreceptor. See *receptor.*

exuviae. The shed cuticle left behind after molting.

flagelliform spermatophore. One of two basic types of spermatophore: in this type, the distal end is produced into a long, slender flagellum. The term also applies to the hemispermatophore, but the flagellum is usually coiled inside the paraxial organ. Compare *lamelliform spermatophore.*

fossorial. Adapted for digging or for life in a burrow, or for both.

fundamental trichobothrial number. The basic number of trichobothria associated with each type (A, B, or C) of fundamental trichobothrial pattern. Compare *accessory trichobothrium.*

fundamental trichobothrial pattern. Any of the three basic patterns of trichobothrial distribution (designated A, B, and C) on the pedipalps of scorpions. See also *accessory trichobothrium, fundamental trichobothrial number.*

ganglion. A grouping of nerve cells in the peripheral nervous system, joined to similar adjacent groupings by connectives.

generalized biogeographic track. A similar pattern of disjunct distribution shared by a varied array of taxa.

Gondwanaland. In the theories of plate tectonics and continental drift, the southern supercontinent, resulting from the breakup of Pangaea in the Jurassic Period (ca. 150–200 M.Y.B.P.) and including an aggregate land mass that now forms South America, Africa, India, Australia, and Antarctica. Compare *Laurasia.*

guild. A group of species that share a common resource (e.g., food).

hemispermatophore. Either of the two halves of the spermatophore, which develop inside the paraxial organ in the male scorpion's mesosoma. During courtship, the two halves are extruded from the genital aperture, cemented together, and deposited on the substrate to form the spermatophore.

indeterminate growth. Growth characterized by increase in body size throughout the entire life of an organism. Compare *determinate growth.*

instar. The period between two successive molts; also, the organism itself during such a period.

interference competition. Direct antagonism between organisms that results in decreased access to resources for the loser. See also *exploitation competition.*

interoreceptor. See *receptor*.

intraguild predation. Predation among species that use common resources and are potential competitors.

inverted life cycle. Temporal displacement of important biological activities (e.g., feeding, reproduction) to periods when other, similar species exhibit minimal activity.

iteroparity. Repeated reproduction during the lifetime of a female.

K-selected species. See *equilibrium species*.

katoikogenic development. The type of embryonic development in which the fertilized ova develop in numerous diverticula branching from the ovariuterus. Compare *apoikogenic development*.

kinesthetic orientation. The ability of an organism to orient itself in space by relying on information from various movement-detecting receptors.

lamelliform spermatophore. One of two basic types of spermatophore: in this type, the distal end is produced into a broad lamina, or blade. The term also applies to the hemispermatophore. Compare *flagelliform spermatophore*.

lamina. The broad, flat distal blade of the lamelliform spermatophore (or hemispermatophore).

Laurasia. In the theories of plate tectonics and continental drift, the northern supercontinent, resulting from the breakup of Pangaea in the Jurassic Period (ca. 150–200 M.Y.B.P.) and including an aggregate land mass that now forms North America and Eurasia. Compare *Gondwanaland*.

lithophilic. Adapted for living in rocky environments, especially in cracks and crevices in boulders or vertical cliff faces.

littoral. Adapted for living in the intertidal zone.

lyriform organ. A lyre-shaped grouping of slit sensilla seen on the cuticle of some arachnids.

mechanoreceptor. A receptor cell that is sensitive to mechanical disturbances such as touch, vibration, or deformation. See also *proprioceptor*, *scolopale*, *slit sense organ*, *trichobothrium*.

mesosoma. The seven-segmented major midbody region, posterior to the cephalothorax, covered dorsally by tergites and ventrally by sternites. On its ventral surface it bears the genital opercula, pectines, and book-lung spiracles; internally, it bears the bulk of the visceral organs. Compare *metasoma*, *opisthosoma*.

metasoma. The five-segmented "tail," forming the posteriormost major body region; the intestine passes through it and terminates posteriorly and ventrally to the fifth metasomal segment. Compare *mesosoma, opisthosoma, telson.*

molting. See *ecdysis.*

monophyletic group. A genealogical group, defined by one or more synapomorphies, containing an ancestral species and all its descendants.

monotypic taxon. A higher taxon containing only one lower taxon (e.g., a family with only one genus, or a genus with only one species).

motoneuron. A neuron that carries impulses to activate muscle cells.

nerve cord. A long collection of ganglia and connectives, usually running from the brain of an animal and down the long axis of its body. In arachnids, the nerve cord extends from the cephalothoracic mass ventrally toward the posterior end of the animal.

neuromere. Any segment of neural tissue found early in development and representing some future portion of the adult nervous system.

neuropile. The central region of a ganglion (as opposed to its outer area, or rind), consisting of a complicated network of axons and dendrites whose cell bodies lie to the outside of the ganglion, in the rind.

ontogenetic allometry. See *allometry.*

opisthosoma. Collectively, the metasoma and mesosoma together. See also *tagma;* compare *prosoma.*

opportunistic species. Species that live in a relatively unstable environment, whose populations are characterized by large, erratic changes in density; also *r-selected species.* Compare *equilibrium species.*

ovariuterus. The major part of the female reproductive system, which includes the longitudinal and transverse ovarian tubes; these form a reticulate network that functions as both ovaries and oviducts. See also *apoikogenic development, katoikogenic development.*

ovoviviparous. Maintaining the fertilized ova inside the maternal body without providing maternal nourishment as they develop (i.e., the ova depend on yolk reserves); the ova hatch in the female's body, resulting in live birth. Compare *viviparous.*

Pangaea. In the theories of plate tectonics and continental drift, the original supercontinent, whose breakup in the Jurassic Period (ca. 150–200 M.Y.B.P.) resulted in the formation of Laurasia and Gondwanaland.

paraxial organ. An organ that produces a hemispermatophore: it is composed of an ejaculatory sac, seminal vesicle, and one or more accessory glands that produce the components of the hemispermatophore and provide the sperm for it.

pecten. Either of the paired comblike appendages making up the second mesosomal segment in scorpions, which function in mechanoreception and, probably, chemoreception; also *pectine*.

pedipalp. Either of the paired chelate appendages comprising the fourth somite in scorpions, which function in prey capture, defense, and sensory perception.

peg sensillum. A sensory structure found on the anteroventral margin of each pectinal tooth and functioning in mechanoreception and, probably, chemoreception.

peripheral inhibition. The occurrence of motoneurons that can inhibit the electrical activity (and hence the tension) of muscle fibers.

phasically sensitive. With reference to receptors, one that usually responds to a stimulus with only a brief period of electrical activity. Compare *tonically sensitive*.

phenology. The cycle of activity of an organism during the year.

photoreceptor. A cell capable of transducing light energy. See also *retinula cell*.

plesiomorphy. A primitive character state. See also *symplesiomorphy*; compare *apomorphy*.

poikilothermic. Exhibiting a body temperature that is not constant, but rather fluctuates in response to environmental temperatures.

postsynaptic potential (*PSP*). An electrical potential induced in a neuron or in a muscle cell by the activity of a cell that synapses upon it: typically the presynaptic cell releases a chemical that binds to the postsynaptic cell, causing the electrical potential.

prepectinal plate. A small ventral sclerotized plate located just anterior to the pectines in fossil scorpions; it is vestigial in extant Neotropical buthids.

proprioceptor. A slowly adapting mechanoreceptor that provides essentially continuous information about the position of various body structures in relation to one another and to distortions of the body.

prosoma. See *cephalothorax*.

protocerebrum. The first of the three major brain portions in arthropods, typically containing centers from the eyes, and important integration and association regions. Compare *deuterocerebrum, tritocerebrum*.

psammophilic. Adapted for living in sandy environments.

r-selected species. See *opportunistic species*.

receptor. Any cell that transduces a specific stimulus, whether from the environment (*exteroreceptor*) or from within the body (*interoreceptor*). See also *chemoreceptor, mechanoreceptor, photoreceptor, receptor organ, sensilla basonica, thermoreceptor*.

receptor organ. A group of tissues that in combination with receptor cells responds to specific stimuli.

retinula cell. Any of a group of retinal cells wherein the specialized membrane region (rhabdomere) of each cell is connected to similar adjacent cells, forming a rhabdom.

rhabdom. In certain visual organs (e.g., the eyes of some arthropods), a complex membrane region where light energy is transduced into electrical impulses: in the scorpion eye, the rhabdom consists of the rhabdomeres of numerous adjacent retinula cells.

rhabdomere. The specialized membrane region of a retinula cell, consisting of complex microvilli (compacted tubes) and presumed to be the site where light is transduced in the eye.

sarcomere. The structural and functional unit of striated muscle: a highly organized subunit of a single muscle cell, containing contractile proteins in an ordered array.

sclerite. Any of the hardened plates into which the exoskeleton is divided. The separation of the sclerites by soft membranes or sutures allows movement while retaining necessary rigidity.

scolopale. A specialized receptor cell in certain mechanoreceptor organs (e.g., the chordotonal organs of insects) that is modified to transduce mechanical energy. See also *scolopophorous organ*.

scolopophorous organ. A sensory organ containing a scolopalous element.

sensilla basonica. Receptors comprised of short, thick hairs innervated by a variable number of sensory neurons and having any of several possible functions, including chemoreception, mechanoreception, and thermoreception.

serosa. The first of the two extraembryonic membranes appearing during embryonic development in apoikogenic scorpions, derived from proliferation of the blastoderm. Compare *amnion*.

seta. Any of the various movable hairlike projections of the scorpion integument that arise from an areolar cup. See also *trichobothrium*.

sexual allometry. See *allometry*.

sister group. A taxon that is the closest genealogical relative of some other taxon (i.e., the two share an immediate common ancestor not shared with any other taxon).

slit sense organ. A very small sensory structure in the cuticle of an arthropod, consisting of a tiny cuticular thinning (slit) innervated by a sensory cell that is activated by cuticular stress; also *slit sensillum*. See also *lyriform organ*.

slit sensillum. See *slit sense organ*.

somite. Any of the body segments in those animals that are composed of a linear series of similar body segments.

spermatophore. A sclerotized, sperm-carrying structure composed of two hemi-spermatophores, which is extruded from the male scorpion's body and deposited on the substrate. After its deposition, the male positions the female over it for her retrieval of the sperm. See also *flagelliform spermatophore, lamelliform spermatophore*.

sternite. Strictly, any of the four ventral mesosomal plates bearing pairs of ventrolateral spiracles leading into book-lung chambers in orthostern (Modern) scorpions. (The fifth ventral mesosomal plate, although it does not bear spiracles, is also referred to as a sternite by neontologists.) Compare *abdominal plate*; see also *sternum*.

sternum. The ventral surface of any body segment (compare *tergum*); also, collectively, the variously shaped fused sterna of somites VI–VIII, located between the coxae of legs III and IV.

stilting. In birth behavior, the raising of the anterior portion of the female scorpion's body prior to and during parturition; also, in response to high ambient temperatures, the raising of a scorpion's body above the substrate, with the posterior portion held highest.

stridulatory organ. A sound-producing organ used to warn potential predators. In *Parabuthus*, it consists of roughened or granular surfaces on the dorsal aspect of metasomal segments I and II, which are vigorously rubbed with the tip of the stinger; in some scopionids, it consists of a granular surface and a patch of stiff setae located on the opposing surfaces of the coxae of the pedipalps and the first pair of legs.

subesophageal ganglion. The ventral portion of the cephalothoracic mass, lying below the gut tube (esophagus); in arachnids, it is often very large and contains major neural centers for the appendages. Compare *circumesophageal region, supraesophageal ganglion*.

supercooling. Lowering the temperature of an organism's body fluids below their ideal freezing point without solidification taking place.

supraesophageal ganglion. The dorsal portion of the cephalothoracic mass, lying above the gut tube (esophagus) and receiving inputs from the eyes and other sensory regions in the head. Compare *circumesophageal region, subesophageal ganglion*.

sympatric distribution. The distribution of two or more populations or taxa with overlapping ranges and thus occupying essentially the same geographic area. Compare *allopatric distribution, disjunct distribution*.

symplesiomorphy. A primitive character state (plesiomorphy) shared between taxa and therefore not useful in determining phylogenetic relationships. Compare *synapomorphy*.

synapomorphy. A derived character state (apomorphy) shared by the taxa of a monophyletic group. Compare *symplesiomorphy*.

synaptoid. A synapselike contact between adjacent neural or sensory cells.

tagma. A union of two or more somites into a structurally distinct group specialized to perform some function or functions; in scorpions, two major tagmata may be recognized, the cephalothorax and the opisthosoma.

telolecithal ovum. An ovum in which the yolk is concentrated to one side (the vegetal side). Compare *alecithal ovum*.

telson. The sting, consisting of the vesicle (containing the paired venom glands) and the sharp, curved aculeus. It forms the terminal portion of the scorpion body, occurring posterior to the fifth metasomal segment.

tergum. The dorsal surface of a body segment. Compare *sternum*.

thermoreceptor. A receptor cell that is sensitive to thermal stimuli.

time minimizer. An animal that maximizes survival and reproduction by minimizing the time spent foraging: such a low-risk, low-reproductive-gain strategy is found in animals subject to high rates of predation.

tonically sensitive. With reference to receptors, one that usually responds to a stimulus with a prolonged period of electrical activity. Compare *phasically sensitive*.

trichobothrial territory. The predictable boundary within which a particular trichobothrium may be found in a given taxon.

trichobothrium. Any of the long, slender, hairlike sensilla arising from cup-shaped areolae on the scorpion pedipalp femur, patella, and chela, and functioning in the perception of airborne vibrations. The trichobothrium is distinguished from typical setae in that the base of its shaft does not completely fill the areola. See also *fundamental trichobothrial pattern*.

tritocerebrum. The third of the three major brain portions in arthropods, typically containing nerves to the anterior regions of the gut and structures adjacent thereto. Compare *deuterocerebrum, protocerebrum*.

troglobitic. Adapted for living exclusively in caves and exhibiting such features as eyelessness, elongation of body parts, and/or depigmentation. Compare *troglophilic*.

troglophilic. Adapted for living in caves and in other habitats, and generally not exhibiting features of cave-adaptedness. Compare *troglobitic*.

trophamnion. The second of the three extraembryonic membranes appearing during embryonic development in katoikogenic scorpions, formed by the proliferation of polar-body cells. Compare *embryonal capsule, embryonal envelope.*

truncal flexure. On the lamelliform spermatophore, the area between the trunk and the distal lamina, which will flex as the female lowers her body over the spermatophore: the flexing pushes the sperm packet into the female's genital atrium.

unguis. Either of the paired lateral claws of the apotele of each leg. Compare *dactyl.*

vicariance event. In historical biogeography, any historical event (often of a geological nature) that fragments an ancestral population into two or more disjunct populations. Each resulting population may evolve in isolation into new taxa. Compare *dispersal event.*

viviparous. Giving birth to live young, which as embryos received at least some nourishment directly from the mother. Compare *ovoviviparous.*

Zeitgeber. The forcing oscillation that entrains a biological oscillation (e.g., the environmental cycle of light or temperature).

Bibliography

Abalos, J. W. 1955. *Bothriurus bertae* sp. n. (Bothriuridae, Scorpiones). *Anales del Instituto de medicina regional, Universidad nacional de Tucumán* 4(2): 231–39.

Abalos, J. W., and C. B. Hominal. 1974. La transferencia espermática en *Bothriurus flavidus* Kraepelin, 1910. *Acta Zoologica Lilloana* 31(5): 47–56.

Abd-el-Wahab, A. 1952. Some notes on the segmentation of the scorpion, *Buthus quinquestriatus* (H.&E.) *Proceedings of the Egyptian Academy of Sciences* 7: 75–91.

———. 1954. The formation of the germ layers and their embryonic coverings in the scorpion *Buthus quinquestriatus* (H.&E.). *Proceedings of the Egyptian Academy of Sciences* 10: 110–18.

———. 1957. The male genital system of the scorpion, *Buthus quinquestriatus*. *Quarterly Journal of Microscopical Science* 98: 111–22.

Abdullah, S. 1957. Calcium chloride for scorpion bites. *British Medical Journal* 1: 1366.

Abushama, F. T. 1962. Bioclimate, diurnal rhythms and water-loss in the scorpion *Leiurus quinquestriatus* (H.&E.). *The Entomologist's Monthly Magazine* 98: 216–24.

———. 1964. On the behaviour and sensory physiology of the scorpion *Leiurus quinquestriatus* (H.&E.). *Animal Behaviour* 12(1): 140–53.

———. 1968. Observations on the mating behaviour and birth of *Leiurus quinquestriatus* (H.&E.), a common scorpion species in the central Sudan. *Revue de zoologie et de botanique africaines* 77: 37–43.

Adam, K. R., H. Schmidt, R. Stämpfli, and C. Weiss. 1966. The effect of scorpion venom on single myelinated nerve fibres of the frog. *British Journal of Pharmacology* 26: 666–77.

Adam, K. R., and C. Weiss. 1958. The occurrence of 5-hydroxytryptamine in scorpion venom. *Journal of Experimental Biology* 35: 39–42.

———. 1959. Actions of scorpion venom on skeletal muscle. *British Journal of Pharmacology* 14: 334–39.

Adler-Graschinsky, E., and S. Z. Langer. 1978. Mechanism of the enhancement in transmitter release from central and peripheral noradrenergic nerve terminals induced by the purified scorpion venom, Tityustoxin. *Naunyn-Schmiedeberg's Archives of Pharmacology* 303: 243–49.

Ahearn, G. A., and N. F. Hadley. 1976. Functional roles of luminal sodium and potassium in water transport across desert scorpion ileum. *Nature* 261: 66–68.

———. 1977. Water transport in perfused scorpion ileum. *American Journal of Physiology* 233: R198–R207.

Alberti, G. 1983. Fine structure of scorpion spermatozoa (*Buthus occitanus* (Amoreux), Buthidae, Scorpiones). *Journal of Morphology* 177: 205–12.

Alexander, A. J. 1956. Mating in scorpions. *Nature* 178: 867–68.

———. 1957. The courtship and mating of the scorpion, *Opisthophthalmus latimanus*. *Proceedings of the Zoological Society* (London) 128: 529–44.

———. 1958. On the stridulation of scorpions. *Behaviour* 12: 339–52.

———. 1959a. Courtship and mating in the buthid scorpions. *Proceedings of the Zoological Society* (London) 133: 145–69.

———. 1959b. A survey of the biology of scorpions of South Africa. *African Wildlife* 13: 99–106.

———. 1967. Problems of limb extension in the scorpion, *Opisthophthalmus latimanus* Koch. *Transactions of the Royal Society* (South Africa) 37(3): 165–81.

———. 1972. Feeding behaviour in scorpions. *South African Journal of Science* 68(9): 253–56.

Alexander, A. J., and D. W. Ewer. 1957. A chemo-receptor in the scorpion, *Opisthophthalmus*. *South African Journal of Science* 53: 421–22.

———. 1958. Temperature adaptive behaviour in the scorpion, *Opisthophthalmus latimanus* Koch. *Journal of Experimental Biology* 35: 349–59.

Allred, D. M. 1973. Scorpions of the National Reactor Testing Station, Idaho. *Great Basin Naturalist* 33: 251–54.

Almassy, R. J., J. C. Fontecilla-Camps, F. L. Suddath, and C. E. Bugg. 1983. Structure of Variant-3 scorpion neurotoxin from *Centruroides sculpturatus* Ewing, refined at 1.8 Å resolution. *Journal of Molecular Biology* 170: 497–527.

Amitai, P. 1980. *Scorpions*. Tel Aviv: Masada Press. 98 pp. [In Hebrew]

Amoreux, P. J. 1762. Tentamen de noxa animalium. Thèse de l'Université Montpellier. Avignon. 59 pp.

Anderson, D. T. 1973. *Embryology and phylogeny in annelids and arthropods*. International Series of Monographs in Pure and Applied Biology, Zoology Division, vol. 50. New York: Pergamon. 495 pp.

Anderson, J. D. 1956. A blind snake preyed upon by a scorpion. *Herpetologica* 12: 327.

Anderson, J. F. 1970. Metabolic rates of spiders. *Comparative Biochemistry and Physiology* 33: 51–72.

Anderson, R. C. 1975. Scorpions of Idaho. *Tebiwa* (Miscellaneous Papers of the Idaho State University Museum of Natural History) 18(1): 1–17.

Angermann, H. 1955. Indirekte Spermatophorenübertragung bei *Euscorpius italicus* Hbst. (Scorpiones, Chactidae). *Naturwissenschaften* 42: 323.

———. 1957. Über Verhalten, Spermatophorenbildung und Sinnesphysiologie von *Euscorpius italicus* Hbst. und verwandten Arten (Scorpiones, Chactidae). *Zeitschrift für Tierpsychologie* 14: 276–302.

Araujo, R. L., and M. V. Gomez. 1976. Potentiation of bradykinin action on smooth muscle by a scorpion venom extract. *General Pharmacology* 7: 123–26.

de Armas, L. F. 1973. Escorpiones del archipiélago cubano. I, Nuevo género y nuevas especies de Buthidae. *Poeyana* 114: 1–28.

———. 1974. Escorpiones del archipiélago cubano. III, Género *Tityus* C. L. Koch, 1836 (Scorpionida: Buthidae). *Poeyana* 135: 1–15.

———. 1975. Un notable caso alimentario en los escorpiones (Arachnida: Scorpionida). *Academia de ciencias de Cuba, Instituto de zoología,* 1: 2–3.

———. 1976. Escorpiones del archipiélago cubano. VI, Familia Diplocentridae (Arachnida: Scorpionidae). *Poeyana* 147: 1–35.

———. 1977. Anomalías en algunos Buthidae (Scorpionida) de Cuba y Brasil. *Poeyana* 176: 1–6.

———. 1980. Aspectos de la biología de algunos escorpiones cubanos. *Poeyana* 211: 1–28.

———. 1982a. Algunos aspectos zoogeográficos de la escorpiofauna antillana. *Poeyana* 238: 1–17.

———. 1982b. Desarrollo postembrionario de *Didymocentrus trinitarius* (Franganillo) (Scorpiones: Diplocentridae). *Academia de ciencias de Cuba, Miscelanea zoologica,* 16: 3–4.

———. 1984. Escorpiones del archipiélago cubano. VII, Adiciones y enmiendas (Scorpiones: Buthidae, Diplocentridae). *Poeyana* 275: 1–37.

———. 1986. Biología y morfometria de *Rhopalurus garridoi* Armas (Scorpiones: Buthidae). *Poeyana* 333: 1–27.

———. 1987. Cópula múltiple en escorpiones (Arachnida: Scorpiones). *Academia de ciencias de Cuba, Miscelanea zoologica,* 30: 1–2.

de Armas, L. F., and H. Contreras. 1981. Gestación y desarrollo postembrionario en algunos *Centruroides* (Scorpionida: Buthidae) de Cuba. *Poeyana* 217: 1–10.

Arseniev, A. S., V. I. Kondakov, V. N. Maiorov, and V. F. Bystrov. 1984. NMR solution spatial structure of "short" scorpion insectotoxin I_5A. *FEBS Letters* 165: 57–62.

el-Asmar, M. F., M. Ismail, K. Ghoneim, and O. H. Osman. 1977. Scorpion (*Buthus minax*, L. Koch) venom fractions with anticholinesterase activity. *Toxicon* 15: 63–69.

el-Asmar, M. F., O. H. Osman, and M. Ismail. 1973. Fractionation and lethality of venom from the scorpion *Buthus minax* (L. Koch). *Toxicon* 11: 3–7.

Auber, M. 1959. Observations sur le biotope et la biologie du scorpion aveugle, *Belisarius xambeui* E. Simon. *Vie et milieu* 10: 160–67.

———. 1963a. Reproduction et croissance de *Buthus occitanus* Amx. *Annales des sciences naturelles: Zoologie* (Paris) 5: 273–85.

———. 1963b. Remarques sur l'ultrastructure des myofibrilles chez des scorpions. *Journal de microscopie* (Paris) 2: 233–36.

———. 1963c. Répartition et ultrastructure des fibres musculaires viscerales de l'intestin môyen de scorpions. *Comptes rendus hebdomadaires des séances de l'Académie des sciences* (Paris), sér. D, 256: 2022–24.

Auber-Thomay, M. 1974. Croissance et reproduction d'*Androctonus australis* (L.) (Scorpions, Buthides). *Annales des sciences naturelles: Zoologie* (Paris) 16: 45–54.

Awati, P. R., and V. B. Tembe. 1956. *Buthus tamulus (Fabr.). The Indian scorpion: Morphology, anatomy and bionomics.* Zoological Monographs, no. 2. University of Bombay. 62 pp.

el-Ayeb, M., M. F. Martin, P. Delori, G. Bechis, and H. Rochat. 1983. Immunochemistry of scorpion α-neurotoxins: Determination of the antigenic site number and isolation of a highly enriched antibody specific to a single antigenic site of Toxin II of *Androctonus australis* Hector. *Molecular Immunology* 20: 697–708.

Babin, D. R., D. D. Watt, S. M. Goos, and R. V. Mlejnek. 1974. Amino acid sequences of neurotoxic protein variants from the venom of *Centruroides sculpturatus* Ewing. *Archives of Biochemistry and Biophysics* 164: 694–706.

———. 1975. Amino acid sequence of neurotoxin I from *Centruroides sculpturatus* Ewing. *Archives of Biochemistry and Biophysics* 166: 125–34.

Babu, K. S. 1961a. Studies on the nervous system of selected arachnids. Ph.D. diss., Sri Venkateswara University, Tirupati, Andhra Pradesh, India.

———. 1961b. Giant fibres in the central nervous system of scorpions. *Journal of Animal Morphology and Physiology* 8(1): 11–18.

———. 1965. Anatomy of the central nervous system of arachnids. *Zoologische Jahrbücher, Abteilung für Anatomie und Ontogenie,* 82(1): 1–154.

———. 1985. Patterns of arrangement and connectivity in the central nervous system of arachnids. In Barth, ed., pp. 3–19.

Babu, K. S., and P. Sanjeeva Reddy. 1967. Unit hair reception activity from the telson of the scorpion *Heterometrus fulvipes. Current Science* 36: 599–600.

Babu, K. S., and S. A. T. Venkatachari. 1966. Certain anatomical features of the ventral nerve cord of the scorpion, *Heterometrus fulvipes*. *Journal of Animal Morphology and Physiology* 13: 22–33.

———. 1972. Activity patterns of interneurons in the ventral nerve cord of the scorpion *Heterometrus fulvipes*. *Indian Journal of Experimental Biology* 10(1): 49–58.

Baccetti, B., and G. Lazzeroni. 1969. The envelopes of the nervous system of pseudo-scorpions and scorpions. *Tissue and Cell* 1(3): 417–24.

Bacon, A. D. 1972. Ecological studies on a population of *Uroctonus mordax* Thorell. Master's thesis, California State University, San Francisco. 54 pp.

Baerg, W. J. 1954. Regarding the biology of the common Jamaican scorpion. *Annals of the Entomological Society of America* 47: 272–76.

———. 1961. *Scorpions: Biology and effects of their venom*. University of Arkansas Agriculture Experimental Station, Bulletin 649. 34 pp.

Balboa, P. F. 1937. Un monstruo aracnológico. *Memorias de la Sociedad cubana de historia natural* 11(1): 55.

Balozet, L. 1971. Scorpionism in the Old World. In Bücherl and Buckley, eds., pp. 349–71.

Bardi, J. K., and C. J. George. 1943. Digestive glands of the scorpion. *Journal of the University of Bombay* 11 (n.s. 3B): 91–115.

Barhanin, J., J. R. Giglio, P. Léopold, A. Schmid, S. V. Sampaio, and M. Lazdunski. 1982. *Tityus serrulatus* venom contains two classes of toxins. Tityus γ toxin is a new tool with a very high affinity for studying the Na^+ channel. *Journal of Biological Chemistry* 257: 12553–58.

Barhanin, J., M. Ildefonse, O. Rougier, S. V. Sampaio, J. R. Giglio, and M. Lazdunski. 1984. Tityus γ toxin, a high affinity effector of the Na^+ channel in muscle, with a selectivity for channels in the surface membrane. *Pflügers Archiv* 400: 22–27.

Barhanin, J., D. Pauron, A. Lombet, R. I. Norman, H. P. M. Vijverberg, J. R. Giglio, and M. Lazdunski. 1983. Electrophysiological characterization, solubilization and purification of the Tityus γ toxin receptor associated with the gating component of the Na^+ channel from rat brain. *EMBO Journal* 2: 915–20.

Barrows, W. M. 1925. Modification and development of the arachnid palpal claw, with especial reference to spiders. *Annals of the Entomological Society of America* 18(4): 483–525.

Barth, F. G. 1972a. Die Physiologie der Spaltsinnesorgane. I, Modellversuche zur Rolle des cuticularen Spaltes beim Reiztransport. *Journal of Comparative Physiology* 78A: 315–36.

———. 1972b. Die Physiologie der Spaltsinnesorgane. II, Funktionelle Morphologie eines Mechanoreceptors. *Journal of Comparative Physiology* 81A: 159–86.

———. 1973. Bauprinzipien und adäquater Reiz bei einem Mechanoreceptor. *Verhandlungen der Deutschen zoologischen Gesellschaft* (Mainz) 66: 25–30.

———. 1978. Slit sense organs: "Strain gauges" in the arachnid exoskeleton. *Symposia of the Zoological Society* (London) 42: 439–48.

———. 1981. Strain detection in the arthropod exoskeleton. In M. S. Laverack and D. Cosens, eds., *Sense organs*, pp. 112–41. Glasgow: Blackie.

———. 1985. Slit sensilla and the measurement of cuticular strains. In Barth, ed., pp. 162–88.

———, ed. 1985. *Neurobiology of arachnids*. Berlin: Springer-Verlag. 385 pp.

Barth, F. G., and J. Bohnenberger. 1978. Lyriform slit sense organs: Thresholds and stimulus intensity ranges in a multiunit mechanoreceptor. *Journal of Comparative Physiology* 125A: 37–43.

Barth, F. G., and P. Picklemann. 1975. Lyriform slit sense organs in spiders: Modelling an arthropod mechanoreceptor. *Journal of Comparative Physiology* 103A: 39–54.

Barth, F. G., and J. Stagl. 1976. The slit sense organs of arachnids: A comparative study of their topography on the walking legs. Chelicerata, Arachnida. *Zoomorphologie* 86(1): 1–23.

Barth, F. G., and M. Wadepuhl. 1975. Slit sense organs on the scorpion leg (*Androctonus australis* L., Buthidae). *Journal of Morphology* 145(2): 209–28.

Bawaskar, H. S. 1984. Scorpion sting. *Transactions of the Royal Society of Tropical Medicine and Hygiene* 78: 414–15.

Bawaskar, H. S., and P. H. Bawaskar. 1986. Prazosin in management of cardiovascular manifestations of scorpion sting. *The Lancet*, no. 8479 (1986 no. 1, March 1): 510–11.

Bechis, G., F. Sampieri, P.-M. Yuan, T. Brando, M.-F. Martin, C. R. Diniz, and H. Rochat. 1984. Amino acid sequence of Toxin VII, a β-toxin from the venom of the scorpion *Tityus serrulatus*. *Biochemical and Biophysical Research Communications* 122: 1146–53.

Beck, E. J. 1885. Description of the muscular and endoskeletal systems of *Scorpio*. *Transactions of the Zoological Society* (London) 11: 339–60.

Beck, L., and K. Görke. 1974. Tagesperiodik, Revierverhalten und Beutefang der Geisselspinne *Admetus pumilio* C. L. Koch im Freiland. *Zeitschrift für Tierpsychologie* 35: 173–86.

Bedini, C. 1967. The fine structure of the eyes of *Euscorpius carpathicus* L. (Arachnida, Scorpiones). *Archives italiennes de biologie* 105: 361–78.

Beer, R. E. 1960. A new species of *Pimeliophilus* (Acarina: Pterygosomidae) parasite on scorpions, with discussion of its postembryonic development. *Journal of Parasitology* 46: 433–40.

Belmonte, C., and L. J. Stensaas. 1975. Repetitive spikes in photoreceptor axons of the scorpion eye: Invertebrate eye structure and tetrodotoxin. *Journal of General Physiology* 66: 649–55.

Benton, C. 1973. Studies on the biology and ecology of the scorpion *Vaejovis carolinianus* (Beauvois). Ph.D. dissertation, University of Alabama. 181 pp.

Bergström, J. 1979. Morphology of fossil arthropods as a guide to phylogenetic relationships. In Gupta, ed., pp. 3–56.

Berkenkamp, S. D. 1973. Observations on the scorpion parasite *Pimeliaphilus joshuae* Newell and Ryckman, 1966 (Acarina: Pterygosomidae). Master's thesis, Arizona State University, Tempe. 81 pp.

Berland, L. 1913. Note sur un scorpion muni de deux queues. *Bulletin de la Société entomologique* (France), 1913: 251–52.

———. 1932. *Les arachnides*. Encyclopédie Entomologique, sér. A, vol. 16. 485 pp.

Bernard, H. M. 1893. The coxal glands of *Scorpio*. *Annals and Magazine of Natural History*, ser. 6, 12: 54–59.

Bernard, P., F. Couraud, and S. Lissitzky. 1977. Effects of a scorpion toxin from *Androctonus australis* venom on action potential of neuroblastoma cells in culture. *Biochemical and Biophysical Research Communications* 77: 782–88.

Bertkau, P. 1878. Versuch einer naturalischen Anordnung der Spinnen nebst Bemerkungen zu einzelnen Gattungen. *Archiv für Naturgeschichte* 44(1): 354–510.

Bettini, S., ed. 1978. *Arthropod venoms*. Handbook of Experimental Pharmacology, vol. 48. Berlin: Springer-Verlag.

Bhaskaran Nair, R., and P. A. Kurup. 1975. Investigations on the venom of the South Indian scorpion *Heterometrus scaber*. *Biochimica et Biophysica Acta* 381: 165–74.

Birula, A. A. B. 1905. Scorpionologische Beiträge. *Zoologischer Anzeiger* 29: 445–50, 621–24.

———. 1917a. *Fauna of Russia and adjacent countries. Arachnoidea*. Vol. 1, *Scorpions*. Tr. B. Munitz. Jerusalem: Israel Program for Scientific Translations, 1965. 154 pp. [Available through U.S. Dept. of Commerce.]

———. 1917b. *Arthrogastric arachnids of Caucasia*. Vol. 1, *Scorpions*. Tr. J. Salkind. Jerusalem: Israel Program for Scientific Translations, 1964. 170 pp. [Available through U.S. Dept. of Commerce.]

Bishop, A. S., and M. G. de Ferriz. 1964. Estudio morfológico, histológico e histoquímico de la glándula venenosa de algunas especies de alacranes de los géneros *Vejovis* C. L. Koch, *Diplocentrus* Peters y *Centruroides* Marx. *Anales del Instituto de biología, Universidad de México* 35: 139–55.

Bohnenberger, J. 1978. The transfer characteristics of a lyriform slit sense organ. *Symposia of the Zoological Society* (London) 42: 449–55.

———. 1981. Matched transfer characteristics of single units in a compound slit sense organ. *Journal of Comparative Physiology* 142A: 391–402.

Bonner, C. 1950. *Studies in magical amulets, chiefly Graeco-Egyptian*. Ann Arbor: University of Michigan Press.

Börner, C. 1903. Die Beingliederung der Arthropoden. 3, Mitteilung: Cheliceraten, Crustaceen und Pantopoden. *Sitzungsberichte der Gesellschaft naturforschender Freunde* (Berlin), 1903: 292–341.

Boudreaux, H. B. 1979. *Arthropod phylogeny with special reference to insects*. New York: Wiley.

Bowerman, R. F. 1972a. A muscle receptor organ in the scorpion postabdomen. I, The sensory system. *Journal of Comparative Physiology* 81A: 133–46.

———. 1972b. A muscle receptor organ in the scorpion postabdomen. II, Reflexes evoked by MRO stretch and release. *Journal of Comparative Physiology* 81A: 147–57.

———. 1975a. The control of walking in the scorpion. I, Leg movements during normal walking. *Journal of Comparative Physiology* 100A: 183–96.

———. 1975b. The control of walking in the scorpion. II, Coordination modification as a consequence of appendage ablation. *Journal of Comparative Physiology* 100A: 197–209.

———. 1976. Electrophysiological survey of joint receptors in walking legs of the scorpion, *Paruroctonus mesaensis*. *Journal of Comparative Physiology* 105A: 353–63.

———. 1977. The control of arthropod walking. *Comparative Biochemistry and Physiology* 56A: 231–47.

———. 1981a. An electromyographic analysis of the elevator/depressor muscle motor programme in the freely-walking scorpion, *Paruroctonus mesaensis*. *Journal of Experimental Biology* 91: 165–77.

———. 1981b. Arachnid locomotion. In C. F. Herreid and C. R. Fourtner, eds., *Locomotion and energetics in arthropods*, pp. 73–102. New York: Plenum.

Bowerman, R. F., and M. Burrows. 1980. The morphology and physiology of some walking leg motor neurons in a scorpion. *Journal of Comparative Physiology* 140A: 31–42.

Bowerman, R. F., and J. A. Butler. 1979. Intrappendage EMG analysis of scorpion walking motor program. [Paper presented at the annual meeting of the Society for Neuroscience, Atlanta.]

Bowerman, R. F., and J. Larimer. 1973. Structure and physiology of the patella-tibia joint receptors in scorpion pedipalps. *Comparative Biochemistry and Physiology* 46A: 139–51.

Bowerman, R. F., and T. M. Root. 1978. External anatomy and muscle morphology of the walking legs of the scorpion *Hadrurus arizonensis*. *Comparative Biochemistry and Physiology* 59A: 57–63.

Bradley, R. 1982. Digestion time and reemergence in the desert grassland scorpion *Paruroctonus utahensis* (Williams) (Scorpionida, Vaejovidae). *Oecologia* 55: 316–18.

———. 1983. Activity and population dynamics of the desert grassland scorpion (*Paruroctonus utahensis*): Does adaptation imply optimization? Ph.D. dissertation, University of New Mexico, Albuquerque. 282 pp.

———. 1984. The influence of the quantity of food on fecundity in the desert grassland scorpion (*Paruroctonus utahensis*) (Scorpionida, Vaejovidae): An experimental test. *Oecologia* 62: 53–56.

———. 1986. The relationship between population density of *Paruroctonus utahensis* (Scorpionida: Vaejovidae) and characteristics of its habitat. *Journal of Arid Environments* 11: 165–72.

Bradley, R., and A. Brody. 1984. Relative abundance of three vaejovid scorpions across a habitat gradient. *Journal of Arachnology* 11: 437–40.

Brauer, A. 1894. Beiträge zur Kenntnis der Entwicklungsgeschichte des Skorpions, I. *Zeitschrift für wissenschaftliche Zoologie* 57: 402–32.

———. 1895. Beiträge zur Kenntnis der Entwicklungsgeschichte des Skorpions, II. *Zeitschrift für wissenschaftliche Zoologie* 59: 351–433.

———. 1917. Über Doppelbildungen des Skorpions *Euscorpius carpathicus* L. *Sitzungsberichte der Deutschen Akademie der Wissenschaften* (Berlin), 1917: 208–21.

Brenner, M. 1972. Evidence for peripheral inhibition in an arachnid muscle. *Journal of Comparative Physiology* 80: 227–31.

Brownell, P. H. 1977a. Compressional and surface waves in sand used by desert scorpions to locate prey. *Science* 197: 479–82.

———. 1977b. Prey-localizing behavior of the nocturnal scorpion, *Paruroctonus mesaensis*. Ph.D. diss., University of California, Riverside. [*Dissertation Abstracts* 38(4): 1509B.]

———. 1984. Prey detection by the sand scorpion. *Scientific American* 251(6): 86–97.

Brownell, P. H., and R. D. Farley. 1979a. Detection of vibrations in sand by tarsal sense organs of the nocturnal scorpion, *Paruroctonus mesaensis*. *Journal of Comparative Physiology* 131A: 23–30.

———. 1979b. Prey localizing behaviour of the nocturnal desert scorpion, *Paruroctonus mesaensis*: Orientation to substrate vibrations. *Animal Behaviour* 27: 185–93.

———. 1979c. Orientation to vibrations in sand by the nocturnal scorpion *Paruroctonus mesaensis*: Mechanism of target localization. *Journal of Comparative Physiology* 131A: 31–38.

Bub, K., and R. F. Bowerman. 1979. Prey capture by the scorpion *Hadrurus arizonensis* Ewing (Scorpiones: Vaejovidae). *Journal of Arachnology* 7: 243–53.

Bücherl, W. 1955/56. Escorpiões e escorpionismo no Brasil. V, Observacões sôbre o aparelho reproductor masculino e o acasalamento de *Tityus trivittatus* e *Tityus bahiensis*. *Memórias do Instituto Butantan* (Sao Paulo) 27: 121–55.

———. 1971. Classification, biology and venom extraction of scorpions. In Bücherl and Buckley, eds., pp. 317–47.

———. 1978. Venoms of Tityinae. A, Systematics, distribution, biology, venomous apparatus, etc., of Tityinae: Venom collection, toxicity, human accidents and treatment of stings. In Bettini, ed., pp. 371–79.

Bücherl, W., and E. E. Buckley, eds. 1971. *Venomous animals and their venoms.* Vol. 3, *Venomous invertebrates.* New York: Academic. 537 pp.

Bulard, M. 1935. *Le scorpion symbole du peuple juif dans l'art religieux des XIVe, XVe, XVIe siècles.* Paris: de Boccard.

Bullock, T. H., and G. A. Horridge. 1965. *Structure and function in the nervous systems of invertebrates,* vol. 2. San Francisco: Freeman.

Burns, M. D. 1973. The control of walking in Orthoptera. I, Leg movements in normal walking. *Journal of Experimental Biology* 58: 45–58.

Buscarlet, L. A., J. Proux, and R. Gerster. 1978. Utilisation du double marquage $HT^{18}O$ dans une étude de bilan métabolique chez *Locusta migratoria migratorioides. Journal of Insect Physiology* 24: 225–32.

Buxton, B. H. 1917. Notes on the anatomy of arachnids. *Zoological Journal of Morphology* 29: 1–31.

Bystrov, V. F., A. S. Arseniev, V. I. Kondakov, V. N. Maiorov, V. V. Okhanov, and Y. A. Ovchinnikov. 1983. NMR conformational study: Polypeptide neurotoxins—Honeybee apamin and scorpion insectotoxin I_5A. In F. Hucho and Y. A. Ovchinnikov, eds., *Toxins as tools in neurochemistry,* pp. 291–309. Berlin: de Gruyter.

Cahalan, M. D. 1975. Modification of sodium channel gating in frog myelinated nerve fibres by *Centruroides sculpturatus* scorpion venom. *Journal of Physiology* 244: 511–34.

Campos, J. A., O. S. Silva, M. Lopez, and L. Freire-Maia. 1980. Signs, symptoms and treatment of severe scorpion poisoning in children. In D. Eaker and T. Wadström, eds., *Natural toxins,* pp. 61–68. New York: Pergamon.

Carbone, E., G. Prestipino, L. Spadavecchia, F. Franciolini, and L. D. Possani. 1987. Blocking of the squid axon K^+ channel by Noxiustoxin: A toxin from the venom of the scorpion *Centruroides noxius. Pflügers Archiv* 408: 423–31.

Carbone, E., G. Prestipino, E. Wanke, L. D. Possani, and A. Maelicke. 1983. Selective action of scorpion neurotoxins on the ionic currents of the squid giant axon. *Toxicon,* Supplement 3: 57–60.

Carbone, E., E. Wanke, G. Prestipino, L. D. Possani, and A. Maelicke. 1982. Selective blockage of voltage-dependent K^+ channels by a novel scorpion toxin. *Nature* 296: 90–91.

Carricaburu, P. 1969. Ocular ditropic of the scorpion *Pandinus imperator. Bulletin de la Société d'histoire naturelle de l'Afrique du Nord* 60(3–4): 91–101.

Carricaburu, P., and M. Cherrak. 1968. Analysis of the electro-retinograms of the scorpion *Androctonus australis. Zeitschrift für vergleichende Physiologie* 61(4): 386–93.

Carricaburu, P., and A. Muñoz-Cuevas. 1986a. Spontaneous electrical activity of the subesophageal ganglion and circadian rhythms in scorpions. *Experimental Biology* 45: 305–10.

———. 1986b. Visual circadian rhythm in the scorpion *Didymocentrus lesuerii* (Gervais) (Arachnida). In *Proceedings of the 10th International congress of arachnology, Jaca,* vol. 1: 7–12.

———. 1987. La modulation du rythme visuel circadien par l'octopamine chez les scorpions et l'adaptation à la vie souterraine. *Comptes rendus hebdomadaires des séances de l'Académie des sciences* (Paris) 305: 285–88.

Carricaburu, P., and R. A. Nait. 1967. Electroretinogramme du scorpion *Androctonus australis* L. *Comptes rendus hebdomadaires des séances de l'Académie des sciences* (Paris), sér. D, 264: 2819–21.

Carthy, J. D. 1966. Fine structure and function of the sensory pegs on the scorpion pecten. *Experientia* 22: 89–91.

———. 1968. The pectines of scorpions. *Symposia of the Zoological Society* (London) 23: 251–61.

Casper, G. S. 1985. Prey capture and stinging behavior in the emperor scorpion, *Pandinus imperator* (Koch) (Scorpiones, Scorpionidae). *Journal of Arachnology* 13: 277–83.

Catterall, W. A. 1975. Cooperative activation of action potential Na^+ ionophore by neurotoxins. *Proceedings of the National Academy of Sciences* (U.S.A.) 72: 1782–86.

———. 1976. Purification of a toxic protein from scorpion venom which activates the action potential Na^+ ionophore. *Journal of Biological Chemistry* 251: 5528–36.

———. 1977. Membrane potential–dependent binding of scorpion toxin to the action potential Na^+ ionophore: Studies with a toxin derivative prepared by lactoperoxidase-catalyzed iodination. *Journal of Biological Chemistry* 252: 8660–68.

———. 1979. Binding of scorpion toxin to receptor sites associated with sodium channels in frog muscle: Correlation of voltage dependent binding with activation. *Journal of General Physiology* 74: 375–91.

———. 1980. Neurotoxins that act on voltage-sensitive sodium channels in excitable membranes. *Annual Review of Pharmacology and Toxicology* 20: 15–43.

Catterall, W. A., R. Ray, and C. S. Morrow. 1976. Membrane potential–dependent binding of scorpion toxin to the action potential Na^+ ionophore. *Proceedings of the National Academy of Sciences* (U.S.A.) 73: 2682–86.

Cekalovic, T. K. 1965–66. Alimentación y habitat de *Centromachetes pococki* (Kraepelin), 1894. *Boletín de la Sociedad de biología* (Concepción) 15: 27–32.

———. 1973. Nuevo caracter sexual secundario en los machos de *Brachistosternus* (Scorpiones, Bothriuridae). *Boletín de la Sociedad de biología* (Concepción) 46: 99–102.

Chandra Sekhara Reddy, D., P. R. Koundihya, R. Ramamurthi, and B. Padmanabha Naidu. 1978. Observations on diurnal rhythmicity in leg, tail and pedipalpal muscles in the scorpion, *Heterometrus fulvipes* C. L. Koch. *Comparative Physiological Ecology* 3: 131–34.

Chandra Sekhara Reddy, D., and B. Padmanabha Naidu. 1977. Diurnal rhythmicity of alkaline and acid phosphatases in the scorpion, *Heterometrus fulvipes* C. L. Koch. *Comparative Physiological Ecology* 2: 217–19.

Chapman, R. F. 1971. *The insects: Structure and function*. London: English University Press. 819 pp.

Chengal Raju, D., M. D. Bashamohideen, and C. Narasimham. 1973. Circadian rhythm in blood glucose and liver glycogen levels of scorpion, *Heterometrus fulvipes*. *Experientia* 29: 964–65.

Ch'eng-Pin, P. 1940. Morphology and anatomy of the Chinese scorpion *Buthus martensi* Karsch. *Peking Natural History Bulletin* 14(2): 103–17.

Cheymol, J., F. Bourillet, M. Roch-Arveiller, and J. Heckle. 1973. Action neuromusculaire de trois venins de scorpions nord-africains (*Leiurus quinquestriatus, Buthus occitanus* et *Androctonus australis*) et de deux toxines extraites de l'un d'entre eux. *Toxicon* 11: 277–82.

Chhatwal, G. S., and E. Habermann. 1981. Neurotoxins, protease inhibitors and histamine releasers in venom of the Indian red scorpion (*Buthus tamulus*): Isolation and partial characterization. *Toxicon* 19: 807–23.

Chicheportiche, R., and M. Lazdunski. 1970. The conformation of small proteins. The state-diagram of a neurotoxin of *Androctonus australis* Hector. *European Journal of Biochemistry* 14: 549–55.

Chippaux, J. P., and M. Goyffon. 1983. Producers of antivenomous sera. *Toxicon* 21: 739–52.

Cloudsley-Thompson, J. L. 1955. On the function of the pectines of scorpions. *Annals and Magazine of Natural History*, ser. 12, 8: 556–60.

———. 1956. Studies in diurnal rhythms. VI, Bioclimatic observations in Tunisia and their significance in relation to the physiology of the fauna, especially woodlice, centipedes, scorpions and beetles. *Annals and Magazine of Natural History*, ser. 12, 9: 305–29.

———. 1957. Studies in diurnal rhythms. V, Nocturnal ecology and water-relations of the British cribellate spiders of the genus *Ciniflo* Bl. *Journal of the Linnean Society (Zoology)* 43: 134–52.

———. 1959. *Spiders, scorpions, centipedes and mites*. Oxford: Pergamon. 278 pp.

———. 1961. *Rhythmic activity in animal physiology and behaviour*. London: Academic. 236 pp.

———. 1962. Lethal temperature of some desert arthropods and the mechanism of heat death. *Entomologia Experimentalis et Applicata* 5: 270–80.

———. 1963. Some aspects of the physiology of *Buthotus minax* (Scorpiones: Buthidae) with remarks on other African scorpions. *Entomologist's Monthly Magazine* 98: 243–46.

———. 1965. The scorpion. *Science Journal* 1(5): 35–41.

———. 1973a. Entrainment of the "circadian clock" in *Buthotus minax* (Scorpiones: Buthidae). *Journal of Interdisciplinary Cycle Research* 4: 119–23.

———. 1973b. Factors influencing the supercooling of tropical Arthropoda, especially locusts. *Journal of Natural History* 7: 471–80.

———. 1975. Entrainment of the "circadian clock" in *Babycurus centrurimorphus* (Scorpiones: Buthidae). *Journal of Interdisciplinary Cycle Research* 6: 185–88.

———. 1978. Biological clocks in Arachnida. *Bulletin of the British Arachnological Society* 4: 184–91.

———. 1981. A comparison of rhythmic locomotory activity in tropical forest Arthropoda with that in desert species. *Journal of Arid Environments* 4: 327–31.

———. 1987. The biorhythms of spiders. In W. Nentwig, ed., *Ecophysiology of spiders*, pp. 371–79. Berlin: Springer-Verlag.

Cloudsley-Thompson, J. L., and C. Constantinou. 1985a. The circadian rhythm of locomotory activity in a neotropical forest scorpion, *Opisthacanthus* sp. (Scorpionidae). *International Journal of Biometeorology* 29: 87–89.

———. 1985b. Diurnal rhythm of activity in the arboreal tarantula *Avicularia avicularia* (L.) (Mygalomorpha: Theraphosidae). *Journal of Interdisciplinary Cycle Research* 16: 113–16.

Cloudsley-Thompson, J. L., and C. S. Crawford. 1970. Lethal temperatures of some arthropods of the southwestern United States. *Entomologist's Monthly Magazine* 106: 26–29.

Constantinou, C. 1980. Entrainment of the circadian rhythm of activity in desert and forest inhabiting scorpions. *Journal of Arid Environments* 3: 133–39.

Constantinou, C., and J. L. Cloudsley-Thompson. 1980. Circadian rhythms in scorpions. In *Proceedings of the 8th International congress of arachnology, Vienna*: 53–55.

———. 1984. Stridulatory structures in scorpions of the families Scorpionidae and Diplocentridae. *Journal of Arid Environments* 7: 359–64.

Coraboeuf, E., E. Deroubaix, and F. Tazieff-Depierre. 1975. Effect of Toxin II isolated from scorpion venom on action potential and contraction of mammalian heart. *Journal of Molecular and Cellular Cardiology* 7: 643–53.

Corda, A. 1835. Über den in der Steinkohlen Formation bei Chomle gefunden fossilen Scorpionen. *Verhandlungen der Gesellschaft des Vaterlands Museums* (Böhme), April: 36.

Corrado, A. P., A. Antonio, and C. R. Diniz. 1968. Brazilian scorpion venom (*Tityus serrulatus*), an unusual sympathetic postganglionic stimulant. *Journal of Pharmacology and Experimental Therapeutics* 164: 253–58.

Corteggiani, E., and A. Serfaty. 1939. Acétylcholine et cholinesterase chez les insects et arachnides. *Comptes rendus hebdomadaires des séances de l'Académie des sciences* (Paris), sér. D, 131: 1124–26.

Couraud, F., E. Jover, J. M. Dubois, and H. Rochat. 1982. Two types of scorpion toxin receptor sites, one related to the activation, the other to the inactivation of the action potential sodium channel. *Toxicon* 20: 9–16.

Couraud, F., H. Rochat, and S. Lissitzky. 1976. Stimulation of sodium and calcium uptake by scorpion toxin in chick embryo heart cells. *Biochimica et Biophysica Acta* 433: 90–100.

———. 1980. Binding of scorpion neurotoxins to chick embryonic heart cells in culture and relationship to calcium uptake and membrane potential. *Biochemistry* 19: 457–62.

Coutinho-Netto, J., A. S. Abdul-Ghani, P. J. Norris, A. J. Thomas, and H. F. Bradford. 1980. The effects of scorpion venom toxin on the release of amino acid neurotransmitters from cerebral cortex *in vivo* and *in vitro*. *Journal of Neurochemistry* 35: 558–65.

Couzijn, H. W. C. 1976. Functional anatomy of the walking-legs of Scorpionida with remarks on terminology and homologization of leg segments. *Netherlands Journal of Zoology* 26(4): 453–501.

———. 1981. Revision of the genus *Heterometrus* Hemprich & Ehrenberg (Scorpionidae: Arachnida). *Zoologische verhandelingen* 184: 1–196.

Crawford, C. S., and R. C. Krehoff. 1975. Diel activity in sympatric populations of the scorpions *Centruroides sculpturatus* (Buthidae) and *Diplocentrus spitzeri* (Diplocentridae). *Journal of Arachnology* 2: 195–204.

Crawford, C. S., and W. A. Riddle. 1974. Cold hardiness in centipedes and scorpions in New Mexico. *Oikos* 25: 86–92.

———. 1975. Overwintering physiology of the scorpion *Diplocentrus spitzeri*. *Physiological Zoology* 48: 84–92.

Crawford, C. S., and R. C. Wooten. 1973. Water relations in *Diplocentrus spitzeri*, a semimontane scorpion from the southwestern United States. *Physiological Zoology* 46: 218–29.

Creighton, T. E. 1984. Evolutionary and genetic origins of protein sequences. In *Proteins: Structures and molecular principles*, pp. 93–131. New York: Freeman.

Cumont, F. 1896–99. *Textes et monuments figurés relatifs aux mystères de Mithra*. 2 vols. Brussels: Lamertin.

Cunha-Melo, J. R., L. Freire-Maia, W. L. Tafuri, and T. A. Maria. 1973. Mechanism of action of purified scorpion toxin on the isolated rat intestine. *Toxicon* 11: 81–84.

Cunliffe, F. 1949. *Pimeliaphilus isometri*, a new scorpion parasite from Manila, P.I. (Acarina, Pterygosomidae). *Proceedings of the Entomological Society* (Washington) 51: 123–25.

Curry, S. C., M. V. Vance, P. J. Ryan, D. B. Kunkel, and W. T. Northey. 1983–84. Envenomation by the scorpion *Centruroides sculpturatus*. *Journal of Toxicology—Clinical Toxicology* 21(4–5): 417–49.

Dahl, F. 1883. Uber die Hörhaare bei den Arachniden. *Zoologischer Anzeiger* 6: 267–70.

Dai, M. E. M., and M. V. Gomez. 1978. The effect of Tityustoxin from scorpion venom on the release of acetylcholine from subcellular fractions of rat brain cortex. *Toxicon* 16: 687–90.

D'Ajello, V., E. Zlotkin, F. Miranda, S. Lissitzky, and S. Bettini. 1972. The effect of scorpion venom and pure toxins on the cockroach central nervous system. *Toxicon* 10: 399–404.

Dales, R. P. 1970. *Practical invertebrate zoology*. Seattle: University of Washington Press. 356 pp.

Dalingwater, J. E. 1980. SEM observations on the cuticles of some chelicerates. In *Proceedings of the 8th International congress of arachnology, Vienna*: 285–89.

Darbon, H., E. Zlotkin, C. Kopeyan, J. van Rietschoten, and H. Rochat. 1982. Covalent structure of the insect toxin of the North African scorpion *Androctonus australis* Hector. *International Journal of Peptide and Protein Research* 20: 320–30.

Delori, P., F. Miranda, and H. Rochat. 1976. Some news and comments about a rational and efficient antivenomous serotherapy. In A. Ohsaka et al., eds., *Animal, plant and microbial toxins*, vol. 2, *Chemistry, pharmacology and immunology*, pp. 407–20. New York: Plenum.

Delori, P., J. van Rietschoten, and H. Rochat. 1981. Scorpion venoms and neurotoxins: An immunological study. *Toxicon* 19: 339–407.

Dennell, R. 1975. The structure of the cuticle of the scorpion *Pandinus imperator* (Koch). *Journal of the Linnean Society (Zoology)* 46: 249–54.

Deonna, W. 1959. *Mercure et le scorpion*. Collection Latomus, no. 37. Brussels.

Devarajulu Naidu, V., and B. Padmanabha Naidu. 1976. Diurnal rhythmic activity in the heart-beat of the scorpion *Heterometrus fulvipes* C. L. Koch. *Indian Journal of Experimental Biology* 14: 1–5.

Devarajulu Naidu, V., and S. A. T. Venkatachari. 1974. Succinate dehydrogenase activity in the heart muscle of the scorpion, *Heterometrus fulvipes* C. Koch. *Indian Journal of Experimental Biology* 12: 539–42.

Diniz, C. R. 1971. Chemical and pharmacological properties of *Tityus* venoms. In Bücherl and Buckley, eds., pp. 311–15.

Diniz, C. R., and J. M. Gonçalves. 1956. Some chemical and pharmacological properties of Brazilian scorpion venoms. In E. Buckley and N. Poges, eds., *Venoms*, pp. 131–39. Publications of the American Association for the Advancement of Science, 44. Washington, D.C.

Diniz, C. R., A. F. Pimenta, J. Coutinho-Netto, S. Pompolo, M. V. Gomez, and G. M. Böhm. 1974. Effect of scorpion venom from *Tityus serrulatus* (Tityustoxin) on the acetylcholine release and fine structure of the nerve terminals. *Experientia* 30: 1304–5.

Diniz, C. R., and J. M. Torres. 1968. Release of an acetylcholine-like substance from guinea pig ileum by scorpion venom. *Toxicon* 5: 277–81.

Dori Raj, B. S., and M. J. Cohen. 1964. Structural and functional correlations in crab muscle fibers. *Naturwissenschaften* 51: 224–25.

Dresco-Derouet, L. 1961. Le métabolisme respiratoire des scorpions. I, Existence d'un rythme nycthéméral de la consommation d'oxygène. *Bulletin du Muséum national d'histoire naturelle* (Paris), sér. 2, 32: 553–57.

———. 1964a. Le métabolisme respiratoire des scorpions chez quelques espèces à différentes températures. *Bulletin du Muséum national d'histoire naturelle* (Paris), sér. 2, 36: 97–99.

———. 1964b. Le métabolisme respiratoire des scorpions. II, Mesures de l'intensité respiratoire chez quelques espèces à différentes températures. *Bulletin du Muséum national d'histoire naturelle* (Paris), sér. 2, 36: 533–57.

Dubale, M. S., and A. B. Vyas. 1968. The structure of the chela of *Heterometrus* sp. and its mode of operation. *Bulletin of the Southern California Academy of Sciences* 67(4): 240–44.

Due, A. D., and G. A. Polis. 1985. Biology of the intertidal scorpion, *Vaejovis littoralis*. *Journal of Zoology* (London) 207: 563–80.

———. 1986. Trends in scorpion diversity along the Baja California peninsula. *American Naturalist* 128: 460–68.

Dufour, L. 1858. Anatomie, physiologie et histoire naturelle de *Galeodes*. *Mémoires présentés à l'Académie des sciences* (Paris) 46: 1247–53.

Dufton, M. J., and H. Rochat. 1984. Classification of scorpion toxins according to amino acid composition and sequence. *Journal of Molecular Evolution* 20: 120–27.

Dumortier, B. 1963. Morphology of sound emission apparatus in Arthropoda. In R.-G. Busnel, ed., *Acoustic behavior of animals*, pp. 319–21. Amsterdam: Elsevier.

Dyar, H. 1890. The number of molts of lepidopterous larvae. *Psyche* 5: 420–22.

Eastwood, E. B. 1978a. Notes on the scorpion fauna of the Cape. III, Some observations on the distribution and biology of scorpions on Table Mountain. *Annals of the South African Museum* 74: 229–48.

———. 1978b. Notes on the scorpion fauna of the Cape. IV, The burrowing activities of some scorpionids and buthids (Arachnida, Scorpionida). *Annals of the South African Museum* 74: 249–55.

———. 1978c. First record of mites on a scorpion in southern Africa. *Journal of the Entomological Society* (South Africa) 41(1): 159.

Edgar, A. L., and H. A. Yuan. 1968. Daily locomotory activity in *Phalangium opilio* and seven species of *Leiobunum* (Arthropoda: Phalangida). *Bios* (Paris) 39: 167–76.

Edney, E. B. 1977. *Water balance in land arthropods*. Berlin: Springer-Verlag. 282 pp.

Edney, E. B., S. Haynes, and D. Gibo. 1974. Distribution and activity of the desert cockroach *Arenivaga investigata* (Polyphagidae) in relation to microclimate. *Ecology* 55: 420–27.

Efrati, P. 1978. Venoms of Buthinae. B, Epidemiology, symptomatology and treatment of scorpion stings. In Bettini, ed., pp. 312–17.

Einhorn, V. F., and R. C. Hamilton. 1977. Action of venom from the scorpion *Leiurus quinquestriatus* on release of noradrenaline from sympathetic nerve endings of the mouse vas deferens. *Toxicon* 15: 403–12.

Ennik, F. 1972. A short review of scorpion biology, management of stings, and control. *California Vector Views* 19(10): 69–80.

Erickson, B. W. 1978. Sequence homology of snake, scorpion, and bee toxins. In P. Rosenberg, ed., *Toxins: Animal, plant and microbial*, pp. 1071–86. New York: Pergamon.

Fabre, J. H. 1907. *Souvenirs entomologiques, 9me série*. Édition définitive. Paris: Délagrave. 229 pp.

———. 1923. *The life of the scorpion*. Tr. A. Teixeira de Mattos. New York: Dodd, Mead. 221 pp.

Farley, R. D. 1987. Postsynaptic potentials and contraction pattern in the heart of the desert scorpion, *Paruroctonus mesaensis*. *Comparative Biochemistry and Physiology* 86A: 121–31.

Farzanpay, R., and M. Vachon. 1979. Contribution à l'étude des caractères sexuels secondaires chez les scorpions Buthidae (Arachnida). *Revue arachnologique* 2(4): 137–42.

Fet, V. Y. 1980. Ecology of the scorpions (Arachnida, Scorpiones) of the southeastern Kara-Kum. *Entomologicheskoe obozrenie* 59(1): 165–70.

Filshie, B. K., and N. F. Hadley. 1979. Fine structure of the cuticle of the desert scorpion, *Hadrurus arizonensis*. *Tissue and Cell* 11(2): 249–62.

Finlayson, L. H. 1976. Abdominal and thoracic receptors in insects, centipedes and scorpions. In P. J. Mill, ed., *Structure and function of proprioceptors in the invertebrates*, pp. 153–211. London: Chapman and Hall.

Firstman, B. 1973. The relationship of the chelicerate arterial system to the evolution of the endosternite. *Journal of Arachnology* 1: 1–54.

Fleissner, G. 1968. Untersuchungen zur Sehphysiologie der Skorpione. *Verhandlungen der Deutschen zoologischen Gesellschaft* (Innsbruck) 61: 375–80.

———. 1972. Circadian sensitivity changes in the median eyes of the North African scorpion, *Androctonus australis* L. In R. Wehner, ed., *Information processing in the visual systems of arthropods*, pp. 133–39. Berlin: Springer-Verlag.

———. 1974. Circadiane Adaptation und Schirmpigmentverlagerung in den Sehzellen der Medianaugen von *Androctonus australis* L. (Buthidae, Scorpiones). *Journal of Comparative Physiology* 91A: 399–416.

———. 1975. A new biological function of the scorpion's lateral eyes as receptors of Zeitgeber stimuli. In *Proceedings of the 6th International congress of arachnology, Amsterdam*: 176–82.

———. 1977a. Entrainment of the scorpion's circadian rhythm via the median eyes. *Journal of Comparative Physiology* 118A: 93–99.

———. 1977b. Scorpion lateral eyes: Extremely sensitive receptors of Zeitgeber stimuli. *Journal of Comparative Physiology* 118A: 101–8.

———. 1977c. The absolute sensitivity of the median and lateral eyes of the scorpion, *Androctonus australis* L. (Buthidae, Scorpiones). *Journal of Comparative Physiology* 118A: 109–20.

———. 1977d. Differences in the physiological properties of the median and lateral eyes and their possible meaning for the entrainment of the scorpion's circadian rhythm. *Journal of Interdisciplinary Cycle Research* 8(1): 15–26.

———. 1983. Efferent neurosecretory fibres as pathways for circadian clock signals in the scorpion. *Naturwissenschaften* 70: 366–67.

———. 1985. Intracellular recordings from spiking and non-spiking cells in the scorpion eye. *Naturwissenschaften* 72: 46–47.

———. 1986. Die innere Uhr und der Lichtsinn von Skorpionen und Käfern. *Naturwissenschaften* 73: 78–88.

Fleissner, G., and G. Fleissner. 1978. The optic nerve mediates the circadian pigment migration in the median eyes of the scorpion. *Comparative Biochemistry and Physiology* 61A: 69–71.

———. 1986. Neurobiology of a circadian clock in the visual system of scorpions. In Barth, ed., pp. 351–75.

Fleissner, G., and S. Heinrichs. 1982. Neurosecretory cells in the circadian clock system of the scorpion *Androctonus australis*. *Cell and Tissue Research* 124: 233–38.

Fleissner, G., and M. Schliwa. 1977. Neurosecretory fibres in the median eyes of the scorpion, *Androctonus australis* L. *Cell and Tissue Research* 178(2): 189–98.

Fleissner, G., and W. Siegler. 1978. Arhabdomeric cells in the retina of the median eyes of the scorpion. *Naturwissenschaften* 65(4): 210–11.

Foelix, R. F. 1976. Rezeptoren und periphere synaptische Verhaltungen bei verschiedenen Arachnida. *Entomologica Germanica* 3(1–2): 83–87.

———. 1982. *Biology of spiders*. Cambridge, Mass.: Harvard University Press. 306 pp.

———. 1985. Mechano- and chemoreceptive sensilla. In Barth, ed., pp. 118–37.

Foelix, R. F., and G. Müller-Vorholt. 1983. The fine structure of scorpion sensory organs. II, Pecten sensilla. *Bulletin of the British Arachnology Society* 6(2): 68–74.

Foelix, R. F., and J. Schabronath. 1983. The fine structure of scorpion sensory organs. I, Tarsal sensilla. *Bulletin of the British Arachnology Society* 6(2): 53–67.

Fontecilla-Camps, J. C., R. J. Almassay, F. L. Suddath, and C. E. Bugg. 1982. The three-dimensional structure of the scorpion neurotoxins. *Toxicon* 20: 1–7.

Fontecilla-Camps, J. C., R. J. Almassay, F. L. Suddath, D. D. Watt, and C. E. Bugg. 1980. Three-dimensional structure of a protein from scorpion venom: A new structural class of neurotoxins. *Proceedings of the National Academy of Sciences* (U.S.A.) 77: 6496–6500.

Fontecilla-Camps, J. C., C. Habersetzer-Rochat, and H. Rochat. 1988. Orthorhombic crystals and three-dimensional structure of the potent Toxin II from the scorpion *Androctonus australis* Hector. *Proceedings of the National Academy of Sciences* (U.S.A.) 85: 7443–47.

Fontecilla-Camps, J. C., F. L. Suddath, C. E. Bugg, and D. D. Watt. 1978. Crystals of a toxic protein from the venom of the scorpion *Centruroides sculpturatus* Ewing: Preparation and preliminary X-ray investigation. *Journal of Molecular Biology* 123: 703–5.

Fouchard, R., and P. Carricaburu. 1970a. Quelques aspects de la physiologie visuelle chez le scorpion *Buthus occitanus*: Étude électrorétinographique. *Bulletin de la Société d'histoire naturelle de l'Afrique du Nord* 61(1/2): 57–68.

———. 1970b. L'électrorétinogramme des yeux lateraux du scorpion *Buthus occitanus* en fonction de l'adaptation à l'obscurité. *Comptes rendus hebdomadaires des séances de l'Académie des sciences* (Paris), sér. D, 272: 849–51.

Fountain, R. L. 1970. Neuromuscular properties of scorpion walking leg muscles. Master's thesis, University of Miami.

Fourtner, C. R., and R. A. Pax. 1972. Chelicerate neuromuscular systems: The distal flexor of the mero-carpopodite of *Limulus polyphemus* (L.). *Comparative Biochemistry and Physiology* 41A: 617–27.

Fourtner, C. R., and R. G. Sherman. 1973. Chelicerate skeletal neuromuscular systems. *American Zoologist* 13: 271–89.

Fox, W. K. 1975. Bionomics of two sympatric scorpion populations (Scorpionida: Vaejovidae). Ph.D. diss., Arizona State University, Tempe. 85 pp.

Fraenkel, G. 1929. Der Atmungsmechanismus des Skorpions. Ein Beitrag zur Physiologie der Tracheenlunge. *Zeitschrift für vergleichende Physiologie* 2: 656–61.

Francke, O. F. 1975. A new species of *Diplocentrus* from New Mexico and Arizona (Scorpionida, Diplocentridae). *Journal of Arachnology* 2: 107–18.

———. 1976a. Redescription of *Parascorpiops montana* Banks (Scorpionida, Vaejovidae). *Entomological News* 87: 75–85.

———. 1976b. Observations on the life history of *Uroctonus mordax* Thorell (Scorpionida, Vaejovidae). *Bulletin of the British Arachnology Society* 3: 254–60.

———. 1977a. Taxonomic observations on *Heteronebo* Pocock (Scorpionida, Diplocentridae). *Journal of Arachnology* 4: 95–113.

———. 1977b. The genus *Diplocentrus* in the Yucatán peninsula with description of two new troglobites. *Bulletin of the Association of Mexican Cave Studies* 6: 49–61.

———. 1978a. Systematic revision of diplocentrid scorpions from circum-Caribbean lands. *Special Publications of the Museum, Texas Tech University*, 14: 1–92.

———. 1978b. Redescription of *Centruroides koesteri* Kraepelin (Scorpionida, Buthidae). *Journal of Arachnology* 6: 65–71.

———. 1978c. New troglobite scorpion of genus *Diplocentrus*. *Entomological News* 89: 39–45.

———. 1979a. Spermatophores of some North American scorpions (Arachnida, Scorpiones). *Journal of Arachnology* 7: 19–32.

———. 1979b. Observations on the reproductive biology and life history of *Megacormus gertschi* Diaz (Scorpiones: Chactidae: Megacorminae). *Journal of Arachnology* 7: 223–30.

———. 1981a. A new genus of troglobitic scorpion from Mexico (Chactoidea, Megacorminae). *Bulletin of the American Museum of Natural History* 170: 23–28.

———. 1981b. Birth behavior and life history of *Diplocentrus spitzeri* Stahnke (Scorpiones: Diplocentridae). *Southwestern Naturalist* 25: 517–23.

———. 1982a. Studies on the scorpion subfamilies Superstitioninae and Typhlochactinae, with description of a new genus (Scorpiones, Chactoidea). *Bulletin of the Association of Mexican Cave Studies* 8: 51–61.

———. 1982b. Are there any bothriurids (Arachnida, Scorpiones) in southern Africa? *Journal of Arachnology* 10: 35–39.

———. 1982c. Birth behavior in *Diplocentrus bigbendensis* Stahnke (Scorpiones, Diplocentridae). *Journal of Arachnology* 10: 157–64.

———. 1982d. Parturition in scorpions (Arachnida, Scorpiones): A review of the ideas. *Revue arachnologique* 4: 27–37.

———. 1984. The life history of *Diplocentrus bigbendensis* Stahnke (Scorpiones, Diplocentridae). *Southwestern Naturalist* 29: 387–93.

———. 1985. Conspectus genericus scorpionum, 1758–1982 (Arachnida, Scorpiones). *Occasional Papers of the Museum, Texas Tech University*, 98: 1–32.

———. 1986. A new genus and a new species of troglobite scorpion from Mexico (Chactoidea, Superstitioninae, Typhlochactini). In J. R. Riddell, ed., *Studies on the cave and endogean fauna of North America*, Texas Memorial Museum, Speleological Monographs, 1: 5–9. Austin.

Francke, O. F., and S. K. Jones. 1982. The life history of *Centruroides gracilis* (Scorpiones, Buthidae). *Journal of Arachnology* 10: 223–39.

Francke, O. F., and W. D. Sissom. 1984. The life history of *Vaejovis coahuilae* Williams (Scorpiones, Vaejovidae), with comparison of the methods used to determine the number of molts to maturity. *Journal of Arachnology* 12: 1–20.

Francke, O. F., and M. E. Soleglad. 1981. The family Iuridae Thorell. *Journal of Arachnology* 9: 233–58.

Freire-Maia, L., and J. A. Campos. 1987. On the treatment of the cardiovascular manifestations of scorpion envenomation. *Toxicon* 25: 125–30.

Freire-Maia, L., J. R. Cunha-Melo, H. A. Futuro-Neto, A. D. Azevedo, and J. Weinberg. 1976. Cholinergic and adrenergic effects of Tityustoxin. *General Pharmacology* 7: 115–21.

Frelin, C., A. Lombet, P. Vigne, G. Romey, and M. Lazdunski. 1981. The appearance of voltage-sensitive Na^+ channels during the *in vitro* differentiation of embryonic chick skeletal muscle cells. *Journal of Biological Chemistry* 256: 12355–61.

Fujimoto, Y., T. Kaku, A. Sato, K. Hoshi, and S. Fujino. 1979. Some properties of the crude extract of scorpion telsons from *Heterometrus gravimanus* and presence of acetylcholine- and histamine-like substances. *Journal of Pharmacy Dynamics* 2: 27–36.

Gajalakshmi, B. S. 1978. Role of lytic cocktail and atropine in neutralizing scorpion venom effects. *Indian Journal of Medical Research* 67: 1038–44.

Galván Cervantes, S. 1966. La picadura de alacrán en la ciudad de Durango. *Salud pública de México* 8(2): 251–80. [Cited in Delori et al. 1976.]

Garnier, G., and R. Stockmann. 1972. Étude comparative de la pariade chez différentes espèces de scorpions et chez *Pandinus imperator*. *Annales de l'Université d'Abidjan*, sér. E (Écologie), 5: 475–97.

Gaskell, W. H. 1902. The origin of vertebrates, deduced from the study of Ammocoetes, Part X. *Journal of Anatomy* (London) 36: 164–208.

Gaubert, P. 1890. Notes sur les organes lyriformes des arachnides. *Bulletin de la Société Philomaire* (Paris) 8(2): 47–53.

———. 1892. Recherches sur les organes de sens et sur les systèmes tégumentaires, glandulaires et musculaires des appendices des arachnides. *Annales des sciences naturelles*, sér. 7, 13: 57–90.

Geetha Bali. 1974. Interaction of photic and non-photic input in scorpion. [Paper presented at the 26th Congress of Physiological Science.]

———. 1975. After discharge in the ventral nerve cord of the scorpion *Heterometrus*. *Current Science* 44(19): 705–6.

———. 1976a. Motor excitation with reference to neural photoreception in scorpions. *Life Science* 18(9): 1009–12.

———. 1976b. Central course of photic input in the ventral nerve cord of scorpion (*Heterometrus fulvipes*). *Experientia* 32(3): 345–47.

———. 1976c. Multimodal neurons in the central nervous system of the scorpion. *Vignana bharathi* 2(1): 93–96.

———. 1977. Photic behavior of the scorpion *Heterometrus fulvipes*. *Indian Journal of Experimental Biology* 15(5): 384–85.

———. 1979. Distribution of photic units in the ventral nerve cord of scorpion. *Vignana bharathi* 5(1–2): 36–38.

———. 1980. Directional sensitivity of certain mechanoreceptive hairs. [Paper presented at the 8th International Congress of Arachnology, Vienna.]

Geetha Bali, and K. Pampathi Rao. 1973. A metasomal neural photoreceptor in the scorpion. *Journal of Experimental Biology* 58(1): 189–96.

Gertsch, W. J. 1974. Scorpionida. In *Encyclopaedia Britannica,* 15th ed., vol. 16: 401–3.

Gertsch, W. J., and D. M. Allred. 1965. Scorpions of the Nevada test site. *Brigham Young University Science Bulletin, Biology Series,* 6(4): 1–15.

Gertsch, W. J., and M. Soleglad. 1966. The scorpions of the *Vejovis boreus* group (subgenus *Paruroctonus*) in North America (Scorpionida, Vejovidae). *American Museum Novitates* 2278: 1–54.

———. 1972. Studies of North American scorpions of the genera *Uroctonus* and *Vejovis* (Scorpionida, Vejovidae). *Bulletin of the American Museum of Natural History* 148: 549–608.

Ghazal, A., M. Ismail, A. A. Abd-el-Rahman, and M. F. el-Asmar. 1975. Pharmacological studies of scorpion (*Androctonus amoreuxi* Aud. & Sav.) venom. *Toxicon* 13: 253–59.

Gilai, A., and I. Parnas. 1970. Neuromuscular physiology of the closer muscles in the pedipalp of the scorpion, *Leiurus quinquestriatus. Journal of Experimental Biology* 52: 325–44.

———. 1972. Electromechanical coupling in tubular muscle fibers. I, The organization of tubular muscle fibers in the scorpion, *Leiurus quinquestriatus. Journal of Cell Biology* 52: 626–38.

Gillespie, J. I., and H. Meves. 1980. The effect of scorpion venoms on the sodium currents of the squid giant axon. *Journal of Physiology* 308: 479–99.

Gilly, W. F., and T. Scheuer. 1984. Contractile activation in scorpion striated muscle fibers. *Journal of General Physiology* 84: 321–45.

Glover, F. A. 1953. Summer foods of the burrowing owl. *Condor* 55: 275.

Gomez, M. V., M. E. M. Dai, and C. R. Diniz. 1973. Effect of scorpion venom, Tityustoxin, on the release of acetylcholine from incubated slices of rat brain. *Journal of Neurochemistry* 20: 1051–61.

Gomez, M. V., C. R. Diniz, and T. S. Barbosa. 1975. A comparison of the effects of scorpion venom Tityustoxin and ouabain on the release of acetylcholine from incubated slices of rat brain. *Journal of Neurochemistry* 24: 331–36.

González-Sponga, M. A. 1970. I, Record del género *Microtityus* para Venezuela. II, *Microtityus biordi,* nueva especie para el sistema de la costa en Venezuela. *Monografías científicas, "Agusto pi Suñer," Instituto universitario pedagógico* (Caracas) 1: 1–18.

———. 1974. Dos nuevas especies de alacranes del género *Tityus,* en las cuevas venezolanas (Scorpionida: Buthidae). *Boletín de la Sociedad venezolana de espeleología* 5(1): 55–72.

———. 1977. Rectificación del caracter "ojos laterales" en varios géneros de la familia Chactidae (Scorpionida) en Venezuela. *Acta Biologica Venezuelica* 9(3): 303–15.

———. 1978a. *Escorpiofauna de la región oriental del estado Bolívar, en Venezuela.* Caracas: Trabajos científicos meritorios del Consejo nacional de investigaciones científicas y tecnológicas. 217 pp.

———. 1978b. *Chactas oxfordi*, nueva especie de la Sierra Nevada de Santa Marta, Colómbia. *Monografías científicas, "Agusto pi Suñer," Instituto universitario pedagógico* (Caracas) 9: 1–20.

———. 1978c. *Chactas choroniensis* (Scorpionida: Chactidae), nueva especie del Parque nacional "Henry Pittier" Edo, Aragua, Venezuela. *Monografías científicas, "Agusto pi Suñer," Instituto universitario pedagógico* (Caracas) 10: 1–18.

Gotlieb, A. 1966. Changes in toxic effect of the venom of the scorpion, *Leiurus quinquestriatus*, as a result of injection of atropine or morphine. *Harefuah* 71: 132–34 [*Chemical Abstracts* 66: 2570, abst. 27303A.]

Gottlieb, K. 1926. Über das Gehirn des Skorpions. *Zeitschrift für wissenschaftliche Zoologie* 127: 185–243.

Govindarajan, S., and G. S. Rajulu. 1974. Presence of resilin in a scorpion *Palamnaeus swammerdami* and its role in the food-capturing and sound-producing mechanisms. *Experientia* 30(8): 908–9.

Goyffon, M. 1978. Biogenic amines and spontaneous electrical activity of the prosomal nervous system of the scorpion. *Comparative Biochemistry and Physiology* 59C: 65–74.

Goyffon, M., G. Chovet, R. Deloines, and M. Vachon. 1973a. Étude du caryotype de quelques scorpions Buthides. In *Proceedings of the 5th International congress of arachnology, Brno*: 23–27.

Goyffon, M., J. Drouet, and J.-M. Francaz. 1980. Neurotransmitter amino acids and spontaneous electrical activity of the prosomian nervous system of the scorpion (*Androctonus australis*). *Comparative Biochemistry and Physiology* 66C: 59–64.

Goyffon, M., and J. Kovoor. 1978. Chactoid venoms. In Bettini, ed., pp. 395–418.

Goyffon, M., J. Luyckx, and M. Vachon. 1975. Sur l'existence d'une activité électrique rythmique spontanée du système nerveux céphalique de scorpion. *Comptes rendus hebdomadaires des séances de l'Académie des sciences* (Paris), sér. D, 280: 873–76.

Goyffon, M., and P. Niaussat. 1975. Interrelationships between cholinergic and monoaminigeric mechanisms in the determinism of spontaneous electrical activity in the central nervous system of scorpion. *Annales d'endocrinologie* (Paris) 36: 101–2.

Goyffon, M., M. Richard, and R. Vennet. 1974. Activité électrique cérébrale spontanée et comportement moteur du scorpion: Intérêt pharmocologique. *Comptes rendus des séances de la Société de biologie* 168(10): 1239.

Goyffon, M., R. Stockmann, and J. Lamy. 1973b. Valeur taxonomique de l'électrophorèse en disques des proteines de l'hémolymphe chez le scorpion: Étude du genre *Buthotus* (Buthidae). *Comptes rendus hebdomadaires des séances de l'Académie des sciences* (Paris), sér. D, 277: 61–63.

Goyffon, M., and M. Vachon. 1976. Activité électrique spontanée des centres nerveux prosomiens du scorpion. [Paper presented at the Congrès du Centenaire de la Société Zoologique de France, Paris.]

Goyffon, M., M. Vachon, and N. Broglio. 1982. Epidemiological and clinical characteristics of scorpion envenomation in Tunisia. *Toxicon* 20: 337–44.

Graber, V. 1879. Über das unicorneale Tracheatenauge. *Archive der mikroskopischen Anatomie* 17: 58–93.

Graham, D. 1972. A behavioral analysis of the temporal organization of walking movements in the first instar and adult stick insect (*Carausius morosus*). *Journal of Comparative Physiology* 81: 23–52.

Grandjean, F. 1952. Sur les articles des appendices chez les acariens actinochitineux. *Comptes rendus hebdomadaires des séances de l'Académie des sciences* (Paris), sér. D, 235: 560–64.

Greenberg, L., and J. W. Ingalls. 1963. Effect of drugs on survival time from scorpion envenomation. *Journal of Pharmaceutical Sciences* 52: 159–61.

Grégoire, J., and H. Rochat. 1983. Covalent structure of Toxins I and II from the scorpion *Buthus occitanus tunetanus*. *Toxicon* 21: 153–62.

Grishin, E. V. 1981. Structure and function of *Buthus eupeus* scorpion neurotoxins. *International Journal of Quantum Chemistry* 19: 291–98.

Grupp, I. L., G. Grupp, M. Gueron, R. Adolph, and N. O. Fowler. 1980. Effects of the venom of the yellow scorpion (*Leiurus quinquestriatus*) on the isolated work-performing guinea pig heart. *Toxicon* 18: 261–70.

Gueron, M., and I. Ovsyshcher. 1987. What is the treatment for the cardiovascular manifestations of scorpion envenomations? *Toxicon* 25: 121–24.

Gupta, A. P., ed. 1979. *Arthropod physiology*. New York: Van Nostrand Reinhold. 762 pp.

Habersetzer-Rochat, C., and F. Sampieri. 1976. Structure-function relationships of scorpion neurotoxins. *Biochemistry* 15: 2254–61.

Habibulla, M. 1970. Neurosecretion in the scorpion *Heterometrus swammerdami*. *Journal of Morphology* 131(1): 1–15.

———. 1971a. Neurosecretion in the brain of a scorpion *Heterometrus swammerdami*—A histochemical study. *General and Comparative Endocrinology* 17(2): 253–55.

———. 1971b. Effects of temperature on proteins and amino acids of the cephalothoracic nerve mass of the scorpion *Heterometrus swammerdami*. *Comparative Biochemistry and Physiology* 39: 499–502.

Hadley, N. F. 1970a. Water relations of the desert scorpion, *Hadrurus arizonensis*. *Journal of Experimental Biology* 53: 547–58.

———. 1970b. Micrometeorology and energy exchange in two desert arthropods. *Ecology* 51: 434–44.

———. 1971. Water uptake by drinking in the scorpion, *Centruroides sculpturatus* (Buthidae). *Southwestern Naturalist* 15: 495–505.

———. 1974. Adaptational biology of desert scorpions. *Journal of Arachnology* 2: 11–23.

Hadley, N. F., and B. K. Filshie. 1979. Fine structure of the epicuticle of the desert scorpion, *Hadrurus arizonensis*, with reference to location of lipids. *Tissue and Cell* 11(2): 263–75.

Hadley, N. F., and R. L. Hall. 1980. Cuticular lipid biosynthesis in the scorpion, *Paruroctonus mesaensis*. *Journal of Experimental Zoology* 212: 373–79.

Hadley, N. F., and R. D. Hill. 1969. Oxygen consumption of the scorpion *Centruroides sculpturatus*. *Comparative Biochemistry and Physiology* 29: 217–26.

Hadley, N. F., and L. L. Jackson. 1977. Chemical composition of the epicuticular lipids of the scorpion, *Paruroctonus mesaensis*. *Insect Biochemistry* 7: 85–89.

Hadley, N. F., J. Machin, and M. C. Quinlan. 1986. Cricket cuticle water relations: Permeability and passive determinants of cuticular water content. *Physiological Zoology* 59: 84–94.

Hadley, N. F., and M. C. Quinlan. 1987. Permeability of arthrodial membrane to water: A first measurement using *in vivo* techniques. *Experientia* 43: 164–66.

Hadley, N. F., J. L. Stuart, and M. C. Quinlan. 1982. An air-flow system for measuring total transpiration and cuticular permeability in arthropods: Studies on the centipede *Scolopendra polymorpha*. *Physiological Zoology* 55: 393–404.

Hadley, N. F., and S. R. Szarek. 1981. Productivity of desert ecosystems. *BioScience* 31: 747–53.

Hadley, N. F., and S. C. Williams. 1968. Surface activities of some North American scorpions in relation to feeding. *Ecology* 49: 726–34.

Hall, R. L., and N. F. Hadley. 1982. Incorporation of synthesized and dietary hydrocarbons into the cuticle and hepatopancreas of the scorpion *Paruroctonus mesaensis*. *Journal of Experimental Zoology* 224: 195–203.

Haller, B. 1912. Über das Zentralnervensystem des Skorpions und der Spinnen. *Archive der mikroskopischen Anatomie* 79: 504–24.

Halse, S. A., P. L. Prideaux, A. Cockson, and K. T. Zwicky. 1980. Observations on the morphology and histochemistry of the venom glands of a scorpion, *Urodacus novaehollandiae* (Scorpionidae). *Australian Journal of Zoology* 28(2): 185–94.

Hansen, H. J. 1893. Organs and characters in different orders of arachnids. *Entomologiske meddelelser* 4: 137–44.

———. 1930. *Studies on Arthropoda*. Vol. 3, *On the comparative morphology of the appendages in the Arthropoda*. Part B, *Crustacea (supplement), Insecta, Myriapoda and Arachnida*. Copenhagen: Gyldendalske Boghandel. 376 pp.

Hanstrom, B. 1923. Further notes on the central nervous system of arachnids: Scorpions, phalangids and trap-door spiders. *Journal of Comparative Neurology* 35: 249–74.

———. 1928. *Vergleichende Anatomie des Nervensystems der wirbellosen Tiere*. Berlin: Springer-Verlag. 628 pp.

Haradon, R. M. 1972. Birth behavior of the scorpion *Uroctonus mordax* Thorell (Vaejovidae). *Entomological News* 83: 218–21.

Harington, A. 1978. Burrowing biology of the scorpion *Cheloctonus jonesii* (Arachnida: Scorpionida: Scorpionidae). *Journal of Arachnology* 5: 243–49.

Harrington, C. H., and T. M. Root. 1983. Scorpion pectines: Sensors of sand texture? [Paper presented to the annual meeting of the Society for Neuroscience, Boston.]

Heatwole, H. 1967. Defensive behavior of some Panamanian scorpions. *Caribbean Journal of Science* 7: 15–17.

Heinrichs, S., and G. Fleissner. 1987. Neural components of the circadian clock in the scorpion *Androctonus australis*: Central origin of the efferent neurosecretory elements projecting to the median eyes. *Cell and Tissue Research* 250: 277–85.

Hennig, W. 1950. *Grundzüge einer Theorie der phylogenetischen Systematik*. Berlin: Deutscher Zentralverlag. 285 pp.

———. 1966. *Phylogenetic systematics*. Urbana: University of Illinois Press. 263 pp.

Henriques, M. C., and M. V. Gomez. 1975. Effect of scorpion venom, Tityustoxin, on the uptake of calcium by isolated nerve ending particles from brain. *Brain Research* 93: 182–87.

Henry, L. 1949. The nervous system and the segmentation of the head in a scorpion (Arachnida). *Microentomology* 14(4): 121–26.

Hershkovich, Y., Y. Elitsur, C. Z. Margolis, N. Barak, S. Sofer, and S. W. Moses. 1985. Criteria map audit of scorpion envenomation in the Negev, Israel. *Toxicon* 23: 845–54.

Hibner, A. 1971. The scorpions of Joshua Tree National Monument. Master's thesis, California State College, Long Beach. 65 pp.

Hjelle, J. T. 1972. Scorpions of the northern California coast ranges. *Occasional Papers of the California Academy of Sciences* 92: 1–59.

———. 1974. Observations on the birth and post-birth behavior of *Syntropis macrura* Kraepelin (Scorpionida: Vaejovidae). *Journal of Arachnology* 1: 221–27.

Hodgkin, A. L., and A. F. Huxley. 1952. A quantitative description of membrane current and its application to conduction and excitation in nerve. *Journal of Physiology* 117: 500–544.

Hoffmann, C. 1964. Zur Funktion der kammformigen Organe von Skorpionen. *Naturwissenschaften* 51: 172.

———. 1965. Die Trichobothrien der Skorpione. *Naturwissenschaften* 52: 436–37.

———. 1967. Bau und Funktion der Trichobothrien von *Euscorpius carpathicus* L. *Zeitschrift für vergleichende Physiologie* 54: 290–352.

Hoffmann, C. C. 1931. Los scorpiones de México. I, Diplocentridae, Chactidae, Vejovidae. *Anales del Instituto de biología* (Mexico) 2: 291–408.

———. 1932. Los scorpiones de México. II, Buthidae. *Anales del Instituto de biología* (Mexico) 3: 243–361.

Holmgren, N. 1916. *Zur vergleichenden Anatomie des Gehirns von Polychaeten, Onychophoren, Xiphosuren, Arachniden, Crustaceen, Myriapoden und Insekten.* Kunglige Svenska Vetenskapsakademiens Handlingar, 56. Stockholm. 303 pp.

Horne, F. R. 1969. Purine excretion in five scorpions, a uropygid and a centipede. *Biological Bulletin* 137: 155–60.

Horner, E., J. Taylor, and H. Padykula. 1965. Food habits and gastric morphology of the grasshopper mouse. *Journal of Mammalogy* 45: 513–35.

Hoyle, G. 1978. Distributions of nerve and muscle fibre types in locust jumping muscle. *Journal of Experimental Biology* 73: 205–33.

Hoyle, G., and M. Burrows. 1973. Correlated physiological and ultrastructural studies on specialized muscles. *Journal of Experimental Zoology* 185: 83–96.

Hu, S. L., H. Meves, N. Rubly, and D. D. Watt. 1983. A quantitative study of the action of *Centruroides sculpturatus* Toxins III and IV on the Na^+ currents of the node of Ranvier. *Pflügers Archiv* 397: 90–99.

Huey, R., E. Pianka, M. Egan, and L. Coons. 1974. Ecological shifts in sympatry: Kalahari fossorial lizards (*Typhlosaurus*). *Ecology* 55: 304–16.

Huffaker, C. B., and R. L. Rabb. 1984. *Ecological entomology*. New York: Wiley. 844 pp.

Huxley, J. S. 1932. *Problems of relative growth*. London: Methuen. 276 pp.

Ignat'yev, A. M., V. P. Ivanov, and Y. S. Balashov. 1976. The fine structure and function of the trichobothria in the scorpion *Buthus eupeus*, Scorpiones, Buthidae. *Entomological Reviews* 55(3): 12–18. [English translation of *Entomologicheskoe obozrenie* 55(3): 525–32.]

Ismail, M., M. F. el-Asmar, and O. H. Osman. 1975. Pharmacological studies with scorpion (*Palamnaeus gravimanus*) venom: Evidence for the presence of histamine. *Toxicon* 13: 49–56.

Ismail, M., A. Ghazal, and E. E. el-Fakahany. 1980. Cardiovascular effects of venom from the scorpion *Buthus occitanus* Amoreux. *Toxicon* 18: 327–37.

Ismail, M., G. Kertesz, O. H. Osman, and M. S. Sidra. 1974a. Distribution of ^{125}I labelled scorpion (*Leiurus quinquestriatus* H.&E.) venom in rat tissues. *Toxicon* 12: 209–11.

Ismail, M., O. H. Osman, and M. F. el-Asmar. 1973. Pharmacological studies of the venom from the scorpion *Buthus minax* (L. Koch). *Toxicon* 11: 15–20.

Ismail, M., O. H. Osman, K. A. Gumaa, and M. A. Karrar. 1974b. Some pharmacological studies with scorpion (*Pandinus exitialis*) venom. *Toxicon* 12: 75–82.

Ivanov, O. C. 1981. The evolutionary origin of toxic proteins. *Toxicon* 19: 171–78.

Ivanov, V. P., and Y. S. Balashov. 1979. The structural and functional organization of the pecten in a scorpion *Buthus eupeus* Koch (Scorpiones, Buthidae) studied by electron microscopy. In *The fauna and ecology of Arachnida*, Trudy Leningradskogo Obshchestva Estestvoispytatelei, 85, pp. 73–87. Leningrad.

Jacques, Y., G. Romey, M. T. Cavey, B. Kartalovski, and M. Lazdunski. 1980. Interaction of pyrethroids with the Na^+ channel in mammalian neuronal cells in culture. *Biochimica et Biophysica Acta* 600: 882–97.

Jahromi, S. S., and H. L. Atwood. 1969. Correlation of structure, speed of contraction, and total tension in fast and slow abdominal muscle fibers of the lobster (*Homarus americanus*). *Journal of Experimental Zoology* 171: 25–37.

Jaimovich, E., M. Ildefonse, J. Barhanin, O. Rougier, and M. Lazdunski. 1982. *Centruroides* toxin, a selective blocker of surface Na^+ channels in skeletal muscle: Voltage-clamp analysis and biochemical characterization of the receptor. *Proceedings of the National Academy of Sciences* (U.S.A.) 79: 3896–3900.

Jayaram, V., D. Chandra Sekhara Reddy, and B. Padmanabha Naidu. 1978. Circadian rhythmicity in phosphorylase activity and glycogen content in the heart muscle of the scorpion *Heterometrus fulvipes* C. L. Koch. *Experientia* 34(9): 1184–85.

Jones, C., J. Nolte, and J. E. Brown. 1971. The anatomy of the median ocellus of *Limulus*. *Zeitschrift für Zellforschung und mikroskopischen Anatomie* 118: 297–309.

Jones, J. C. 1962. Current concepts concerning insect hemocytes. *American Zoologist* 2: 209–46.

Jover, E., N. Martin-Moutot, F. Couraud, and H. Rochat. 1980. Binding of scorpion toxins to rat brain synaptosomal fraction: Effects of membrane potential, ions and other neurotoxins. *Biochemistry* 19: 463–67.

Junge, D. 1976. Voltage clamping. In *Nerve and muscle excitation*, pp. 57–76. Sunderland, Mass.: Sinauer Associates.

Karlsson, E. 1973. Chemistry of some potent animal toxins. *Experientia* 29: 1319–27.

Karsch, F. 1879a. Scorpionologische Beiträge, I. *Mitteilungen des Münchener entomologischen Vereins* 3: 6–22.

———. 1879b. Scorpionologische Beiträge, II. *Mitteilungen des Münchener entomologischen Vereins* 3: 97–136.

Kästner, A. 1940. Ordnung der Arachnida: Scorpiones. In W. Kükenthal and

T. Krumbach, eds., *Handbuch der Zoologie*, Band 3, Hälfte 2, Teil 1, *Chelicerata*, pp. 117–240. Berlin: de Gruyter.

———. 1968. *Invertebrate zoology*. Vol. 2, *Arthropod relatives: Chelicerata, Myriapoda*. Tr. H. W. Levi and L. R. Levi. New York: Interscience. 472 pp.

Katz, N. L., and C. Edwards. 1972. The effect of scorpion venom on the neuromuscular junction of the frog. *Toxicon* 10: 133–37.

Kaur, G., and M. S. Kanungo. 1967. Metabolism of the (pedipalp) muscle of the scorpion. *Life Science* 6(2): 113–17.

Keegan, H. L. 1980. Scorpions of medical importance. Jackson: University Press of Mississippi. 140 pp.

Keegan, H. L., and W. R. Lockwood. 1971. Secretory epithelium in venom glands of two species of scorpion of the genus *Centruroides* Marx. *American Journal of Tropical Medicine and Hygiene* 20: 770–85.

Keimer, L. 1929. Bemerkungen und Lesefrüchte zur altägyptischen Naturgeschichte. *Kêmi* 2: 84–106.

Keister, M., and J. Buck. 1973. Respiration: Some exogenous and endogenous effects on rate of respiration. In M. Rockstein, ed., *The physiology of Insecta*, 2d ed., vol. 6: 469–509. New York: Academic.

Keller, O. 1913. Skorpione. In *Die antike Tierwelt*, vol. 2, pp. 470–79. Leipzig: Engelmann.

Kennaugh, J. 1959. An examination of the cuticles of two scorpions, *Pandinus imperator* and *Scorpiops hardwickii*. *Quarterly Journal of Microscopical Science* 100: 41–50.

Kennedy, D., and W. J. Davis. 1977. Organization of invertebrate motor systems. In S. R. Geiger et al., eds., *Handbook of physiology*, sec. 1, vol. 1, pt. 2, pp. 1023–89. Bethesda, Md.: American Physiology Society.

King, W. W., and N. F. Hadley. 1979. Water flux and metabolic rates of free-roaming scorpions using the doubly labeled water technique. *Physiological Zoology* 52: 176–89.

Kinzelbach, R. 1970. Skorpione als Strandbewohner. *Natur und Museum* 100(8): 351–55.

Kjellesvig-Waering, E. N. 1966a. Silurian scorpions of New York. *Journal of Paleontology* 40(2): 359–75, 3 pls.

———. 1966b. The scorpions of Trinidad and Tobago. *Caribbean Journal of Science* 6(3–4): 123–35.

———. 1969. Scorpionida: The holotype of *Mazonia woodiana* Meek and Worthen, 1868. *Fieldiana Geology* 12: 171–90.

———. 1972. *Brontoscorpio anglicus*: A gigantic Lower Paleozoic scorpion from central England. *Journal of Paleontology* 46(1): 39–42.

———. 1986. *A restudy of the fossil Scorpionida of the world*. Paleontographica Americana, no. 55. Ithaca, N.Y.: Paleontological Research Institute. 287 pp.

Knowlton, G. F., and W. L. Thomas. 1936. Food habits of Skull Valley lizards. *Copeia* 1936: 64–66.

Koch, C. L. 1836–45. *Die Arachniden*. 12 vols. Nuremberg.

Koch, L. E. 1977. The taxonomy, geographic distribution and evolutionary radiation of Australo-Papuan scorpions. *Records of the Western Australian Museum* 5(2): 83–367.

———. 1978. A comparative study of the structure, function and adaptation to different habitats of burrows in the scorpion genus *Urodacus* (Scorpionida, Scorpionidae). *Records of the Western Australian Museum* 6(2): 119–46.

———. 1981. The scorpions of Australia: Aspects of their ecology and zoogeography. In A. Keast, ed., *Ecological biogeography of Australia*, pp. 875–84. The Hague: Junk.

Kock, D. 1969. Ein Skorpion frisst an einer Zwergmaus. *Natur und Museum* 99: 107–12.

Koehler, F. H., Jr. 1979. Physical characteristics of the fluorescence spectra of scorpions. Master's thesis, California State University, San Francisco.

Kopeyan, C., G. Martinez, S. Lissitzky, F. Miranda, and H. Rochat. 1974. Disulfide bonds of Toxin II of the scorpion *Androctonus australis* Hector. *European Journal of Biochemistry* 47: 483–89.

Kopeyan, C., G. Martinez, and H. Rochat. 1978. Amino acid sequence of neurotoxin V from the scorpion *Leiurus quinquestriatus quinquestriatus*. *FEBS Letters* 89: 54–58.

———. 1979. Amino acid sequence of neurotoxin III of the scorpion *Androctonus australis* Hector. *European Journal of Biochemistry* 94: 609–15.

———. 1985. Primary structure of Toxin IV of *Leiurus quinquestriatus quinquestriatus*: Characterization of a new group of scorpion toxins. *FEBS Letters* 181: 211–17.

Koppenhöfer, E., and H. Schmidt. 1968a. Die Wirkung von Skorpiongift auf die Ionenströme des ranvierschen Schnürrings. *Pflügers Archiv* 303: 133–49.

———. 1968b. Incomplete sodium inactivation in nodes of Ranvier treated with scorpion venom. *Experientia* 24: 41–42.

Kovoor, J., W. R. Lourenço, and A. Muñoz-Cuevas. 1987. Conservation des spermatozoïdes dans les voies génitales des femelles et biologie de la reproduction des scorpions (Chélicérates). *Comptes rendus de l'Académie des sciences* (Paris), sér. 3, 10: 259–64.

Kraepelin, K. 1891. *Revision der Skorpione*. Part 1, *Die Familie der Androctonidae*. Jahrbuch der Hamburgischen wissenschaftlichen Anstalten, 8. 144 pp., 2 pls.

———. 1894. *Revision der Scorpione*. Part 2, *Scorpionidae und Bothriuridae*. Jahrbuch der Hamburgischen wissenschaftlichen Anstalten, 11. 248 pp., 3 pls.

———. 1899. *Scorpiones und Pedipalpi*. In F. Schulze, ed., *Das Tierreich*, Lieferung 8. Berlin: Friedländer. 265 pp.

———. 1905. Die geographische Verbreitung der Skorpione. *Zoologische Jahrbücher* 22: 321–64.

———. 1907. Die sekundären Geschlechtscharaktere der Skorpione, Pedipalpen und Solifugen. *Mitteilungen aus dem Naturhistorischen Museum in Hamburg* 25: 181–225.

Krafft, B. 1970. Les rythmes d'activité d'*Agelena consociata* Denis: Activité de tissage et activité locomotrice. *Revue Biologica Gabonica* 6: 99–130.

Krapf, D. 1986a. Predator–prey relations in diurnal *Scorpio maurus* L. In *Proceedings of the 10th International congress of arachnology, Jaca,* vol. 1: 133.

———. 1986b. Contact chemoreception of prey in hunting scorpions (Arachnida, Scorpiones). *Zoologischer Anzeiger* 217: 119–29.

Kraus, O. 1976. Zur phylogenetischen Stellung und Evolution der Chelicerata. *Entomologica Germanica* 3: 1–12.

Labeyrie, V. 1978. The significance of the environment in the control of insect fecundity. *Annual Review of Entomology* 23: 69–89.

Lamoral, B. H. 1971. Predation on terrestrial molluscs by scorpions in the Kalahari Desert. *Annals of the Natal Museum* 21(1): 17–20.

———. 1976. *Akentrobuthus leleupi,* a new genus and species of humicolous scorpion from eastern Zaire, representing a new subfamily of the Buthidae. *Annals of the Natal Museum* 22(3): 681–91.

———. 1979. The scorpions of Namibia (Arachnida, Scorpionida). *Annals of the Natal Museum* 23(3): 497–784.

———. 1980. A reappraisal of suprageneric classification of Recent scorpions and of their zoogeography. In *Proceedings of the 8th International congress of arachnology, Vienna*, pp. 439–44.

Lamoral, B. H., and S. Reynders. 1975. A catalogue of the scorpions described from the Ethiopian faunal region. *Annals of the Natal Museum* 22(2): 489–576.

Land, M. F. 1972. Stepping movements made by jumping spiders during turns mediated by the lateral eyes. *Journal of Experimental Biology* 57: 15–40.

Lane, N. J., and J. B. Harrison. 1980. An unusual form of tight junction in the nervous system of the scorpion. *European Journal of Cell Biology* 22: 244.

Lankester, E. R. 1883. Notes on the habits of the scorpions *Androctonus funestus* Ehr. and *Euscorpius italicus* Roes. *Journal of the Linnean Society (Zoology)* 16: 455–62.

———. 1884. On the skeleto-trophic tissues and coxal glands of *Limulus, Scorpio,* and *Mygale. Quarterly Journal of Microscopical Science* 24: 129–62.

———. 1885a. Notes on certain points in the anatomy and generic characters of scorpions. *Transactions of the Zoological Society* (London) 11: 372–84, pls.

———. 1885b. Comparison of the muscular and endoskeletal systems of *Limulus* and *Scorpio*, and considerations of the morphological significance of the facts recorded. *Transactions of the Zoological Society* (London) 11: 361–72.

———. 1904. The structure and classification of Arthropoda. *Quarterly Journal of Microscopical Science* 47: 523.

Lankester, E. R., W. B. S. Benham, and E. J. Beck. 1885. On the muscular and endoskeletal systems of *Limulus* and *Scorpio,* with some notes on the anatomy and generic characters of scorpions. *Transactions of the Zoological Society* (London) 11: 311–60.

Lankester, E. R., and A. G. Bourne. 1883. The minute structure of the lateral and central eyes of *Scorpio* and of *Limulus. Quarterly Journal of Microscopical Science* 23: 177–212.

Larrouy, G., M. C. Signorel, and Y. Cambefort. 1973. Comportement en captivité de *Pandinus imperator* C. L. Koch et naissance des jeunes. *Bulletin de la Société d'histoire naturelle* (Toulouse) 109(3–4): 346–50.

Latreille, P. A. 1810. *Considérations générales sur l'ordre naturel des animaux composants les classes des Crustacés, des Arachnides et des Insectes.* Paris: Schoell.

Laurie, M. 1890. The embryology of a scorpion (*Euscorpius italicus*). *Quarterly Journal of Microscopical Science* 31: 105–41.

———. 1891. Some points in the development of *Scorpio fulvipes. Quarterly Journal of Microscopical Science* 32: 587–97.

———. 1896a. Notes on the anatomy of some scorpions, and its bearing on the classification of the order. *Annals and Magazine of Natural History,* ser. 6, 17: 185–93.

———. 1896b. Further notes on the anatomy and development of scorpions, and their bearing on the classification of the order. *Annals and Magazine of Natural History,* ser. 6, 18: 121–33.

Laverack, M. S. 1966. Observations on a proprioceptive system in the legs of the scorpion *Hadrurus hirsutus. Comparative Biochemistry and Physiology* 19: 241–51.

Lawrence, R. F. 1953. *The biology of the cryptic fauna of forests, with special reference to the indigenous forest of South Africa.* Cape Town: Balkena. 408 pp.

———. 1954. Fluorescence in Arthropoda. *Journal of the Entomological Society* (South Africa) 17: 167–70.

———. 1969. A new genus of psammophile scorpion and new species of *Opisthophthalmus* from the Namibia desert. *Scientific Papers of the Namib Desert Research Station* 48: 105–16.

Lazarovici, P., P. Yanai, M. Pelhate, and E. Zlotkin. 1982. Insect toxin components from the venom of a chactoid scorpion, *Scorpio maurus palmatus* (Scorpionidae). *Journal of Biological Chemistry* 257: 8397–8404.

Lazarovici, P., and E. Zlotkin. 1982. A mammal toxin derived from the venom of a chactoid scorpion. *Comparative Biochemistry and Physiology* 71C: 177–81.

Le Berre, M. 1979. Analyse sequentielle du comportement alimentaire du scorpion *Buthus occitanus* (Amor.) (Arach. Scorp. Buth.). *Biology of Behaviour* 4: 97–122.

Lefèbvre, P. J., A. S. Luyckx, E. Moerman, and M. Bogaert. 1978. Scorpion venom–induced release of noradrenalin does not modify glucagon-output from the isolated perfused dog stomach. *Hormone and Metabolic Research* 10: 80–81.

Legendre, R. 1968. Morphologie et développement des Chélicérates. *Fortschritte der Zoologie* 19(1): 1–50.

———. 1985. The stomatogastric nervous system and neurosecretion. In Barth, ed., pp. 38–49.

Le Guelte, L., and R. Ramousse. 1979. Effets de facteurs environnementaux sur le rythme du comportement constructeur chez l'araignée *Araneus diadematus* L. *Biology of Behaviour* 4: 289–302.

Leneveu, E., and M. Simonneau. 1986. Scorpion venom inhibits selectively Ca^{2+}-activated K^{+} channels *in situ*. *FEBS Letters* 209: 165–68.

Le Pape, G. 1974. Sur quelques aspects des relations mère-jeunes chez trois espèces de scorpions Buthidae. *Revue de comportement animal* 8: 261–64.

Le Pape, G., and M. Goyffon. 1975. Accouplement interspécifique suivi de parturition dans le genre *Androctonus* (Scorpionida, Buthidae). *Comptes rendus hebdomadaires des séances de l'Académie des sciences* (Paris), sér. D, 280: 2005–8.

Lester, D., P. Lazarovici, M. Pelhate, and E. Zlotkin. 1982. Purification, characterization and action of two insect toxins from the venom of the scorpion *Buthotus judaicus*. *Biochimica et Biophysica Acta* 701: 370–81.

Levy, G., and P. Amitai. 1980. *Fauna Palaestina. Arachnida.* Vol. 1, *Scorpiones.* Jerusalem: Israel Academy of Sciences and Humanities. 130 pp.

Likes, K., W. Banner, Jr., and M. Chavez. 1984. *Centruroides exilicauda* envenomation in Arizona. *Western Journal of Medicine* 141: 634–37.

Linden, C. D., and M. A. Raftery. 1976. Isolation of a scorpion toxin for use as a probe of the electrically excitable sodium channel. *Biochemical and Biophysical Research Communications* 72: 646–53.

Linnaeus, C. 1758. *Systema Naturae.* 10th Edition. Stockholm.

Linsenmair, K. E. 1968. Anemomenotaktische Orientierung bei Skorpionen (Chelicerata, Scorpiones). *Zeitschrift für vergleichende Physiologie* 60: 445–49.

———. 1972. Anemomenotactic orientation in beetles and scorpions. In K. Schmidt-Koenig, ed., *Animal orientation and navigation,* pp. 501–10. Washington, D.C.: National Aeronautics and Space Administration.

Lin Shiau, S. Y., W. C. Tseng, and C. Y. Lee. 1975. Pharmacology of scorpion Toxin II in the skeletal muscle. *Naunyn-Schmiedeberg's Archives of Pharmacology* 289: 359–68.

Locket, N. A. 1986. Albinism and eye structure in an Australian scorpion, *Urodacus yaschenkoi* (Scorpiones, Scorpionidae). *Journal of Arachnology* 14: 101–15.

Longenecker, G. L., H. E. Longenecker, Jr., D. D. Watt, C. G. Huggins, and

C. A. Peterson. 1980. Inhibition of the angiotensin converting enzyme by venom of the scorpion *Centruroides sculpturatus*. *Toxicon* 18: 667–70.

Lourenço, W. R. 1974. Relações entre alguns Aracnídeos e Quilópodos que habitam os cupinzeiros. *Cerrado* 6(26): 24–25.

———. 1975. Étude préliminaire sur les scorpions du District Fédéral (Brasil). *Revista brasileira de biologia* 35(4): 679–82.

———. 1976. Sur *Bothriurus asper araguayae* (Vellard, 1934) (Scorpiones, Bothriuridae). *Revista brasileira de biologia* 36(4): 911–18.

———. 1978. Étude sur les scorpions appartenants au "complexe" *Tityus trivittatus* Kraepelin, 1898, et, en particulier, de la sous-espèce *Tityus trivittatus fasciolatus* Pessôa, 1935 (Buthidae). Ph.D. diss., L'Université Pierre et Marie Curie. Paris. 128 pp.

———. 1979a. Le scorpion Buthidae: *Tityus mattogrossensis* Borelli, 1901 (Morphologie, écologie, biologie et développement post-embryonnaire). *Bulletin du Muséum national d'histoire naturelle* (Paris), sér. 4, 1A: 95–117.

———. 1979b. La biologie sexuelle et le développement post-embryonnaire du scorpion Buthidae: *Tityus trivittatus fasciolatus* Pessôa, 1935. *Revista de nordeste biologia* 2: 49–96.

———. 1981a. Scorpions cavernicoles de l'équateur *Tityus demangei* n. sp. et *Ananteris ashmolei* n. sp. (Buthidae); *Troglotayosicus vachoni* n. gen., n. sp. (Chactidae), scorpion troglobie. *Bulletin du Muséum national d'histoire naturelle* (Paris), sér. 4, 3: 635–62.

———. 1981b. Sur la distribution géographique et l'écologie de *Opisthacanthus cayaporum* Vellard, 1932 (Scorpiones, Scorpionidae). *Revista brasileira de biologia* 41(2): 343–49.

———. 1981c. Estudo da variabilidade do caráter número de dentes dos pentes nos escorpiões *Tityus cambridgei* Pocock, 1897, e *Rhopalurus laticauda* Thorell, 1876. *Revista brasileira de biologia* 41(3): 545–48.

———. 1982. Notes sur quelques acariens parasites de scorpions. *Acarologia* 23: 245–47.

———. 1983. La faune des Scorpions de Guyane française. *Bulletin du Muséum national d'histoire naturelle* (Paris), sér. 4, 5A: 771–808.

———. 1985. Essai d'interprétation de la distribution du genre *Opisthacanthus* (Arachnida, Scorpiones, Ischnuridae) dans les régions Néotropicale et Afrotropicale. Étude taxinomique, biogéographique, évolutive et écologique. Thèse de Doctorat d'état en Sciences naturelles, L'Université Pierre et Marie Curie. Paris. 287 pp.

Lourenço, W. R., and O. F. Francke. 1985. Revision des connaissances sur les scorpions cavernicoles (troglobies) (Arachnida, Scorpiones). *Mémoires de biospéléologie* 12: 3–7.

Lower, H. F. 1956. The terminology of the insect cuticle. *Nature* 178: 1355–56.

Lucas, S., and W. Bücherl. 1971. Aparelhos estriduladores do escorpião, *Rho-*

palurus iglesiasi dorsomaculatus (Prado) 1938, e da *Aranha caranguejeira, Theraphosa blondi* (Latreille) 1804. *Ciência e cultura* 23(5): 635–37.

MacArthur, R. H. 1972. *Geographical ecology*. New York: Harper and Row. 269 pp.

MacArthur, R. H., and E. O. Wilson. 1967. *The theory of island biogeography*. Princeton, N.J.: Princeton University Press. 203 pp.

Maccary, A. 1810. *Mémoire sur le scorpion qui se trouve sur la montagne de Cette*. Paris: Gabon. 48 pp.

Macedo, T. M., and M. V. Gomez. 1982. Effects of Tityustoxin (TsTX) from scorpion venom on the release and synthesis of acetylcholine in brain slices. *Toxicon* 20: 601–6.

Machan, L. 1966. Studies on structure, electroretinogram, and spectral sensitivity of the median and lateral eyes of the scorpion. Ph.D. diss., University of Wisconsin, Madison. [*Dissertation Abstracts* 27: 1329B.]

———. 1967. Studies on the structure and electrophysiology of scorpion eyes. *South African Journal of Science* (Dec.): 512–20.

———. 1968a. The effect of prolonged dark adaptation on sensitivity and the correlation of shielding pigment position in the median and lateral eyes of the scorpion. *Comparative Biochemistry and Physiology* 26: 365–68.

———. 1968b. Spectral sensitivity of scorpion eyes and the possible role of shielding pigment effect. *Journal of Experimental Biology* 49: 95–105.

Maier, L., T. Root, and E.-A. Seyfarth. 1987. Heterogeneity of spider leg muscle: Histochemistry and electrophysiology of identified fibers in the claw levator. *Journal of Comparative Physiology* 157: 285–94.

Main, B. Y. 1956. Taxonomy and biology of the genus *Isometroides* Keyserling (Scorpionida). *Australian Journal of Zoology* 4: 158–64.

Makioka, T., and K. Koike. 1984. Parthenogenesis in the viviparous scorpion, *Liocheles australasiae*. *Proceedings of the Japanese Academy*, ser. B, 60(9): 374–76.

———. 1985. Reproductive biology of the viviparous scorpion, *Liocheles australasiae* (Fabricius) (Arachnida, Scorpiones, Scorpionidae). I, Absence of males in two natural populations. *International Journal of Invertebrate Reproduction and Development* 8(6): 317–23.

Malek, S. R. A. 1964. A study of a scorpion cuticle. I, The structure and staining reactions of the fully formed cuticle of *Buthus quinquestriatus* (H.&E.). *Proceedings of the Linnean Society* (London) 175: 101–16.

Mandell, F., and J. W. Graef. 1974. General care of the patient. In J. W. Graef and T. E. Cone, eds., *Manual of pediatric therapeutics*, p. 12. Boston: Little, Brown.

Mani, M. S. 1968. *Ecology and biogeography of high altitude insects*. The Hague: Junk. 538 pp.

Manton, S. M. 1958. Hydrostatic pressure and leg extension in arthropods, with special reference to arachnids. *Annals and Magazine of Natural History*, ser. 13, 1(3): 161–82.

Marei, Z. A., and S. A. Ibrahim. 1979. Stimulation of rat uterus by venom from the scorpion *Leiurus quinquestriatus*. *Toxicon* 17: 251–58.

Maretic, Z. 1983. Latrodectism: Variations in clinical manifestations provoked by *Latrodectus* species of spiders. *Toxicon* 21: 457–66.

Marples, T. G., and D. J. Shorthouse. 1982. An energy and water budget for a population of the arid zone scorpion *Urodacus yaschenkoi* (Birula 1903). *Australian Journal of Ecology* 7: 119–27.

Martin, M. F., L. G. Garcia y Perez, M. el-Ayeb, C. Kopeyan, G. Bechis, E. Jover, and H. Rochat. 1987. Purification and chemical characterizations of seven toxins from the Mexican scorpion, *Centruroides suffusus suffusus*. *Journal of Biological Chemistry* 262: 4452–59.

Martin, M. F., and H. Rochat. 1984a. Purification and amino acid sequence of Toxin I″ from the venom of the North African scorpion *Androctonus australis* Hector. *Toxicon* 22: 695–703.

———. 1984b. Purification of thirteen toxins active on mice from the venom of the North African scorpion *Buthus occitanus tunetanus*. *Toxicon* 22: 279–91.

Mastanaiah, S., D. Chengal Raju, D. Ravi Varma, and K. S. Swami. 1979. Rhythmic variations in the activities of aldolase and isocitrate dehydrogenase in the heart muscle of the scorpion, *Heterometrus fulvipes* (C. Koch). *Experientia* 35: 69–70.

Mastanaiah, S., D. Chengal Raju, and K. S. Swami. 1977. Rhythmic variation in the isocitrate dehydrogenase activity in the scorpion *Heterometrus fulvipes* (C. Koch). *Experientia* 33: 1051–52.

Mastanaiah, S., V. Doraswamy Reddy, D. Chengal Raju, and K. S. Swami. 1978. Circadian rhythmic activity of lipase in the scorpion, *Heterometrus fulvipes* (C. Koch). *Current Science* 47: 130–32.

Mathew, A. P. 1948. Nutrition in the advanced embryos of *Palamnaeus scaber*. *Proceedings of the National Academy of Sciences* (India), sec. B (Biological Sciences), 27(4): 111–18.

———. 1956. Embryology of *Heterometrus scaber*. *Bulletin of the Central Research Institute* (University of Travancore) 1: 1–96.

———. 1957. Mating in scorpions. *Journal of the Bombay Natural History Society* 54: 853–57.

———. 1960. Embryonic nutrition in *Lychas tricarinatus*. *Journal of the Zoological Society of India* 12: 220–28.

———. 1962. Changes in the structure of the ovarian and diverticular mucosa in *Heterometrus scaber* (Thorell) during a reproductive cycle. In *Proceedings of the 1st All-India congress of zoology (1959)*, pt. 2, pp. 100–111.

———. 1965. On the movable claw of the pedipalp in the scorpion *Heterometrus scaber*. *Journal of Animal Morphology and Physiology* 12(2): 271–75.

———. 1966. The movable claw of the pedipalp of the scorpion, *Heterometrus scaber*. In *Proceedings of the 2nd All-India congress of zoology (1962)*, pt. 2, pp. 411–14.

———. 1968. Embryonic nutrition in *Urodacus abruptus* Pocock. *Journal of Animal Morphology and Physiology* 15(1/2): 162–67.

Matthiesen, F. A. 1960. Sõbre o acasalamento de *Tityus bahiensis* (Perty, 1834). *Revista de agricultura* (Piracicaba) 35(4): 341–46.

———. 1961. Notas sõbre escorpiões. *Revista de agricultura* (Piracicaba) 36(3): 129–47.

———. 1962. Parthenogenesis in scorpions. *Evolution* 16: 255–56.

———. 1969–70. Le développement post-embryonnaire du scorpion Buthidae: *Tityus bahiensis* (Perty, 1834). *Bulletin du Muséum national d'histoire naturelle* (Paris), sér. 2, 41(6): 1367–70.

———. 1970. Reproductive system and embryos of Brazilian scorpions. *Anais da Academia brasileira de ciências* 42(3): 627–32.

———. 1971. The breeding of *Tityus serrulatus* Lutz and Mello, 1927, in captivity. *Revista brasileira de pesquisas médicas e biológicas* 4(4–5): 299–300.

de Maupertuis, P.-L. 1733. Expériences sur les scorpions. *Histoire de l'Académie royale des sciences* (Paris), *Mémoires*, 1731: 223–29.

Maury, E. A. 1968a. Aportes al conocimiento de los escorpiones de la República Argentina. I, Observaciones biológicas sobre *Urophonius brachycentrus* (Thorell) Bothriuridae. *Physis* (Buenos Aires) 27(75): 407–18.

———. 1968b. Aportes al conocimiento de los escorpiones de la República Argentina. II, Algunas consideraciones sobre el género *Bothriurus* en la Patagonia y Tierra del Fuego con la descripción de una nueva especie (Bothriuridae). *Physis* (Buenos Aires) 28(76): 149–64.

———. 1969. Observaciones sobre el ciclo reproductivo de *Urophonius brachycentrus* (Thorell, 1877) (Scorpiones, Bothriuridae). *Physis* (Buenos Aires) 29(78): 131–39.

———. 1971. Essai d'une classification des sous-familles de scorpions, Bothriuridae. In *Proceedings of the 5th International congress of arachnology, Brno*, pp. 29–36.

———. 1973a. Las tricobotrias y su importancia en la sistemática del género *Brachistosternus* Pocock 1894 (Scorpiones, Bothriuridae). *Physis* (Buenos Aires) 32(85): 247–54.

———. 1973b. Los escorpiones de los sistemas serranos de la Provincia de Buenos Aires. *Physis* (Buenos Aires) 32(85): 351–71.

———. 1975a. Sobre el dimorphismo sexual de la pinza de los pedipalpos en los escorpiones Bothriuridae. *Bulletin du Muséum national d'histoire naturelle* (Paris), sér. 3, 305: 765–71.

———. 1975b. Escorpiofauna patagónica. I, Sobre una nueva especie del género *Timogenes* Simon, 1880 (Bothriuridae). *Physis* (Buenos Aires) 34(88): 65–74.

———. 1975c. La estructura del espermatoforo en el género *Brachistosternus* (Scorpiones, Bothriuridae). *Physis* (Buenos Aires) 34(89): 179–82.

———. 1975d. Escorpiones y escorpionismo en el Perú. V, *Orobothriurus*, un nuevo género de escorpiones altoandino (Bothriuridae). *Revista peruana de entomología* 18: 14–25.

———. 1977. Un nuevo *Brachistosternus* de los medanos costeros bonaerenses (Scorpiones, Bothriuridae). *Physis* (Buenos Aires) 37(93): 169–76.

———. 1978. Escorpiofauna patagónica. II, *Urophonius granulatus* Pocock 1898 (*Bothriuridae*). *Physis* (Buenos Aires) 38(94): 57–68.

———. 1980. Usefulness of the hemispermatophore in the systematics of the scorpion family Bothriuridae. In *Proceedings of the 8th International congress of arachnology, Vienna*, pp. 335–39.

———. 1983. Singular anomalía sexual en un ejemplar de *Brachistosternus pentheri* Mello-Leitão 1931 (Scorpiones, Bothriuridae). *Revista de la Sociedad de entomología de Argentina* 42(1/4): 155–56.

Mazurkiewicz, J. E., and E. M. Bertke. 1972. Ultrastructure of the venom gland of the scorpion, *Centruroides sculpturatus* (Ewing). *Journal of Morphology* 137: 365–84.

Mazzotti, L., and M. A. Bravo-Becherelle. 1963. Scorpionism in the Mexican Republic. In H. L. Keegan and W. W. McFarland, eds., *Venomous and poisonous animals and noxious plants of the Pacific area*, pp. 119–31. New York: Pergamon.

McAlister, W. H. 1960. Early growth rates in offspring from two broods of *Vejovis spinigerus* Wood. *Texas Journal of Science* 12: 158–62.

———. 1965. The mating behaviour of *Centruroides vittatus* Say (Arachnida: Scorpionida). *Texas Journal of Science* 17: 307–12.

———. 1966. The aggregating tendency of *Centruroides vittatus* Say (Arachnida: Scorpionida). *Texas Journal of Science* 18: 80–84.

McClendon, J. F. 1904. On the anatomy and embryology of the nervous system of the scorpion. *Biological Bulletin* 8: 38–55.

McCormick, S. J., and G. A. Polis. 1982. Arthropods that prey on vertebrates. *Biological Reviews* 57: 29–58.

———. 1986. Comparison of the diet of *Paruroctonus mesaensis* at two sites. In *Proceedings of the 9th International congress of arachnology, Panama City*, pp. 167–71.

McDaniels, M. M. 1968. Notes on the biology of California scorpions. *Entomological News* 79: 278–84.

McElhinny, M. W., N. S. Haile, and A. R. Crawford. 1974. Palaeomagnetic evidence shows Malay peninsula was not a part of Gondwanaland. *Nature* 252: 641–45.

McGeer, E. G., and P. L. McGeer. 1981. Neurotoxins as tools in neurobiology.

In J. R. Smythies and R. J. Bradley, eds., *International review of neurobiology*, vol. 22, pp. 173–204. New York: Academic.

McIndoo, E. 1911. The lyriform organs and tactile hairs of araneids. *Proceedings of the Academy of Natural Sciences* (Philadelphia) 63: 375–418.

McIntosh, M. E. 1969. Biochemical and immunochemical studies on the venom from the scorpion *Centruroides sculpturatus* Ewing. Ph.D. diss., University of Kansas, Lawrence. 179 pp.

McIntosh, M. E., and D. D. Watt. 1967. Biochemical-immunochemical aspects of the venom from the scorpion *Centruroides sculpturatus*. In F. E. Russell and P. R. Saunders, eds., *Animal toxins*, pp. 47–58. New York: Pergamon.

———. 1972. The purification of toxins from the North American scorpion *Centruroides sculpturatus*. In A. DeVries and E. Kochva, eds., *Toxins of animal and plant origin*, vol. 2, pp. 529–44. New York: Gordon and Breach.

Mello-Leitão, C. 1945. *Escorpiões sul-americanos*. Arquivos do Museu Nacional (Rio de Janeiro) 40. 468 pp.

Mercier, J., and S. Dessaigne. 1970. Influence of some psycholeptic drugs upon the behavior of the scorpion *Androctonus australis*. *Comptes rendus des séances de la Société de biologie et de ses filiales* 164(2): 341–44.

———. 1971. Action of gallamine on the motility and reflexes of the praying mantis and the scorpion. *Comptes rendus des séances de la Société de biologie et de ses filiales* 165(6): 1368–71.

Meves, H., N. Rubly, and D. D. Watt. 1982. Effect of toxins isolated from the venom of the scorpion *Centruroides sculpturatus* on the Na^+ currents of the node of Ranvier. *Pflügers Archiv* 393: 56–62.

———. 1983. Potential-dependent action of a scorpion toxin of the sodium current of the frog myelinated nerve fibre. *Journal of Physiology* 345: 1–77.

———. 1984a. Voltage-dependent effect of a scorpion toxin on sodium-current inactivation. *Pflügers Archiv* 402: 24–33.

Meves, H., J. M. Simard, and D. D. Watt. 1984b. Biochemical and electrophysiological characteristics of toxins from the venom of the scorpion *Centruroides sculpturatus*. *Journal de physiologie* (Paris) 79: 185–91.

———. 1986. Interactions of scorpion toxins with the sodium channel. *Annals of the New York Academy of Sciences* 479: 113–32.

Mill, P. J. 1972. *Respiration in the invertebrates*. New York: Macmillan. 201 pp.

Miller, C., E. Moczydlowski, R. Latorre, and M. Phillips. 1985. Charybdotoxin, a protein inhibitor of single Ca^{2+}-activated K^+ channels from mammalian skeletal muscle. *Nature* 313: 316–18.

Millot, J., and R. Paulian. 1943. Valeur fonctionnelle des poumons des scorpions. *Bulletin of the Zoological Society* (France) 68: 97–98.

Millot, J., and M. Vachon. 1949. Ordre des scorpions. In P.-P. Grassé, ed., *Traité de zoologie*, vol. 6, pp. 386–436. Paris: Masson.

Miranda, F., C. Kopeyan, H. Rochat, C. Rochat, and S. Lissitzky. 1970. Purification of animal neurotoxins: Isolation and characterization of eleven neurotoxins from venoms of the scorpions *Androctonus australis* Hector, *Buthus occitanus tunetanus* and *Leiurus quinquestriatus quinquestriatus*. *European Journal of Biochemistry* 16: 514–23.

Mitchell, R. W. 1968. *Typhlochactas*, a new genus of eyeless cave scorpion from Mexico (Scorpionida, Chactidae). *Annales de spéléologie* 23(4): 753–77.

———. 1971. *Typhlochactas elliotti*, a new eyeless cave scorpion from Mexico (Scorpionida, Chactidae). *Annales de spéléologie* 26(1): 135–48.

Mitchell, R. W., and S. B. Peck. 1977. *Typhlochactas sylvestris*, a new eyeless scorpion from montane forest litter in Mexico (Scorpionida, Chactidae, Typhlochactinae). *Journal of Arachnology* 5(2): 159–68.

Moffett, S., and G. S. Doell. 1980. Alteration of locomotor behavior in wolf spiders carrying normal and weighted egg cocoons. *Journal of Experimental Zoology* 213: 219–26.

Moss, J., R. W. Colburn, and I. J. Kopin. 1974a. Scorpion toxin–induced catecholamine release from synaptosomes. *Journal of Neurochemistry* 22: 217–21.

Moss, J., N. B. Thoa, and I. J. Kopin. 1974b. On the mechanism of scorpion toxin–induced release of norepinephrine from peripheral adrenergic neurons. *Journal of Pharmacology and Experimental Therapeutics* 190: 39–48.

Mozhayeva, G. N., A. P. Naumov, E. D. Nosyreva, and E. V. Grishin. 1980. Potential-dependent interaction of toxin from venom of the scorpion *Buthus eupeus* with sodium channels in myelinated fibre. *Biochimica et Biophysica Acta* 597: 587–602.

Narahashi, T., B. I. Shapiro, T. Deguchi, M. Scuka, and C. M. Wang. 1972. Effects of scorpion venom on squid axon membranes. *American Journal of Physiology* 222: 850–57.

Nayar, K. K. 1966. Endocrine nature of the lymphoid organs of the scorpion, *Heterometrus scaber* Thor. In *Proceedings of the 2nd All-India congress of zoology (1962)*, pt. 2, pp. 173–74.

Nemenz, H., and J. Gruber. 1967. Experimente und Beobachtungen an *Heterometrus longimanus petersii* (Thorell) (Scorpiones). *Verhandlungen der Zoologisch-botanischen Gesellschaft* (Vienna) 107: 5–24.

Newell, I. M., and R. E. Ryckman. 1966. Species of *Pimeliaphilus* (Acari: Pterygosomidae) attacking insects, with particular reference to the species parasitizing Triatominae (Hemiptera: Reduviidae). *Hilgardia* 37(12): 403–36.

Newlands, G. 1969a. Scorpion defensive behavior. *African Wildlife* 23(2): 147–53.

———. 1969b. Scorpion preparation for scientific study and display. *Journal of the Entomological Society* (South Africa) 32(2): 491–93.

———. 1972a. A description of *Hadogenes lawrenci* sp. nov. (Scorpiones) with a checklist and key to the South-West African species of the genus *Hadogenes*. *Madoqua*, ser. 2, 1(54–62): 133–40.

———. 1972b. Ecological adaptations of Kruger National Park scorpionids (Arachnida: Scorpiones). *Koedoe* 15: 37–48.

———. 1972c. Notes on psammophilous scorpions and a description of a new species (Arachnida: Scorpionida). *Annals of the Transvaal Museum* (Pretoria) 27(12): 241–57.

———. 1974a. The venom-squirting ability of *Parabuthus* scorpions (Arachnida: Buthidae). *South African Journal of Medical Science* 39(4): 175–78.

———. 1974b. Transvaal scorpions. *Fauna and Flora* 25: 3–7.

———. 1978a. Arachnida (except Acari). In M. J. A. Werger, ed., *Biogeography and ecology of southern Africa*, pp. 677–84. The Hague: Junk.

———. 1978b. Review of southern African scorpions and scorpionism. *South African Medical Journal* 54: 613–15.

———. 1985. A re-appraisal of the rock scorpions (Scorpionidae: *Hadogenes*). *Koedoe* 28: 35–45.

Newlands, G., and C. Martindale. 1980. The buthid scorpion fauna of Zimbabwe-Rhodesia with checklists and keys to the genera and species, distributions and medical importance (Arachnida: Scorpiones). *Zeitschrift für angewandte Zoologie* 67: 51–77.

Newport, G. 1843. On the structure, relations, and development of the nervous and circulatory systems, and on the existence of a complete circulation of the blood in vessels, in Myriapoda and macrourous Arachnida. *Philosophical Transactions of the Royal Society* (London) 133: 243–302.

Okamoto, H. 1980. Binding of scorpion toxin to sodium channels *in vitro* and its modification by β-bungarotoxin. *Journal of Physiology* 299: 507–20.

Okamoto, H., K. Takahashi, and N. Yamashita. 1977. One-to-one binding of a purified scorpion toxin to Na^+ channels. *Nature* 266: 465–68.

Oliveira, Z. E. G., I. F. Heneine, and M. V. Gomez. 1976. The effect of sulfhydryl reagents on the Tityustoxin-induced acetylcholine release in rat brain slices. *Journal of Neurochemistry* 27: 43–46.

Osman, O. H., M. Ismail, M. F. el-Asmar, and S. A. Ibrahim. 1972. Effect on the rat uterus of the venom from the scorpion *Leiurus quinquestriatus*. *Toxicon* 10: 363–66.

Palka, J., and K. S. Babu. 1967. Toward the physiological analysis of defensive responses of scorpions. *Zeitschrift für vergleichende Physiologie* 55: 286–98.

Pampathi Rao, K. 1963. Some aspects of the electrical activity in the central nervous system of the scorpion, *Heterometrus swammerdami*. In *Proceedings of the 16th International congress of zoology, Washington, D.C.*, vol. 2, p. 69.

———. 1964. Neurophysiological studies of an arachnid, the scorpion *Heterometrus fulvipes*. *Journal of Animal Morphology and Physiology* 11: 133–42.

Pampathi Rao, K., and T. Gropalakrishnareddy. 1967. Blood borne factors in circadian rhythms of activity. *Nature* 213: 1047–48.

Pampathi Rao, K., and M. Habibulla. 1973. Correlation between neurosecretion and some physiological functions of the scorpion *Heterometrus swammerdami*. *Proceedings of the Indian Academy of Sciences*, sec. B, 77(4): 148–55.

Pampathi Rao, K., and V. S. R. Murthy. 1966. Oscillographic analysis of some proprioceptors in the scorpion *Heterometrus fulvipes*. *Nature* 212: 520–21.

Pansa, M. C., G. M. Natalizi, and S. Bettini. 1973. Effect of scorpion venom and its fractions on the crayfish stretch receptor organ. *Toxicon* 11: 283–86.

Parker, G. H. 1886. The eyes in scorpions. *Bulletin of the Museum of Comparative Zoology* (Harvard) 13: 173–208.

Parnas, I., D. Avgar, and A. Shulov. 1970. Physiological effects of venom of *Leiurus quinquestriatus* on neuromuscular systems of locust and crab. *Toxicon* 8: 67–79.

Parnas, I., and F. E. Russell. 1967. Effects of venoms on nerve, muscle and neuromuscular junction. In F. E. Russell and P. R. Saunders, eds., *Animal toxins*, pp. 401–15. New York: Pergamon.

Parrish, C. 1966. The biology of scorpions. *Pacific Discovery* (Mar.–Apr.): 2–11.

Parry, D. A., and R. H. Brown. 1959a. The hydraulic mechanism of the spider leg. *Journal of Experimental Biology* 36: 423–33.

———. 1959b. The jumping mechanism of salgicid spiders. *Journal of Experimental Biology* 36: 654–64.

Pasantes, H., R. Tapia, B. Ortega, and G. Massieu. 1965. Free amino acids and activity of some pyridoxal phosphate-dependent enzymes in the nervous system of three arthropod species. *Comparative Biochemistry and Physiology* 16: 523–29.

Patten, W. 1890. On the origin of vertebrates from arachnids. *Quarterly Journal of Microscopical Science* 31: 2.

Paulus, H. F. 1979. Eye structure and the monophyly of the Arthropoda. In Gupta, ed., pp. 299–383.

Pavan, M. 1956. Studi sugli Scorpioni. II, Contributo citometrico alla conoscenza delle cellule epidermiche e dei loro prolungamenti. *Bollettino del Laboratorio di zoologia F. Silvestri* (Portici) 33: 586–93.

———. 1958. Studi sugli Scorpioni. IV, Sulla birifrangenza e sulla fluorescenza dell'epicuticola. *Bollettino della Società entomologica italiana* 87(1–2): 23–26.

Pavesi, P. 1881. Teratologia: Toradelfia di uno scorpione. *Rendiconti dell'Istituto lombardo di scienze e lettere* (Milan) 14: 329–32.

Pavlovsky, E. N. 1913. Scorpiotomische Mitteilungen. I, Ein Beitrag zur Morphologie der Giftdrüsen der Skorpione. *Zeitschrift für wissenschaftlichen Zoologie* 105: 157–77.

———. 1917. Sur l'appareil genital mâle et sur un cas d'anomalie de cet appareil chez *Isometrus maculatus* (Scorpionides, fam. Buthidae). *Comptes rendus de la Société de biologie* (Paris) 80: 502–5.

———. 1921. Sur l'appareil genital mâle chez *Scorpio maurus* L. *Bulletin de la société d'histoire naturelle* (Algiers) 12: 194–98.

———. 1924a. Skorpiotomische Mitteilungen, III, IV. *Zoologische Jahrbücher, Abteilung für Anatomie* 46: 473–508.

———. 1924b. Studies on the organization and development of scorpions. *Quarterly Journal of Microscopical Science* 68: 615–40.

———. 1924c. On the morphology of the male genital apparatus in scorpions. *Trudy leningradskogo obshchestva estestvoispytatelei* 53(2): 17–86.

———. 1925. Zur Morphologie des weiblichen Genitalapparates und zur Embryologie der Skorpione. *Annuaire du Musée zoologique de l'Académie des sciences* (Leningrad) 26: 137–205.

———. 1926. Studies on the organization and development of scorpions. V, The lungs. *Quarterly Journal of Microscopical Science* 70: 221–61.

Pelhate, M., and E. Zlotkin. 1982. Actions of insect toxin and other toxins derived from the venom of the scorpion *Androctonus australis* on isolated giant axons of the cockroach (*Periplaneta americana*). *Journal of Experimental Biology* 97: 67–77.

Peters, W. 1861. Über eine neue Eintheilung der Skorpione. *Monatsberichte der Königlichen preussischen Akademie der Wissenschaften* (Berlin) 1861: 507–16.

Petrunkevitch, A. 1922. The circulatory system and segmentation in Arachnida. *Journal of Morphology* 36: 157–85.

———. 1953. Paleozoic and Mesozoic Arachnida of Europe. *Geological Society of America, Memoirs,* 53: 1–128, 58 pls.

———. 1955. Arachnida. In R. C. Moore, ed., *Treatise on invertebrate paleontology,* pt. P, *Arthropoda,* vol. 2, pp. 42–162. Lawrence: Kansas University Press and the Geological Society of America.

Pflugfelder, O. 1930. Zur Embryologie des Skorpions *Hormurus australasiae* (F.). *Zeitschrift für wissenschaftliche Zoologie* 137: 1–29.

Phillips, J. E. 1977. Excretion in insects: Function of gut and rectum in concentrating and diluting the urine. *Federation Proceedings* 36: 2480–86.

Pianka, E. R. 1970. On r- and K-selection. *American Naturalist* 104: 592–97.

di Pirajno, A. D. 1955. *A cure for serpents.* Tr. K. Naylor. London: Andre Deutsch.

Piza, S. de Toledo, Jr. 1939a. Observacões sôbre o aparelho reprodutor do *Tityus bahiensis. Jornal de agronomia* 2: 49–56.

———. 1939b. Estudos anatômicos em escorpiões brasileiros. I, O aparelho reprodutor feminino e o inicio do desenvolvimento do embriao de *Tityus serrulatus. Jornal de agronomia* 2: 273–76.

Pocock, R. I. 1893a. Notes on the classification of scorpions, followed by some observations upon synonymy, with descriptions of new genera and species. *Annals and Magazine of Natural History,* ser. 6, 12: 303–30.

———. 1893b. Notes upon the habits of some living scorpions. *Nature* 48: 104–7.

———. 1894. Scorpions and their geographical distribution. *Natural Science* 4(24): 353–64.

———. 1896. How and why scorpions hiss. *Natural Science* 9: 17–25.

———. 1900. *Arachnida*. In W. T. Blanford, ed., *The fauna of British India, including Ceylon and Burma*. London: Taylor & Francis. 279 pp.

———. 1901. The Scottish Silurian scorpion. *Quarterly Journal of Microscopical Science* 44: 291–311.

———. 1902a. Studies on the arachnid endosternite. *Quarterly Journal of Microscopical Science* 46: 225–62.

———. 1902b. *Arachnida: Scorpiones, Pedipalpi, et Solifugae*. In *Biologia Centrali-Americana (Zoologia)*. London: Taylor & Francis. 71 pp.

———. 1904. On a new stridulating organ in scorpions discovered by W. J. Burchell in Brazil in 1828. *Annals and Magazine of Natural History*, ser. 7, 13: 56–62.

———. 1911. *A monograph of the terrestrial Carboniferous Arachnida of Great Britain*. Monographs of the Paleontological Society (London), 44. 84 pp.

Police, G. 1901. Sui centri nervosi sottointestinali dell'*Euscorpius italicus*. *Bollettino della Società dei naturalisti* (Naples) 15: 1–24.

———. 1902. Sul sistema nervoso stomatogastrico dello scorpione. *Archives italiennes de zoologie* (Naples) 1: 179–98.

———. 1904. Sui centri nervosi dei cheliceri e del rostro nello scorpione. *Bollettino della Società dei naturalisti* (Naples) 18: 130–35.

———. 1907. Sugli occhi dello scorpione. *Zoologische Jahrbücher, Abteilung für Anatomie und Ontogenie* 25: 1–70.

Polis, G. A. 1979. Prey and feeding phenology of the desert sand scorpion *Paruroctonus mesaensis* (Scorpionidae: Vaejovidae). *Journal of Zoology* (London) 188: 333–46.

———. 1980a. Seasonal patterns and age-specific variation in the surface activity of a population of desert scorpions in relation to environmental factors. *Journal of Animal Ecology* 49: 1–18.

———. 1980b. The effect of cannibalism on the demography and activity of a natural population of desert scorpions. *Behavioral Ecology and Sociobiology* 7: 25–35.

———. 1981. The evolution and dynamics of intraspecific predation. *Annual Review of Ecology and Systematics* 12: 225–51.

———. 1984. Age structure component of niche width and intraspecific resource partitioning by predators: Can age groups function as ecological species? *American Naturalist* 123: 541–64.

———. 1986. Sexual variation in the feeding ecology of the scorpion *Paruroctonus mesaensis*. In *Proceedings of the 9th International congress of arachnology, Panama City*, pp. 193–96.

———. 1988. Trophic and behavioral response of desert scorpions to harsh environmental periods. *Journal of Arid Environments* 14(2): 123–34.

Polis, G. A., and R. D. Farley. 1979a. Behavior and ecology of mating in the cannibalistic scorpion. *Paruroctonus mesaensis* Stahnke (Scorpionida: Vaejovidae). *Journal of Arachnology* 7: 33–46.

———. 1979b. Characteristics and environmental determinants of natality, growth and maturity in a natural population of the desert scorpion, *Paruroctonus mesaensis* (Scorpionida: Vaejovidae). *Journal of Zoology* (London) 187(4): 517–42.

———. 1980. Population biology of a desert scorpion: Survivorship, microhabitat, and the evolution of a life history strategy. *Ecology* 61(3): 620–29.

Polis, G. A., and W. R. Lourenço. 1986. Sociality among scorpions. In *Proceedings of the 10th International congress of arachnology, Jaca,* vol. 1: 111–15.

Polis, G. A., and S. J. McCormick. 1986a. Patterns of resource use and age structure among a guild of desert scorpions. *Journal of Animal Ecology* 55: 59–73.

———. 1986b. Scorpions, spiders and solpugids: Predation and competition among distantly related taxa. *Oecologia* 71: 111–16.

———. 1987. Competition and predation among species of desert scorpions. *Ecology* 68: 332–43.

Polis, G. A., C. N. McReynolds, and G. Ford. 1985. Home range geometry of the desert scorpion, *Paruroctonus mesaensis. Oecologia* 67: 273–77.

Polis, G. A., C. A. Myers, and R. Holt. 1989. The ecology and evolution of intraguild predation: Potential competitors that eat each other. *Annual Review of Ecology and Systematics* 20. In press.

Polis, G. A., C. A. Myers, and M. A. Quinlan. 1986. Burrowing biology and spatial distribution of desert scorpions. *Journal of Arid Environments* 10: 137–46.

Polis, G. A., W. D. Sissom, and S. J. McCormick. 1981. Predators of scorpions: Field data and a review. *Journal of Arid Environments* 4: 309–26.

Pomeranz, A., P. Amitai, I. Braunstein, Y. Reichenberg, M. Finelt, and A. Drukker. 1984. Scorpion sting: Successful treatment with nonhomologous antivenin. *Israel Journal of Medical Science* 20: 451–52.

Pool, W. R. 1979. Characterization of components of the venom of *Centruroides sculpturatus* Ewing. Ph.D. diss., Creighton University, Omaha, Nebraska.

Possani, L. D., M. A. R. Dent, B. M. Martin, A. Maelicke, and I. Svendsen. 1981a. The amino terminal sequence of several toxins from the venom of the Mexican scorpion *Centruroides noxius* Hoffmann. *Carlsberg Research Communications* 46: 207–14.

Possani, L. D., P. L. Fletcher, Jr., A. B. C. Alagõn, A. C. Alagõn, and J. Z. Juliã. 1980. Purification and characterization of a mammalian toxin from venom of the Mexican scorpion, *Centruroides limpidus tecomanus* Hoffmann. *Toxicon* 18: 175–83.

Possani, L. D., B. M. Martin, J. Mochca-Morales, and I. Svendsen. 1981b. Purification and chemical characterization of the major toxins from the venom of the Brazilian scorpion *Tityus serrulatus* Lutz and Mello. *Carlsberg Research Communications* 46: 195–205.

Possani, L. D., B. M. Martin, and I. Svendsen. 1982. The primary structure of Noxiustoxin: A K^+ channel blocking peptide purified from the venom of the scorpion, *Centruroides noxius* Hoffmann. *Carlsberg Research Communications* 47: 285–89.

Possani, L. D., B. M. Martin, I. Svendsen, G. S. Rode, and B. W. Erickson. 1985. Scorpion toxins from *Centruroides noxius* and *Tityus serrulatus*: Primary structures and sequence comparison by metric analysis. *Biochemical Journal* 229: 739–50.

Possani, L. D., W. E. Steinmetz, M. A. R. Dent, A. C. Alagõn, and K. Würthrich. 1981c. Preliminary spectroscopic characterization of six toxins from Latin American scorpions. *Biochimica et Biophysica Acta* 669: 183–92.

Pringle, J. W. S. 1938. Proprioception in insects. II, The action of the campaniform sensilla on the legs. *Journal of Experimental Biology* 15: 114–31.

———. 1955. The function of lyriform organs of arachnids. *Journal of Experimental Biology* 32: 270–78.

———. 1961. Proprioception in arthropods. In J. A. Ramsay and V. B. Wigglesworth, eds., *The cell and the organism*, pp. 256–82. Cambridge: Cambridge University Press.

———. 1963. The proprioceptive background to mechanisms of orientation. *Ergebnisse der Biologie* 26: 1–11.

Probst, P. J. 1972. Zur Fortpflanzungsbiologie und zur Entwicklung der Giftdrüsen beim Skorpion *Isometrus maculatus* (DeGeer, 1778) (Scorpiones: Buthidae). *Acta Tropica* 29(1): 1–87.

Przibram, H., and F. Megusâr. 1912. Wachstummessungen an *Sphodromantis bioculata* Burm. I, Lange und Masse. *Archiv für Entwicklungsmechanik der Organismen* (Wilhelm Roux) 34: 680–741.

Pyke, G. H. 1984. Optimal foraging theory: A critical review. *Annual Review of Ecology and Systematics* 15: 523–75.

Rack, M., D. Richter, and N. Rubly. 1987. Purification and characterization of a β-toxin from the venom of the African scorpion *Leiurus quinquestriatus*. *FEBS Letters* 214: 163–66.

Raghavaiah, K., K. Satyanarayanam, R. Ramamurthi, and V. Chandrasekharam. 1978. Hyperglycemic principle in the cephalothoracic ganglionic mass in the Indian scorpion *Heterometrus fulvipes*. *Indian Journal of Experimental Biology* 16(8): 944–46.

Raghavaiah, K., M. Sheeramachandramurthy, R. Ramamurthi, P. Satyam, and V. Chandrasekharam. 1977. The effect of epinephrine and the hyperglycemic factor of the scorpion's cephalothoracic ganglionic mass on the phosphorylase activity of the hepato-pancreas of the scorpion *Heterometrus fulvipes*. *Experientia* 33(5): 690–91.

Ramakrishna, T. 1977. Is light receptor a dipole? *Neuroscience Letters* 5: 51–55.

Randall, W. C. 1966. Microanatomy of the heart and associated structure of two scorpions, *Centruroides sculpturatus* Ewing and *Uroctonus mordax* Thorell. *Journal of Morphology* 119: 161–80.

Rao, V. P. 1963. Studies on the peripheral nervous system of the scorpion, *Heterometrus fulvipes*. Ph.D. diss., Sri Venkateswara University, Tirupati, Andhra Pradesh, India.

Rathmayer, W., C. Walther, and E. Zlotkin. 1977. The effect of different toxins from scorpion venom on neuromuscular transmission and nerve action potentials in the crayfish. *Comparative Biochemistry and Physiology* 56C: 35–39.

Ravindranath, M. H. 1974. The hemocytes of a scorpion *Palamnaeus swammerdami*. *Journal of Morphology* 144: 1–10.

Ray, R., C. S. Morrow, and W. A. Catterall. 1978. Binding of scorpion toxin to receptor sites associated with voltage-sensitive sodium channels in synaptic nerve ending particles. *Journal of Biological Chemistry* 253: 7307–13.

Reddy, C. R., K. V. Kumari, and K. S. Babu. 1980. Effect of scorpion (*Heterometrus fulvipes*) venom on succinate dehydrogenase activity of the cockroach *Periplaneta americana*. *Indian Journal of Experimental Biology* 18: 867–69.

Reddy, C. R. R. M., G. Suvarnakumari, C. S. Devi, and C. N. Reddy. 1972. Pathology of scorpion venom poisoning. *Journal of Tropical Medicine and Hygiene* 75: 98–100.

Redi, F. 1671. *Experimenta circa generationem insectorum*. Amsterdam: Frisius. 330 pp.

Reissland, A., and P. Görner. 1985. Trichobothria. In Barth, ed., pp. 138–61.

Remy, P., and P. Leroy. 1933. Présence de scorpions dans la zone subterrestre du littoral marin. *Bulletin Mens Société Linné* (Lyon) 11: 39–42.

Richards, A. G. 1951. *The integument of arthropods*. Minneapolis: University of Minnesota Press. 411 pp.

Riddle, W. A. 1978. Respiratory physiology of the desert grassland scorpion *Paruroctonus utahensis*. *Journal of Arid Environments* 1: 243–51.

———. 1979. Metabolic compensation for temperature change in the scorpion *Paruroctonus utahensis*. *Journal of Thermal Biology* 4: 125–28.

———. 1981. Cuticle water activity and water content of beetles and scorpions from xeric and mesic habitats. *Comparative Biochemistry and Physiology* 68A: 231–35.

Riddle, W. A., C. S. Crawford, and A. M. Zeitone. 1976. Patterns and hemolymph osmoregulation in three desert arthropods. *Journal of Comparative Physiology* 112: 295–305.

Riddle, W. A., and S. Pugach. 1976. Cold hardiness in the scorpion, *Paruroctonus aquilonalis*. *Cryobiology* 13: 248–53.

Rimsza, M. E., D. R. Zimmerman, and P. S. Bergeson. 1980. Scorpion envenomation. *Pediatrics* 66: 298–302.

Robertson, H. G., S. W. Nicolson, and G. N. Louw. 1982. Osmoregulation and temperature effects on water loss and oxygen consumption in two species of African scorpion. *Comparative Biochemistry and Physiology* 71A: 605–9.

Rochat, H., P. Bernard, and F. Couraud. 1979. Scorpion toxins: Chemistry and mode of action. In B. Ceccarelli and F. Clementi, eds., *Advances in cytopharmacology*, vol. 3, *Neurotoxins: Tools in neurobiology*, pp. 325–34. New York: Raven.

Rochat, H., C. Kopeyan, L. G. Garcia, G. Martinez, J. P. Rosso, A. Pakaris, M. F. Martin, A. Garcia, N. Martin-Moutat, J. Grégoire, and F. Miranda. 1976. Recent results on the structure of scorpion and snake toxins. In A. Ohsaka et al., eds., *Animal, plant and microbial toxins*, vol. 2, *Chemistry, pharmacology and immunology*, pp. 79–87. New York: Plenum.

Roig-Alsina, A. 1978. Aspectos ecológicos de tres escorpiones del piedemonte precordillerano de Mendoza. *Ecosur* 5(10): 183–90.

Roig-Alsina, A., and E. A. Maury. 1981. Consideraciones sistemáticas y ecológicas sobre *Brachistosternus* (*Leptosternus*) *borelii* Kraepelin 1911 (Scorpiones, Bothriuridae). *Physis* (Buenos Aires) 39(97): 1–9.

Rolfe, W. D. 1980. Early invertebrate terrestrial faunas. In A. Panchen, ed., *The terrestrial environments and the origin of land vertebrates*, pp. 117–57. Systematics Association Special Volume No. 15. London: Academic.

———. 1985. Early invertebrate arthropods: A fragmentary record. *Philosophical Transactions of the Royal Society* (London), sec. B, 309: 207–18.

Rolfe, W. D., and E. C. Beckett. 1984. Autecology of Silurian Xiphosurida, Scorpionida, Cirripedia, and Phyllocarida. *Special Papers in Palaeontology* 32: 27–37.

Romey, G., J. P. Abita, R. Chicheportiche, H. Rochat, and M. Lazdunski. 1976. Scorpion neurotoxin: Mode of action on neuromuscular junctions and synaptosomes. *Biochimica et Biophysica Acta* 448: 607–19.

Romey, G., R. Chicheportiche, M. Lazdunski, H. Rochat, F. Miranda, and S. Lissitzky. 1975. Scorpion neurotoxin—A presynaptic toxin which affects both Na^+ and K^+ channels in axons. *Biochemical and Biophysical Research Communications* 64: 115–21.

Romine, W. O., Jr., G. M. Schoepfle, J. R. Smythies, G. al-Zahid, and R. J. Bradley. 1974. Pharmacological evidence related to the existence of two sodium channel gating mechanisms. *Nature* 248: 797–99.

Root, T. M. 1979. The structure and neuromuscular physiology of the scorpion leg motor system. Ph.D. diss., University of Wyoming, Laramie.

———. 1980. Motor innervation of the trochanter-femur elevator and depressor muscles in the scorpion. *American Zoologist* 20(4): 941.

———. 1985. Central and peripheral organization of scorpion locomotion. In Barth, ed., pp. 337–47.

———. 1988. Histochemistry and neuromuscular physiology of some leg muscles of the scorpion *Paruroctonus mesaensis*. In preparation.

Root, T. M., and R. F. Bowerman. 1978. Intra-appendage movements during walking in the scorpion *Hadrurus arizonensis*. *Comparative Biochemistry and Physiology* 59A: 49–56.

———. 1979. Neuromuscular physiology of scorpion leg muscles. *American Zoologist* 19(3): 993.

———. 1981. The scorpion walking leg motor system: Muscle fine structure. *Comparative Biochemistry and Physiology* 69A: 73–78.

Root, T. M., and S. C. Gatwood. 1981. Histochemistry of scorpion leg muscles. *American Zoologist* 21(4): 147.

Rosin, R. 1965. A new type of poison gland found in the scorpion *Nebo hierichonticus* (E. Sim.) (Diplocentridae, Scorpiones). *Rivista di parassitologia* 26: 111–22.

Rosin, R., and A. Shulov. 1961. Sound production in scorpions. *Science* 133: 1918–19.

———. 1963. Studies on the scorpion *Nebo hierichonticus*. *Proceedings of the Zoological Society* (London) 140(4): 547–75.

Ross, R. H., Jr., and R. E. Monroe. 1970. Utilization of acetate-1-^{14}C by the tarantula, *Aphonepelma* sp., and the scorpion, *Centruroides sculpturatus*, in lipid synthesis. *Comparative Biochemistry and Physiology* 36: 765–73.

Rosso, J. P., and H. Rochat. 1985. Characterization of ten proteins from the venom of the Moroccan scorpion *Androctonus mauretanicus mauretanicus*, six of which are toxic to the mouse. *Toxicon* 23: 113–25.

Roth, V., and W. Brown. 1980. Arthropoda: Arachnida. In R. C. Brusen, ed., *Common intertidal invertebrates of the Gulf of California*, pp. 347–55. Tucson: University of Arizona Press.

Ruhland, M. 1976. Untersuchungen zur neuromuskulären Organisation eines Muskels aus Laufbeinregeneraten einer Vogelspinne (*Dugesiella hentzi* Ch.). *Verhandlungen der Deutschen zoologischen Gesellschaft* (Hamburg) 69: 238.

Ruhland, M., E. Zlotkin, and W. Rathmayer. 1977. The effect of toxins from the venom of the scorpion *Androctonus australis* on a spider nerve-muscle preparation. *Toxicon* 15: 157–60.

Saint-Remy, G. 1886a. Recherches sur la structure du cerveau du scorpion. *Comptes rendus hebdomadaires des séances de l'Académie des sciences* (Paris), sér. D, 102: 1492–94.

———. 1886b. Sur la structure des centres nerveux chez le scorpion. *Bulletin de la Société des sciences* (Nancy) (2)8(20): xxix.

———. 1887. Structure du cerveau chez le scorpion et la scolopendre. *Bulletin de la Société des sciences* (Nancy) (2)9(21): xxxi–xxxii.

Sampieri, F., and C. Habersetzer-Rochat. 1978. Structure-function relationships in scorpion neurotoxins: Identification of the superreactive lysine residue in Toxin I of *Androctonus australis* Hector. *Biochimica et Biophysica Acta* 535: 100–109.

Sanjeeva Reddy, P. 1969. Studies on the input from a hair sensillum into the central nervous system of the scorpion, *Heterometrus fulvipes*. Ph.D. diss., Sri Venkateswara University, Tirupati, Andhra Pradesh, India.

———. 1971. Function of the supernumerary sense cells and the relationship between modality of adequate stimulus and innervation pattern of the scorpion hair sensillum. *Journal of Experimental Biology* 54: 233–38.

Sanjeeva Reddy, P., and K. Pampathi Rao. 1970a. The central course of the hair afferents and the pattern of contralateral activation in the central nervous system of the scorpion, *Heterometrus fulvipes*. *Journal of Experimental Biology* 53: 165–70.

———. 1970b. Tactile input through conducting pathways and behavior of the scorpion *Heterometrus fulvipes*. *Indian Journal of Physiology and Pharmacology* 15(2): 39–40.

San Martín, P. R. 1961. Observaciones sobre la ecología y distribución geográfica de tres especies de escorpiones en el Uruguay. *Revista de la Facultad de humanidades y ciencias, Universidad de la República* (Montevideo) 19: 1–42.

———. 1968. *Bothriurus vachoni*, n. sp. del Brasil (Scorpionida, Bothriuridae). *Acta Biologica Venezuelica* 6(2): 38–51.

———. 1969. Estudio sobre la compleja estructura esqueleto esclerificado del organo paraxial del género *Brachistosternus* (Bothriuridae—Scorpionida). *Boletín de la Sociedad de biología* (Concepción) 41: 13–30.

———. 1972. Fijación de los caracteres sistemáticos en los Bothriuridae (Scorpiones). I, Queliceros: Morfología y nomenclatura. *Boletín de la Sociedad de biología* (Concepción) 44: 47–55.

San Martín, P. R., and T. K. Cekalovic. 1972. Fijación de los caracteres sistemáticos en los Bothriuridae (Scorpiones). II, Queliceros: Estudio de diferenciación a nivel genérico. *Boletín de la Sociedad de biología* (Concepción) 44: 57–71.

San Martín, P., and L. Gambardella. 1966. Nueva comprobación de la partenogénesis en *Tityus serrulatus* Lutz y Mello-Campos 1922. *Revista de la Sociedad entomológica Argentina* 28(1–4): 79–84.

———. 1967. Descripción del espermatoforo de *Bothriurus bucherli* San Martín 1963 (Scorpiones—Bothriuridae). *Revista de la Sociedad entomológica Argentina* 24: 17–20.

Santhanakrishnan, B. R., and V. B. Raju. 1974. Management of scorpion stings in children. *Journal of Tropical Medicine and Hygiene* 77: 133–35.

Satyanarayanam, K., V. R. Selvarajan, and K. S. Swami. 1975. Some biochemical studies of denervated pedipalpal muscles of scorpion, *Heterometrus fulvipes*. *Comparative Biochemistry and Physiology* 52B: 383–86.

Savory, T. H. 1935. *The Arachnida*. London: Edward Arnold. xi + 218 pp.

———. 1961. *Spiders, men, and scorpions*. London: University of London Press. 191 pp.

———. 1977. *The Arachnida*. 2d ed. London: Academic. 340 pp.

Schantz, R., and M. Goyffon. 1970. Neurotransmitter amino acids and spontaneous electrical activity of the prosomian nervous system of the scorpion. *Comptes rendus des séances de la Société de biologie* (Paris) 164: 1225–27.

Schatz, L. 1952. The development and differentiation of the arthropod procuticle: Staining. *Annals of the Entomological Society of America* 45: 678–86.

Schawaller, W. 1979. Erst Nachweis eines Skorpions im dominikanischen Bernstein (Stuttgarter Bernsteinsammlung: Arachnida, Scorpionida). *Stuttgarter Beiträge zur Naturkunde*, ser. B (Geologie und Palaeontologie), 45: 1–15.

———. 1981. Zwei weitere Skorpione im dominikanischen Bernstein (Stuttgarter Bernsteinsammlung: Arachnida, Scorpionida). *Stuttgarter Beiträge zur Naturkunde*, ser. B (Geologie und Palaeontologie), 82: 1–14.

Scheuring, L. 1913. Die Augen der Arachniden. *Zoologische Jahrbücher, Abteilung für Anatomie und Ontogenie*, 33: 553–636.

Schimkewitsch, W. 1893. Sur la structure et sur la signification de l'endosternite des arachnides. *Zoologischer Anzeiger* 16: 300–308.

———. 1894. Über Bau und Entwicklung des Endosternites der Arachniden. *Zoologische Jahrbücher, Abteilung für Anatomie und Ontogenie*, 8: 191–216.

Schliwa, M. 1979. The retina of the phalangid, *Opilio ravennae*, with particular reference to arhabdomeric cells. *Cell and Tissue Research* 204: 473–95.

Schliwa, M., and G. Fleissner. 1977. Efferente neurosekretorische Fasern in den Medianaugen und im Sehnerv des Skorpions *Androctonus australis*. *Verhandlungen der Deutschen zoologischen Gesellschaft* (Erlangen) 70: 230.

———. 1979. Arhabdomeric cells of the median eye retina of scorpions. I, Fine structural analysis. *Journal of Comparative Physiology*, ser. A, 130(3): 265–70.

———. 1980. The lateral eyes of the scorpion, *Androctonus australis*. *Cell and Tissue Research* 206(1): 95–114.

Schmitt, O., and H. Schmidt. 1972. Influence of calcium ions on the ionic currents of nodes of Ranvier treated with scorpion venom. *Pflügers Archiv* 333: 51–61.

Schneider, A. 1892. Sur le système arteriel du scorpion. *Tablettes zoologiques* 2: 157–98.

Schoener, T. W. 1971. Theory of feeding strategies. *Annual Review of Ecology and Systematics* 2: 369–404.

Schroeder, O. 1908. Die Sinnesorgane der Skorpionskamme. *Zeitschrift für wissenschaftlichen Zoologie* 90: 436–44.

Schultze, W. 1927. Biology of the large Philippine forest scorpion. *Philippine Journal of Science* 32: 375–89.

Scorza, J. V. 1954. Sistemática, distribución geográfica y observaciones ecológicas de algunos alacranes encontrados en Venezuela. *Memorias de la Sociedad de ciencias naturales La Salle* (Caracas) 14(38): 179–214.

Scriber, J. M., and F. Slansky. 1981. The nutritional ecology of immature insects. *Annual Review of Entomology* 29: 183–211.

Selvarjan, V. R. 1976. Studies on the electrical stimulation of the central nervous system of implanted scorpions *Heterometrus fulvipes*. *Marathwada University Journal of Science* 15(8): 107–10.

Seyfarth, E. A. 1978. Lyriform slit sense organs and muscle reflexes in the spider leg. *Journal of Comparative Physiology* 125: 45–57.

———. 1980. Daily patterns of locomotor activity in a wandering spider. *Physiological Entomology* 5: 199–206.

Seyfarth, E. A., and F. G. Barth. 1972. Compound slit sense organs on the spider leg: Mechanoreceptors involved in kinesthetic orientation. *Journal of Comparative Physiology* 78: 176–91.

Seyfarth, E. A., and J. Bohnenberger. 1980. Compensated walking of tarantula spiders and the effect of lyriform slit sense organ ablation. In *Proceedings of the 8th International congress of arachnology, Vienna*, vol. 1, pp. 249–55.

Shachak, M. 1980. Energy allocation and life history strategy of the desert isopod *H. reaumuri*. *Oecologia* 45: 404–13.

Shachak, M., and S. Brand. 1983. The relationship between sit-and-wait foraging strategy and dispersal in the desert scorpion *Scorpio maurus palmatus*. *Oecologia* 60: 371–77.

Shear, W. 1982. Remarkable new fossils of terrestrial arthropods from the Devonian (Middle Givetian) of Gilboa, New York, and thoughts on arachnomorph phylogeny. *American Arachnology* 26: 11.

Shorthouse, D. J. 1971. Studies on the biology and energetics of the scorpion, *Urodacus yaschenkoi* (Birula, 1904). Ph.D. diss., Australian National University, Canberra. 163 pp.

Shorthouse, D. J., and T. Marples. 1980. Observations on the burrow and associated behavior of the arid-zone scorpion *Urodacus yaschenkoi* (Birula). *Australian Journal of Zoology* 28: 581–90.

———. 1982. The life stages and population dynamics of an arid zone scorpion *Urodacus yaschenkoi* (Birula, 1903). *Australian Journal of Ecology* 7: 109–18.

Shulov, A. 1955. On the poison of scorpions in Israel, II. *Harefuah* 49: 131–33. [Cited in Shulov and Levy 1978.]

———. 1958. Observations on the mating habits of two scorpions, *Leiurus quinquestriatus* H.&E. and *Buthotus judaicus*. In *Proceedings of the 10th International congress of entomology (1956), Montreal*: 877–80.

Shulov, A., and P. Amitai. 1958. On the mating habits of three scorpions: *Leiurus quinquestriatus* H.&E., *Buthotus judaicus* E. Sim. and *Nebo hierichonticus* (E. Sim.). *Archives de l'Institut Pasteur d'Algérie* 36: 351–69.

———. 1960. Observations sur les scorpions: *Orthochirus innesi* E. Sim., 1910, ssp. *negebensis* nov. *Archives de l'Institut Pasteur d'Algérie* 38: 117–29.

Shulov, A., and G. Levy. 1978. Venoms of Buthinae. A, Systematics and biology of Buthinae. In Bettini, ed., pp. 309–12.

Shulov, A., R. Rosin, and P. Amitai. 1960. Parturition in scorpions. *Bulletin of the Research Council of Israel*, sec. B (Zoology), 9: 65–69.

Simard, J. M., H. Meves, and D. D. Watt. 1986. Effects of Toxins VI and VII from the scorpion *Centruroides sculpturatus* on the sodium currents of the frog node of Ranvier. *Pflügers Archiv* 406: 620–28.

Simard, J. M., and D. D. Watt. 1983. Spontaneous, rhythmic myoelectric activity induced by a scorpion neurotoxin, reduced [calcium]$_0$ and cadmium in chick striated muscle. *Comparative Biochemistry and Physiology* 75A: 413–19.

Simon, E. 1879. Ordre Scorpiones Th. In *Arachnides de France* 7: 79–115. Paris: Roret.

Sinha, R. C. 1966. Effect of starvation on the concentration of ascorbic acid in the pedipalp muscle of the scorpion (*P. bengalensis*). *Experientia* 22(4): 221–22.

Sinha, R. C., and M. S. Kanungo. 1967. Effects of starvation on the scorpion *Palamnaeus bengalensis*. *Physiological Zoology* 40: 386–90.

Sissom, W. D. 1986. Description of the male of *Vaejovis gracilis* Gertsch and Soleglad (Scorpiones: Vaejovidae), with a clarification of the identity of the species. In J. R. Riddell, ed., *Studies on the cave and endogean fauna of North America*, Texas Memorial Museum, Speleological Monographs, 1: 11–16.

———. 1989. *Typhlochactas mitchelli*, a new species of eyeless, montane forest litter scorpion from northeastern Oaxaca, Mexico (Chactidae, Superstitioninae, Typhlochactini). *Journal of Arachnology* 16: 365–71.

Sissom, W. D., and O. F. Francke. 1981. Scorpions of the genus *Paruroctonus* from New Mexico and Texas (Scorpiones, Vaejovidae). *Journal of Arachnology* 9: 93–108.

———. 1983. Post-birth development of *Vaejovis bilineatus* Pocock (Scorpiones: Vaejovidae). *Journal of Arachnology* 11: 69–75.

———. 1985. Redescriptions of some poorly known species of the *nitidulus* group of the genus *Vaejovis* (Scorpiones, Vaejovidae). *Journal of Arachnology* 13: 243–66.

Smith, G. T. 1966. Observations on the life history of the scorpion *Urodacus abruptus* Pocock (Scorpionida), and an analysis of its home sites. *Australian Journal of Zoology* 14: 383–98.

Snodgrass, R. E. 1952. *A textbook of arthropod anatomy.* Ithaca, N.Y.: Comstock. 363 pp.

Soleglad, M. E. 1973. Scorpions of the *mexicanus* group of the genus *Vejovis* (Scorpionida, Vejovidae). *Wassmann Journal of Biology* 31(2): 351–72.

———. 1975. The taxonomy of the genus *Hadrurus* based on chela trichobothria (Scorpionida: Vejovidae). *Journal of Arachnology* 3(2): 113–34.

———. 1976. A revision of the scorpion subfamily Megacorminae (Scorpionida: Chactidae). *Wassmann Journal of Biology* 34(2): 251–303.

Southcott, R. V. 1955. Some observations on the biology, including mating and other behavior, of the Australian scorpion *Urodacus abruptus* Pocock. *Transactions of the Royal Society* (South Australia) 78: 145–54.

Sreenivasa Reddy, R. P. 1959. A contribution towards the understanding of the functions of the pectines of scorpions. *Journal of Animal Morphology and Physiology* 6(2): 75–80.

———. 1963. Biology of scorpions, with special reference to the pectines. Ph.D. diss., Sri Venkateswara University, Tirupati, Andhra Pradesh, India.

Srivastava, D. S. 1955. Maxillary processes and mechanism of feeding in scorpions. *Saugar University Journal*, pt. 2, 1(4): 85–91.

Stahnke, H. L. 1945. Scorpions of the genus *Hadrurus* Thorell. *American Museum Novitates* 1298: 1–9.

———. 1966. Some aspects of scorpion behavior. *Bulletin of the Southern California Academy of Sciences* 65: 65–80.

———. 1970. Scorpion nomenclature and mensuration. *Entomological News* 81: 297–316.

———. 1972a. UV light, a useful field tool. *BioScience* 22: 604–7.

———. 1972b. A key to the genera of Buthidae (Scorpionida). *Entomological News* 83: 121–33.

———. 1974. Revision and keys to the higher categories of Vejovidae. *Journal of Arachnology* 1: 107–41.

Stahnke, H. L., and M. Calos. 1977. A key to the species of *Centruroides* Marx (Scorpionida: Buthidae). *Entomological News* 88: 111–20.

Stahnke, H. L., and A. H. Dengler. 1964. The effect of morphine and related substances on the toxicity of venoms. I, *Centruroides sculpturatus* Ewing scorpion venom. *American Journal of Tropical Medicine and Hygiene* 13: 346–51.

Stallcup, W. B. 1977. Comparative pharmacology of voltage-dependent sodium channels. *Brain Research* 135: 37–53.

Stauffer, P. H. 1974. Malaya and Southeast Asia in the pattern of continental drift. *Bulletin of the Geological Society of Malaysia* 7: 89–138.

Stockmann, R. 1979. Développement postembryonnaire et cycle d'intermue chez un scorpion Buthidae *Buthotus minax occidentalis* (Vachon et Stockmann). *Bulletin du Muséum national d'histoire naturelle* (Paris), sér. 4, A1(2): 405–20.

Størmer, L. 1963. *Gigantoscorpio willsi, a new scorpion from the Lower Carboniferous of Scotland and its associated preying microorganisms.* Skrifter av det Norske Videnskaps-Akademi, Oslo I, Matematisk-Naturvidenskapelig Klasse, n.s., 8. 171 pp.

———. 1970. Arthropods from the Lower Devonian (Lower Emsian) of Alken an der Mosel, Germany. I, Arachnida. *Senckenbergiana Lethaea* 51: 335–69.

Subramanyam Reddi, G., V. R. Selvarajan, and K. S. Swami. 1976. Studies on amino transferase activity of de-nervated pedipalpal muscle of scorpion *Heterometrus fulvipes*. *Marathwada University Journal of Science* 15(8): 149–51.

Swoveland, M. C. 1978. External morphology of scorpion pectines. Master's thesis, California State University, San Francisco.

Taylor, R. W. 1971. Observations on birth and postpartum behavior in the scorpion *Vejovis carolinus* Koch. *Transactions of the Kentucky Academy of Sciences* 32: 80–82.

Tazieff-Depierre, F. 1968. Propriétés pharmacologiques des toxines du venin de scorpion (*Androctonus australis*). *Comptes rendus hebdomadaires des séances de l'Académie des sciences* (Paris), sér. D, 267: 240–43.

———. 1972. Venin de scorpion, calcium et émission d'acétylcholine par les fibres nerveuses dans l'iléon de cobaye. *Comptes rendus hebdomadaires des séances de l'Académie des sciences* (Paris), sér. D, 275: 3021–24.

Teissier, G. 1960. Relative growth. In T. H. Waterman, ed., *The physiology of Crustacea*, vol. 1. London: Academic.

Tembe, V. B., and P. R. Awati. 1942. External morphology and anatomy of scorpion (*Buthus tamulus* Fabr.). *Journal of the University of Bombay* 11 (n.s. 3B): 54–76.

———. 1944. External morphology and anatomy of scorpion (*Buthus tamulus* Fabr.). *Journal of the University of Bombay* 12 (n.s. 5B): 1–13.

Thorell, T. 1876a. Études scorpiologiques. *Atti della Società italiana di scienze naturali* 19: 75–272.

———. 1876b. On the classification of scorpions. *Annals and Magazine of Natural History*, ser. 4, 17: 1–15.

Thornton, I. W. B. 1956. Notes on the biology of *Leiurus quinquestriatus* (H.&E. 1829). *British Journal of Animal Behaviour* 4: 92–93.

Tikader, B. K., and D. B. Bastawade. 1983. *Fauna of India*. Vol. 3, *Scorpions*. Calcutta: Zoological Survey of India. 670 pp.

Tilak, R. 1970. On An interesting observation on the living habits of *Hormurus nigripes* Pocock (Arachnida: Ischnuridae). *Science and Culture* 36: 174–76.

Tintpulver, M., T. Zerachia, and E. Zlotkin. 1976. The action of toxins derived from scorpion venom on the ileal smooth muscle preparation. *Toxicon* 14: 371–77.

Tjonneland, A., S. Okland, and B. Midttun. 1985. Myocardial ultrastructure in five species of Scorpions (Chelicerata). *Zoologischer Anzeiger* 214 (1/2): 7–17.

Tongiorgi, P. 1959. Effects of the reversal of the rhythm of nycthemeral illumination on astronomical orientation and diurnal activity in *Arctosa variana* C. L. Koch (Araneae—Lycosidae). *Archive of Italian Biology* 97: 251–65.

Toolson, E. C. 1985. Uptake of leucine and water by *Centruroides sculpturatus* (Ewing) embryos (Scorpiones, Buthidae). *Journal of Arachnology* 13: 303–10.

Toolson, E. C., and N. F. Hadley. 1977. Cuticular permeability and epicuticular lipid composition in two Arizona vejovid scorpions. *Physiological Zoology* 50: 323–30.

———. 1979. Seasonal effects on cuticular permeability and epicuticular lipid composition in *Centruroides sculpturatus* Ewing 1928 (Scorpiones: Buthidae). *Journal of Comparative Physiology* 129: 319–25.

Toren, T. J. 1973. Biology of the California coast range scorpion *Vaejovis gertschi striatus* Hjelle. Master's thesis, California State University, San Francisco. 74 pp.

Torres, F., and H. Heatwole. 1967a. Factors influencing behavioral interaction of female parent and offspring in scorpions. *Caribbean Journal of Science* 7(1/2): 19–22.

———. 1967b. Orientation of some scorpions and tailless whip-scorpions. *Zeitschrift für Tierpsychologie* 24: 546–57.

Tourtlotte, G. 1974. Studies on the biology and ecology of the northern scorpion, *Paruroctonus boreus* (Girard). *Great Basin Naturalist* 34(3): 167–79.

Toye, S. A. 1970. Some aspects of the biology of two common species of Nigerian scorpions. *Journal of Zoology* (London) 162: 1–9.

Tu, A. T. 1977. Scorpion venoms. In *Venoms: Chemistry and molecular biology*, pp. 459–83. New York: Wiley.

von Ubisch, M. 1922. Über eine neue *Iurus*-Art aus Kleinasien nebst einigen Bemerkungen über die Funktion der Kamme der Scorpione. *Zoologische Jahrbücher, Abteilung für Systematik, Ökologie und Geographie*, 44: 503–16.

Ugolini, A., I. Carmignani, and M. Vannini. 1986. Mother-young relationship in *Euscorpius*: Adaptive value of the larva permanence on the mother's back (Scorpiones, Chactidae). *Journal of Arachnology* 14: 43–46.

Uthaman, M., and Y. Srinivasa Reddy. 1979. Bimodal oscillation in the oxygen consumption of the isolated tissues of the scorpion *Heterometrus fulvipes*. *Journal of Arid Environments* 2: 347–53.

———. 1985. Rhythmic changes in the activity of neurosecretory cells of the scorpion *Heterometrus fulvipes* (C. Koch). *Chronobiology* 12: 331–38.

Vachon, M. 1940. Sur la systématique des scorpions. *Mémoires du Muséum national d'histoire naturelle* (Paris), 13(2): 241–60.

———. 1951. Sur quelques scorpions "halophiles" (*Microbuthus fagei*, *Mesobuthus confucius* et *Euscorpius flavicaudis*). *Bulletin du Muséum national d'histoire naturelle* (Paris), sér. 2, 23(3): 256–60.

———. 1952. *Études sur les scorpions*. Algiers: Publications de l'Institut Pasteur d'Algérie. 482 pp.

———. 1953. The biology of scorpions. *Endeavour* 12(46): 80–89.

———. 1956. Sur des nouveaux caractères familiaux et génériques chez les scorpions. In *Proceedings of the 14th International congress of zoology (1953), Copenhagen*, pp. 471–74.

———. 1957. La régénération appendiculaire chez les scorpions (Arachnides). *Comptes rendus hebdomadaires des séances de l'Académie des sciences* (Paris), sér. D, 244: 2556–59.

———. 1963. De l'utilité, en systématique, d'une nomenclature des dents des chélicères chez les scorpions. *Bulletin du Muséum national d'histoire naturelle* (Paris), sér. 2, 35: 161–66.

———. 1973. Étude des caractères utilisés pour classer les familles et les genres des scorpions (Arachnides). I, Les trichobothriaux et types de trichobothriotaxie chez les scorpions. *Bulletin du Muséum national d'histoire naturelle* (Paris), sér. 3, 104: 857–958.

———. 1975. Sur l'utilisation de la trichobothriotaxie du bras des pédipalpes des scorpions (Arachnides) dans le classement des genres de la famille des Buthidae Simon. *Comptes rendus hebdomadaires des séances de l'Académie des sciences* (Paris), sér. D, 281: 1597–99.

———. 1980. Essai d'une classification sous-générique des scorpions du genre *Scorpiops* Peters, 1861 (Arachnida, Scorpionida, Vaejovidae). *Bulletin du Muséum national d'histoire naturelle* (Paris), sér. 4, 2: 143–60.

Vachon, M., and W. R. Lourenço. 1985. Scorpions cavernicoles du Sarawak (Borneo): *Chaerilus chapmani* sp. n. (Chaerilidae), scorpion troglobie. *Mémoires de biospéléologie* 12: 9–18.

Vachon, M., R. Roy, and M. Condamin. 1970. Le développement postembryonnaire du scorpion *Pandinus gambiensis* Pocock. *Bulletin de l'Institut français d'Afrique noire*, sér. A, 32: 412–32.

Vachon, M., and R. Stockmann. 1968. Contribution à l'étude des scorpions africains appartenants au genre *Buthotus* Vachon 1949 et étude de la variabilité. *Monitore zoologico italiano*, n.s., 2 (suppl.): 81–149.

Van der Hammen, L. 1977. A new classification of Chelicerata. *Zoologische mededelingen* 51: 307–19.

———. 1978. The evolution of the chelicerate life-cycle. *Acta Biotheoretica* 27(1/2): 44–60.

Vannini, M., M. Balzi, A. Becciolini, I. Carmignani, and A. Ugolini. 1985a. Water exchange between mother and larvae in scorpions. *Experientia* 41: 1620–21.

Vannini, M., and A. Ugolini. 1980. Permanence of *Euscorpius carpathicus* (L.) larvae on the mother's back (Scorpiones, Chactidae). *Behavioral Ecology and Sociobiology* 7: 45–47.

Vannini, M., A. Ugolini, and I. Carmignani. 1985b. Mother-young relationships in *Euscorpius* (Scorpiones): Trophic exchange between mother and larvae. *Monitore zoologico italiano*, n.s., 19: 172.

Vannini, M., A. Ugolini, and C. Marucelli. 1978. Notes on the mother-young relationship in some *Euscorpius* (Scorpiones, Chactidae). *Monitore zoologico italiano*, n.s., 12: 143–54.

Varela, J. C. 1961. Gestación, nacimiento y eclosión de *Bothriurus bonariensis* var. *bonariensis* (Koch, 1842). *Revista de la Facultad de agronomía, Universidad de la República* (Montevideo) 19: 225–44.

Vargas, O., M. F. Martin, and H. Rochat. 1987. Characterization of six toxins from the venom of the Moroccan scorpion *Buthus occitanus mardochei*. *European Journal of Biochemistry* 162: 589–99.

Vasantha, N., S. A. T. Venkatachari, P. Murali Mohan, and K. S. Babu. 1975. On the acetylcholine content in the scorpion, *Heterometrus fulvipes* C. Koch. *Experientia* 31: 451.

———. 1977. On the possible mode of action of neurohormones on cholinesterase activity in the ventral nerve cord of scorpion *Heterometrus fulvipes*. *Experientia* 33: 238–39.

Venkatachari, S. A. T. 1971. Analysis of spontaneity on the nervous system of the scorpion *Heterometrus fulvipes*. *Indian Journal of Experimental Biology* 9(3): 338–44.

———. 1975. Physiological analysis of through-conducting systems in the ventral nerve cord of the scorpion, *Heterometrus fulvipes* C. Koch. *Biologisches Zentralblatt* 94: 409–22.

Venkatachari, S. A. T., and K. S. Babu. 1970. Activity of motor fibers in the scorpion, *Heterometrus fulvipes*. *Indian Journal of Experimental Biology* 8: 102–11.

Venkatachari, S. A. T., and V. Devarajulu Naidu. 1969. Choline esterase activity in the nervous system and the innervated organs of the scorpion, *Heterometrus fulvipes*. *Experientia* 25: 821–22.

Venkatachari, S. A. T., and P. Muralikrishna Dass. 1968. Choline esterase activity rhythm in the ventral nerve cord of scorpion. *Life Sciences* 7: 617–21.

Venkateswara Rao, P. 1963. Studies on the peripheral nervous system of the scorpion, *Heterometrus fulvipes*. Ph.D. diss., Sri Venkateswara University, Tirupati, Andhra Pradesh, India.

Venkateswara Rao, P., and S. Govindappa. 1967. Dehydrogenase activity and its diurnal variations in different muscles of the scorpion *Heterometrus fulvipes*. *Proceedings of the Indian Academy of Sciences*, ser. B, 66: 243–49.

Vereshchagin, S. M., V. P. Lapitskii, and V. P. Tyshchenko. 1971. Functional characteristics of neurons of the central nervous system of scorpion *Buthus caucasicus*. *Biologicheskie nauki* 14(12): 25–30.

Vermaseren, M. J. 1978. *Mithriaca IV*. Études Préliminaires aux Religions Orientales, vol. 16. Leiden: Brill.

Vijayalakshimi, N. R., and P. A. Kurup. 1976. Free amino acids and proteins of the haemolymph, hepatopancreas and muscle of the scorpion, *Heterometrus scaber* (Thor.). *Indian Journal of Experimental Biology* 14: 236–38.

Vital Brazil, O., A. C. Neder, and A. P. Corrado. 1973. Effects and mechanism of action of *Tityus serrulatus* venom on skeletal muscle. *Pharmacological Research Communications* 5: 137–50.

Volkova, T. M., I. E. Dulubova, I. N. Telezhinskaya, and E. V. Grishin. 1984a. Toxic components of the Central Asian scorpion *Orthochirus scrobiculosus*. *Bioorganicheskaya khimiya* (USSR) 10: 1100–1108.

Volkova, T. M., A. F. Garcia, I. N. Telezhinskaya, N. A. Potapenko, and E. V. Grishin. 1984b. Amino acid sequence of two neurotoxins from the venom of the Central Asian scorpion *Buthus eupeus*. *Bioorganicheskaya khimiya* (USSR) 10: 979–82.

Vyas, A. B. 1971. Adaptive diversification of the prosomatic appendages in the scorpion *Heterometrus fulvipes*. *Annals of Zoology* 7(3): 65–80.

———. 1974a. The cheliceral muscles of the scorpion *Heterometrus fulvipes*. *Bulletin of the Southern California Academy of Sciences* 73(1): 9–13.

———. 1974b. Scanning electron microscopy of book lungs of the scorpion *Heterometrus fulvipes*. *Entomological News* 85: 88–91.

Vyas, A. B., and S. M. Laliwala. 1972a. Microanatomy of the book lungs of scorpion and the mechanism of respiration. *Vidya* (Gujarat University Journal) 15(1): 122–28.

———. 1972b. Certain noteworthy features of the circulatory system of *Heterometrus fulvipes*. *Proceedings of the National Academy of Sciences* (India) 42(B): 267–71.

Walckenaer, C. A. 1844. *Histoire naturelle des insects: Aptères*. 4 vols. Paris: Roret.

Walcott, C. 1969. A spider's vibration receptor: Its anatomy and physiology. *American Zoologist* 9: 133–44.

Wang, G. K., and G. Strichartz. 1982. Simultaneous modifications of sodium channel gating by two scorpion toxins. *Biophysical Journal* 40: 175–79.

———. 1983. Purification and physiological characterization of neurotoxins from venoms of the scorpions *Centruroides sculpturatus* and *Leiurus quinquestriatus*. *Molecular Pharmacology* 23: 519–33.

Wanless, F. R. 1977. On the occurrence of the scorpion *Euscorpius flavicaudis* (DeGeer) at Sheerness Port, Isle of Shepey, Kent. *Bulletin of the British Arachnological Society* 4: 74–76.

Warburg, M. R. 1968. Behavioral adaptations of terrestrial isopods. *American Zoologist* 8: 545–59.

Warburg, M. R., and A. Ben-Horin. 1978. Temperature and humidity effects on scorpion distribution in northern Israel. *Symposia of the Zoological Society* (London) 42: 161–69.

———. 1979. Thermal effect on the diel activity rhythm of scorpions from mesic and xeric habitats. *Journal of Arid Environments* 2: 339–46.

———. 1981. The response to temperature gradients of scorpions from mesic and xeric habitats. *Comparative Biochemistry and Physiology* 68A: 277–79.

Warburg, M. R., S. Goldenberg, and A. Ben-Horin. 1980a. Scorpion species diversity and distribution within the Mediterranean and arid regions of northern Israel. *Journal of Arid Environments* 3: 205–13.

———. 1980b. Thermal effect on evaporative water loss and haemolymph osmolarity in scorpions at low and high humidities. *Comparative Biochemistry and Physiology* 67A: 47–57.

Warburton, C. 1909. Arachnida embolobranchiata: Scorpions, spiders, mites, etc. In S. F. Harmer and A. E. Shipley, eds., *The Cambridge natural history*, vol. 4, *Crustacea and Arachnida*, pp. 295–473. London: Macmillan.

Warnick, J. E., E. X. Albuquerque, and C. R. Diniz. 1976. Electrophysiological observations on the action of the purified scorpion venom, Tityustoxin, on nerve and skeletal muscle of the rat. *Journal of Pharmacology and Experimental Therapeutics* 198: 155–67.

Waterman, J. A. 1950. Scorpions in the West Indies, with special reference to *Tityus trinitatis*. *Caribbean Medical Journal* 12: 167–77.

Watt, D. D. 1964. Biochemical studies of the venom from the scorpion, *Centruroides sculpturatus*. *Toxicon* 2: 171–80.

Watt, D. D., D. R. Babin, and R. V. Mlejnek. 1974. The protein neurotoxins in scorpion and elapid snake venoms. *Agricultural and Food Chemistry* 22: 43–51.

Watt, D. D., and M. E. McIntosh. 1972. Effects on lethality of toxins in venom from the scorpion *Centruroides sculpturatus* by group specific reagents. *Toxicon* 10: 173–81.

Watt, D. D., J. M. Simard, D. R. Babin, and R. V. Mlejnek. 1978. Physiological characterization of toxins isolated from scorpion venom. In P. Rosenberg, ed., *Toxins: Animal, plant and microbial*, pp. 647–60. New York: Pergamon.

Watt, D. D., J. M. Simard, and P. M. Mancuso. 1982. Oscillation in resting tension of chick skeletal muscle treated with Toxin V from scorpion venom, reduced $[\text{calcium}]_o$ and cadmium. *Comparative Biochemistry and Physiology* 71A: 375–82.

Weltzin, R. 1981. Dorsal leg nerve joint receptors in restrained and freely walking scorpions. Master's thesis, University of Wyoming, Laramie.

Weltzin, R., and R. F. Bowerman. 1980. Scorpion walking leg proprioceptors. *American Zoologist* 20(4): 940.

Werner, F. 1934. *Scorpiones, Pedipalpi*. In H. G. Bronn, ed., *Klassen und Ordnungen des Tierreichs*, Band 5, Abteilung 4, Buch 8. Leipzig: Akademische Verlagsgesellschaft M.B.H. 316 pp.

Weygoldt, P. 1969. *The biology of pseudoscorpions*. Cambridge, Mass.: Harvard University Press. 145 pp.

———. 1979. Significance of later embryonic stages and head development in arthropod phylogeny. In Gupta, ed., pp. 107–35.

Weygoldt, P., and H. Paulus. 1979a. Untersuchungen zur Morphologie, Taxonomie und Phylogenie der Chelicerata. I, Morphologische Untersuchungen. *Zeitschrift für zoologische Systematik und Evolutionsforschung* 17(3): 85–116.

———. 1979b. Untersuchungen zur Morphologie, Taxonomie und Phylogenie der Chelicerata. II, Cladogramme und die Entfaltung der Chelicerata. *Zeitschrift für zoologische Systematik und Evolutionsforschung* 17(3): 177–200.

Wheeler, K. P., J. Barhanin, and M. Lazdunski. 1982. Specific binding of Toxin II from *Centruroides suffusus suffusus* to the sodium channel in electroplaque membranes. *Biochemistry* 21: 5628–34.

Wheeler, K. P., M. Lazdunski, and D. D. Watt. 1983. Classification of Na^+ channel receptors specific for various scorpion toxins. *Pflügers Archiv* 397: 164–65.

Whitmore, D. H., R. Gonzalez, and J. G. Baust. 1985. Scorpion cold hardiness. *Physiological Zoology* 58: 526–37.

WHO [World Health Organization]. 1981. *Progress in the characterization of venoms and standardization of antivenoms: Report of WHO coordination meeting on venoms and antivenoms, Zürich, 24–27 September 1979*. WHO Offset Publication no. 58. Geneva.

Wigglesworth, V. B. 1947. The insect cuticle. *Biological Reviews* 23: 408–51.

Williams, G. 1962. Seasonal and diurnal activity of harvestmen (Phalangida) and spiders (Araneida) in contrasted habitats. *Journal of Animal Ecology* 31: 23–42.

Williams, S. C. 1963. Feeding ecology of the scorpion *Anuroctonus phaeodactylus*, in a chaparral community recovering from fire. Master's thesis, San Diego State College. 374 pp.

———. 1966. Burrowing activities of the scorpion *Anuroctonus phaeodactylus* (Wood) (Scorpionida: Vaejovidae). *Proceedings of the California Academy of Sciences* 34: 419–28.

———. 1968. Scorpion preservation for taxonomic and morphological studies. *Wasmann Journal of Biology* 26(1): 133–36.

———. 1969. Birth activities of some North American scorpions. *Proceedings of the California Academy of Sciences* 37(1): 1–24.

———. 1970a. A systematic revision of the giant hairy-scorpion genus *Hadrurus* (Scorpionida: Vejovidae). *Occasional Papers of the California Academy of Sciences* 87: 1–62.

———. 1970b. Three new scorpions of *Vejovis* from Death Valley, California (Scorpionida: Vejovidae). *Pan-Pacific Entomologist* 46(1): 1–11.

———. 1970c. Coexistence of desert scorpions by differential habitat preference. *Pan-Pacific Entomologist* 46(4): 254–67.

———. 1970d. Scorpion fauna of Baja California, Mexico: Eleven new species of *Vaejovis* (Scorpionida, Vaejovidae). *Proceedings of the California Academy of Sciences* 37: 275–332.

———. 1971a. Developmental anomalies in the scorpion *Centruroides sculpturatus*. *Pan-Pacific Entomologist* 47(1): 76–77.

———. 1971b. Birth behavior in the South African scorpion *Hadogenes*. *Pan-Pacific Entomologist* 47(1): 79–80.

———. 1971c. In search of scorpions. *Pacific Discovery* 24(3): 1–10.

———. 1976. The scorpion fauna of California. *Bulletin of the Society of Vector Ecology* 3: 1–4.

———. 1980. Scorpions of Baja California, Mexico, and adjacent islands. *Occasional Papers of the California Academy of Sciences* 135: 1–127.

———. 1987. Scorpion bionomics. *Annual Review of Entomology* 32: 274–96.

Williams, W. 1962. Seasonal and diurnal activity of harvestmen (Phalangida) and spiders (Araneida) in contrasted habitats. *Journal of Animal Ecology* 31: 23–42.

Wills, L. J. 1947. *A monograph of British Triassic scorpions*. Monographs of the Palaeontological Society (London), 100/101. 137 pp.

———. 1959. The external anatomy of some Carboniferous "scorpions," part 1. *Palaeontology* 1(4): 261–82.

———. 1960. The external anatomy of some Carboniferous "scorpions," part 2. *Palaeontology* 3(3): 276–332.

Wilson, D. M. 1967. Stepping patterns in tarantula spiders. *Journal of Experimental Biology* 47: 133–51.

Wilson, E. O. 1975. *Sociobiology: The new synthesis*. Cambridge, Mass.: Harvard University Press. 697 pp.

Wuttke, W. 1966. Untersuchungen zur Aktivitätsperiodik bei *Euscorpius carpathicus* L. (Chactidae). *Zeitschrift für vergleichende Physiologie* 53: 405–48.

Yarom, R. 1970. Scorpion venom: A tutorial review of its effects in men and experimental animals. *Clinical Toxicology* 3: 561–69.

Yellamma, K., P. Murali Mohan, and K. S. Babu. 1980. Morphology and physiology of giant fibres in the seventh abdominal ganglion of the scorpion *Heterometrus fulvipes*. *Proceedings of the Indian Academy of Sciences* 89(1): 29–38.

Yellamma, K., K. Subhashini, M. Mohan, and K. S. Babu. 1982. Microanatomy of the seventh abdominal ganglion and its peripheral nerves in the scorpion *Heterometrus fulvipes*. *Proceedings of the Indian Academy of Sciences* 91(3): 225–34.

Yokota, S. D. 1979. Water, energy and nitrogen metabolism in the desert scorpion *Paruroctonus mesaensis*. Ph.D. diss., University of California, Riverside. 316 pp.

Yokota, S. D., and V. H. Shoemaker. 1981. Xanthine excretion in a desert scorpion, *Paruroctonus mesaensis*. *Journal of Comparative Physiology* 142: 423–28.

Yoshikura, M. 1975. Comparative embryology and phylogeny of Arachnida. *Kumamoto Journal of Science* 12(2): 71–142.

Zavaleta, A., J. Navarro, and R. Castro de la Mata. 1981. Pharmacological effects of a Peruvian scorpion (*Hadruroides lunatus*) venom. *Toxicon* 19: 906–9.

Zell, A., S. E. Ealick, and C. E. Bugg. 1985. Three-dimensional structures of scorpion neurotoxins. In R. A. Bradshaw and J. Tang, eds., *Molecular architecture of proteins and enzymes*, pp. 65–97. New York: Academic.

Zinner, H., and P. Amitai. 1969. Observations on hibernation of *Compsobuthus acutecarinatus* Simon and *C. schmiedeknechti* Vachon (Scorpionidea, Arachnida) in Israel. *Israel Journal of Zoology* 18: 41–47.

Zlotkin, E., G. Martinez, H. Rochat, and R. Miranda. 1976. A protein from scorpion venom toxic to crustaceans. In A. Ohsaka et al., eds., *Animal, plant and microbial toxins*, vol. 1, *Biochemistry*, pp. 73–80. New York: Plenum.

Zlotkin, E., F. Miranda, C. Kopeyan, and S. Lissitzky. 1971. A new toxic protein in the venom of the scorpion *Androctonus australis* Hector. *Toxicon* 9: 9–13.

Zlotkin, E., F. Miranda, and S. Lissitzky. 1972. A factor toxic to crustaceans in the venom of the scorpion *Androctonus australis* Hector. *Toxicon* 10: 211–16.

Zlotkin, E., F. Miranda, and H. Rochat. 1978. Venoms of Buthinae. C, Chemistry and pharmacology of Buthinae scorpion venoms. In Bettini, ed., pp. 317–69.

de Zolessi, L. D. 1956. Observaciones sobre el comportamiento sexual de *Bothriurus bonariensis* (Koch) (Scorpiones, Bothriuridae). *Boletín de la Facultad de agronomía, Universidad de la República* (Montevideo) 35: 1–10.

Zwicky, K. T. 1968. A light response in the tail of *Urodacus*, a scorpion. *Life Sciences* 7: 257–62.

———. 1970a. The spectral sensitivity of the tail of *Urodacus*, a scorpion. *Experientia* 26: 317.

———. 1970b. Behavioral aspects of the extraocular light sense of *Urodacus*, a scorpion. *Experientia* 26: 747–48.

Index of Taxa

In this index, "k" after a page number signifies a key to the identification of a taxon; "f" means a second mention on the next page, and "ff" means separate references on the next two pages. Two numbers with a dash between them mark a discussion spanning two or more pages, and *passim* denotes separate mentions on three or more pages in close but not necessarily consecutive sequence.

Index of Subjects

In this index, "f" means a second mention on the next page; "ff" means separate references on the next two pages; two numbers with a dash between them mark a discussion spanning two or more pages; and *passim* denotes separate mentions on three or more pages in close but not necessarily consecutive sequence.

Library of Congress Cataloging-in-Publication Data

The biology of scorpions

Bibliography: p.
Includes index.
1. Scorpions. I. Polis, Gary A., 1946–
QL458.7.B56 1989 595.4'6 84-40330
ISBN 0-8047-1249-2

∞ This book is printed on acid-free paper